Springer Complexity

Springer Complexity is an interdisciplinary program publishing the best research and academic-level teaching on both fundamental and applied aspects of complex systems – cutting across all traditional disciplines of the natural and life sciences, engineering, economics, medicine, neuroscience, social and computer science.

Complex Systems are systems that comprise many interacting parts with the ability to generate a new quality of macroscopic collective behavior the manifestations of which are the spontaneous formation of distinctive temporal, spatial or functional structures. Models of such systems can be successfully mapped onto quite diverse "real-life" situations like the climate, the coherent emission of light from lasers, chemical reaction-diffusion systems, biological cellular networks, the dynamics of stock markets and of the internet, earthquake statistics and prediction, freeway traffic, the human brain, or the formation of opinions in social systems, to name just some of the popular applications.

Although their scope and methodologies overlap somewhat, one can distinguish the following main concepts and tools: self-organization, nonlinear dynamics, synergetics, turbulence, dynamical systems, catastrophes, instabilities, stochastic processes, chaos, graphs and networks, cellular automata, adaptive systems, genetic algorithms and computational intelligence.

The three major book publication platforms of the Springer Complexity program are the monograph series "Understanding Complex Systems" focusing on the various applications of complexity, the "Springer Series in Synergetics", which is devoted to the quantitative theoretical and methodological foundations, and the "SpringerBriefs in Complexity" which are concise and topical working reports, case-studies, surveys, essays and lecture notes of relevance to the field. In addition to the books in these two core series, the program also incorporates individual titles ranging from textbooks to major reference works.

Springer Series in Synergetics

Founding Editor: H. Haken

The Springer Series in Synergetics was founded by Herman Haken in 1977. Since then, the series has evolved into a substantial reference library for the quantitative, theoretical and methodological foundations of the science of complex systems.

Through many enduring classic texts, such as Haken's *Synergetics and Information and Self-Organization*, Gardiner's *Handbook of Stochastic Methods*, Risken's *The Fokker Planck-Equation* or Haake's *Quantum Signatures of Chaos*, the series has made, and continues to make, important contributions to shaping the foundations of the field.

The series publishes monographs and graduate-level textbooks of broad and general interest, with a pronounced emphasis on the physico-mathematical approach.

More information about this series at http://www.springer.com/series/712

Till Frank

Determinism and Self-Organization of Human Perception and Performance

 Springer

Till Frank
Dept. Psychology and Physics
University of Connecticut
Storrs, CT, USA

ISSN 0172-7389   ISSN 2198-333X   (electronic)
Springer Series in Synergetics
ISBN 978-3-030-28820-4   ISBN 978-3-030-28821-1   (eBook)
https://doi.org/10.1007/978-3-030-28821-1

This Springer imprint is published by the registered company Springer Nature Switzerland AG.
The registered company address is: Gewerbestrasse 11, 6330 Cham, Switzerland

To Sofia

Foreword

Frank's monograph focuses on two main aspects:

1. Mathematical models of a number of experiments in psychophysics, movement science, behavioral science and related fields.
2. What Frank calls the "physics perspective."

Specific kinds of behavior in humans and animals that can be described verbally (e.g. walking/running) can also be conceived of as patterns (e.g. the coordination of limbs). Frank describes them by means of a few appropriately chosen variables, and in line with his previous papers, formulates differential equations of a type he denotes as Lotka-Volterra-Haken (LVH) equations. He also studies their steady state and oscillatory solutions, which depend on a specific "bifurcation" parameter.

When such a parameter exceeds a critical value, the type of solution and the related behavioral pattern change qualitatively, e.g. from walking to running when the required speed is increased, or from grasping an object with one hand or with both hands, depending on the object's size.

Using bifurcation theory (including the "attractor" concept), he presents a number of interesting examples both verbally and in terms of equations. His detailed elaborations are supported by numerous figures.

In this way, students and scientists can learn how to model further observations. Also, this approach may serve as a framework (or point of departure) for the development of more detailed microscopic treatments.

When applying bifurcation theory to real-world phenomena, typically a basic problem arises: at the bifurcation point the formerly stable fixed point (attractor) becomes unstable and new fixed points (some of which are attractors) become possible. The question: what "real world" mechanism drives the system from the unstable fixed point into the new attractors and, in particular, which one?

In the context of human behavior this means that a choice between two (or more) behavioral patterns must be made: a problem of decision-making.

Frank addresses this problem, and seeks to leave out any "spiritualism" using his concept of the "physics perspective." In this context, he adopts the view of "Skinner's determinism" with its central concept of the role of causal chains. Here, chance events play a minor role. Frank develops this purely deterministic approach quite rigorously, and invoking his "physics perspective" dismisses all hitherto used concepts and denotations (e.g. decisions, memory, forgetting, fixation, knowledge, ambiguity, learning, deficits, failures, impairment, information), replacing them with his own interpretations.

Summing up: in physics there are numerous instances in which chance effects play a major role. They become especially pronounced as "critical fluctuations" at phase-transitions of systems in thermal equilibrium (e.g. ferromagnets) and nonequilibrium (e.g. lasers).

In many cases of practical interest, it doesn't matter what they "ontologically" are—though there is a deep "philosophical" interest (cf. the debate on quantum theory). But what does all this have to do with human behavior in relation to Frank's "physics perspective?"

Or from a more practical point of view: how well can we predict a person's decision at a "bifurcation" point? And, what influences his/her final decision?

These are surely tantalizing questions. I won't attempt to answer them. All I want to say is this: many changes in behavioral patterns resemble nonequilibrium phase-transitions with their critical fluctuations (often connected with symmetry breaking). One example is Kelso's experiment on the coordination of the movement of the two index fingers, where the transition from parallel to symmetric motion is accompanied by critical fluctuations.

Is quantum physics relevant for organisms? On the microscopic level, certainly: it governs the formation of chemical bonds, the opening of ion channels, the light-absorption of molecules. Rhodopsin is decomposed by only a few quanta of light, a probabilistic effect. The subsequent chain of reactions leads to visual perception—a "macroscopic" event.

Is this example an exception to the rule, or can the amplification of quantum, i.e., nondeterministic processes also give rise to macroscopic effects in other cases?

I believe only experiments can answer that question, but not pure "thought experiments." But irrespective of the "nature" of the chance event (quantum or classical) a human or animal must rely on it when she/it has to quickly respond to a novel situation that implies different choices. But what happens if the situation isn't novel? In this case, in our discussion we have to realize that there is a fundamental difference between living beings and physical systems: humans and animals can learn, physical systems cannot.

In this way, humans learn to deal with their surroundings, including other humans. Thus, their behavior becomes more reliable, more predictable—more deterministic. (Which doesn't rule out, however, that occasionally humans (and animals) act in quite an unexpected manner).

I am sure that Frank's excellent treatise will give rise to a further discussion on just what it means to be human.

University of Stuttgart, Germany   Hermann Haken

Preface

Humans and animals move and position their bodies in numerous ways. Humans walk, run, crawl, roll, sway, tiptoe, balance, jump, climb, reach, dance, tap, clap their hands, smile, sing, and make other noises, just to name a few examples. Humans stand, lie, sit, knee, crouch, and lean. Animals walk, run, trot, gallop, bound and hop, fly and swim in various ways. Humans not only can form dynamic and static patterns with their bodies and limbs but also form a rich variety of internal states. Humans think, imagine, have inner speech, dream, hallucinate, and feel. In general, humans and animals exhibit a huge variety of possible brain activity states and brain activity patterns. Moreover, there are a plenitude of ways how humans and animals interact with their environments and react to certain circumstances of their surroundings. Humans drink and eat, grasp, hold, push, pull, kick, throw, touch, rub, point to, draw, read and write, kiss, make love, and talk to each other. Humans listen, judge, estimate, and identify objects, other humans, and animals.

When considering humans and animals as systems, then we are puzzled with the richness of the system dynamics of those human and animal systems. Let us compare human and animal systems with mechanical systems such as a wheel or a seesaw. A wheel can roll or fall. A seesaw can stand still or move up and down. Mechanical systems lack the richness of human and animal systems. In view of the richness of the dynamics of human and animal systems, the question arises whether or not it is possible to address the variety of possible brain and body activity patterns produced by humans and animals from a unified scientific point of view—just as classical mechanics addresses mechanical systems like the wheel and the seesaw from a unified point of view.

This book provides an answer to this question. Accordingly, humans and animals are considered as self-organizing, pattern formation systems. The dynamics of human and animal systems or, more precisely, the formation of human and animal brain and body activity patterns is understood from a unifying physics perspective. The appropriate theoretical frameworks are the theory of pattern formation, as developed by various research groups, and synergetics, a theory of self-organization founded by Professor Hermann Haken.

The aim of this book is to provide an introduction to the self-organization, pattern formation perspective of human beings and animals. It is highlighted what humans and animals have in common with pattern formation systems of the inanimate world such as cloud patterns in the sky, hurricane spirals, and tornado funnels. It is shown that the aforementioned richness of possible body patterns, internal states, and reaction patterns of humans and animals can be explained within the theory of pattern formation and synergetics. In line with Skinner's famous conception of humans as machines, it is demonstrated that all human and animal aspects related to body patterns, internal states, and reaction patterns are determined by laws. In doing so, it is shown that determinism of human life is a built-in feature of the self-organization perspective. In particular, it is shown that the self-organization perspective is sufficient to explain various experimental findings ranging from gross motor movements to event-related neuronal firing patterns and psychological diseases without the need to mystify humans and animals.

Theoretical concepts will be introduced and explained in detail using graphs and examples. For this reason, this book is designed for undergraduate and graduate students and experimental researchers studying human perception, cognition, and behavior. Mathematical material is presented in separate sections. Therefore, the book can also serve as a guideline for theoretical researchers, computer scientists, and modelers that come with an advanced quantitative background.

Acknowledgments

I am indebted to Professor H. Haken for inviting me to write this book for the Springer Series of Synergetics. I wish to thank Dr. C. Caron and Ms. G. Hakuba of the Springer-Verlag for their help in preparing this manuscript.

Storrs, CT, USA   Till Frank
May 2019

Contents

Chapter 1
Introduction

1.1 What Am I?

The question "What am I" is a question that many of us have asked themselves at some point in time. This question is not about what we are doing for a living. We are factory workers, shop assistants, teachers, and so on. Somebody who asks the question "What am I" wants to know, what he or she is as an entity. Am I what my body is? Am I what I am thinking? Answering the question "What am I" is important for individuals in order to gain a satisfactory perspective of life. The question "What am I" has been asked from a bird's eye view as the question "What is a human?". Religion, philosophy, and the sciences have made various efforts to answer this question. In particular, the natural sciences such as biology, neuroscience, chemistry, and physics and the social sciences such as anthropology, sociology, linguistics, and psychology have attempted to draft a scientific picture of what a human being is. The scientific view of human beings thus obtained is also of importance for Western medicine. In the Western world, the diagnosis of a disease frequently relies on the scientific concept of human beings. Likewise, the treatment of a patient using therapy, drug medication, or surgery is conducted in line with the scientific view of human beings. This scientific view is supported by experimental studies and theoretical works and is permanently evolving. Not only does Western medicine profit from a proper understanding of the nature of human beings. The way people live together within a society depends on what the people living in that society think about the human nature. Likewise, the way different societies live together depends on how human nature is conceived across different societies. In summary, answering the question "What am I" benefits individuals, communities, societies, and different societies interacting with each other. The nature of humans is in part revealed by the experiences that humans make and the actions that humans perform. Therefore, the question about the human nature may be answered by addressing a related question "Why do humans behave, think, feel, and perceive as

© Springer Nature Switzerland AG 2019
T. Frank, *Determinism and Self-Organization of Human Perception and Performance*, Springer Series in Synergetics,
https://doi.org/10.1007/978-3-030-28821-1_1

they do?" The deterministic perspective proposed by Skinner [300] and the concepts of synergetics, a theory of self-organization and pattern formation proposed by Haken [152–154, 160, 162], allow us to address both questions, the related and the original one. In this book the attempt is made to bring Skinner's perspective and Haken's theoretical framework together in order to obtain a comprehensive view of human "perception", "cognition", and "behavior" and the nature of human beings in general.

1.2   A Note on Terminology

Throughout this book, the author will use a particular terminology. The specific terminology used in this book is not crucial for the understanding of this text [112]. What is at issue is that the reader gains an understanding of the meaning of the concepts that will be introduced by the author. Given that understanding, the reader may re-name the concepts and calls them differently. The reader may think of a dictionary that translates words between languages. For example, the English word "Apple" translates to "Apfel" in German, "Pomme" in French, "Mela" in Italian, "Manzana" in Spanish, and so on. Table 1.1 provides names of some of the concepts that will be used in the remainder of this first chapter and throughout the book. The reader should feel free to substitute the suggested names by phrases that he or she is more comfortable with.

1.3   Skinner's Perspective

Skinner and his co-workers conducted a series of experiments with animals in which they showed that animals could learn to react with different behaviors to objects exerting different forces on them. For example, pigeons were trained (i.e., restructured, see Chaps. 2 and 9) with the help of two signs such that the pigeons eventually reacted to one sign with a pecking behavior and to the other sign by rotating around their own axes like a spinning dancer. The reactions were considered

Table 1.1 Notations of some technical terms used in this book

Author	Reader
Brain and body activity patterns	
External forces	
Forces	
Physics-based determinism	
Physics perspective	
Self-organization	
Structure	

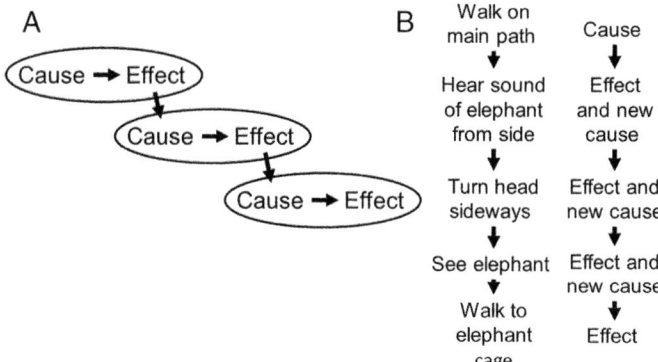

Fig. 1.1 Cause-and-effect relationships. Panel (**a**): A chain of cause-and-effect relationships. Panel (**b**): Example for a visitor walking through a zoo

cause-and-effect relationships. The signs were the causes, while the observed reactions were the effects. In fact, Skinner suggested that all human behavior can be explained in terms of cause-and-effect relationships. Importantly, an effect can be the cause of another effect. In doing so, sequences of experiences and actions can be explained as chains of cause-and-effect relationships, see Fig. 1.1a. Skinner states that "A response may produce or alter some of the variables which control another response. The result is a chain. ... One episode in our behavior generates conditions responsible for another. We look to one side and are stimulated by an object with causes us to move in its direction. ... Chaining need not be the result of movement in space. We wander ... verbally, for example, in a causal conversation or when we speak our thoughts" [300, p. 224].

Let us dwell on Skinner's example, see Fig. 1.1b. Let us consider a walk through a zoo. A visitor walks on the main path. Suddenly, he or she hears the sound of elephants from the side. The person turns the head sideways in the direction of the sound and sees the elephants. Subsequently, our visitor leaves the main path and walks to the elephant cage. Let us reformulate this little story in terms of cause-and-effect relationships. Walking through the zoo on the main path that passes by the elephant cage makes that the visitor comes close enough to the elephants to hear them. The elephant noises cause a head turn of the visitor. The head turn makes it possible for the visitor to see the elephants. Seeing the elephants makes the visitor to leave the main path and walk towards the elephant cage. The sequence of reactions or actions performed by the visitors can be described within the framework of synergetics as will be shown in later chapters. For modeling purposes, we may simplify the chain of events as shown in Fig. 1.2a. The walking dynamics of the hypothesized zoo visitor is modeled by means of two behaviors: walking on the main path and walking towards the cage. To mathematize the example, the variable W is introduced and is put equal to 1 to denote walking on the main path and equal to 2 to denote walking towards the cage. Likewise, the head orientation is described in terms of two distinct states. The head either points in the direction of the walking

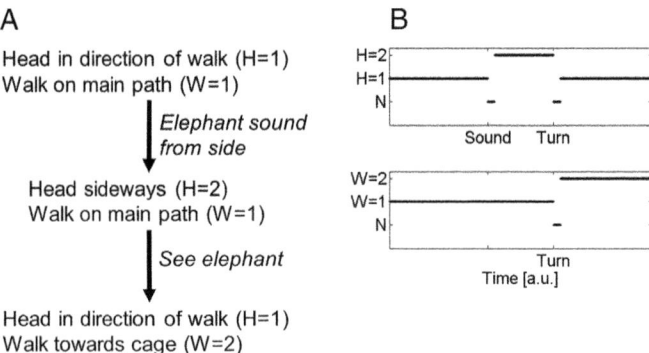

Fig. 1.2 Simplified version of the zoo visitor example. Panel (**a**): Sequence of events. Panel (**b**): Simulation results from the human pattern formation model that will be introduced below and in later chapters

path or is oriented sideways. Again, a variable is introduced, the variable H. H assumes the value 1 to describe head orientation in the direction of walk and 2 to describe that the head is turned sideways. Figure 1.2a presents the short story about the zoo visitor as a three-events sequence. The visitor walks on the main path with the head in the direction of the walk. This brings him or her close to the elephant cage and makes that he or she hears elephant sounds from the side. Hearing the sound causes the visitor to turn the head sideways while walking on the main path. The visitor sees the elephants and, as a result, walks towards the cage. The head turns back such that it is oriented in walking direction. Figure 1.2b presents results from a simulation of the self-organization, pattern formation model that will be introduced in later chapters. In this simulation not only two states for walking and head dynamics but three states have been used. It is taken into account that when the visitor switches between categorically different states (e.g., from walking on the main path to walking towards the elephant cage) that there is a short period in which the dynamics cannot be classified as belonging to either of the two states. In Fig. 1.2b these periods are marked as "N" states, where "N" stands for "nothing".

Let us return to Skinner's main argument about cause-and-effect relationships determining human experiences and behaviors. Accordingly, whatever an individual experiences and does follows a sequence that is determined by laws. Consequently, things of the animate (humans, animals, etc.) and inanimate (pieces of ice, pieces of iron, etc.) worlds are treated on an equal footing. They are described by states that evolve in time and satisfy certain evolution equations. Within the framework of synergetics such laws in terms of evolution equations can be derived from bottom-up, mechanistic (e.g., neuronal) approaches [38, 39, 107, 158, 159, 215], see also Sect. 4.5. In addition, laws for human experiences and behavior may be formulated irrespective of any explicit mechanistic perspective using a top-down modeling approach [158, 215]. In fact, the model that underlies Fig. 1.2b was derived with the help of the latter approach.

1.4   Physics Perspective

Skinner's point of view that there are laws that determine human experiences and behavior is consistent with the fundamental assumption of physics about the lawfulness of nature. To look at the world from a physics perspective means to assume that all observations satisfy lawful relationships. In this context, the author defines the physics perspective as the assumption that the laws of physics determine all events and processes in the animate and inanimate worlds. In particular, the life of any human individual is determined by the laws of physics.

   Physics-based determinism is the application of the physics perspective to understand the nature of humans and animals and a narrowing of the perspective in a particular sense. Accordingly, physics-based determinism comes with the assumption that deterministic laws (i.e., laws that do not involve a random component) are in a large number of cases sufficient to describe human experiences and human and animal behavior. A more detailed discussion about these issues will be presented in Chap. 2. Since the laws of physics describe cause-and-effect relationships, the physics perspective of the universe and the concept of physics-based determinism imply both that humans and animals are physical systems, whose experiences and behaviors are determined by chains of cause-and-effect relationships. In doing so, Skinner's point of view about the chaining of cause-and-effect relationships is re-obtained.

1.5   Self-Organization, Pattern Formation, and Synergetics

Let us clarify concepts and theoretical schools that contribute to synergetics as a theoretical framework. Figure 1.3 provides a schematic overview in this regard. First of all, as mentioned above, applying the physics perspective to understand human and animal experiences and behavior (and adding some extras concerning randomness) leads to the concept of physics-based determinism. There are two theories, the theory of self-organization and dynamical systems theory, that both are tailored to address deterministic systems as defined by the concept of physics-based determinism. The theory of self-organization and dynamical systems theory are two related theories without clear boundaries such that various elements of dynamical system theory can be found in the theory of self-organization and vice versa. Throughout this book, self-organization refers to mechanisms that require that systems are composed of several components (at least two). Spatially extended systems such as chemical substances solved in a liquid in a small container, ants of an ant colony, or neurons belonging to a particular cortical brain area belong to the class of multi-component systems as well. Roughly speaking, self-organization means that there is a multi-component system whose components interact with each other and, in doing so, produce a certain outcome. That is, the mechanism of self-organization puts emphasis on interactions between components. Self-organization

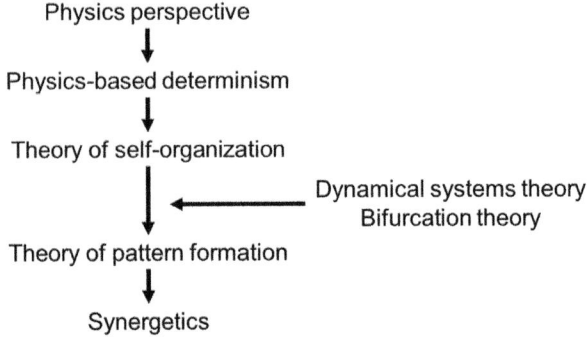

Fig. 1.3 Overview of various theoretical frameworks and relations between them

describes all kind of mechanisms that involve interactions between components (or parts) such that the interactions lead to the observation of interest. Within the theory of self-organization there are various distinct research areas.

A particular research area is the theory of pattern formation, see Fig. 1.3 again. A precise definition of the concept "pattern" will be given in Chap. 4. For the moment, only some examples of patterns will be named. Geometric patterns such as stripes may correspond to spatial patterns in the sense the phrase "pattern" is understood in this book. For example, the stripes of a zebra make up a spatial pattern [243]. The key issue is that the pattern is produced in a particular way, namely, by means of a self-organization of the components of the system under consideration. The beats produced by a drummer, who hits a drum with a particular frequency (say ten beats per second), constitute a temporal pattern. Again, the actual temporal structure is not the key issue here. As we will see in Chap. 4 the issue at hand is how the temporal structure emerges. If it emerges in a certain way, then it qualifies as a pattern. There are not only spatial and temporal patterns. Rather, there are spatiotemporal patterns as well. For example, hot air steaming up at some locations and flowing down as cold air at other locations can produce cloud patterns in the air that form rolls. These rolls are patterns that extend in space. In addition, they are temporal structures because if a small volume of air molecules would be colored somehow, then that marked volume would change its position in time. It would travel downwards for a while with the cold air and upwards for a while with the hot air. Patterns may be grouped into categories. For example, with respect to human locomotion there is the category of walking patterns and the category of running patterns. A human being, who is walking, performs one pattern out of the category of walking patterns.

The theory of pattern formation is concerned with the formation of patterns due to self-organization. For example, the objective is to explain how a spatially extended chemical system that shows initially no spatial structure can turn into a system that shows a stripe pattern. The theory of pattern formation is also concerned with the question which pattern of a category emerges. For example, in the context of the aforementioned chemical system, the question arises, why the emerging stripe pattern exhibits a particular spatial period, for example, a period of 1 centimeter.

Finally, if there are different types (categories) of patterns, the theory of pattern formation is concerned with the transitions between those types of patterns. For example, the theory of pattern formation may be applied to study gait transitions in humans such as transitions from walking to running. The theory of pattern formation makes extensive use of concepts of bifurcation theory, which in turn can be considered as part of dynamical systems theory.

Bifurcations will be introduced in detail in Sect. 3.2. At this stage, a bifurcation should be understood as a transition between qualitatively different states, that is, a transition in which a system switches between categories.

Finally, there are several scientific schools that have suggested in the literature different approaches to describe and understand pattern formation. Synergetics is one of those schools. Just as the theory of pattern formation, synergetics makes extensive use of the concepts of bifurcation theory, see Fig. 1.3 again.

The key concepts of synergetics will be discussed in later chapters. At this stage, only one of the most important features of synergetics for our understanding of human and animal behavior (and experiences) should be mentioned. Within the framework of synergetics, transitions between categorically distinct experiences and behaviors are discussed in analogy to phase transitions of matter. An important class of such phase transitions of matter are the transitions between the aggregate states of matter: solid, liquid, and gas. Figure 1.4 illustrates the analogy between behavioral transitions and phase transitions of matter. Panel (a) illustrates phase transitions between aggregate states of water. When starting at a sufficiently low temperature and increasing the temperature gradually, phase transitions from ice to water and water to gas can be observed. Panel (b) illustrated transitions between postural states and gaits. A sitting person may stand up and start to walk. The ice-water-gas sequence of aggregate states is obtained by phase transitions. The sitting-standing-walking sequence is assumed to exhibit transitions that are of a similar kind to the phase transitions of matter. Just like ice turns into water and into gas under appropriate conditions, a person under consideration turns from a sitting person into a standing person and from a standing person into a walking person under appropriate conditions, see Fig. 1.4 again. Due to this feature, synergetics as a conceptual framework arrives again at Skinner's conception of human beings as physical entities determined by cause-and-effect relationships. There are certain conditions that make us to do the things we do (just as there are certain conditions that make that ice turns into water).

However, in contrast to Skinner's perspective and in contrast to the afore-mentioned, general physics perspective, the framework of synergetics provides

Fig. 1.4 Analogy between phase transitions between aggregate states (panel **a**) and behavioral transitions (panel **b**)

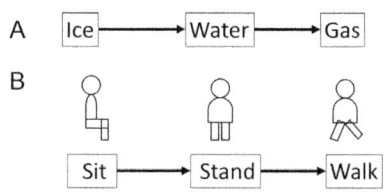

additional explanatory power. While from the physics perspective the conclusion can be drawn that human experiences and behaviors are determined by laws (see above and see more in this regard in Chap. 2), this conclusion does not readily provide us with mechanisms that explain how experiences and behaviors emerge and how and why transitions between different types of experiences and behaviors take place. Synergetics describes the emergence of experiences and behaviors and transitions between them as the formation of patterns and transitions between patterns due to self-organization. In doing so, synergetics features principles and mechanisms on a certain level of description that will be discussed in detail in later chapters. These mechanisms of self-organization apply not only to living systems such as humans and animals but also to inanimate pattern formation systems such as lasers and hurricanes. Synergetics is concerned with self-organizing pattern formation systems of the animate and inanimate worlds and treats them on an equal footing. Therefore, according to synergetics, the mechanisms of self-organization that determine human experiences and behaviors (and animal experiences and behaviors) belong to a general class of mechanisms and exhibit a wide scope of applications.

1.6   Human Pattern Formation Reaction Model

The model used throughout this book is referred to as human pattern formation reaction model. Below only a short introduction will be presented. The model will be worked out in detail in Chap. 5. In Sect. 1.7.2 below and in Chap. 5 it will also be motivated why the model is referred to as reaction model. At this stage, human reactions should be considered as the way how humans react to (the forces exerted by) objects and events in their environments. For example, a driver reacts to a stop sign at the road by pushing the brake pedal. As such the human pattern formation reaction model applies to animals just as it applies to humans. To simplify the presentation, in the sections below, it will be introduced for humans.

1.6.1   Component and System Dynamics Descriptions

Let us introduce the basic model for understanding human reactions and the dynamics of isolated human systems (see Chaps. 5 and 6) within the framework of pattern formation and synergetics. The model will be referred to as human pattern formation reaction model. At this stage, only a brief introduction will be given. The model will be refined in the following chapters.

The human pattern formation reaction model involves two descriptive levels that are illustrated in Fig. 1.5. Panels (a)–(c) introduce in several steps the components of the model and the interactions between the components. Panel (d) describes how patterns evolve in the components shown in panels (a)–(c). In this sense, panel (d) is

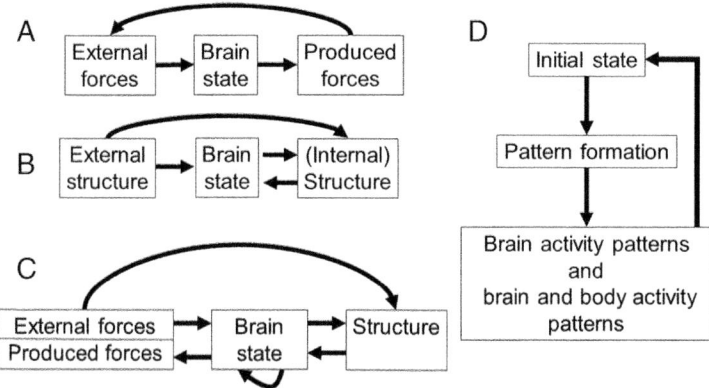

Fig. 1.5 Component descriptions (panels **a**, **b**, **c**) and system dynamics description of the human pattern formation reaction model (panel **d**)

about the dynamics of the formation of patterns. Let us start with the components, see panels (a)–(c). Panels (a) and (b) are for didactical purposes only. They will be merged to panel (c). As illustrated in panel (a), the model considers forces acting on human individuals and forces produced by individuals. In doing so, the model takes a classical physics approach as starting point. Accordingly, systems interact via forces. Forces acting on individuals correspond to forces of the environments of the individuals of interest. Therefore, they are considered as external forces. Examples of external forces will be given below. Humans produce forces to move things around (e.g., to displace a cup). In the context of our model, all kind of muscle forces are considered as produced forces. For example, the forces that result in a grasping movement are considered as produced forces. In certain applications of the model it might be useful to consider not only the forces but also the consequences of the forces with respect to the body parts. For example, the position and orientation of the left upper and lower arms might be described in terms of the positions of three markers placed on shoulder, elbow, and wrist. These marker variables do not describe forces but they describe parts of the individual under consideration that do not belong to the nervous system and the human brain, in particular. Therefore, these variables might be put in the same class of variables as the variables that describe the produced forces. In other words, the model considers a class of variables that describe the produced forces of an individual and variables that change in time due to those forces but are not related to the nervous system and the human brain, in particular.

A key component of the model is the brain state of an individual of interest. This state describes the brain activity as measured in various ways (e.g., electric potentials measured by electroencephalography over the scalp or firing rates of neurons). The sensory system of the human body acts as an interface between external forces and the brain state. Due to this interface, external forces affect the dynamics of the brain state. Likewise, the muscular system of the human body (in combination with other body systems) can be considered as interface between the brain state and the produced forces. With the help of this second interface, various relationships between brain states and produced forces can be described (i.e., it can be described how produced forces depend on brain states). Finally, produced forces may affect and change environmental circumstances such that the forces exerted by the environment and acting on an individual change as well (see the curved arrow in panel (a) pointing from the right to the left).

As shown in panel (b), the model takes structural components of individuals and their environments into account. The structure of the environment is referred to as external structure, while the structure of individuals is called the internal structure. However, in what follows the phrase "internal" will be dropped. Since interactions between individuals and their environments take place by means of forces, the external structure of a person may be described by the external forces acting on that person. Therefore, external structure and external forces are two aspects of the same system. In what follows, we will not distinguish between these two aspects and, for the sake of simplicity, use only the phrase "external forces" to refer to the environmental conditions of an individual. Consistent with panel (a), the arrow in panel (b) pointing from external structure to the brain state points out that in general external forces (i.e., external structure) act on the brain state of an individual. Moreover, the structure of a human individual (i.e., the internal structure) addressed in panel (b) refers to the properties of the human brain and body that can be considered to a first approximation as being fixed on the time scale on which the brain state evolves. In particular, the structural components of the brain are defined as those components that determine the evolution of the brain state (see the arrow pointing to the left from the structure to the brain state) but do not vary in time on the time scale on which the brain dynamics of interest takes place or they vary only to a negligibly small amount. Having said that in order to take plasticity of the neuronal system into account it is assumed that the brain state can affect the structure and induce changes on relative slow time scales (see the arrow pointing to the right from the brain state to the structure). Finally, panel (b) also indicates that external forces may have a direct impact on the structure (see the curved arrow pointing to the right from the external structure to the internal structure).

Panel (c) merges panels (a) and (b). In addition, panel (c) highlights that the brain is a dynamical system. That is, the evolution of human and animal brain activity depends on the ongoing brain activity. This is illustrated by the curved arrow that points from the brain state to the brain state. Panel (c) presents the component description of the human pattern formation reaction model. Table 1.2 summarizes various features of the components shown in panel (c) as discussed so far.

Table 1.2 allows us to summarize the component description of the model. Accordingly, the model distinguishes between the individual and its environment. The individual is described in terms of three (internal) components: produced forces, the brain state, and structure. The environment is described in terms of one external component: the external forces. The three internal components are divided into two groups. The first group is composed of two components that describe variables that change in time: the brain state and the produced forces. The second group consists

Table 1.2 Components and their characteristic properties

Component	External versus internal	Force versus brain activity	Constant versus time-dependent
External forces/external structure	External	Force	Const. or vary
Produced forces	Internal	Force	Const. or vary
Brain state	Internal	Brian activity	Varies with time
(Internal) Structure	Internal	Related to both	Const. or slowly varying

of a single component. This component describes properties that do not change over time (at least to a first approximation) and is called the structure. As far as the time varying components are concerned, the model distinguishes between neuronal activity and muscle activity (and its consequences in terms of body movements). The former is captured by the brain state, while the latter is given in terms of the forces produced by an individual (and the positions and orientations of the body parts of that individual). The structure describes fixed properties both of the nervous system and the muscular system. In doing so, structure relates both to brain activity and produced forces (as indicated in Table 1.2). Note that the components shown in panel (c) are sufficient to describe most of the phenomena that will be presented in the following chapters of this book. Without loss of generality, further components might be added if needed. That is, the pattern formation reaction model is not restricted to the use of the components shown in panel (c) of Fig. 1.5.

The structure mentioned above should be considered as an effective structure because typically a part of the human body or brain is composed of smaller parts. For example, a synapse, which is a particular connection between two neurons, might be considered as a basic structural entity. However, when having a closer look, a synapse might be described in terms of three structural subparts: the pre-synaptic and post-synaptic parts and the synaptic cleft.

Panel (c) illustrates possible interactions between the components that have been discussed already in the context of panels (a) and (b). The human brain may transmit signals to the muscular system. The signals may lead to the production of certain forces that result in certain body movements. The produced forces may also be involved in maintaining certain postural arrangements of the body. Produced forces and in particular body movements may affect the environment and, in doing so, the external forces. The environment as described in the model by means of external forces may affect both the brain state and the structure. The brain state is assumed to evolve in time. The dynamics of the brain state in general depends on the current structure of the brain. In this sense, structure affects the dynamics or evolution of the brain state. Structure was introduced above as the fixed properties of the human body, in general, and the human brain, in particular. However, structure properties may exhibit slow drifts induced by brain activity. If so, brain states affect structure. By taking into consideration that there might be an effect of the brain state on its structure, the model can capture neuronal plasticity induced by the activity of the brain itself.

Let us turn next to panel (d). Panel (d) describes schematically the formation of patterns as relevant for our understanding of human experiences and "behaviors". As shown in panel (d), pattern formation starts with an initial state. This initial state may be affected by prior patterns, which leads to a circular causality loop that will be discussed below.

While the notion of patterns has been introduced above for all kinds of systems of the animate and inanimate worlds, in the context of the human pattern formation reaction model, patterns are defined as special configurations of the components involved in the pattern formation. Brain activity (BA) patterns are either special brain states or special states describing the human brain and its structure. The latter case applies when the brain structure is slowly changing. Brain and body activity (BBA) patterns are all remaining cases that involve either the production of forces or external forces acting on the individual of interest or both. More precise definitions regarding patterns will be given in Chaps. 4 and 5. Patterns emerge and disappear. These transient processes are referred to as pattern formation.

Patterns may form (i.e., emerge) only in a subset of the components shown in panel (c). For example, pattern formation may be restricted to the brain state or it may involve the brain state and its structure. Pattern formation may involve produced forces and the brain state but not external forces and structure. A systematic discussion in this regard will be given in Chap. 5.

With respect to the terminology "perception", "cognition" and "behavior" that has been frequently used in the literature, BA patterns may be related to "end-products of cognitive activities". Likewise, BBA patterns may be related to "percepts" and "behaviors". The emergence and disappearance of patterns may regarded as "cognitive activity" in the case of the formation of BA patterns or as "perception" or the buildup of "behavior" in the case of the formation of BBA patterns. However, as will be shown in later chapters the terminology "perception", "cognition", and "behavior" is not needed for our understanding of human reactions and the dynamics of human isolated systems.

1.6.2 Ordinary State and Grand State

As discussed in the previous section, it is useful to distinguish between structure and state. In this context, the brain state, the produced forces, and the limb and body positions taken together describe the ordinary state of a human being or animal. As illustrated in Fig. 1.6 the ordinary state evolves on a time scale on which the structure is assumed to be constant or varies only slowly. The variables of the ordinary state and the structure taken together describe the individual of interest. In contrast, the environment (as discussed above) is described in terms of external forces. Taking the variables of the individual and the environment together we obtain a complete description of the overall system. We refer to the state of the overall system as grand state, see Fig. 1.6.

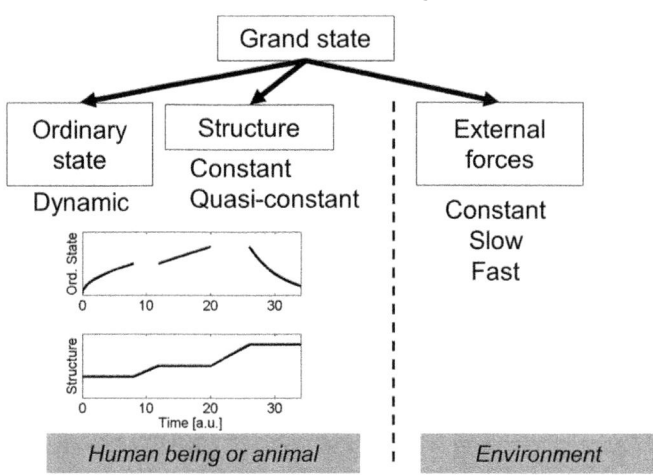

Fig. 1.6 Illustration of the distinction between ordinary state and grand state

In the previous section, Sect. 1.6.1, we also introduced the notion of an initial state. In this context, note that frequently we will used the phrase initial conditions rather than the phrase initial state. As such, the phrase initial conditions takes a broader perspective and refers to the values of the ordinary state variables and the values of the structure variables at a particular reference time point. The initial conditions even refer to the circumstances of the environment at that time point. However, if we describe the system on the level of the grand state then the initial grand state just describes what we have summarized before: the initial ordinary state, the initial structure, and the initial external forces. For this reason, the phrases initial state and initial conditions are to some extent interchangeable.

1.6.3   Circular Causality

The pattern formation reaction model exhibits various circular causality loops on the level of the component description (see Fig. 1.5c) and on the level of the system dynamics (Fig. 1.5d). These loops indicate that pattern formation may be subjected to circular causality. Three of the circular causality loops will be discussed in what follows.

As mentioned above, the formation of patterns as a dynamical process (panel (d) of Fig. 1.5) features a circular causality loop that links initial states of pattern formation with final states and vice versa final states of pattern formation with initial states. Given the formation of a particular pattern of interest at a particular time point of interest, the initial state relevant for that process may be affected by the BA or BBA pattern that has emerged or disappeared prior to the time point of interest. As

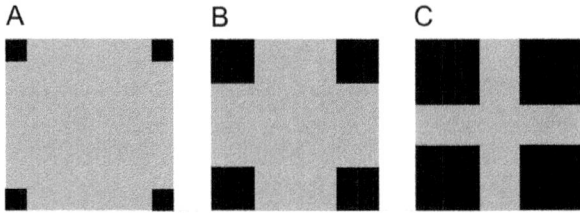

Fig. 1.7 Three images that have been used in a particular figure-ground experiment

a result, the pattern formation model presented in panel (d) of Fig. 1.5 features a circular causality loop. On the one hand, pattern formation leads to the emergence of BA or BBA patterns. On the other hand, the buildup and decay of patterns at a given time point may depend on the pattern that has emerged or disappeared prior to that time point. Figure 1.7 exemplifies this feature of pattern formation for a particular experiment on "figure-ground perception" [139, 187]. Figure 1.7 shows three images. The image in panel (a) is typically "perceived" as four squares that are located in the four corners of a light-colored square. In contrast, the image in panel (c) is typically "perceived" as a light-colored cross on a dark background. Experimental research has shown that the image presented in the middle (panel b) can be "perceived" both as four squares or as a light-colored cross [139, 187]. It has been shown that when image (a) is shown first and image (b) second, then image (b) is typically "perceived" as four squares. In contrast, when image (c) is shown first and subsequently image (b), then image (b) is typically "perceived" as a light-colored cross on a dark background. This illustrates that "perception" depends on what was "perceived" previously. From the perspective of pattern formation, the example illustrates that the pattern formation of BA/BBA patterns at a given time point depends on what BA/BBA pattern was present prior to that time point. We will return to this example in Sect. 6.5.

Note that the loop described in Fig. 1.5d can involve various components of the human pattern formation reaction model shown in Fig. 1.5c. In the simplest case the circular causality loop only takes place on the level of the brain state. The curved arrow that points from the brain state to the brain state in Fig. 1.5c indicates such a feedback loop.

Let us return to the component description sketched in panel (c) of Fig. 1.5. BA and BBA pattern do not only affect the pattern formation of subsequent patterns as discussed above. Rather, they may change the structure (i.e., the internal structure) on time scales slower than the time scales that determine the emergence and disappearance of patterns. In this case, the two arrows pointing in panel (c) of Fig. 1.5 from structure to the brain state and from the brain state to the structure establish a circular causality loop. More precisely, the arrow pointing to the left from structure to brain state describes the situation in which a given structure supports the emergence of a particular pattern. In the circular causality scenario, the emerging pattern is assumed to change the structure that determined the pattern formation process in the first place. This restructuring process is indicated by the arrow pointing to the right from the brain state to the structure. As a result of the structural change, the pattern that just has emerged may disappear again.

In the simplified scenario that will be considered throughout this book, pattern formation systems exhibit two different time scales. According to Fig. 1.6a, the structure can be regarded as constant (or slowly evolving) over the time period of the buildup and disappearance for a particular pattern, say pattern A associated with a particular "percept", "cognitive end-product", or "behavior". Subsequently, the structure changes, which is assumed to trigger the disappearance of pattern A and buildup of another pattern, say pattern B. In general, pattern B will be associated with a "percept", "cognitive end-product", or "behavior" different form the one associated with pattern A. Therefore, the right hand side of Fig. 1.5c involving the two components, brain state and structure, describes a loop that can lead to sequences of "perceptual experiences", "chains of cognitive activities and their end-products", or "behavioral sequences".

The left hand side of Fig. 1.5c describes a similar loop that involves time-varying external forces. This loop has been address several times above. For example, a human may perform a particular "behavior" that is associated with a certain BBA pattern, say pattern A. Due to that "behavior" that we may label as "behavior A" the person of interest may be exposed to an external force that was not present before. That force, in turn, may affect the formation of BBA patterns such that the BBA pattern A disappears and another BBA pattern, say pattern B, emerges. Pattern B is assumed to be associated with a different "behavior". Consequently, the person under consideration makes a transition from one "behavior" to another one. In the end, the left hand side describes a loop that can lead to sequences of "behavioral activities" due to the fact that humans with the help of their produced forces can change the external circumstances under which their "behaviors" emerged in the first place. In doing so, the effect A of a cause A can change (e.g., remove) its cause A and replace it by a cause B, which leads to an effect B.

Both loops of the component level may act together. In fact, the human reaction model that was used to simulate the hypothetical zoo visitor on its walk through the zoo (see above) was set up in a way that it exploits both loops at the same time. Accordingly, the chaining of the cause-and-effect relationships presented in Fig. 1.2 is a consequence of the two aforementioned loops on the component level.

1.7 Forces

1.7.1 Fundamental and Auxiliary Forces

Humans produce forces and are subjected to forces. Therefore, it is worthwhile to clarify what kind of forces are relevant to understand human reactions and the dynamics of the human isolated system. There are four fundamental forces in

Table 1.3 Fundamental and
auxiliary forces

Fundamental forces	Auxiliary forces
Gravitational force	Mechanical forces
Electromagnetic force	Chemical driving forces
Strong force	
Weak force	

physics: the gravitational force, the electromagnetic force, the weak force and the strong force. They are listed in Table 1.3. The gravitational force acts on objects that have mass and pulls masses together. Gravitation is important for human body movements and posture. Walking, running, sitting, lying, etc. is possible due to the gravitational force of the earth that pulls us down to the floor. In outer space the gravitational pull of the earth is greatly reduced. For astronauts it becomes difficult to walk. The electromagnetic force acts between charged particles and determines the dynamics of electromagnetic waves. In particular, light waves of the visual spectrum acting on an observer's eyes and the electric signals in a human brain are determined by the electromagnetic force. As we will see in a moment, the electromagnetic force plays a key role for human "perception" and "behavior" in other contexts as well. The strong force holds the nucleus of atoms together. The nucleus of atoms is composed of positively charged particles. Due to the electromagnetic force these particles push each other away. Therefore, as a consequence of the electromagnetic force any nucleus would immediately fall apart. The strong force holds the particles together despite the electromagnetic force. To the best of the author's "knowledge", the strong force does not have any relevance for understanding the human nature (except for the fact that all atoms and molecules humans are composed of only exist because of the strong force). The weak force plays like the strong force only a role in atomic physics. The weak force determines a particular decay of atoms, the so-called beta decay.

In addition to these four fundamental forces, classical mechanics introduces another force, the mechanical force, as a force that is exerted by an object and acts on another object. For example, if a metal ball rolls on a plain surface and hits another metal ball that does not move, then the first ball exerts a mechanical force on the second ball. Frequently, mechanical forces are in fact electromagnetic forces. In the ball collision example the atoms in each ball are held together by electromagnetic forces. The first ball does not destroy the second and, vice versa, the second ball does not destroy the first ball because the electromagnetic forces hold the metal atoms together. When the two balls collide, at the collision interface the balls interact via electromagnetic forces. For this reason, the mechanical force exerted by the first ball on the second actually is an electromagnetic force. Mechanical forces are important. It would be inconvenient to describe many problems in classical mechanics only in terms of the four fundamental forces. Mechanical forces play an important role for understanding human movements and posture and all kind of "activities" that involve force production via the muscular system. For example, if we push a box with both hands along a plain surface then we produce a force that can conveniently be described by a mechanical force acting on the box. However, as it has been argued in the example of the two colliding balls, the force actually is an electromagnetic force that acts on the interface between our hands and the surface of the box. In this sense, electromagnetic forces do not play only a role for understanding light waves or brain activity but they play also a role for our understanding of all kind of circumstances in which humans produce or are subjected to mechanical forces. Table 1.3 lists mechanical forces as auxiliary forces. There is another force listed in Table 1.3 in this class of auxiliary forces: the chemical driving force. Chemical reactions can conveniently be described by chemical reaction equations and chemical driving forces. However, as such, chemical reactions happen due to the electromagnetic forces between atoms and molecules. Chemical driving forces that drive chemical reactions are in fact electromagnetic forces.

Again, using electromagnetic forces to describe chemical reactions is for practical applications inconvenient. Therefore, chemical driving forces just as mechanical forces are useful forces although they do not exist in the sense of a fundamental force. Chemical driving forces determine all biochemical reactions in the human body. In particular, they determine cell signaling in neurons by means of biochemical reactions. Moreover, they determine the transmission of electromagnetic forces from one neuron to another neuron at the synaptic gap by means of biochemical reactions. In doing so, the dynamics of brain activity is not only determined by electromagnetic forces but also by chemical driving forces (which, however, in the end are electromagnetic forces again). Table 1.4 shows which forces are relevant for our understanding of brain activity, reactions to external forces as mediated by the sensory systems (vision, hearing, touch/haptics, taste and smell), and force production.

Table 1.4 Applications of forces (for body movements and posture only produced muscle forces are considered, force production and descending/ascending signaling to/from the muscular system is ignored)

Application	Forces involved
Brain activity	Electromagnetic forces and chemical driving forces
Vision	Electromagnetic forces and chemical driving forces (at receptors in the human eye)
Hearing	Mechanical forces (producing sound waves), chemical driving forces (at receptors in the ear), electromagnetic forces (at receptors in the ear)
Touch/haptics	Mechanical forces (producing deformations), chemical driving forces (at skin receptors), and electromagnetic forces (at skin receptors)
Taste and smell	Chemical driving forces and electromagnetic forces (at receptors)
Body movements and posture	Gravitational forces and mechanical forces

1.7.2 Reaction to Forces Versus "Stimulus-Response Theory"

Humans and animals live in environments that determine to some extent what they do and what processes take place in their neuronal systems. For example, consider a hypothetical father-daughter situation as shown in Fig. 1.8. Let us assume the father is looking at his daughter. Light waves are reflected from the daughter towards the eyes of the father. As a result, light enters the eyes of the father and the light of the visual spectrum exerts a forces on light-sensitive molecules of receptor cells of the eyes. This leads to biochemical reactions and the production of electric forces that make that electrical signals are transmitted to certain areas in the brain. The electrical signals that in turn come with their own electric forces can affect the formation of BBA patterns in various ways. At this stage, it is important to notice that the daughter can be considered as the origin of the forces acting on the visual system of the father or on the father as a whole, when he is considered as a pattern formation system. In general, we say that when a living thing or an object is the origin or source of a force then that living thing or object exerts a force. In this sense, the daughter exerts a force on the father. The daughter does not only exert a force on the father in the visual world when the father is looking at the daughter. Rather, the daughter has various other possibilities to exert forces on her father. For example, as indicated in Fig. 1.8 the daughter may talk to the father (e.g., calls him "Dad"). In doing so, the daughter produces sound waves that hit the mechanoreceptors of the ears of her father. The sound waves exert mechanical forces on the receptors. They are transformed again into electric forces that can impact the formation of BBA patterns in certain areas of the father's brain. When father and daughter are holding hands, as they do in Fig. 1.8, then the daughter exerts forces on the skin receptors of the father. In addition, the daughter may want the father to go into a particular direction and may pull on his hand. If so, the daughter exerts a force on the father again. In general, the human senses of vision, hearing, touch, smell, and taste are subsystems of the human body that are affected by forces that come from the environment. Depending on what is appropriate for a proper description of the situation under consideration, we may locate the origin of forces in another person. If so that other person exerts a force on the person of interest. Objects (e.g., a red traffic light, a loudspeaker from which music is played, a closing door

Fig. 1.8 Father-daughter situation in which the daughter exerts different kind of forces on the father (and vice versa the father exerts forces on the daughter)

that hits a person) can exert forces on humans too. Finally, we may not locate the origin of a force in a living thing or object but rather see the origin of the force in a physical phenomenon. For example, we may consider the light waves hitting the father's eyes as the source of the forces from the visual world surrounding the father. Likewise, we may consider the soundwave as the source of the forces that act on the mechanoreceptors in the father's ears. In summary, living things, objects, and physical phenomena can be the origin of forces that act on humans (and animals) and, in doing so, can affect (and determine to some degree) the formation of BA and BBA patterns. In other words, living things, objects, and physical phenomena exert forces on humans (and animals). Humans (and animals), in turn, have various ways to react to those forces. Therefore, the pattern formation model outlined in Fig. 1.5 is referred to as a reaction model of humans (and animals) as seen from the perspective of the theory of pattern formation and the theory of synergetics.

These considerations are in analogy to consideration in classical mechanics on particle systems. Particles can be subjected to forces. If so, particles react to the forces applied to them. This fundamental perspective is taken in Fig. 1.5a. Figure 1.5a should been seen in analogy to the classical mechanics perspective of a particle that is subjected to forces and reacts to the forces by producing forces. All what has been changed in Fig. 1.5 with respect to this classical mechanics perspective is that the particle has been replaced by the brain state. While in classical mechanics a particle is subjected to forces and produces forces on other particles, the human pattern formation reaction model takes that the brain state is subjected to forces and is involved in the production of forces.

In the literature on human (and animal) "behavior" frequently the term "stimulus" has been used. This term is neither used in classical mechanics nor in the theory of pattern formation. A ball on an inclined ramp rolls down the ramp because the gravitational force acts on the ball and makes it to roll down. The ball does not roll down because of any "stimulus" acting on it. A hurricane in the ocean is a self-organizing, spatio-temporal pattern. The hurricane forms when the conditions over the ocean are appropriate. In particular, the spiral form of the hurricane is a consequence of the forces that act on air and liquid molecules (in part due to the earth rotation). The hurricane pattern does not form because there is any "stimulus" acting on it. The transition from a clear sky pattern to a hurricane pattern is induced by the forces acting on the liquid/air system over the ocean. As mentioned above, transitions between qualitatively distinct states of pattern formation self-organizing systems are seen in analogy to phase transitions. In the context of phase transitions, we note that ice turns into water (as indicated in Fig. 1.4a) under appropriate external conditions such as sufficient heating. Phase transitions are not affected or induced by "stimuli". Likewise, when a sitting person stands up then this is due to the external and internal conditions of that person (see Fig. 1.4b). In particular, let us imagine there is an alarm clock ringing and indicating that it is time to leave and to go to catch the bus. The alarm clock exerts forces on the human system and, in doing so, induces a transition from sitting to standing. This situation is just like the situations described earlier: there are appropriate conditions that make a person to make a transition from sitting to standing, there are appropriate circumstances under which hurricanes emerge, and there are appropriate conditions under which phase transitions happens. The concept of a "stimulus" is not needed when describing human reactions from a physics perspective. Moreover, it is not recommended to use the concept of a "stimulus" because if the concept of a stimulus is used in studies on humans and animals for pattern formation phenomena but the concepts of forces and external conditions are used to describe aggregate phase transitions and pattern formation in the inanimate world, then we would use two different descriptions for the same basic phenomenon. We would indicate that there is a fundamental difference between pattern formation in humans and animals, on the one hand, and pattern formation and phase transitions in the inanimate world, on the other hand. However, the principles of physics apply to systems of the animate and inanimate worlds in the same way. There is no fundamental difference between pattern formation in the inanimate and animate worlds. Note that the aforementioned appropriate conditions frequently can be described by certain parameters that will be introduced in Sect. 3.2.2.

Let us turn to the hypothetical zoo visitor that is described by the sequence of three BBA patterns shown in panel (a) of Fig. 1.2. First, the visitor forms a BBA pattern of walking on the main path with the head in direction of the walk. Subsequently, the visitor forms a BBA pattern with the head turned sideways still walking on the main path. Finally, the visitor forms a BBA pattern of walking towards the elephant cage with the head oriented in the direction of the walk. The first transition between BBA patterns is induced by forces exerted by the elephants in terms of sound waves that act on the visitor. The second transition between BBA patterns is induced by forces exerted by the elephants in terms of electromagnetic waves of the visual spectrum that act on the visitor. As mentioned above, the theory of pattern formation, in general, and synergetics, in particular, are theoretical frameworks that address self-organizing, pattern formation systems of the animate and inanimate worlds on an equal footing. Within these frameworks self-organizing, pattern formation systems (and in particular humans and animals) are affected by forces.

1.8   Levels of Analysis

1.8.1   Psychophysics

In the literature, certain relationships between the components presented in panel (c) of Fig. 1.5 have been studied. First of all, the reactions of humans and animals to external forces have been studied in the field of psychophysics. In this regard, a goal of psychophysics is to determine quantitative laws or cause-and-effect relationships. That is, the objective is to determine mappings between physical quantities describing external forces acting on a person, on the one hand, and the consequences induced by those forces, on the other hand. Benchmark examples of psychophysical laws are laws that describe the relationship between the value of a physical quantity and the graded (i.e., quantitative) reaction of a person to that quantity. The graded reaction is frequently considered as "perceived" value. For example, various studies have examined the relationship between the size of objects (i.e., the "actual" size) and the size of objects reported by human observers (i.e, the "perceived" size). With respect to the human pattern formation reaction model, psychophysics attempts to identify the relationship between external forces and produced forces, see panel (c) of Fig. 1.5. Interestingly, psychophysics is interested in a short-cut that points from external forces as cause to produced forces as effect. In doing so, brain state and structure are not explicitly taken into account.

Let us illustrate the relationship between psychophysics and the human pattern formation reaction model in somewhat more detail. When we pick up and lift an object, then we must overcome the gravitational force that pushes the object downwards. In other words, due to their heaviness objects produce forces when they are picked up. In this context, the question arises about how the human system reacts to those forces. In general, the answer to that question depends on the context. In some scenarios, humans actually pick up objects and, in doing so, produce forces that counterbalance to some extent the gravitational forces. In other scenarios humans produce internal states of brain activity about the heaviness of the objects without picking the objects up. For example, as a result of such a brain state associated with object heaviness, an individual may not pick up an object and lift it from location A to location B. Instead, the individual may push the object from A to B. In experimental studies about internal brain states associated with the heaviness of objects, typically, a test person will be asked to estimate the heaviness of an object rather than the magnitude of the force necessary to lift the object. Let us take estimated or reported heaviness as a proxy for "perceived" force. From a naive point of view one might expect that the "perceived" force equals or approximates the force required to pick up an object (i.e, the gravitational force pulling the object down). In particular, when the weight of an object is doubled then the "perceived" heaviness should increase by a factor two as well. Experimental research has shown that this is not the case [310]. In fact, the "perceived" force is related to the "actual" force by a monotonically increasing function whose increase speeds up [310] as shown in panel (a) of Fig. 1.9. The function is also called a monotonically accelerating function. This function is a special case of a power law function. In general, power law functions exhibit an exponent. For accelerating functions the exponent is larger than 1. Therefore, gravitational forces in this kind of experimental paradigms result in human reactions involving power laws with exponents larger than 1. If the exponent equals 1, then the function becomes a straight line. Size "perception" typically satisfies a straight line function [310]. Finally, if the exponent is smaller than 1, then power law functions are monotonically increasing functions whose increase slows down. That is, they are monotonically increasing but de-accelerating functions. When light increases in brightness then this implies that the electromagnetic force of the light acting on an observer's eye increases in strength. In this context, the human reaction to the force exerted by the light of the visual spectrum (i.e., the "perceived" brightness) can be described by such a monotonically

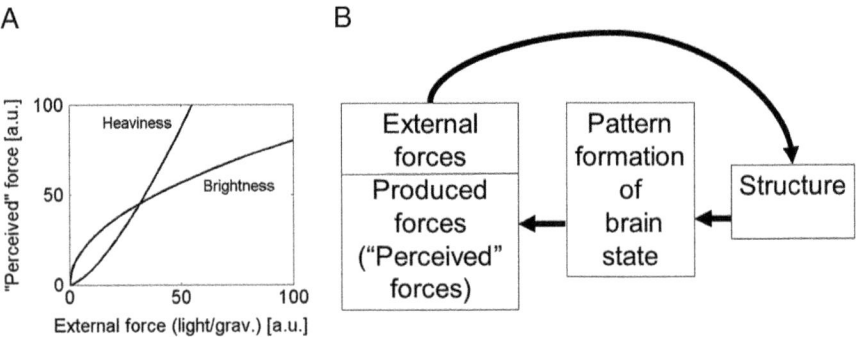

Fig. 1.9 Psychophysical laws given in terms of power laws as observed in experiments (panel **a**) and explanation of psychophysical laws in terms of the human pattern formation model (panel **b**)

increasing but de-accelerating function of the electromagnetic force. That is, the "perceived" strength of the electromagnetic force exerted by light satisfies a power law with an exponent smaller than 1 [310]. Panel (a) illustrates this kind of relationship as well. As will be shown in Chap. 8, power laws that describe human reactions to external forces can be explained within the self-organization, pattern formation framework. In fact, the pattern formation approach includes power laws (as observed in psychophysical experiments) as an in-built property. In the case of the aforementioned psychophysical magnitude estimation experiments, pattern formation is assumed to take place in the brain state only. However, the structure that determines the pattern formation is assumed to be affected by the external forces. That is, the component model shown in panel (c) of Fig. 1.5 reduces to the simplified component model shown in panel (b) of Fig. 1.9.

1.8.2 Neuroscience Research

Second, the impact of external forces on the formation of brain activity patterns has been studied in its own merit. Typically, this kind of research is conducted in the field of neuroscience. For example, taking a pattern formation perspective, a moving object that exerts a force on a human observer by sending or reflecting light to the observer's eyes is assumed to induce the emergence of a brain activity pattern related to the movement of the object. In this example, neuroscientists would like to quantify the brain activity pattern specific to the moving object or its force and would like to determine how the pattern depends on the properties (such as speed and direction) of the object movement. In general, a part of neuroscience research is about the connection between external forces and brain states. In the component model presented in panel (c) of Fig. 1.5 there is a direct link from external forces to the brain state and an indirect link via the structure of the brain. Neuroscience research highlights the importance of those two connections for understanding the nature of humans.

Neuroscience research is not only concerned with the relationship between external forces and the emergence and disappearance of certain brain states (or patterns of brain activity). Neuroscience research is also concerned with the interpretation of brain states. Is a burst of brain activity in a certain brain area related to a particular movement (e.g., a grasping movement)? For example, a certain neuronal activity observed in a test person at a particular location may be related to descending signals to motor neurons in the neck of that person such that he or she turns his or her head in a particular direction (e.g., in order to move the head in the direction of a sound source; recall our zoo visitor example).

Let us turn to an experiment on monkeys in which the relationship between neuronal activity and "behavior" (body movement; here: eye movements) was examined. Panels (a) and (b) of Fig. 1.10 illustrate some basic physiological facts that are needed to understand the experiment. Panel (a) shows a typical structure of a neuronal network involving two neurons. Neuron A sends signals to neuron B. Neuron A can be regarded as sender, whereas neuron B plays to role of a receiver. Neuron A generates signals in terms of pulses. The pulses travel along the so-called axon, see panel (a). The pulses reach an interface unit that connects neuron A with neuron B, which is called the synapse. The synapse transforms the pulse-like signals into electric currents. The currents travel along the so-called dendrite to neuron B. In fact, neurons typically exhibit various dendrites and, consequently, are able to receive signals from more than one neuron. For our purposes, the key issue is that

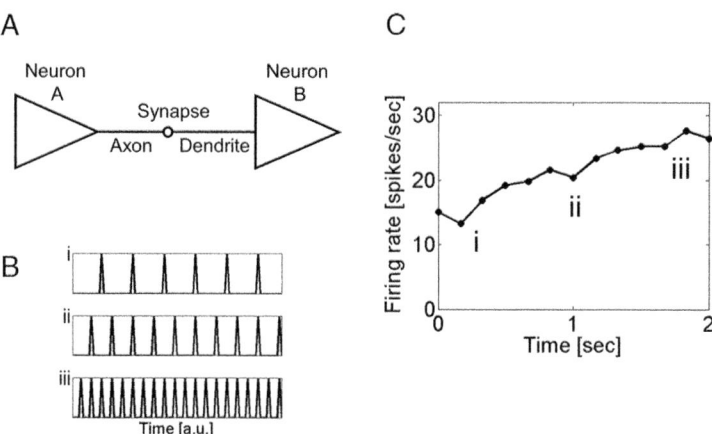

Fig. 1.10 On the experiment by Shadlen and Newsome on reactions of monkeys to moving objects. Panel (**a**) shows a scheme of two connected neurons. Pulse-like signals are generated by neuron A and run through the axon towards neuron B. Panel (**b**) shows pulse-like signals with low, medium, and high firing rates (from top to bottom). Panel (**c**) shows the firing rate measured in certain neurons as a function of time during a period in which a monkey was exposed to moving dots

the pulses in the axon are sent with a particular frequency as measured in the number of spikes that pass a particular location in the axon per second. That is, the neuronal activity of neuron A can be quantified in terms of the firing rate as measured in spikes per unit. Panel (b) illustrates schematically three qualitatively different firing rates. From top to bottom the number of spikes that can be counted in a time unit increases. Therefore, the firing rate increases when looking from subpanel (i), to subpanel (ii), to subpanel (iii). In an experiment reported by Shadlen and Newsome [296, 297] visual "perception" of adult rhesus monkeys was examined. A group of moving dots were displays in the middle of a screen and were shown to the monkeys. In doing so, the light pattern caused by the dots generated an external force field that impacted the monkeys' eyes. In some of the trials the dots moved leftwards, in others they moved rightwards. The monkeys were "trained" (i.e., restructured) such that they reacted to the different kind of force fields generated by the moving dots with eye movements. More precisely, the monkeys were "trained" (i.e., restructured) to gaze at a particular target point on the left of the screen when they "perceived" the dots as moving leftwards. Likewise, the monkeys were "trained" to look at a particular point on the right of the screen, when they "perceived" the dots as moving rightwards. The monkeys had surgically implanted measuring devices that recorded the neural activity of neurons in a certain brain area, called the lateral intraparietal (LIP) area. Firing rates were calculated from measurements of those LIP neurons. Panel (c) shows the firing rate of LIP neurons as a function of time. The zero time point indicates the onset of the dot movements. The firing rate increased more or less monotonically as a function of time.

Shadlen and Newsome [296, 297] argued that this characteristic increase was correlated with the eye movement reaction of the monkeys and, in doing so, with the "perception" of the dots as moving in a particular direction. As will be shown in Sects. 6.3.5 and 7.5.5 the monotonic increase of neuronal activity as shown in panel (c) is a key feature of the formation of brain activity patterns. Within the framework of synergetics and the pattern formation approach to the dynamics of human systems, panel (c) illustrates on the level of neuronal firing rates the buildup of a brain activity pattern. This example illustrates that within the framework of synergetics human reactions can be studied on a level that is typically used in neuroscience research.

1.9 Why Synergetics and the Theory of Pattern Formation?

Although Skinner's point of view is at the heart of all sciences and, in particular, echoes the fundamental point of view of physics, namely, that all natural processes are lawful, Skinner's perspective exhibits limitations in several ways. For example, no explicit mechanism is given that explains human reactions and the dynamics of the isolated human system (i.e., "perception", "cognition", and "behavior"). Skinner's perspective does not provide a theoretical framework that allows the explicit formulation of laws. Moreover, physiological findings and research in neuroscience are ignored (for a reason that will be briefly reviewed in Chap. 2).

What is needed is a theoretical framework that addresses the features listed in Table 1.5. First of all, the framework should be able to address observations made in human and animal studies from Skinner's general perspective, that is, in terms of cause-and-effect relationships and cause-and-effect chains. Moreover, the approach should provide a general mechanism that explains those observations and why humans and animals react in the way they do. Furthermore, the framework should be quantitative. That is, the objective is to mathematize the laws that describe human and animal reactions and the evolution of brain activity related to such reactions. Likewise, the objective is to formulate laws that describe the evolution of brain activity independent from external forces (i.e., brain activity of the isolated human system, see Chaps. 5 and 6). As a result of its quantitative nature, the framework should allow us to formulate laws that determined how external circumstances (i.e., external forces) are related to human reactions. These laws should be derived from the microscopic structural properties of the systems at hand (bottom-up approach). As far as humans and animals are concerned, the laws should follow from neuronal network theory, the neurophysiology of the human and animal bodies, and the structural properties of their environments. This implies that it should be possible to include findings from neuroscience research into the theoretical framework of interest. However, due to the fact that the framework should be quantitative and that the laws can be formulated in mathematical equations, the approach should also yield psychophysical laws. From our discussions provided so far, it is clear that the theory of pattern formation, in general, and synergetics, in particular, as theoretical frameworks feature the aforementioned properties (i.e., the features listed in Table 1.5).

Furthermore, the scope of synergetics and the theory of pattern formation is not limited to living systems such as humans and animals. Synergetics is concerned with pattern formation systems of the animate and inanimate worlds and treats them on an equal footing [152–154, 160, 162]. Due to this property synergetics is tailored

Table 1.5 Features of the pattern formation approach and synergetics

Main feature	Derived feature
Physics perspective	Cause-and-effect relationships
	Cause-and-effect chains
General mechanism	
Quantitative	
Bottom-up approaches [160], Sect. 4.5	Neuroscience applications [38, 39, 107, 158, 159]
Top-down approaches	Psychophysics applications [158, 215]
Broad scope [97, 160, 215, 227, 243, 356], Sect. 4.5	
Not subjected to mystification of humans (see this book)	

Table 1.6 Synergetics as a theoretical framework addressing various levels relevant for understanding human beings and animals as individuals and as part of interacting human and animal many body systems (e.g., "societies", "swarms")

Level	Key mechanism
Cellular level	Self-organization of biochemical reactions involved in cell signaling [99, 101, 111]
Neuronal level	Self-organization and pattern formation of brain activity [38, 39, 107, 158, 159]
Body level	Self-organization of body movements and formation of body patterns [177, 182]
Level of interacting individuals	Self-organization of "relations" between life partners, among family members, and in small groups [115, 198], Chap. 11
	Self-organization and pattern formation in large scale human and animal many body systems (e.g., "societies", "swarms") [352]

to help researchers with developing a view of the nature of humans and animals that is not biased by non-scientific (e.g., religious) perspectives. That is, since the aim of a researcher in synergetics is to determine laws that hold both for humans and animals, on the one hand, and for non-living matter, on the other hand, such a researcher will less likely mystify the nature of humans and animals and will less likely give humans and animals properties and powers that they do not have.

Finally, as far as the application of synergetics and the theory of pattern formation to human beings and animals is concerned, another benefit is that the theoretical framework applies to various levels of interest: the cellular level, neuronal level, body level, and "social" level, see Table 1.6.

1.10   Organization of This Book

This book is organized in four parts, as shown in Table 1.7. Part I of this book consists of two chapters presented at the beginning (Chap. 2) and at the end of this book (Chap. 12). In the following chapter, Chap. 2, we will dwell on the topic of determinism. The starting point will be the fundamental assumption underlying physics: lawfulness of all phenomena of the inanimate and animated worlds in the sense that the laws of physics apply both to living and non-living things. In this context, it will be pointed out that the physics perspective is a particular perspective out of several possible perspectives that are not necessarily inconsistent with each other. In the final chapter, Chap. 12, we will return to this issue about the relationship between the physics perspective taken in this book and other perspectives of the world (e.g., religious perspectives). In this context, some thoughts about ethics and the purpose of life will be developed.

Table 1.7 Topics overview

Part	Topic	Chapter
	Introduction	1
I	Determinism	2
II	Basics of pattern formation theory and synergetics	
	(a) Self-organizing system $\rightarrow$ pattern formation systems	3
	(b) Pattern formation	4
III	Human and animal pattern formation	
	("Perception", "cognition", "behavior")	
	(a) Categorical reactions: general case	5
	(b) Ordinary states pattern formation	6
	(c) Grand states pattern formation	7
	(d) Continuous reactions	8
IV	Special applications	
	(a) Restructuring humans and animals ("conditioning")	9
	(b) Clinical psychology	10
	(c) Life trajectories	11
I	Ethics and purpose of life	12
	Appendix	13

Part II of this book introduces the basic concepts of the theory of pattern formation and synergetics. These concepts will be introduce in two steps. In Chap. 3 it will be shown how pattern formation systems can be defined as special cases of self-organizing systems. In this context, a key role will play the notion of attractors. In Chap. 4 pattern formation as such will be discussed. Part III shows how to apply the concepts introduced in part II to pattern formation in humans and animals. In doing so, part III shows how to address "perception", "cognition", and "behavior" from the physics perspective, in general, and the perspective of self-organizing pattern formation systems, in particular. While Chaps. 5, 6, and 7 will be devoted to discuss categorical reactions and the formation of categorical brain activity patterns, Chap. 8 will be about continuous reactions.

Part IV presents a few special topics. In Chap. 9 some experimental paradigms about the restructuring of humans and animals as used in studies by Skinner and Pavlov will be addressed. Chapter 10 presents applications in clinical psychology. Finally, Chap. 11 is about chains of cause-and-effect relationships that determine the entire daily routines of people and even span over the whole lifetime of human beings. In this context also chains of cause-and-effect relationships involving two people will be briefly addressed.

1.11   Mathematical Notes

1.11.1   Human Pattern Formation Reaction Model

Let us introduce variables to describe the human pattern formation reaction model in mathematical terms. Let q_b denote the brain state and q_f describe produced forces (including body positions if necessary, see above). Taking the brain state and produced forces together, we obtain the ordinary state $q = (q_b, q_f)$. The internal structure determining the evolution of the ordinary state is denoted by N_q. External structure and external forces are denoted by the variable α. The grand state is given by everything and corresponds to the triplet (q, N_q, α). Figure 1.11a presents the variables in the component level of the human pattern formation reaction model that was introduced in Sect. 1.6.

Pattern formation can involve various components. In general, pattern formation takes place in a state space described by the state X. Moreover, pattern formation depends on an initial state S. Let t_0 denote the initial time point of interest. Then $X(t_0) = S$. The evolution of the state X is determined by

$$\frac{d}{dt} X = N(X) , \tag{1.1}$$

where N denotes a non-linear operator and t denotes time. N should not be confused with N_q. Only in a particular special case (see below), we have $N = N_q$. Equation (1.1) describes the evolution of the state and in the context of the human pattern formation reaction model the formation of a pattern. Note that the term "pattern" will be defined in Chap. 4. As indicated in the third box from top of Fig. 1.11b, eventually, the state will converge to an attractor. Mathematically speaking,

$$X(t \to \infty) \in W , \tag{1.2}$$

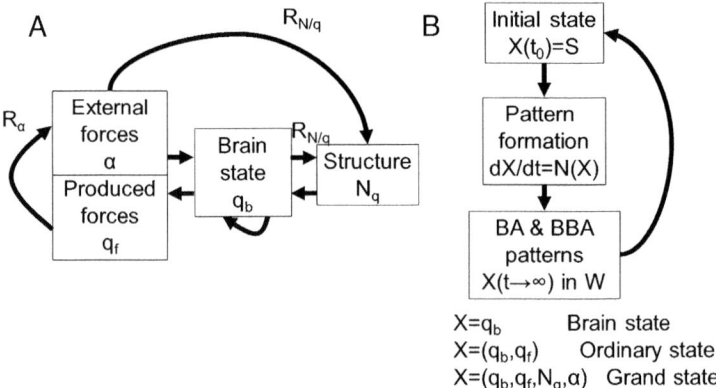

Fig. 1.11   Variables of the human pattern formation reaction model

where W describes an attractor (see Chap. 3). As mentioned above, pattern formation as described in Fig. 1.11b can involve various components shown in Fig. 1.11a. Let us briefly mentioned a few special cases.

Evolution of Brain States and Formation of BA Patterns

First, let us assume that pattern formation only involves the brain state. Accordingly, we put $X = q_b$. In this case, Eq. (1.1) reads

$$\frac{d}{dt}q_b = N_{q/b}(q_b) \, , \tag{1.3}$$

where $N_{q/b}$ is the relevant substructure of N_q describing the evolution of brain activity. Note that in $N_{q/b}$ the subindex q/b is used rather than q_b to avoid second layer subindices. If a stationary state $q_{b,st}$ emerges, in the sense of $q_b(t \to \infty) = q_{b,st}$, then this state is an example of a brain activity pattern (BA pattern), see Chap. 4.

Not all BA patterns emerge exclusively on the brain state. In general, BA patterns are special grand states that emerge in state spaces spanned by the brain state variable q_b and the brain structure variable $N_{q/b}$, when assuming that $N_{q/b}$ changes slowly. For example, if we consider the grand state X given by the pair of variables $(q_b, N_{q/b})$, then the evolution of that state is given by $dq_b/dt = N_{q/b}(q_b)$ and $dN_{q/b}/dt = R_{N/q/b}(q_b, N_{q/b})$, where the function $R_{N/q/b}$ determines the evolution of $N_{q/b}$ and is the relevant special case of the function $R_{N/q}$. Again, the subindices N/q and $N/q/b$ have been used rather than N_q and $N_{q/b}$ in order to avoid second layer subindices. The first equation, $dq_b/dt = N_{q/b}(q_b)$, describes the evolution of the brain activity. The second equation, $dN_{q/b}/dt = R_{N/q/b}(q_b, N_{q/b})$, describes the change of the brain structure, that is, brain plasticity. A pattern emerging in the state $(q_b, N_{q/b})$ is a BA pattern that involves structural components.

Equation (1.3) describes the special case of the formation of BA patterns in which brain structure is fixed. The more general case of the formation of BA patterns is that the structure $N_{q/b}$ may evolve in time.

Pattern Formation of Ordinary States Under the Impact of External Forces

In this case, pattern formation involves the ordinary state $q = (q_b, q_f)$. Accordingly, we put $X = q$ and Eq. (1.1) becomes

$$\frac{d}{dt}q = N_q(q, \alpha) \, . \tag{1.4}$$

This is the aforementioned special case in which $N = N_q$. If a state becomes stationary like $q(t \to \infty) = q_{st}$ then q_{st} is an example of a BBA pattern (brain and body activity pattern) because it involves a force production component, which is related to body activity.

Pattern Formation of Grand States: General Case

In general, changes taking place on the external and internal structure levels can be part of the formation of a pattern of interest. In order to describe this scenario, we consider the grand state and put $X = (q, N_q, \alpha)$ with $q = (q_b, q_f)$. This implies that Eq. (1.1) corresponds to three evolution equations that can be written like

$$\frac{\mathrm{d}}{\mathrm{d}t} q = N_q(q, \alpha) ,$$

$$\frac{\mathrm{d}}{\mathrm{d}t} N_q = R_{N/q}(q, N_q, \alpha) ,$$

$$\frac{\mathrm{d}}{\mathrm{d}t} \alpha = R_\alpha(q, N_q, \alpha) . \tag{1.5}$$

The operators $R_{N/q}$ and R_α describe the evolution of the (internal) structure N_q and external structure α, respectively. The impact of those operators is schematically indicted in Fig. 1.11a. Note that in $R_{N/q}$ the subindex N/q is used rather than N_q to avoid second layer subindices.

Pattern Formation Model for Discrete Systems

Equations (1.1)–(1.5) are mathematically incomplete. They should be considered in a symbolic way. What is missing is the precise definition of the type of the state variable X. A human individual under consideration may be described in terms of a set of scalar variables (e.g., in terms of a number of electrical signals over the scalp measured by an electroencephalography). If a human individual is described in terms of a set of scalar variables, then the individual is considered to be a discrete system. In this case X, q_b, q_f, α are vectors. $N, N_{q/b}, N_q, R_{N/q}$ and R_α are vector-valued functions. Vectors and vector-valued functions will be denoted by bold-face print. Equations (1.1)–(1.5) become

$$\frac{\mathrm{d}}{\mathrm{d}t} \mathbf{X} = \mathbf{N}(\mathbf{X}) , \tag{1.6}$$

$$\mathbf{X} = \mathbf{q}_b \;\Rightarrow\; \frac{\mathrm{d}}{\mathrm{d}t} \mathbf{q}_b = \mathbf{N}_{q/b}(\mathbf{q}_b) , \tag{1.7}$$

$$\mathbf{X} = \mathbf{q} \;\Rightarrow\; \frac{\mathrm{d}}{\mathrm{d}t} \mathbf{q} = \mathbf{N}_q(\mathbf{q}, \tilde{\alpha}) , \tag{1.8}$$

$$\mathbf{X} = (\mathbf{q}, \mathbf{N}_q, \tilde{\alpha}) \;\Rightarrow\; \frac{\mathrm{d}}{\mathrm{d}t} \mathbf{q} = \mathbf{N}_q(\mathbf{q}, \tilde{\alpha}) , \tag{1.9}$$

$$\frac{d}{dt}\mathbf{N}_q = \mathbf{R}_{N/q}(\mathbf{q}, \mathbf{N}_q, \tilde{\alpha}) ,\qquad(1.10)$$

$$\frac{d}{dt}\tilde{\alpha} = \mathbf{R}_\alpha(\mathbf{q}, \mathbf{N}_q, \tilde{\alpha}) .\qquad(1.11)$$

Note that in the case of α the notation $\tilde{\alpha}$ is used to denote a vector. Equations (1.6)–(1.11) are the mathematically complete forms of Eqs. (1.1)–(1.5) in the discrete variable case.

As an example, consider a brain structure $\mathbf{N}_{q/b}$ that involves a single scalar variable β that changes along with the formation of the pattern of interest. Let us assume that external forces do not vary and that force production can be neglected. In this case, Eqs. (1.9)–(1.11) reduced to

$$\frac{d}{dt}\mathbf{q}_b = \mathbf{N}_{q/b}(\mathbf{q}_b, \beta, \alpha) ,\qquad(1.12)$$

$$\frac{d}{dt}\beta = R_\beta(\mathbf{q}_b, \beta, \alpha).\qquad(1.13)$$

We will return to this example in applications in Chap. 7 when studying the pattern formation of grand states.

Pattern Formation Model for Spatially Extended Systems Described by Field Variables

Human "perception", "cognition", and "behavior" may require a description in terms of field variables. For example, the brain activity in a certain area of the brain may be considered as a field (rather than a discrete set of signals as picked up by electroencephalography). Let $\mathbf{x}$ denote the vector describing the area of interest. For example, we have x for a one-dimensional, (x, y) for a two-dimensional, and (x, y, z) for a three-dimensional area. In this case X, q_b, q_f, α are field variables that depend on $\mathbf{x}$ and time t. Therefore, changes of the field with time are described by the partial time derivative $\partial/\partial t$. The quantities $N, N_{q/b}, N_q, R_{N/q}$ and R_α are operators that involve $\mathbf{x}$ but also differential operators (i.e., gradients) that will be denoted by the nabla symbol: ∇. The operators may also involve integrals. If so, integral operators are considered as inverse differential operators (i.e., inverse nabla operators). Therefore, in the description below no separate symbol for integral operators will be used. For pattern formation involving field variables Eqs. (1.1)–(1.5) read

$$\frac{\partial}{\partial t}X(\mathbf{x}, t) = N(\nabla, \mathbf{x}, X) ,\qquad(1.14)$$

$$X = q_b \implies \frac{\partial}{\partial t}q_b(\mathbf{x}, t) = N_{q/b}(\nabla, \mathbf{x}, q_b) ,\qquad(1.15)$$

$$X = q \; \Rightarrow \; \frac{\partial}{\partial t} q(\mathbf{x}, t) = N_q(\nabla, \mathbf{x}, q, \alpha) \, , \tag{1.16}$$

$$X = (q, N_q, \alpha) \; \Rightarrow \; \frac{\partial}{\partial t} q(\mathbf{x}, t) = N_q(\nabla, \mathbf{x}, q, \alpha) \, , \tag{1.17}$$

$$\frac{\partial}{\partial t} N_q = R_{N/q}(q, N_q, \alpha) \, , \tag{1.18}$$

$$\frac{\partial}{\partial t} \alpha(\mathbf{x}, t) = R_\alpha(q, N_q, \alpha) \, . \tag{1.19}$$

Equations (1.14)–(1.19) are the complete forms of Eqs. (1.1)–(1.5) in terms of field variables describing a human individual.

Pattern Formation Involving Several Field Variables

In the case of the grand state description defined by Eqs. (1.17)–(1.19) pattern formation involves several interacting fields. However, it might be necessary to describe, for example, the brain state q_b with several field variables like $q_{b,1}(\mathbf{x}, t)$, $q_{b,2}(\mathbf{x}, t)$, ... For example, $q_{b,1}$ may be used to describe axonal pulse rates in a certain cortical area, while $q_{b,2}$ describes the dendritic currents in that area. In this case, the components of the human pattern formation reaction model become vector-valued field variables. For example, we have

$$\mathbf{q}_b(\mathbf{x}, t) = (q_{b,1}(\mathbf{x}, t), q_{b,2}(\mathbf{x}, t), \dots) \, . \tag{1.20}$$

In this case, a mathematically description of the human pattern formation reaction model can be achieved by merging the notations of Eqs. (1.6)–(1.11) and Eqs. (1.14)–(1.19).

1.11.2 Miscellaneous

The mathematical details of human pattern formation reaction models that describe cause-and-effect chains as in the zoo visitor example discussed in Sect. 1.3 will be presented in Chap. 7. The mathematical model describing the figure-ground experiment presented in Sect. 1.6.3 will be addressed in more detail in Sects. 6.5.3 and 6.5.4. Mathematical details will be presented in this regard in Sect. 6.6. The mathematical details of psychophysical power laws addressed in Sect. 1.8.1 as well as the modeling of the power laws in terms of the human pattern formation reaction model will be presented in Chap. 9. Modeling the firing rates in the experiment on the reactions of monkeys to moving dots as reviewed in Sect. 1.8.2 will be discussed in more detail in Sects. 6.3.5 and 7.5.5. Mathematical details will be presented there.

Chapter 2
Determinism

This chapter introduces the concept of determinism. In the first part of this chapter two related learning paradigms will be presented: "classical conditioning" and "operant conditioning". These paradigms give experimental demonstrations for the determinism of "behavior", that is, the notion that "behavior" is determined by cause-and-effect relationships. In the second part of this chapter, the concept of determinism will be introduced in the context of the physics perspective of the nature of humans and animals. In order to demonstrate the gravity of the concept of determinism, the "Which shoes should I put on?" example [112] will be discussed. The example will make clear that determinism implies that humans and animals do not make "choices". In that context, it will then also become clear that everyday language sometimes is inappropriate to describe phenomena from a scientific, physics perspective. In the third part of this chapter, the relevance to study processes that occur inside humans and animals, for example, on the brain level, will be addressed. While Skinner argues—in some sense in line with the physics perspective of the world—that the study of such processes is not necessary for understanding humans and animals, there are very good reasons why such internal processes should be studied. In the final part of this chapter, the physics perspective will be put in the context of other perspectives (e.g., religious perspectives). Some comments on the scope of the physics perspective and on the relationship between physics and religion will be made.

2.1 "Classical Conditioning"

Causes and effects in the inanimate world can often be identified with little effort. For example, when a guitarist plucks a guitar string, then the string produces a particular sound. In this example, the guitarist applies a certain force on the string and the string reacts by vibrating and producing a force that generates a sound wave

© Springer Nature Switzerland AG 2019
T. Frank, *Determinism and Self-Organization of Human Perception and Performance*, Springer Series in Synergetics,
https://doi.org/10.1007/978-3-030-28821-1_2

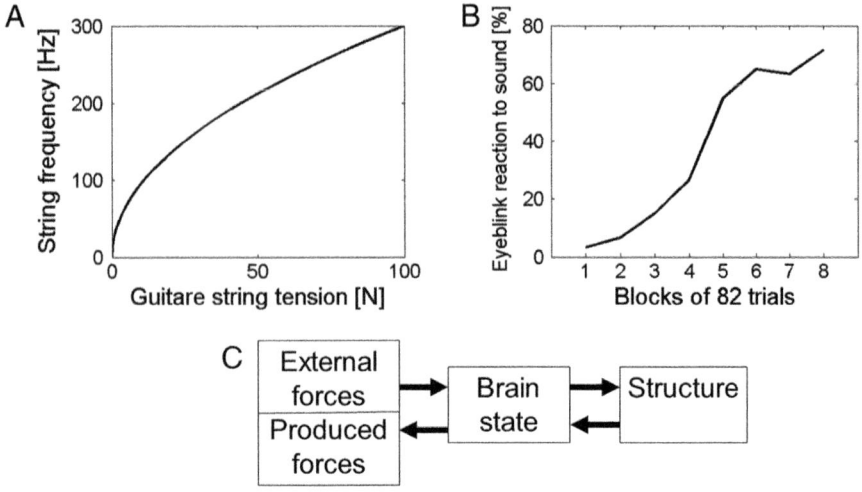

Fig. 2.1 Examples and theory of reactions to forces that change due to changes on the structural level. Panel (**a**): Frequency-tension relation of a guitar string indicating that the produced frequency (reaction to a plucking force) depends on the string structure (i.e., tension) such that when the structure is changed the frequency changes. Panel (**b**): Relationship between eye blink reactions to sounds in rabbits and days of "training" or restructuring indicating that the eye blink reactions to sounds become more frequent due to structural changes (putatively on the brain level) induced by the restructuring trials (data from Schneiderman et al. [284]). Panel (**c**): Simplified human pattern reaction model that can account for variations in the relationship between produced forces (i.e., reactions) and the applied external forces. For the original model see Fig. 1.5

with a particular frequency. The force generated by the guitar player is the cause. The string vibrations and the force produced by the vibrating string are the effects or the reactions to the guitar player's force. These reactions depend on the tension of the string. A string with a high tension produces a sound with a relatively high frequency, whereas a string with a low tension produces a sound with a relatively low frequency. Panel (a) of Fig. 2.1 presents a typical frequency-tension relationship of vibrating strings. The graph in panel (a) reveals that how the string reacts to an applied force can be changed by changing the tension (i.e., the structure) of the string. Pavlov's experiments on animals allow us to discuss animal "behavior" (and human "behavior") in a similar way.

Pavlov's experiments with dogs showed that animals can be manipulated (like the tension of a string can be manipulated) such that they react to certain external forces in the same way as they do to certain other external forces. Pavlov examined the salivation reaction of dogs [51, 288]. He and his co-workers placed meat powder on the tongue of a dog and recorded the amount of saliva produced by the dog. In doing so, they were able to quantify the salivation reaction of the dog to the chemical driving forces of the meat powder acting on the taste receptors of the dog. As such, the measurement procedure could be used to determine the salivation reaction to any external force affecting the dog of interest. In fact, Pavlov and his co-workers studied the salivation reaction to the mechanic force of a sound wave that acts on the mechanoreceptors in the ears of a dog. The sound wave was the sound wave produced by a ringing bell. First, the baseline reaction was determined. Under baseline conditions, the dog did not produce saliva when the bell was ringing. Subsequently, the dog was exposed to several "training" trials (i.e., restructuring trials). In each "training" trial, the experimenter rang the bell while the meat powder was in the mouth of the dog. That is, the dog was exposed to both external forces at the same time: the chemical driving force of the meat powder and the mechanical force of the soundwave of the ringing bell. After a sufficient large number of such trials, the dog produced saliva when the bell was ringing even when there was no meat powder involved. That is, the dog showed a salivation reaction to the force of a sound wave of the ringing of a bell [51, 288]. In this experiment, there were two causes: the chemical driving forces of the meat powder and the mechanical force of the soundwave of the ringing bell. The effect was the production of forces in the so-called parasympathetic system that lead to the secretion of saliva. While under baseline conditions the mechanical force of the soundwave did not cause a secretion reaction, after the "training" period this particular mechanical force caused a salivation reaction.

Pavlov's conditioning experiments with animals demonstrate a qualitative aspect of animal (and human) reactions to the forces in the world: Pavlov's experiments demonstrate that cause-and-effect relationships can be created. In addition to this qualitative aspect, "classical conditioning" experiments typically reveal a quantitative aspect of the animal (and human) reaction model. The effect of restructuring becomes stronger when the experiment is conducted over longer periods. Panel (b) of Fig. 2.1 shows data from an eyeblink experiment reported by Schneiderman et al. [284] conducted with rabbits. Rabbits were exposed to air puffs to the eye. The mechanical forces exerted by the air puffs acting on the eye caused eyeblink reactions of the rabbits. The air puffs were combined with 800 Hz tones that were played for 600 ms. In doing so, the rabbits were exposed to mechanical forces to the eyes and ears coming in terms of air streams and sound waves. The animals were "trained" in blocks of 82 trials. Panel (b) shows the percentage of eyeblink reactions to the 800 Hz tones over the course of the administered blocks. The percentage of eyeblink reactions increased. That is, first, a cause-and-effect relationship between sound and eyeblinking was established and, second, the established cause-and-effect relation became stronger or more pronounced over "training" trials. The second feature is a typical quantitative aspect of "classical conditioning" experiments.

It is nowadays assumed that in "classical conditioning" experiments such as Pavlov's experiment on dogs or the experiment by Schneiderman et al. on rabbits the structure on the neuronal level is changed [288] leading to the creation of novel cause-and-effect relationships. We will return to this issue in Chap. 9. In particular, in Chap. 9 we will apply the human pattern formation reaction model shown in panel (c) of Fig. 2.1 to animal "conditioning" experiments. Accordingly, the paired external forces from two sources (meat powder and sound or air puffs and sound) induce the emergence of appropriate BBA patterns that also occur in pairs in the animal brains. The fact that the patterns occur in pairs is assumed to change the structure of the animal brains. This change in structure makes that animals exhibit novel cause-and-effect relationships. For example, they react to a sound by salivation or eyeblink reactions. For more details see Chap. 9.

Let us dwell on the quantitative aspect of conditioning experiments. Let us compare panels (a) and (b) of Fig. 2.1 in this regard. Panel (a) shows how a guitar string reacts (in terms of its sound frequency) to somebody plucking the string when the string is put under a particular tension. Panel (b) shows eyeblink reactions of rabbits (reported as percentage values) to 800 Hz tones when the rabbits have been exposed to a number of experimental trials. Comparing the graphs of panels (a) and (b), we see that both graphs have in common that they describe quantitative changes of cause-and-effect relationships. The graph in panel (b) tells us how the impact of the (mechanic force exerted by the) 800 Hz tones changes when the number of experimental trials is increased. When increasing the number of trials the impact becomes stronger in the sense that the percentage of reaction to the (800 Hz tone induced) mechanical forces increases. From a physiological perspective, the graph in panel (b) indicates to what extent the neuronal structure changes when the number of "conditioning" (i.e., restructuring) events is increased (see [288] and Chap. 9). The graph in panel (a) tells us how the impact of the force of the guitarist changes when the tension of a guitar string is increased. When the tension is increased, the plucking force (produced by the guitar player) leads to the production of soundwaves with higher frequencies. From a material physics perspective, the graph in panel (a) indicates to what extent there are changes in the atomic and molecular structure of the string when the string tension is increased that lead to changes in the way the string material reacts to forces. From a physics perspective, "classical conditioning" experiments are about changing the structure of a material, which happens to be an animal if the experiments involve animals or a human if the experiments involve human participants. Changing the structure of a material, in general, changes how the material reacts to forces. When talking about science, "conditioning experiments" should be seen in the first place from this perspective and should not be over-interpreted using mystic concepts.

At a first glance, the important aspect about Pavlov's experiment seems to be that the brain structure of humans and animals can be changed by applying appropriately paired forces. However, having a second look at this argument, it becomes clear that in fact it happens very often that we use several forces (i.e., not a single one) in a coordinated and appropriate way in order to change the structure of a material. For example, in order to make a small hole in a piece of paper, we need to hold the paper with one hand and at the same time punch the hole through the paper with a sharp object that we hold in the other hand. The structure of the paper material is changed by applying simultaneously and in an appropriate way two forces to the material.

2.2 "Operant Conditioning"

Thorndike conducted early experiments on what is nowadays called "operant conditioning" [13, 288]. These experiments were conducted in a more systematic fashion by Thorndike's followers, in particular, by Skinner [13, 192, 288]. Just like "classical conditioning" experiments, "operant conditioning" experiments are nowadays assumed to change the neuronal structure of an animal such that the animal reacts in a certain way (as "desired" by the experimenter) to a certain situation (characterized by a particular set of external forces). The main difference between "classical and operant conditioning" is in the experimental procedure used to produce changes in the neuronal structure of animals.

Thorndike conducted experiments with cats that were put into so-called puzzle boxes. A hungry cat was put in a cage-like box and food was placed outside of the box such that the cat could see and smell the food. That is, the cat was exposed to the electromagnetic forces of the light waves coming from the food and the chemical driving forces of the food acting on the smell receptors of the cat. In addition, there were the mechanical forces from the box confining the cat within the box and the electromagnetic forces of the visual world around the cat coming among other things from the box. The box exhibited a door that was closed but could be opened when the cat hit or moved a particular bar or latch. Typically, the cat moved around in the box for a while and sooner or later by accident hit/moved the latch. When hitting the latch, the door opened, the cat escaped and snatched the food. After some break, the same cat was put in the box, again, with the door closed and new food placed outside. In this second trial, at some point in time the cat hit the latch again and in doing so escaped to eat the food. This procedure was repeated several times. With an increasing number of repetitions, the time for the cat to hit the latch and open the door decreased. In view of this observation, Thorndike hypothesized that when a reaction to an external force is rewarded (i.e., a particular effect is rewarded), then a cause-and-effect relationship is established between the external force as cause and the reaction as effect. Consequently, in Thorndike's cat experiment, the electromagnetic forces of the light waves from the door in combination with the forces of the food cause that the cat produces muscle forces that move her body and limbs in a certain way and eventually open the door.

Thorndike hypothesized further that in general during the maturation of humans and animals pleasurable experiences lead to a change of the structure of humans and animals such that in the end humans and animals satisfy the so-called law of effect. The law of effect states that humans and animals react more frequently in a certain way to external forces when the effect of that type of reaction is pleasurable [13, 288].

As mentioned above, while Thorndike conducted benchmark experiments in the field of "operant conditioning", more systematic experiments were carried out by Thorndike's followers. For example, Tolman studied cause-and-effect relationships of rats in mazes and how such relationships changed due to "operant conditioning" [319]. In one of his experiments, a mace was used with a starting position and two doors. The doors were marked differently. Roughly speaking, they differed in color: the frame of one door was painted in white, the frame of the other door was painted in black. When a rat passed through one door, say, the white door, the animal received a food reward. In the case of the other door (the black door), it did not receive a reward. Accordingly, the experimental procedure was relatively simple. A rat was placed on the starting position. The rat walked through one of the doors and (depending on the door) was rewarded or not. This experimental procedure was repeated several times over a period of several days. Tolman recorded the number of runs, in which rats passed through the door with the reward, which will be referred to as runs through the food door. He found that in the beginning the probability of such runs was 50%. However, over a period of 20 days the probability increased to almost 100%, see Fig. 2.2a. Figure 2.2a indicates that a cause-and-effect relationship was established during the course of the experiment. After 20 days, the forces exerted by the food door on the animals (in terms of the electromagnetic forces of the light acting on the animals' eyes) caused the animals to go through the food door.

Let us discuss Tolman's experiment with respect to the human pattern formation reaction model shown in a simplified version in panel (c) of Fig. 2.2. Accordingly, in the beginning of the experiment the electromagnetic force of the light waves from the door frames did not affect the emergence of brain and body activity (BBA) patterns relevant for directional movements of the animals towards or away from the doors. At the beginning of each trial, that is, when the rat was put on the starting position, the rat's brain state was in a "random" initial condition. "Random" means the brain state in terms of the ongoing brain activity was dependent on the prior

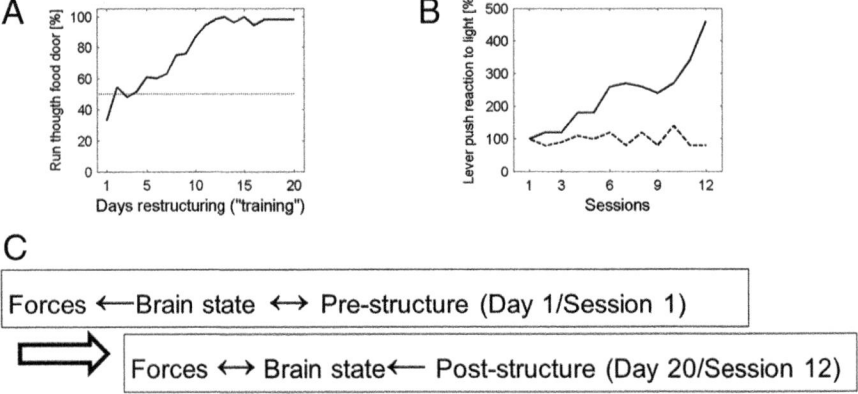

Fig. 2.2 Examples and theory of restructuring experiments by means of paired forces involving rewards. Panels (**a**) and (**b**) refer to animal experiments by Tolman [319] and Guerra-Narbone et al. [150]. Panel (**a**): Rats were restructured such that eventually they reacted to the food door by running through it. Panel (**b**): Mice were restructured such that eventually they reacted to a light signal by pushing a lever. Panel (**c**): Human and animal pattern formation reaction model explaining brain restructuring phenomena as demonstrated in panels (**a**) and (**b**) by assuming that the brain structure changes due to reward-related external forces

history of the rat and was likely to be different for each trial. Likewise, while the rat was put in each trial on the starting position, the precise location within that starting position and the orientation of the animal differed from trial to trial. Therefore, initially the external forces of the visual world differed from trial to trial as well. On day one, these two "random" elements ("random" brain state and "random" external forces) affected the emergence of movement-related BBA patterns. Consequently, in some trials BBA patterns of animals moving through the food door emerged and in other trials BBA patterns of animals moving through the other door emerged. The probability for the emergence of either of the two BBA patterns was the same and, consequently, the probability that animals walked through a particular door was the same for both doors (i.e., 50%).

During the course of the restructuring of the animals (i.e., the "training"), the neuronal structure of the rats changed such that the electromagnetic force of the light waves from the door frames started to affect crucially the formation of BBA patterns. In Fig. 2.2c this change is indicated by a change from a pre-structure to a post-structure. Likewise, it is indicated that for animals with post-structure external forces from the door frames impacted the brain state (while this was not the case for animals with the pre-structure). In our example about differently colored doors, the color of the doors, that is, spectral properties of the electromagnetic forces of the light waves, played a role. Accordingly, it is assumed that light waves from the door frames biased the formation of BBA patterns in post-structure animals. Light waves from the food door induced the emergence of a BBA pattern related to the movement through the food. In contrast, light waves from the other door induced the emergence of a BBA pattern associated with a movement away from that door. Panel (c) indicates that after completion of brain restructuring, that is, when the post-structure was formed, pattern formation was more or less independent of the ongoing brain activity. That is, with respect to the system dynamics level shown in Fig. 1.5d of Chap. 1, for animals with pre-structure the initial state did affect pattern formation. In contrast, for animals with post-structure the initial state did not affect pattern formation.

In summary, just as the eyeblink experiment with rabbits by Schneiderman et al. [284], Tolman's experiment [319] illustrates that cause-and-effect relationships can be created. If so, such relationships may involve a quantitative aspect that changes gradually during the restructuring process. While in the eyeblink experiment "classical conditioning" was used for restructuring (i.e., pairing of external forces, where one of the two forces induced a so-called "reflex", see Chap. 13), in Tolman's experiment "operant conditioning" was used (i.e., pairing of external forces, where one force was reward related) to restructure animals and change the way BBA patterns are formed under the impact of external forces and conditions.

Just like Tolman, Skinner and his co-workers used rewards and punishment to study the emergence of animal "behavior". In particular, Skinner and his co-workers were able to "train" animals to press a lever in order to obtain food [13]. In what follows, let us briefly review a typical animal experiment in which animals were restructured such that they showed a particular food-rewarding reaction ("behavior") to a particular light signal (i.e., the force of an electromagnetic wave). In the experiment, a light source (an electric bulb) was used that could be switched on and off. When the light was switched on and the animal pressed the lever then a food reward was given. When the lever was pressed while the light was off no reward was given. Using this experimental procedure, it was possible to restructure the animals involved in the experiment: the experimental procedure changed the repertoire of cause-and-effect relationships that determined the reactions ("behaviors") of the animals. In particular, after several experimental sessions, animals pressed the lever more frequently when the light was switched on. That is, the electromagnetic force generated by the light bulb induced the emergence of a particular brain activity and body movement pattern that involved pressing the lever. In terms of a cause-and-effect relationship, the light was the cause and the lever pressing movements the effect.

Panel (b) of Fig. 2.2 shows experimental results from Guerra-Narbone et al. [150]. Mice were "trained" in 12 subsequent sessions using the aforementioned restructuring (i.e., "operant conditioning") procedure. For each session, the number of lever presses during the light on period and the number of lever presses during the light off period was determined and the ratio of the two numbers was calculated. Figure 2.2b shows for two animal groups the ratio values as percentage values as functions of the "training" sessions. The two animal groups differed by certain neurophysiological conditions. The results from the two groups are shown separately in Fig. 2.2b as solid and dashed lines, respectively. In the beginning the ratio was at 100%. Animals of both groups pressed the lever during the light on periods as often as they pressed it during the light off periods. However, for the animals under appropriate neurophysiological conditions (see solid line) from session to session the ratio increased, that is, the animals pressed more frequently the lever when the light was on. At the end of the 12 "training" sessions, those animals pressed the lever four to five times more frequently when the light was on as compared to the condition when the light was off. For the animals that were in the second group the experimental procedure did not have an effect. They pressed the lever with a particular frequency irrespective whether the light was switched on or off even after 12 sessions. For more details (in particular, with respect to the difference between the two groups of animals) the reader is referred to the original study [150].

2.3 Skinner's Conceptualization of Humans and Animals

Skinner emphasized in his works that the observations in animal experiments that involve paired forces, where one force is reward related (i.e., "operant conditioning" experiments), reflect general principles that do not only apply to animals but also to humans [300]. According to Skinner, when an infant grows up, he or she is subjected to restructuring via reward related forces. Being rewarded affects the cause-and-effect relationships of the infant. This also holds for the adult. Moreover, punishment rather than reward has a similar effect. The "behaviors" (i.e., reactions) of an individual can be understood on the basis of the circumstances of the individual. In other words, any "behavior" (reaction) is the effect of a particular cause and the underlying cause-and-effect relationship relevant for the observed "behavior" (reaction) was established by a restructuring process that took place in the past of the individual. In this sense, human "behavior" is determined. Skinner regarded humans as entities that are made of biological material but follow rules like machines. Just as machines are determined, humans are determined. Skinner considered human beings as man-machines [300]. This is not a triviality and the consequences are far-reaching. For example, it implies that the concept of "free will" is a misconception—at least from Skinner's perspective.

2.4 Physics Perspective

The second part of this chapter introduces from a theoretical point of view the physics perspective of humans and animals. Moreover, in that context, the concept of physics-based determinism will be introduced. It will be shown that physics-based determinism is consistent with key elements of Skinner's concept of determinism. However, there are also two peculiarities of physics-based determinism that are opposed to Skinner's lines of thought. They concern the relevance of the study of internal states and the limitations of physics-based determinism as a theoretical framework.

2.4.1 Physics Perspective: Two Remarks

The author distinguishes between the physics perspective of humans and animals and various other perspectives. In this context, two important notes should be made. First, no value is attached to the physics perspective or any perspective of those others perspectives. In particular, the physics perspective is not regarded superior to any of the others perspectives [112].

Second, the physics perspective is a departure point that can be used to draw conclusions about the world. These conclusions can be compared with experimental observations. The observations can be consistent or inconsistent with the conclusions. If they are inconsistence, the current physics perspective is falsified and revised. This procedure leads to a change of the physics perspective over time. Note that this procedure only works within the physics perspective. That is, this procedure is not applicable to falsify other perspectives. As mentioned above, the physics perspective is about the world or the universe. In this sense, the physics perspective is an all-encompassing perspective. Consequently, the physics perspective includes as a special case humans and animals: it describes the way the internal (brain activity) dynamics of humans and animals evolves in time and how humans and animals react to forces acting on them. Having said that the physics perspective is subjected to limitations. This has been anticipated above, when stating that the development of the physics perspective only works within the physics perspective. The physics perspective is limited to its own scope, which means that it makes statements that apply only within the framework of the physics perspective. Although this seems to be a triviality, it is important to be clear about this issue. No claim is made that the physics perspective is the "true" or "correct" perspective of the universe [112]. We will return to this issue at the end of this chapter and discuss the relation between the physics perspective and other perspectives (e.g., religious and ethical perspectives).

2.4.2   Definition of the Physics Perspective

The physics perspective of the universe is defined as the perspective that states that everything that happens in the universe is determined by the first principles of physics. This means that all events in the universe are described by the Schrödinger equation [63] and the four fundamental forces: the gravitational force, the electromagnetic force, the strong force and the weak force (see Chap. 1). In other words,

> the physics perspective states that the laws of physics determine everything that happens in the animate and inanimate worlds.

The laws of physics are formulated on a fundamental level by means of quantum mechanics [63] in terms of the Schrödinger equation and involve the four fundamental forces.

2.4.3   Some Facts About the Schrödinger Equation

In order to understand the basic elements of the physics perspective it is not necessary to understand the explicit details of the Schrödinger equation. Likewise, the reader does not need to "know" the precise definition of the four fundamental forces. The Schrödinger equation is a fundamental equation in quantum mechanics [63] and can be used to define the evolution of a probability function that describes the state of the particles of the universe. Moreover, the state of the universe at any given time point can serve as initial condition to solve the evolution equation. Therefore, the history of the universe is not relevant to determine future events. Only the current state is of importance. In other words, the past is irrelevant for the evolution of the probability function describing the universe as long as the current state of the universe is given. Consequently, whether or not there "exists" a past at all (whatever "exists" means in that context) is irrelevant for the physics perspective. Finally, the Schrödinger equation is a deterministic evolution equation for the probability function. That is, for any initial condition there exists exactly one solution. There is no "arbitrariness" about the solution of the Schrödinger equation. Let us summarize the three aspects of the Schrödinger equation that are relevant to understand the physics perspective of the universe. First, the Schrödinger equation can be used to describe the evolution of a probability function that describes the universe. Second, the state of the universe at any time point can be used as initial condition for solving the Schrödinger equation. Third, for any initial condition there is exactly one solution. In this sense, the Schrödinger equation is a deterministic equation.

2.4.4 Physics-Based Determinism

Let us discuss different routes that lead to physics-based determinism: (a) quantum mechanics and the Schrödinger equation in the classical limit, (b) large scale perspectives in which not only the systems of interest but also other systems are taken into consideration in order account for possible deterministic perturbations produced by those other systems, and (c) systems with observables that correspond to mean values of subsystem observables. These three approaches are illustrated in Fig. 2.3.

Schrödinger Equation and Physics-Based Determinism

In this section arguments will be outlined that have been presented earlier in Frank [112]. As mentioned above, the Schrödinger equation is a quantum mechanical evolution equation that determines the evolution of a probability function. Let us

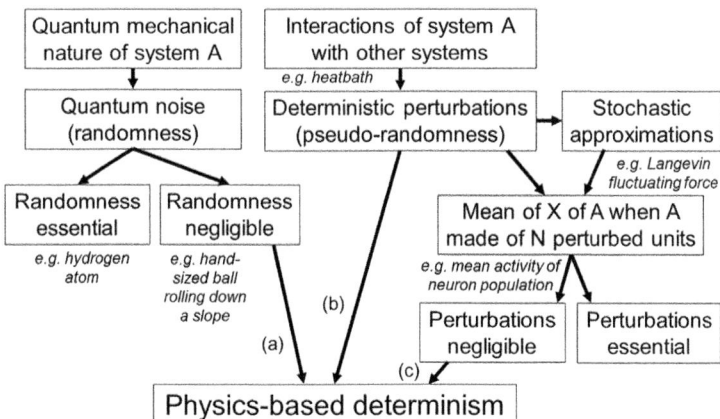

Fig. 2.3 Three routes (a), (b), and (c) within the physics perspective of the universe that lead to physics-based determinism. See text for details

denote the probabilistic aspect as described by the Schrödinger equation as quantum noise. The Schrödinger equation features a fundamental physical parameter: the Planck constant. The Planck constant is relatively small. This implies that many probabilistic aspects of quantum mechanics can only be observed for systems that are very small and/or operate at a very low energy. In other words, quantum noise plays only a role for certain applications and can be neglected in other applications. For many applications quantum noise is irrelevant. This case is known in the literature as the classical limit of quantum mechanics [26, 29, 35, 66, 205, 301]. In this classical limit, the quantum mechanical description via the Schrödinger equation of a system A reduces to the corresponding classical description of the system. In Fig. 2.3 this case is indicated (on the left-hand-side of the figure) by the branch, in which quantum noise is negligible. If quantum noise can be neglected, the Schrödinger equation can be replaced by deterministic equations describing physical and chemical phenomena. For example, mechanical systems can be described by the deterministic evolution equations of Newtonian mechanics. Electromagnetic phenomena can be described by the deterministic Maxwell equations. Chemical reactions can be described by deterministic rate equations. Deterministic in this context means that the evolution equations do not feature a random component. Importantly, just as for the Schrödinger equation, for those deterministic evolution equations there is exactly one solution for any given initial condition.

For example, if we place a hand-sized ball at the top of slope (e.g., a ramp with a particular incline), the ball will roll down the slope due to the gravitational force pulling the ball downwards. While in principle the ball movement can be described within the framework of quantum mechanics, the probabilistic aspect of quantum mechanics is negligible in this application. Consequently, the dynamics of the ball is determined by Newton's equation of classical mechanics. At every point in time, the current state[1] of the ball on the slope can be considered as initial condition of the ball that determines completely all future states of the ball. Various worked-out examples can be found in the literature. For example, the classical limit of quantum mechanics for a confined particle in a box [346, Sec. 5.9] and the quantum mechanical harmonic oscillator [26, Sec. 6.11] have been discussed. More explicitly, it has been argued that quantum mechanical randomness (i.e., quantum noise) can be neglected for a particle with mass of 1 gram that is moving with a speed of 1 m/sec [301, Sec. 9.5] or a 30 gram bullet in flight, which position is measured with an accuracy of 0.01 cm [26, Sec. 6.11].

If the system A of interest is relatively isolated from other systems, then in the aforementioned case, in which quantum noise can be neglected, from the discussion above it follows that the system evolves in a deterministic fashion. That is, our considerations lead to the introduction of the concept of physics-based determinism, as indicated in Fig. 2.3, see arrow (a).

---

[1] Which in this case means ball position and ball velocity.

Statistical Physics and Physics-Based Determinism

Humans and animals are systems that are composed of many subsystems, one the one hand, and, on the other hand, are in contact with their environments that are composed again of many components. Therefore, let us consider next systems that interact with many other systems. In the case of human individuals and animals these other systems may be part of the humans or animals themselves. For example, a particular brain area, labeled A, of a human individual may be active during the dream stage and the activity may be associated with a particular visual experience. In general, on-going brain activity in other brain areas that are not related to that visual experience can impact the activity in the brain area A. In doing so, the dynamics of the brain activity in the system A is perturbed. On an even smaller scale, cell signaling in neurons, the increase and decay of membrane potentials of neurons, and the electrical signaling along axons and dendrites is perturbed by the fact that these processes operate at a finite temperature. The fact that the human brain operates at a finite temperature means that the brain activity is perturbed by thermal micro-vibrations of its bio-molecules. On a larger scale, while a individual is performing a task (e.g., reading a book), he or she typically is subjected to various types of non-task related external forces (e.g., forces exerted by a person, who happens to walk by the reader, or caused by the music to which the reader is listening). These forces can again be regarded as interactions between the system A under consideration (e.g., the reading individual) and other systems. The right-hand-side of Fig. 2.3 addresses the situation in which a system A interacts in various ways with other systems. In this case, randomness originating from the quantum mechanical nature of the system A and the other systems is not addressed because this aspect is addressed on the left-hand-side of Fig. 2.3. Accordingly, it is assumed that for the description of the system A and the other interacting systems the phenomenon of quantum noise can be neglected. Consequently, the interactions and the resulting perturbations are deterministic, as indicated in Fig. 2.3. For example, thermal molecular micro-vibrations that impact system A are typically modeled by means of a large number of oscillators that oscillate with different frequencies and act on the system A. The set or ensemble of oscillators is considered as a heat bath [160]. Often the number of the systems that exert forces on the system A of interest is relatively large. Consequently, the total effect of all forces originating from those systems is similar to a random force acting on the system A. Therefore, we may say that the deterministic perturbations exhibit some kind of pseudo-randomness. As indicated in Fig. 2.3 there are two physical descriptions of such pseudo-random perturbations.

On the one hand, the deterministic nature of the perturbations can be taken into account and the perturbations can be described as a (large) set of deterministic systems interacting with the system A of interest. For example, in order to study the impact of thermal perturbations, a large set of heat bath oscillators that satisfy deterministic evolution equations can be simulated. Likewise, in so-called molecular dynamics simulations the dynamics of many interacting molecules is simulated that satisfy deterministic evolution equations of Newtonian mechanics. In combination with the aforementioned assumption that quantum noise can be neglected, this approach leads to an overall deterministic description of the system A of interest and to the introduction of physics-based determinism. In Fig. 2.3 this is indicated by the arrow (b).

On the other hand, in statistical physics pseudo-randomness can be described by stochastic models, see Fig. 2.3. These models do not account for the deterministic nature of the perturbations and, in doing so, are approximative models. However, they typically simplify the mathematical description of the system A and the other interacting systems and allow researchers to obtain important insights into the problems at hand. A benchmark example in this regard is the introduction of a Langevin fluctuating force that allows us to obtain a comprehensive understanding of systems subjected to pseudo-random perturbations [88, 271].

Physics-Based Determinism When the Observable of Interest Is a Mean Value

Frequently, the observable of interest corresponds to a mean value of some observables that belong to more elementary physical units (or components). For example, it has been argued that it is useful to describe brain activity associated with human activities and experiences on the basis of the mean activity of certain neuron populations (rather than on the basis of the activity of single neurons). Such models take as departure point the mean activity of neuron populations [124, 158, 175, 255, 272, 351]. If mean values are considered then the questions arises to what extent they are affected by the aforementioned pseudo-random perturbations. Assuming that the elementary physical units (e.g., single neurons of a population) are perturbed in an independent fashion, then the effects of those perturbations typically cancel out on the level of the mean value. In particular, let N denote the number of units that contribute to a mean value under consideration. Then, the so-called central limit theorem of statistics [183] states that the impact of those perturbations on the mean value decays like $1/\sqrt{N}$. That is, if the number N of units is relatively large, then the impact is relatively small. Consequently, for mean values X of a system A composed of a relatively large number N of units, perturbation can often be neglected. In this case, our considerations end up with a deterministic system again, see route (c) indicated in Fig. 2.3.

However, if the objective is to study explicitly the variability of a mean value X of a system A, then the deterministic description may be replaced by a stochastic one. For example, so-called critical phenomena (see Sect. 3.2.6 in Chap. 3) that require a probabilistic description have been studied for a neuron population model by considering the impact of a Langevin fluctuating force [116].

Physics-Based Determinism and "Perception", "Cognition", and "Behavior"

Let us define physics-based determinism as the physics perspective of humans and animals when assuming that the probabilistic aspect of quantum mechanics can be neglected for the phenomena of interest. Moreover, we assume that interactions with other systems that are not related to the phenomenon of interest and lead to deterministic perturbations are either included into the deterministic description of the system or can be neglected (e.g., because actually the observable of interest is a mean value that is computed from a relatively large number of units such that fluctuations of that mean value are negligibly small). Consequently, physics-based determinism states that under the two aforementioned assumptions (i.e., when neither quantum noise nor pseudo-randomness due to non-related interactions with other systems play a role) phenomena related to "perception", "cognition", and "behavior" of humans and animals can be described by deterministic evolution equations. According to those deterministic evolution equations, current states of affairs cause future states of affairs in a deterministic sense.

According to physics-based determinism, "perception", "cognition", and "behavior" is determined in just the same way as the dynamics of the ball rolling down the ramp mentioned above. Of course, the equations that determine the reactions of humans and animals and the evolution of the brain activity of humans and animals are in general different from Newton's equation of classical mechanics that determines the dynamics of a ball on a slope. However, as it will be shown in the subsequent chapters, they have the same general property as Newton's equation: for any given initial condition, there is only one, unique solution.

2.5 "Free Will"

2.5.1 Physics-Based Determinism Versus "Free Will"

Let us apply physics-based determinism to understand human and animal "perception", "cognition", and "behavior" and the role of "free will". As already mentioned above, the physics perspective takes the Schrödinger equation as a departure point. In the Schrödinger equation there are no expressions that could be regarded as terms that reflect "free will", which implies that under physics-based determinism the corresponding evolution equations do not feature a property that could be interpreted as "free will". Having said that, one needs to acknowledge that a precise definition of "free will" is difficult to achieve. If so, how can we state that the physics perspective, and, in particular, the Schrödinger equation, does not take "free will" into consideration? The reason for this is that if we attempt to define "free will" in a common-sense way then we see that "free will" is in contradiction with the physics perspective defined above. For example, Chisholm [49, page 32] writes that if we have "free will" then "we cause certain events to happen, and nothing – or no one – causes us to cause these events to happen". This is in contradiction with the aforementioned physics perspective that states that the conditions of the universe at an arbitrary reference point can serve as initial conditions and determine the future states of the universe. What happens in the future is caused by the conditions of the reference state. Consequently, when using the notion of a past for a moment, "perception", "cognitive activity", and "behavior" of a human or animal observed at a particular point in time is a consequence of the state of affairs given at any earlier point in time.

In this context, Fig. 2.4 illustrates some pitfalls of "free will" advocates. Panel (a) describes a so-called "stimulus-response" situation from a "free will" perspective. There is a "stimulus" (e.g., a small red soft ball lying on the floor) and a "response" (a 10 years old child kicking the ball away). Typically, various "responses" can happen given the same "stimulus" (e.g., the 10 years old child could grasp the ball or step on it). Some "free will" advocates postulate that there is "freedom to choose" a "response" (i.e., a reaction) out of the set of possible "responses" (i.e., reactions). While that "choice" can be affected by some factors (e.g., heaviness of the ball), the key issue is that according to "free will" advocates there is an element of "independent will" that independent of all factors affects the "choice" of the "response" (i.e., reaction). The corresponding physics perspective of that situation

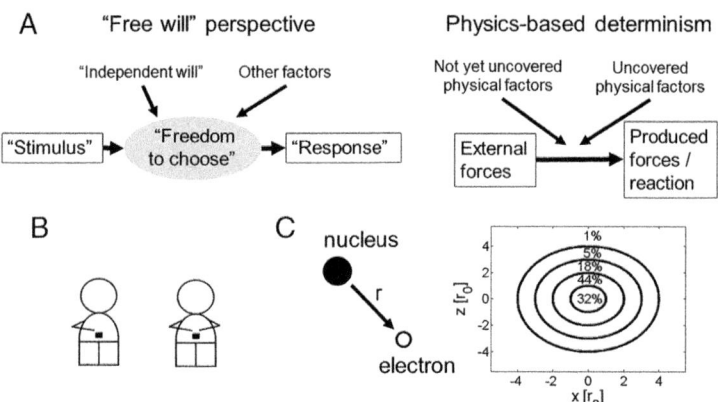

Fig. 2.4 Pitfalls of "free will" advocates. Panel (**a**, left): "Stimulus-response free will" perspective. Panel (**a**, right): When taking all conditions into account (e.g., current structure and brain state) any given external force ("stimulus") results in a unique reaction ("response"). Panel (**b**): Medium sized objects are grasped with one hand or with two hands depending on the current ongoing brain activity. Humans do not "choose" to grasp them with one hand or two hands. Depending on their ongoing brain activity, they are either in the condition to grasp them with one hand or in the condition to grasp them with two hands. Panel (**c**): Probability to find the electron of a hydrogen atom in certain spheres and shells around the nucleus as obtained from quantum mechanical calculations. Here $r_0 = 5.29 \, 10^{-11}$ m is the Bohr radius. "Free will" in humans motivated by the randomness involved in quantum mechanical descriptions would imply that any electron of a hydrogen atom exhibits the same kind of human "free will" power

is shown on the right of panel (a). There are external forces describing the external structure of the person under consideration (e.g., the light reflected from the ball exerting a force on the eyes of the 10 years old child) and there is a reaction in terms of produced forces (e.g., the kicking movement of the child). Depending on the explicit circumstances described by a plenitude of factors (e.g., the brain state of the child at the time when the child is confronted with the ball), there is one unique reaction to the given external forces and circumstances. On the "behavioral" level this reaction will come in terms of certain produced forces. Importantly, some of the factors that affect the reaction to the external forces of interest will be uncovered. Other factors will not yet be uncovered.

A similar situation that has been extensively studied by experimental research is illustrated in panel (b). A person is confronted with a medium sized object. That is, an object that is not too large such that it can be grasped will little effort with one hand (the right or the left hand). In addition, it is assumed that the object is not too small such that it is not too cumbersome to grasp the object with two hands. In experimental studies it has been shown that test persons under such circumstances show both grasping patterns. Sometimes they grasp objects of this kind with one hand. Sometimes they grasp them with two hands. Detailed experimental work has revealed that the reaction in a given test trial to this kind of medium sized objects depends on the grasping pattern that was performed in the prior test trial. If a test person had grasped an object with one hand, then a medium sized object is grasped with one hand. In contrast, if a test person had grasped an object with two hands, then a medium sized object is grasped with two hands [214, 269]. Note that referring to the prior test trial is not a plea for introducing the past in the description of human affairs. Rather, from the physics perspective at issue are the conditions of the test person at the time point of a given test trial. However, these conditions depend in general on what the test person did in the preceding test trial. We will return in Chap. 6 to this example in the context of the human pattern formation reaction model shown in Fig. 1.5 (see Chap. 1). To summarize, in line with the concept of physics-based determinism, in grasping experiments involving objects that can be grasped in different ways, there is no "freedom to choose" how the objects are grasped. The grasping pattern is determined by factors. Among those factors is the state of the person under consideration (i.e., the ordinary brain state and the structure, see Fig. 1.5) at the time point when he or she is confronted with the to-be-grasped object. "Free will" advocates have argued that a situation as shown in panel (b) of Fig. 2.4 with all factors held constant can still be seen in either of two ways. In contrast, in line with the concept of physics-based determinism, any given situation evolves in a unique way. This unique way can be calculated from the relevant laws when taking all relevant factors into account.

Finally, let us return to the probabilistic nature of physical laws in quantum mechanics and to stochastic descriptions of physical systems.

As mentioned above, according to the physics perspective, the Schrödinger equation allows us to formulate equations for human reactions in the first place in terms of evolution equations for probabilities. Can this feature be used to introduce the notion of "free will"? In fact, some "free will" advocates have argued that random elements in the description of human "perception", "cognition", and "behavior" indicate the existence of "free will" [237]. In order to discuss this argument, let us consider a hydrogen atom, which requires a quantum mechanical description (see Fig. 2.3). A hydrogen atom consists of a positively charged nucleus and a negatively charged electron. Using a classical mechanics description, the electron circles on an orbit around the nucleus like the earth is circling around the sun. In contrast, according to the quantum mechanical description of the electron, the electron can be found at any location around the nucleus. What differs from location to location is the probability that the electron can be found at that location. In particular, when the electron is in its so-called quantum mechanical ground state, the probability (more precisely, the probability density) as a function of the distance from the nucleus is a exponentially decaying function (for mathematical details see Sect. 2.8.2). Using that function, the probability to find the electron in certain spheres and shells around the nucleus can be calculated analytically. Panel (c) presents the probabilities thus obtain [12] (see also the mathematical notes in Sect. 2.8.2). For example, within a sphere of one Bohr radius (denoted by r_0) around the nucleus the electron can be found with a probability of 32%. Within a shell that goes from 1 Bohr radius to 2 Bohr radii, the occupation probability is 44%. Note that panel (c) present a two-dimensional cut to the spheres and shells that extent actually in all three dimensions x, y, z (i.e., these spheres and shells are like the layers of an onion). From the graph shown in panel (c) we see that in general close to the nucleus the probability to find the electron is relatively high, whereas further away the probability is relatively low. A supporter of the "free will" concept who relates "free will" to the presence of randomness may point out that quantum mechanics features a built-in random element and that some aspects of human brain functioning require a quantum mechanical description. This probabilistic aspect of quantum mechanics would introduce a random element in the description of human "perception", "cognition", and "behavior". Furthermore, the random element would be interpreted as "free will". If so, any electron spinning around a nucleus as shown in panel (c) would possess the same kind of "free will" power. The "free will" of a human individual would not differ qualitatively from the "free will" of an electron. In fact, since everything in physics (atoms, molecules, etc.) is described in the first place by means of quantum mechanics, every part and piece of the universe would exhibit "free will". Any piece of metal, any chair, any table would have "free will", which is something that supporters of the "free will" concept typically do not want to say. In short, the concepts of quantum mechanics cannot be used to motivate the existence of "free will". Therefore, taking quantum noise (i.e., the probabilistic element of quantum mechanics) into account in order to describe humans makes the description more mathematically involved but does not add anything regarding the discussion about "free will".

A similar argument can be put forward when "free will" advocates argue that randomness in general (not necessarily quantum noise) is evidence or a hint for the presence of "free will". As described on the right-hand-side of Fig. 2.3, a system of interest may be subjected to pseudo-random perturbations due to environmental forces that perturb the system in various ways. In this context, we may consider the forecast of a hurricane. Scientists may calculate for an emerging hurricane the most likely path that it will take and, in addition, other, less likely paths that the hurricane may take as a result of perturbations that will shift the hurricane out of the most likely path. In the end, the hurricane path will be described in a probabilistic sense. In this example, it is clear that the fact that the path of the hurricane is described in a probabilistic sense does not mean that the hurricane has "free will". Likewise, the physics perspective for human and animal "behavior" that takes randomness into account in terms of laws that are probabilistic in nature (e.g., by introducing Langevin fluctuating forces [32, 118, 286], see also Fig. 2.3 on the right) does not imply that the randomness involved in those laws reflects any form of "free will". In other words, if any phenomenon whose description involves an element of randomness involved "free will", then any hurricane (and any stock price jumping up and down on the stock market) would exhibit the "free will" property as well.

2.5.2 We Do Not "Choose". We Are in the Condition of Doing Something: The Shoe Example

In this section, the shoe example will be presented that has been discussed earlier in Frank [112]. Let us illustrate a key issue of physics-based determinism, the absence of "choice", with an example. Let us consider a person who is about to leave for work and is about to put his or her shoes on for leaving. For sake of simplicity, we assume that there are only two pairs of shoes available that are labeled with A and B. Panel (a) of Fig. 2.5 illustrates this situation from a "free will" perspective. According to the "free will" perspective, at issue is to make a "choice" between A and B. The person under consideration asks "What shoes should I put on?" In the psychological literature, it is sometimes assumed that at that point in time it is possible for the person under consideration to choose between A and B. Such a time point is referred to as "choice point". Once the person has asks the question "What shoes should I put on?", he or she makes a "decision" reflecting "free will" and "chooses" either shoes A or B.

Panel (b) of Fig. 2.5 illustrates the situation from the perspective of physics-based determinism. As mentioned above, the deterministic equations that can be derived from the Schrödinger equation exhibit unique solutions for any kind of initial conditions. Consequently, the solutions of those deterministic equations do not describe "choice points". Therefore, according to physics-based determinism, the notion that there is a unique situation that can make that the person of interest either puts on shoes A or puts on shoes B is a misconception. Rather, events that enfold in time are characterized by cause-and-effect relationships that are given in terms of if-then laws as illustrated in panel (b) of Fig. 2.5. According to panel (b), a closer look at the person confronted with the question "What shoes should I put on?", will reveal that in fact there is not a unique situation but there are two different situations. First, there is the situation A in which the person asks "What

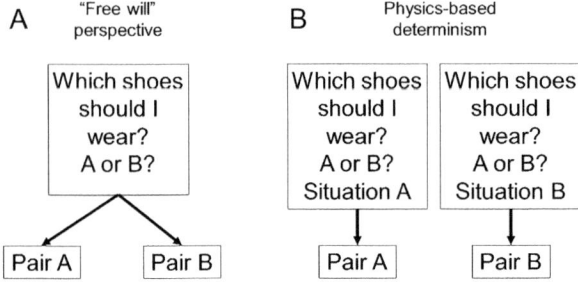

Fig. 2.5 The shoe example from the "free will" perspective (panel **a**) and from the perspective of physics-based determinism (panel **b**)

shoes should I put on"? and in which the initial conditions (in terms of the brain state, the structure, and the external forces, see Fig. 1.5 in Chap. 1) are such that he or she will put on shoes A. Second, there is the situation B in which the person asks "What shoes should I put on?"—like in situation A—but in which the initial conditions are different from situation A, namely, such that he or she will put on shoes B.

We see that the concept "choice" is not needed in the physics perspective of the world. In fact, it reflects a misconception when looking at human "behavior" in the context of physics-based determinism. Consider a speaker X, who tells a friend: "In this shoe shop, I had the choice between two pairs of shoes, a red and a black one, and I took the red shoes." Then, from a physics perspective, the speaker X has a misunderstanding about the situation in which he or she was involved. In contrast, consider a speaker Y, who tells a friend "In this shoe shop, I saw two pairs of shoes, a red and a black one. I was in the condition to take the red shoes. Therefore, I took the red shoes." Then speaker Y describes his or her situation in line with physics-based determinism.

2.5.3   What Do You Want? Chicken or Fish?

From a physics perspective, the question "What do you want? Chicken or fish?" is misleading because it suggests that the person being asked would have a "choice". Rather, consistent with physics-based determinism, we should ask a person "Are you in the condition for eating chicken or are you in the condition for eating fish?"

The considerations made above on physics-based determinism and "free will" support Skinner's point of view. Having pointed out this communality between Skinner's perspective and the physics perspective of the nature of humans, let us address two differences.

2.6   Internal States

In the context of Tolman's experiment on rats (reviewed above) the initial "behavior" of the rats to walk through either of the two doors with equal probability was explained by taking the internal state of the rat into account. Let us highlight the importance to consider internal states or conditions of a system by means of a simple mechanical example. Panel (a) of Fig. 2.6 presents the design of the system. The system is composed of two mass points (e.g., metal balls) labeled 1 and 2, two springs labeled 1 and 2, and a force plate. The mass point 1 is considered to be the internal state of the system. The mass point 2 acts as a sensor, that is, as input interface with the environment. In contrast, the force plate acts as output interface with the environment. That is, the force plate measures the force that the system produces. Note that the plate itself is immobile (i.e., does not move). The mass points 1 and 2 are mobile and can change their positions. Since the system involves two mass points, it might be referred to as two-body system. The question arises how does the two-body system react to external forces acting on the sensor? What force does the two-body system produce on the force plate as a reaction to an external force acting on the sensor mass? The problem can be solved explicitly (see mathematical notes in Sect. 2.8.3). Here, only a qualitative discussion will be given.

As such, the force applied to the sensor mass pushes mass 1 in the direction of mass 2. Due to the spring 2 this pushes mass 1 in the direction of the plate. Spring 1 is squeezed and a force that wants to push the plate to the left is produced. As a result, the external force (input force) that pushes the sensor to the left is transformed into an (output) force that wants to push the plate to the left. However, the internal state can crucially modify this overall relationship between external (input) force and produced (output) force. For example, if the internal mass 1 is initially (i.e., at

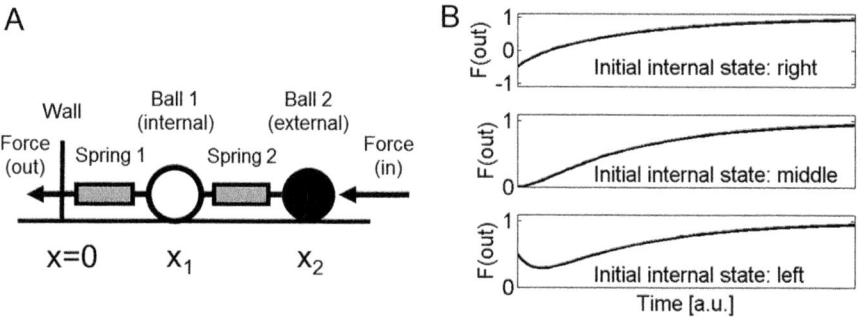

Fig. 2.6 Schema of a two-body system (panel **a**) and the impact of the system's internal state on the system's reaction to an external force (panel **b**). Panel (**a**): The system is composed of two mass points, two springs, and a immobile force plate (shown on the left). An external force, $F(in)$, acts on the mass point 2. Mass point 1 is considered as the internal state of the system. Panel (**b**): Produced forces $F(out)$ over time as measured at the plate for a constant external force but for different initial internal conditions

the time point when the external force is applied to the sensor) close to the sensor mass 2, then spring 1 is stretched. In this case, initially spring 1 pulls at the plate. There is a force that wants to move the plate to the right. This is irrespective of the external force. In contrast, if the internal mass 1 is initially close to the force plate, then spring 1 is squeezed which implies that there is a pushing force on the plate. This is again independent of the external force on the sensor mass. Panel (b) shows the time course of the produced force measured at the force plate for a constant external force for three different initial internal states. In all three conditions the external force is the same. The top panel refers to the case when the internal mass initially is close to the sensor mass. The middle panel shows the produced force for the case when the internal mass is in a neutral position. The bottom panel shows the force produced when the internal mass is initially located close to the output interface (i.e., the plate). Clearly, the produced force as a function of time differs for all three cases.

Internal states can have quantitative and qualitative impacts. For the two-body system, we may consider the external force on the sensor as cause and the produced force on the plate as the effect. If so, then the cause-and-effect relationship depends on the internal state of the two-body system as given by the location of the internal mass. For the relatively simple two-body system the produced force converges to the same value (given a constant external force) irrespective of the initial internal state. Therefore, in this example, the internal state affects quantitatively the cause-and-effect relationship of the two-body system by affecting the transient dynamics of the produced force. In contrast, in the case of the rats in Tolman's two door maze, at least during the first few days, under idealized conditions, the internal states of the rats determine whether they go through the food door or the other door. Therefore, in the case of Tolman's rats the internal states have a qualitative impact on the reactions of the rats to the doors.

Skinner argued that it is not necessary to address internal states in order to understand human and animal reactions [300]. The reason for this is that internal states themselves can be understood as being caused by something. Following Skinner's argument, let us consider the idealized case in which a mapping exists from external forces to internal states. That is, the mapping describes how internal states depend on appropriately defined external forces. In this idealized scenario it is indeed sufficient to study how humans or animals react to external forces that act or have acted on that human beings or animals under consideration. In other words, in the idealized case, for any internal state one can identify a cause-and-effect chain that gave rise to the internal state of interest and can follow this cause-and-effect chain backwards in time until one arrives at a cause that is given by an external force. Therefore, studying internal states does not do any harm but it is not necessary to do so.

From the physics perspective this argument exhibits at least two weak points. It is not clear whether or not the idealized case holds for humans and animals. When following a cause-and-effect chain backwards in time we may never encounter a cause that is entirely given in terms of external forces. That is, internal states may play a crucial role throughout the cause-and-effect chain that describes a human or animal during its whole lifespan, from birth to death. Second, from a practical point of view, the procedure to eliminate internal states by identifying the relevant external forces related to those internal states is often somewhat cumbersome. Consider the two-body system shown in Fig. 2.6. The produced force can be calculated given the position of ball 1, which reflects the internal state. Taking Skinner's point of view, we would need to observe the experimenter who set up the two-body system. When we observe how the experimenter placed ball 1 into the overall design, then we can conclude the position of ball 1. Observing how the experimental design has been assembled is a relatively cumbersome task as compared to measuring the position of ball 1.

Recall that the physics perspective states that the internal and external conditions that define a human and animal at any time point can serve as a reference point and are sufficient to explain future "perceptual", "cognitive", and "behavioral" activities of that human or animal. In view of this perspective, from a theoretical point of view, it is more convenient to address human individuals and animals by taking the relevant internal states into account. If so, the history of the human individuals and animals under consideration can be ignored.

2.7 Scope of Physics

2.7.1 *What Physics Does and What It Does Not Need To Do*

Physics is an all-encompassing theory of the animate and inanimate worlds. Consequently, physics explains all observations made on humans and animals. Physics does not need to address in any way concepts that are outside of the physics perspective. In other words, physics does not need to explain concepts outside the physics perspective in terms of physical concepts.

For example, the three main religions Christianity, Islam, and Hinduism assume that humans have a soul. The soul is not a concept of physics. Physics does not need to explain the soul in terms of physical concepts such as biochemical reactions in human cells and firing rates of neurons. Physics may explain from the physics perspective[2] why humans have believes (e.g., believes may be explained as synchronization phenomena, see Chap. 12). In particular, physics may explain why humans believe in a human soul. However, such an approach is not an explanation of the soul. Moreover, it does not address the soul as seen from religious perspectives such as Christianity, Islam, and Hinduism. For a more detailed discussion in this regard see the discussion on ethics in Chap. 12

Nowadays, in the neuroscience literature studies have been published on neural correlates of concepts that do not exist in physics. Such studies are like studies that

---

[2]I.e., not from a religious perspective.

try to explain the human soul in terms of biochemical reactions in human cells and neuronal firing rates.

2.7.2 Physics and Other Perspectives: Second Look

The physics perspective is about the universe as a whole and, consequently, addresses all affairs that concern the nature of humans and animals. The physics perspective provides a departure point (see Sect. 2.4.1) that can be used to draw conclusions. Although it has not been stated explicitly, the conclusions are drawn using Western logic. Here, Western logic (used in this book as a synonym for Boolean logic) stands among other things for AND and OR operations on statements that lead to new statements. For example, according to Western logic a true statement is obtained if two true statements are combined by an AND operation. However, if one of the two statements or both are false, then the combined statement is false.

Two contradicting hypotheses may be formulated and subsequently tested for consistency with fundamental assumptions of physics using Western logic. In doing so, we will find that at least one of the two hypotheses is incorrect (we may find that both are incorrect, but we will not find that both are correct—by definition of the word "contradicting"). Importantly, this testing of the two hypotheses is conducted within the framework provided by the physics perspective. The physics perspective is made to address hypotheses within its perspective. Although this seems to be a trivial statement, this means that the physics perspective is not made to address issues "outside" the physics perspective. In particular, the physics perspective does not address other perspectives.

Let us illustrate this issue by means of Fig. 2.7. Panel (a) shows a father and his daughter. The father makes the statement "Today is Monday". The daughter makes the statement "Today is Tuesday". In order to compare the statements a theory is needed. Only in the context of a particular theory, the statements can be compared. For example, the theory can be used to figure out whether the statements are different, if only one of the statements can be true, and if so, which statement is true. The need for a theory in order to make this kind of comparisons and conclusions is a crucial issue. As such, two statements that are not identical are not necessarily in contradiction with each other in the sense that only one of the two statements can be true. For example, let us assume the father says "Today is Monday". The daughter says "Today is my birthday". The statements are not necessarily in contradiction with each other although they are not identical.

Panel (b) lists schematically several perspectives about the nature of humans and animals. Among those perspectives is the physics perspective labeled by P9 (note that the label here has no meaning). There are other perspectives. In order to compare those perspectives or theories, we need a theory again. That theory might be used to determine whether the perspectives are different or not, whether or not only one of the perspectives can be true, and if so, which is true. The physics

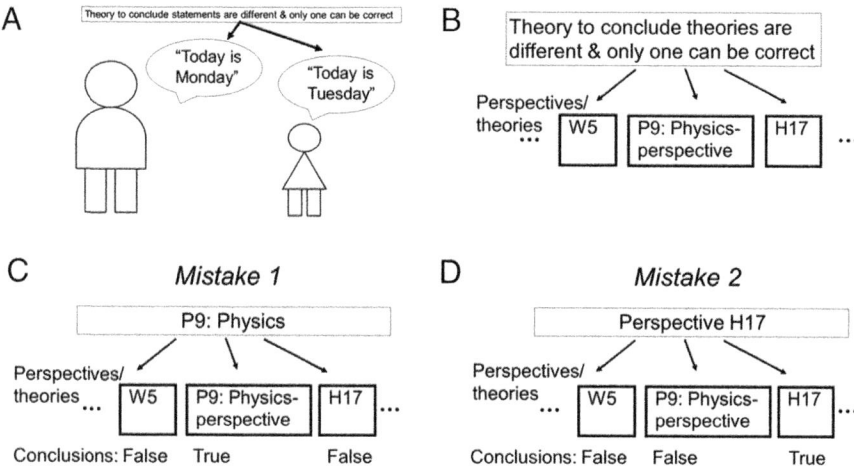

Fig. 2.7 Physics is a particular perspective out of many possible perspectives and is not a perspective "above" other perspectives. Panel (**a**): A theory/perspective is needed to compare the two statements made by the father and the daughter. Panel (**b**): A theory/perspective is need to compare between several theoretical perspectives and frameworks including the physics perspective as one of those perspectives. Panels (**c**) and (**d**): Mistakes frequently made in the literature. Physics is not a perspective "above" other perspective (panel **c**). When putting a particular perspective H17 "above" other perspectives such as physics, one most likely will find that all theoretical frameworks are "incorrect" except for the perspective H17 (panel **d**)

perspective as defined by the author is not that theory that is used to compare between perspectives. Likewise, this book is not about comparing perspectives.

We may compare the situation with the job specialization and job categorization in many modern societies. In many societies, a distinction is made between different jobs such as the job of a baker and the job of a pilot. Among other things, the jobs come with different duties. Importantly, people performing different jobs typically have received different types of "training". The purpose of the "training" is to put the worker of a particular profession in the position to perform his or her job successfully. For example, the purpose of a baker "apprenticeship" is to make that the apprentice eventually will be able to work in a bakery. The purpose of a pilot "training" is to make sure that the trainee will be able to fly a plane. From this it follows that it is not advised that a baker flies a plane or that a pilot works in a bakery.

By analogy, the physics perspective is design to address questions like "What is the physics of the universe?" and "What is the physics of humans?" *While the physics perspective can make statements about all phenomena addressed by any other perspective concerning the universe including humans and animals, the physics perspective is not in the position to make any statements about other perspectives* (e.g., religious perspectives). Importantly, the physics perspective does not negate the existence of "free will" or "the human soul" as such. The physics perspective only makes the statement that within a physics framework the concepts of "free will" and "the human soul" are not needed in order to address human and animal reactions and the dynamics of human and animal brain activity. Just like the baker does not tell the pilot what to do, and the pilot does not tell the baker what to do, the physics perspective does not make statements about other perspectives.

Note also that it is not inconsistent to state that the physics perspective is all-encompassing and addresses all phenomena in the animate and inanimate worlds and to state that the physics perspective does not make statements about other perspectives. The reason for this is that the physics perspective addresses all phenomena in the animate and inanimate worlds within its framework.

These considerations can be applied to clarify the relationship between the physics perspective and religious perspectives. As mentioned above, the physics perspective is not a perspective to determine which perspective or theory of the universe is the "correct" one. Panel (c) of Fig. 2.7 illustrates a mistake frequently made in this context. It has been argued that if we find that physics can explain all phenomena of the world, then this would negate the existence of a God. However, as shown in panel (c), if physics is taken as a theory to decide which perspective is the "correct" one, then this trivially will lead to the conclusion that physics is the "correct" perspective and that all other perspectives are "incorrect". These considerations imply that (1) physics is not a suitable higher-order theory to compare theories and (2) the physics perspective is not tailored to prove the existence or non-existence of divine entities such as God, Allah, Vishnu, and so on. In particular, the relationship between the physics perspective and the religious perspectives of various faith groups is twofold. On the one hand, the physics perspective rejects (as elements of the physics perspective) religious concepts that are inconsistent with the physics perspective. For example, the concept of a "soul" as something that possess "free will" power does not belong to the repertoire of concepts of the physics perspective. On the other hand, *as such* the physics perspective does not speak in favor or against any religious concept.

Finally, note that there is another frequently made mistake that is illustrated in panel (d). Sometimes, a religious perspective is taken to compare between that perspective and the physics perspective and to "prove" that the physics perspective is "false" and the religious perspective of interest is "correct". Such an approach comes with various methodological problems. For example, the question arises whether the religious perspective uses Western logic to draw conclusions. Irrespective of such problems in methodology, if someone starts his or her considerations with assumptions that are "different" from the assumptions of the physics perspective (assuming that a measure can be defined what "different" means), then it is not surprising that that person ends up with the conclusion that the physics perspective is "false". In the end, approaches as illustrated in panels (c) and (d) just tell us that there is someone saying something like "I am a baker and that is why I do not fly a plane" or "I am a pilot and that is why I do not bake bread".

2.8 Mathematical Notes

2.8.1 Schrödinger Equation

The Schrödinger equation describes the evolution of a wave function. Let $\Psi(x, t)$ denote a wave function defined on a one-dimensional space with coordinate x. Here t denotes time. In general, Ψ is a complex-valued function. Let $\hat{H}$ denote the Hamiltonian operator of the system under consideration. i is the imaginary unit (i.e., $i = \sqrt{-1}$). Furthermore, $\hbar$ is the Planck quantum. Then, the Schrödinger equation reads [63]

$$\frac{\partial}{\partial t} \Psi(x, t) = \frac{1}{i\hbar} \hat{H} \Psi . \tag{2.1}$$

Let us consider the special case in which Ψ is a real-valued function. For example, this is the case for the ground state of the hydrogen atom. Then the probability p to find the system in the interval $[A, B]$ is given by the integral [301, Sec. 9.6]

$$p(x \in [A, B], t) = \int_A^B [\psi(x, t)]^2 \, dx . \tag{2.2}$$

The formal solution of Eq. (2.1) reads (assuming $\hat{H}$ does not depend explicitly on time)

$$\Psi(x, t) = \exp\{-\frac{i}{\hbar} \hat{H}(t - t_0)\} \Psi(x, t_0) , \tag{2.3}$$

where $\Psi(x, t_0)$ denotes the wave function at an arbitrary reference time point t_0 that serves as initial condition. Equation (2.3) tells us that for any initial condition there is a unique solution. From this unique solution at any time point $t \geq t_0$ the probability (2.2) can be computed to find the system in any interval $[A, B]$ of interest.

2.8.2 Ground State of the Electron of the Hydrogen Atom

In the stationary case the Schrödinger equation (2.1) reads

$$\hat{H} \Psi = E \Psi , \tag{2.4}$$

where Ψ is the stationary wave function and E is the energy level of the state of interest. The ground state of the electron of the hydrogen atom can be determined by means of the stationary Schrödinger equation (2.4) for

$$\hat{H} = -\frac{\hbar^2}{2\mu}\nabla^2 - \frac{B}{r},$$
(2.5)

where μ is the reduced mass of the two-body system, ∇^2 is the three-dimensional Laplace operator, $B > 0$ is a constant, and r is the distance between the nucleus and the electron. The solution of Eq. (2.4) with $\hat{H}$ defined by Eq. (2.5) reads for the radial component of the wave function [26, 301, 346]

$$\Psi(r) = C \exp\{-r/r_0\},$$
(2.6)

where $r_0 = 5.29 \ 10^{-11}$m is the Bohr radius (or 1 atomic unit) and $C > 0$ is a normalization constant. The corresponding probability density is the square of Ψ and equals $\Psi^2(r) = C^2 \exp\{-2r/r_0\}$ and describes an exponentially decaying function [26, Fig. 6.20 of Sec. 13]. In order to obtain the probability to find the electron in a certain space around the nucleus, we need to integrate the probability density $\Psi^2(r)$ over the respective space (e.g., see Eq. (2.2) for a one-dimensional problem). In doing so, the probabilities to find the electron in certain spheres and shells around the nucleus can be obtained [301, Sec. 9.6]. These probabilities are shown in Fig. 2.4c.

2.8.3 Two-Body System

In what follows the mathematical details of the two-body system discussed in Sect. 2.6 will be presented. In order to demonstrate that the position of the internal mass affects the force produced by the two-body system it is sufficient to focus on the so-called overdamped case of Newtonian mechanics. Accordingly, it is assumed that the two balls of the two-body system are subjected to strong friction. Let $x_1(t)$ and $x_2(t)$ denote the positions of the two balls. Let $\gamma > 0$ denote the friction constant. Let F_{in} and F_{out} denote the input and output forces. Let F_1 and F_2 denote the spring forces of springs 1 and 2, respectively, that push the balls to the right. Then, x_1 and x_2 evolve like

$$\gamma \frac{d}{dt}x_2 = -F_{in} + F_2(t), \quad \gamma \frac{d}{dt}x_1 = -F_2(t) + F_1(t),$$
(2.7)

and we have

$$F_2(t) = \kappa(L - (x_2 - x_1)), \quad F_1(t) = \kappa(L - x_1), \quad F_{out} = -F_1(t),$$
(2.8)

where $\kappa > 0$ is the spring constant and $L > 0$ is the rest length of the springs. Since we are interested in determining F_{out} and F_{out} depends on $x_1(t)$, it is sufficient to solve the model for x_1. A detailed calculation yields

$$F_{out}(t) = \kappa(x_1(t) - L) ,$$

$$x_1(t) = a \exp\{\lambda_1 \kappa t/\gamma\} + b \exp\{\lambda_2 \kappa t/\gamma\} + C , \ C = L - F_{in}/\kappa \quad (2.9)$$

with

$$a = x_1(0) - b - C , \ b = \frac{(\lambda_1 + 2)x_1(0) - x_2(0) - \lambda_1 C}{\lambda_1 - \lambda_2} \quad (2.10)$$

and $\lambda_1 = -(3-\sqrt{5})/2, \lambda_2 = -(3+\sqrt{5})/2$. In the application discussed in Sect. 2.6, the initial value $x_2(0)$ of the external or sensor mass 2 is fixed. However, the internal initial state given in terms of the position $x_1(0)$ of mass 1 is varied. From Eq. (2.10) it follows that this internal state $x_1(0)$ affects the parameters a and b that in turn affect the solution $x_1(t)$, which eventually affects the force output $F_{out}(t)$. The simulations shown in Fig. 2.6b show the three cases with $x_1(0) = L/2$ (to the left of middle position), $x_1(0) = L$ (at middle position), and $x_1(0) = 3L/2$ (to the right of the middle position). Model parameters: $\gamma = \kappa = L = 1$ in a.u. and $F_{in} = 1$ force unit.

2.8.4 Miscellaneous

As mentioned in the Sect. 2.5.1, grasping objects in different ways (e.g., with one hand or two hands) has been studied in various experimental and theoretical studies. Grasping medium sized objects with one hand or two hands (see Fig. 2.4b) will be modeled in Chap. 6, see Sects. 6.5.3, 6.5.4, and 6.6. The shoe example discussed in Sect. 2.5.2 will be modeled explicitly in Sects. 4.6.1 and 6.4.1.

Chapter 3
From Self-Organizing Systems to Pattern Formation Systems

In this chapter and the following chapter fundamental concepts of the theory of pattern formation and synergetics will be introduced. The goal of this chapter is to clarify the notion of self-organizing systems and pattern formation systems. It will be shown that while it is difficult to arrive at a clear definition of self-organizing systems, pattern formation systems can be defined in a precise way. With the definition of pattern formation systems at hand, this chapter provides a basis for understanding humans and animals from a pattern formation perspective.

3.1 Classification of Systems

3.1.1 Thingbeings

Let us clarify the classes of systems relevant for the self-organization and pattern formation approach to understand human and animal reactions and the dynamics of human and animal brain activity.

First of all, all living beings and non-living things in this universe may be taken together to a single group, the group of thingbeings. Having formed this group of thingbeings, we can separate it again into the thingbeings of the inanimate world (things or non-living things) and thingbeings of the animate world (living beings). The latter includes humans and animals. In doing so, we arrive at Fig. 3.1.

For the study of self-organization the distinction between the animate and inanimate worlds is irrelevant because the principles of self-organization hold in both domains [97, 160, 215]. The purpose of Fig. 3.1 is to point out that frequently a distinction is made between things and living beings. However, for our understanding of human and animal "perception", "cognition", and "behavior" from a pattern formation perspective this distinction does not play a crucial role.

© Springer Nature Switzerland AG 2019

T. Frank, *Determinism and Self-Organization of Human Perception and Performance*, Springer Series in Synergetics, https://doi.org/10.1007/978-3-030-28821-1_3

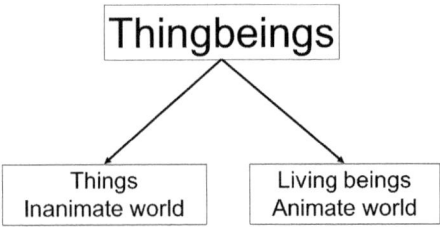

Fig. 3.1 The group of thingbeings and the two sub-groups of things and living things

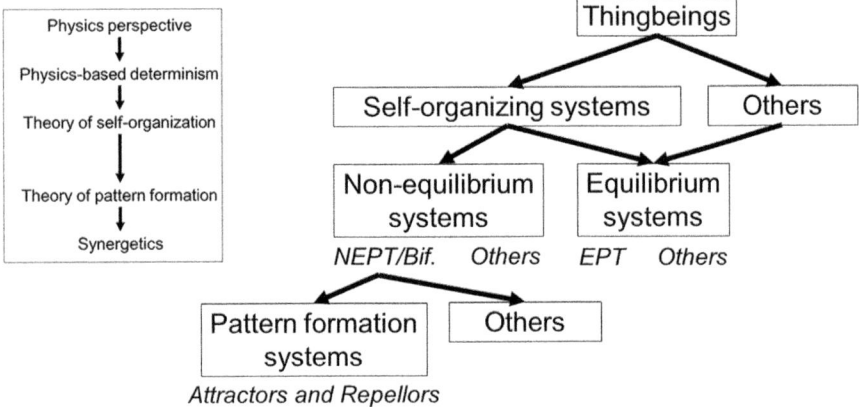

Fig. 3.2 Pattern formation systems in the hierarchy of physical systems (NEPT and EPT are abbreviations for non-equilibrium and equilibrium phase transitions, respectively)

Accordingly, "perception", "cognition", and "behavior" should not be understood as phenomena reserved for living beings.

Let us turn to the classification scheme shown in Fig. 3.2 that is more useful for understanding humans and animals from a self-organization and pattern formation perspective. As an insert in Fig. 3.2, Fig. 1.3 that was discussed in Chap. 1 is shown. In Fig. 3.2 we are interested in clarifying the step from the theory of self-organization to the theory of pattern formation addressed in Fig. 1.3. We will do so in terms of the systems involved in those step.

First, thingbeings can be separated into self-organizing systems and systems that do not qualify as self-organizing systems. A look at the literature on self-organization reveals that different authors have different notions about what self-organization means. Nevertheless, a minimal definition for self-organizing systems will be given below that is consistent with most of the work published in the field of self-organization. Second, self-organizing systems can be grouped into equilibrium and non-equilibrium systems. Equilibrium systems are in thermodynamic equilibrium with their environments. For example, when we put a cup of hot coffee on a table in a room at room temperature and wait for a sufficiently long time, then the temperature of the coffee will drop until it reaches the same temperature as the air in the room. At this point, the cup of coffee is in thermodynamic equilibrium with its environment. Non-equilibrium systems are systems that exhibit a different temperature as their environments and/or exhibit an energy source and a flow of energy through the systems. A pot with boiling water on a stove that is heated from below by a gas fire is an example of a non-equilibrium system. The fire pumps energy into the water. As a result, the water obtains a higher temperature as its environment. The water is not at equilibrium with the environment. For non-equilibrium self-organizing systems the flow of energy typically is needed to maintain the functioning of the systems. An example of such a self-organizing system is the laser. The laser produces a particular form of light, the laser light. The laser must be pumped with energy. Otherwise, laser light cannot be produced. Throughout this book, humans and animals are considered to be non-equilibrium self-organizing systems. Humans and animals need energy to do their businesses. Typically, humans and animals can get this energy by eating and drinking.

As mentioned above, there is not a clear definition of self-organizing systems. In order to be more specific, two steps can be taken as illustrated in Fig. 3.2. First, the subset of equilibrium and non-equilibrium self-organizing systems are considered that exhibit phase transitions or bifurcations. In the case of equilibrium systems, these phase transitions are referred to as equilibrium phase transitions. In the case of non-equilibrium systems, we either have non-equilibrium phase transitions or bifurcations. Phase transitions and bifurcations share a common feature that will be discussed below. Therefore, phase transitions and bifurcations can be used to define a subclass of self-organizing systems. The second step is to focus on the subclass of non-equilibrium self-organizing systems that exhibit non-equilibrium phase transitions or bifurcations. Among those systems are pattern formation systems.

Pattern formation systems will be defined below as the subclass of systems that exhibit bifurcations, which implies the existences of attractors and repellors. Examples of non-equilibrium self-organizing pattern formation systems are lasers, hurricanes, cloud patterns, humans and animals, who are reacting to external forces (e.g., who are "perceiving"), form body movement patterns like walking, or exhibit a particular brain activity dynamics (e.g., humans, who are "thinking" and "dreaming").

3.1.2 Self-Organizing Systems

Self-organization refers to the fact that the components of a system act together to form a whole. The organization or assembly of the components comes from within the system in the sense that it is not dictated from outside. This does not mean that self-organizing systems are isolated from the environment. There might be external forces acting on self-organizing systems. The forces might affect or even trigger a particular self-organization of the systems. However, these external forces do not specify the result of the self-organization. Let us define self-organizing systems as follows:

Self-organizing systems are systems that exhibit a group of mutually interacting components or a field with mutually interacting components at different spatial locations.

A group of mutually interacting components is a group whose components cannot be considered separately of the other components without changing the dynamics of the components. That is, in order to understand the evolution of mutually interacting components, one cannot take out any component of the group and study its dynamics independently of the dynamics of the other components. For example, if the size of the group is two, then the first component acts on the second component and vice versa such that the two components must be studied together in order to understand the dynamics of the two components. An example of a system with two mutually interacting components are two children swinging up and down on a seesaw. When the side of the first child goes down, the child pushes himself or herself up such that the side of the second child will go down. When the second child comes close to the ground, then he or she will push himself or herself up such that the first child goes down again and a new swinging cycle begins. An example of two systems that are not mutually interacting is a caregiver (e.g., a parent) who rocks back and forth a baby cradle. The two components are the caregiver and the cradle. The dynamics of the cradle can only be understood when taking the rhythmic movements of the caregiver into account. However, the rhythmic movements of the caregiver can be understood independently of the dynamics of the cradle (when assuming that the reactive forces from the cradle to the caregiver can be neglected in a first approximation). Such "master-slave" systems or similar two components systems exhibiting a driving force component and a driven component do not belong to the class of self-organizing systems (at least for the purposes of this book).

Note that when talking about a group of components, this means that for systems composed of spatially discrete units there must be at least two components. A single component system cannot be considered as self-organizing system. This consequence also follows from the notion of self-organization as an organization of several components to a whole. If there is only one component then this notion fails.

3.1.3 Pattern Formation Systems

Given the definitions introduced above, in this book pattern formation systems are defined as follows:

> Pattern formation systems are a special case of non-equilibrium self-organizing systems that exhibit bifurcations due to the self-organization of their components and are globally stable.

That is, pattern formation systems are a special case of self-organizing non-equilibrium systems that exhibit bifurcations or non-equilibrium phase transitions that can be described in terms of bifurcations, see Fig. 3.2 again. In addition, as mentioned above, the systems are assumed to possess a special stability property: they are assumed to be globally stable. We will return to this property in Chap. 4. From the definition of self-organizing systems and Fig. 3.2 it follows that pattern formation systems exhibit the following three properties. First, they exhibit a group of mutually interacting components. Second, the systems exhibits bifurcations (i.e., each system exhibits at least one). Third, the bifurcations are consequences of the interactions between the mutually interacting components. This definition of pattern formation systems emphasizes that bifurcations in pattern formation systems arise from mutual interactions between system components. Bifurcations will be discussed in detail in the following sections. Importantly, bifurcations involve the existence of attractors and repellors. This issue is highlighted in Fig. 3.2.

3.2 Phase Transitions and Bifurcations

3.2.1 About Jumps and Kinks

At the heart of the theory of pattern formation are the related concepts of phase transitions and bifurcations. Let us consider equilibrium and non-equilibrium self-organizing systems, see Fig. 3.2. By definition, equilibrium systems are assumed to operate at equilibrium. For non-equilibrium systems, it would be useful to introduce a counterpart to an equilibrium condition. A pseudo-equilibrium condition for non-equilibrium system is that we require that they are either stationary or that they belong to the subclass of pattern formation systems and are located on an attractor. In this context, stationary means that the observable of interest does not vary over time (i.e., is constant), whereas the concept of an attractor will be defined below. At this stage, it is sufficient to note that exhibiting stationarity and being located in an attractor state are two conditions for non-equilibrium systems that have a similar meaning as the equilibrium condition for equilibrium systems.

Having established some sort of equivalent starting point for equilibrium and non-equilibrium systems, it is assumed next that an equilibrium or non-equilibrium system exhibits a real-valued parameter that describes a part of the system structure and exhibits a special property that will be described in what follows. For the moment, let us refer to this parameter as structure parameter. The special property of this structure parameter is the following. First, the values of the structure parameter can be put on a real line such that the real line can be decomposed into intervals that involve or do not involve critical points. Second, when the parameter is varied within an interval without a critical point then physical quantities that depend on the state of the system vary in a smooth fashion as function of the parameter. Third and in contrast to the second property, when the parameter is varied within an interval that exhibits a critical point such that it is varied beyond the critical value then there are physical quantities that depend on the state of the system that either jump from

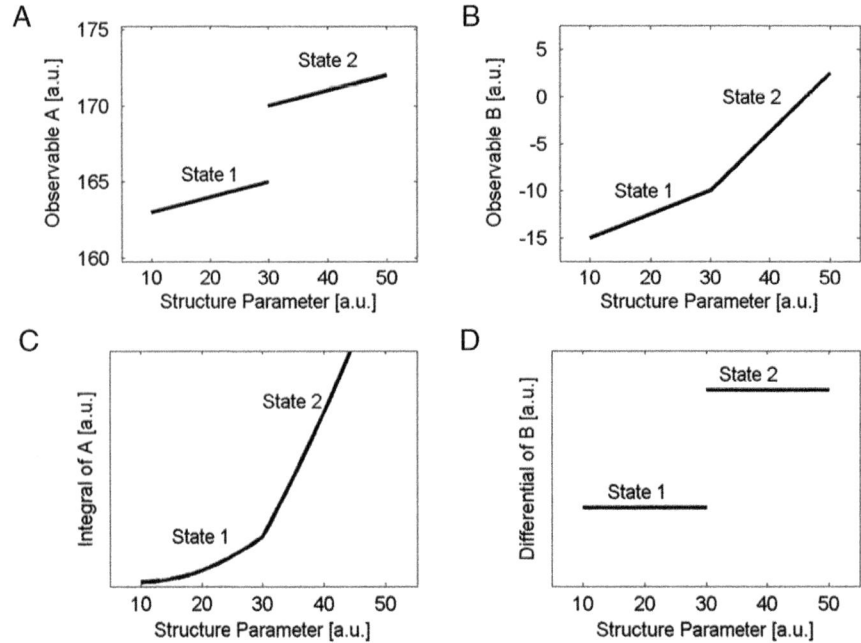

Fig. 3.3 Illustration of bifurcations and phase transitions. An observable related to the state of a system varies in a non-smooth fashion at a critical value of a structure parameter. The relationship between observable and structure parameter exhibits a jump (panel **a**) or kink (panel **b**). Panel (**c**): Integral of graph in panel (**a**). Panel (**d**): Differentiated graph from panel (**d**)

one value to another or change in a non-smooth fashion. Figure 3.3 illustrates both possibilities. Panel (a) shows a jump of an observed state variable (physical quantity) as a function of the structure parameter. Panel (b) present a non-smooth change of the observed state variable. A non-smooth change is also referred to as a kink.

What is the meaning of the smooth and non-smooth changes of physical quantities? The smooth changes indicate that the states within the intervals belong to the same category of states. They differ quantitatively but they are qualitatively the same kind of states. For example, in panels (a) and (b) all states for the structure parameter between 10 and 30 units describe qualitatively the same kind of state (e.g., walking of a human). Likewise, all state for the structure parameter between 30 and 50 describe qualitatively the same kind of state (e.g., running). The jump-like or non-smooth changes at the critical points indicate that the systems under consideration make a transition between qualitatively different states. In other words, smooth changes indicate within-category changes. Jumps and kinks indicate transitions between categories. In the context of equilibrium systems, the categories are called phases or aggregate states. For example, water can be found in solid, liquid, and gaseous aggregate states. For equilibrium systems, the between-category transitions are called equilibrium phase transitions. For non-equilibrium systems the between-category transitions are called non-equilibrium phase transitions or bifurcations. Note however that in the literature sometimes a distinction is made between non-equilibrium phase transitions and bifurcations. The reason for this is that the phrase bifurcation is primarily used in dynamical systems theory and dynamical systems theory often focuses only on a certain aspect of non-equilibrium phase transitions. For our purposes, a non-equilibrium phase transition is a jump-like or non-smooth change of a physical quantity indicating a transition between qualitatively different state. A bifurcation is just the same (i.e., a jump-like or non-smooth change of a physical quantity indicating a transition) and in addition means that the transition can be described in terms of attractors and repellors. If non-equilibrium phase transitions can be described in terms of attractors and repellors, then these phase transitions are both non-equilibrium phase transitions and bifurcations.

Jump-like and non-smooth phase transitions and bifurcations are related to each other. If the graph describing the observable of interest is integrated over the relevant structure parameter, then a jump-like change becomes a non-smooth change. That is, if the original observable is replaced by the integral of the observable with respect to the structure parameter under consideration, then the jump-like transition becomes a non-smooth transition. Panel (c) displays the integral of the observable shown in panel (a) as function of the structure parameter. Clearly, the jump of the observable disappears. However, at the critical point the integrated observable changes in a non-smooth fashion. By analogy, if an observable displaying a non-smooth change is differentiated with respect to the corresponding structure parameter, then the resulting observable exhibits a jump-like change at the critical value. Panel (d) shows the graphs shown in panel (b) differentiated with respect to the structure parameter. Since the observable shown in panel (b) increases linearly with the structure parameter in the states 1 and 2, the differentiated observable is just a constant over the structure parameter. Due to the fact that for state 2 the observable B increases faster than for state 1, this constant is larger for state 2 as compared to state 1. This leads in panel (d) to a jump-like transition at the critical value of the differentiated observable from a low constant value to a high constant value.

Figure 3.4 presents examples of phase transitions and bifurcations. Panel (a) features four examples of jump-like transitions. The two sub-panels on the top illustrate that it does not matter whether graphs describing within-category changes are parallel or not. The sub-panel on the bottom left points out that the direction of change is irrelevant for the existence of a jump-like phase transition or bifurcation. Finally, the sub-panel on the bottom right demonstrates that within-category changes do not need to follow linear relationships. In fact, in general, the within-category state changes are described by nonlinear functions of the structure parameters under consideration. Panel (b) exemplifies transitions involving a kink. The two sub-panels on the top illustrate that the direction of change does not matter for the existence of a non-smooth phase transition or bifurcation. The sub-panels on the bottom illustrate again that the graphs do not need to correspond to linear functions of the structure parameters of interest.

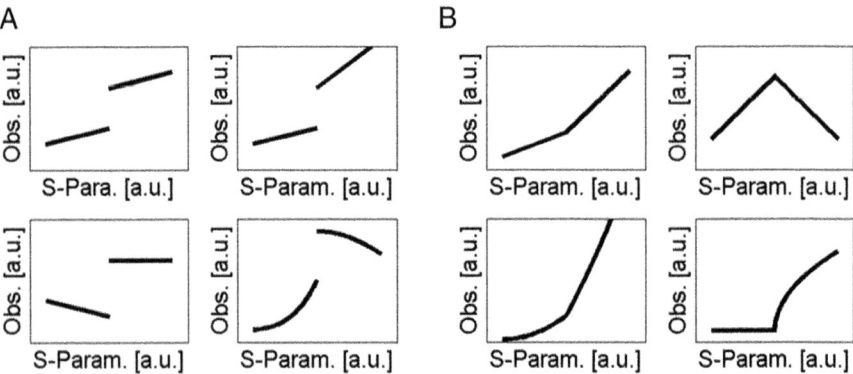

Fig. 3.4 Examples of bifurcations and non-equilibrium phase transitions. Panels (**a**) and (**b**) exemplify transitions with jumps and kinks, respectively

3.2.2 Bifurcation Parameters and Bifurcation Points

The structure parameters that induce phase transitions and bifurcations will be called bifurcation parameters. Likewise, the values of the bifurcation parameters at which phase transitions and bifurcations take place (i.e., jump-like or non-smooth changes of physical quantities occur) will be referred to as bifurcation points or critical bifurcation parameter values. Note that in the literature bifurcation parameters have also been called "control parameters" [160]. From our discussion about "free will" and "choice" in Chap. 1 and Sect. 2.5, it follows that the notion of "control" is not needed. In fact, "control" is a misleading concept when studying humans and animals from a physics perspective. This will become also clear when modeling pattern sequences in Chaps. 6 and 7. Although the phrase "control parameter" is not intended to imply that the thing which exhibits a "control parameter" has "control" over itself. Rather, the idea is that the system under consideration is "controlled" to some extent by the "control parameter". However, to avoid confusion and to help to obtain a scientific picture about the nature of humans and animals free of any mystification, the phrase "control" will be dropped in any context and will not be used in this book.

Note also that typically there is not a single bifurcation parameter. For a given system, various structure parameters of the system can induce phase transitions and bifurcations when they are varied beyond certain critical values. This issue will become in particular important when discussing the relationship between the theoretical framework of synergetics and the theoretical framework of ecological psychology (see Sect. 6.5.5).

3.2.3 Examples of Phase Transitions and Bifurcations

Figure 3.5 illustrates equilibrium and non-equilibrium phase transitions for several systems that belong to different system classes. Panel (a) shows the density of water as a function of the temperature. At zero degrees Celsius the function makes a jump. The reason for this is that at that temperature water changes from its solid aggregate state as ice into its liquid aggregate state commonly referred to as water. In this example, temperature is a bifurcation parameter. Note that the density of water as a function of the temperature in the liquid state seems to be a monotonically decaying function. In fact, due to the so-called anomaly of water, the density increases between 0° and 4° Celsius and decays monotonically only for temperatures larger than 4° Celsius. That is, there is a maximum at 4° Celsius. In this context, note again that for the existence of a phase transition or bifurcation our concern is about the jump-like and non-smooth changes (not so much about the smooth changes).

Panel (b) shows the magnetic field of iron, which is a ferromagnetic material, as a function of the temperature of the material [204]. The magnetic field on the vertical axis is shown in rescaled units as percentage of the absolute maximal magnetic field. The temperature T is shown in Celsius on the horizontal axis. For T below 770 °C the magnetic field is different from zero. Iron is in its ferromagnetic state. When increasing the temperature, the magnetic field decreases monotonically. It drops to zero at $T = 770$ °C and stays at zero for higher temperatures. For

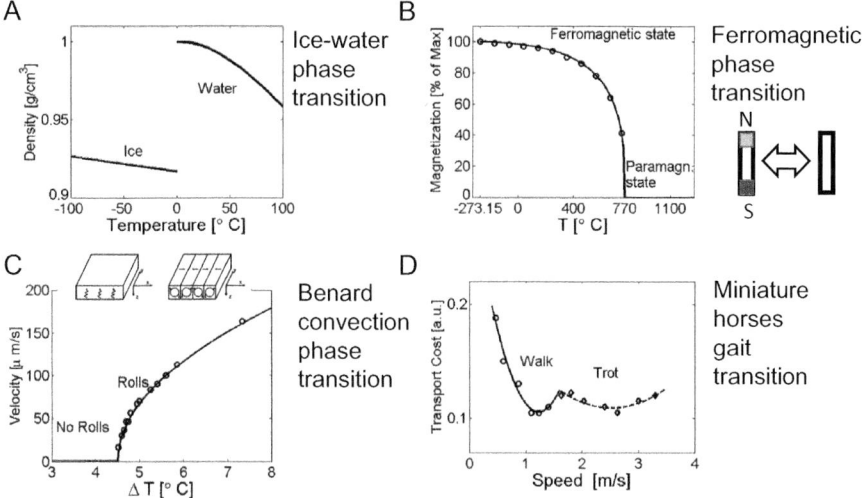

Fig. 3.5 Examples of phase transitions and bifurcations: aggregate phase transition from ice to water (panel **a**), equilibrium phase transition of a ferromagnet from the non-magnetic to the magnetic phase, when lowering the temperature (panel **b**), non-equilibrium phase transition of a heated fluid layer leading to the emergence of roll patterns (panel **c**), bifurcation from walk to trot in miniature horses (panel **d**). Data for panels (**b**)–(**d**) are taken from [204, 349] and [143], respectively

higher temperatures iron is in its paramagnetic state. In this state, it exhibits no magnetic field. In particular, at $T = 770°$ the strength of the magnetic field changes in a non-smooth function. The graph displays a kink. This kink indicates again that at $T = 770\,°C$ there is transition between two qualitatively different states. The ferromagnetic and paramagnetic states are also simply referred to as magnetic and non-magnetic phases. Accordingly, panel (b) illustrates a ferromagnetic phase transition or a phase transition between a magnetic and non-magnetic phase. Temperature is regarded as bifurcation parameter. The temperature of $770\,°C$ is the bifurcation point (also called phase transition point).

Panel (c) shows on the vertical axis the velocity of fluid particles streaming upwards in miniature roll patterns that were produced in a laboratory experiment [349]. More precisely, in the laboratory experiment a fluid was heated from below. For sufficiently high temperatures of the heat source the fluid started to stream upwards at certain locations and downwards at other locations. In doing so, rotating rolls were created. The rotation velocity of the rolls was measured in terms of the velocity of the upstreaming fluid particles. The laboratory experiment was conducted for different temperatures of the heat source. Panel (c) shows roll velocity on the vertical axis and on the horizontal axis the heat source temperature as difference ΔT with respect to the temperature at the top of the fluid layer. The symbol Δ preceding the temperature symbol T indicates that we are dealing with a temperature difference rather than an absolute temperature. For temperature differences ΔT below $4.51°$ Celsius, the system did not produce fluid streams and rolls. Accordingly, the fluid streams had velocity of zero. For temperature differences above $4.51°$ Celsius fluid streams and, consequently, roll patterns could be observed and the fluid stream velocity was different from zero. In particular, when the temperature difference was increased even further, then the fluid stream velocity increased as well. The temperature difference was the bifurcation parameter and the value of $4.51°$ Celsius was the critical temperature value that separated two qualitatively different types of states. Let us dwell on the nature of the two different states. For temperature values below the critical temperature the fluid showed heat transport by diffusion. That is, heat was transported from the bottom to the top of the fluid layer by random collisions of the fluid molecules without mass transport. That is, as mentioned above, there were no fluid streams. In contrast, for temperature values above the critical temperature the fluid showed heat transport by convection. Convection means mass transport. Fluid particles moved in streams from bottom to top and, in doing so, transported heat energy from the hot region to the cold region. Vice versa, fluid particles moved from top to bottom and, in doing so, cold fluid cells were transported from the cold region to the hot region. Convection of this kind is also called Benard convection or Rayleigh-Benard convection [97, 160, 212, 252, 349]. In panel (c), the velocity graph as a function of the temperature of the heat source (more precisely, the temperature difference) displays a kink. The kink just as the transition from zero to non-zero velocity indicates that the experiment involved two qualitatively different states.

Finally, panel (d) shows the transport cost (measured in terms of the oxygen consumption) of miniature horses for different locomotion speeds [143, 166]. In fact, the graph shows the result obtain from a single horse (however, qualitatively similar results were found for a total of three miniature horses). Interpolation lines have been added to help interpret the graph. Following the interpolation lines, the whole graph can be regarded as two U shaped graphs that are merged together at a certain locomotion speed, which is the critical locomotion speed. At that critical point the graph as a whole exhibits a sharp peak. That is, there is a kink or non-smooth change. The kink indicates that the two speed intervals for speeds below and above the critical locomotion speed describe two qualitatively different types of locomotion. In fact, for locomotion speeds smaller than the critical value the miniature horse was walking. In contrast, for speeds larger than the critical value the horse performed a trot. That is, at the critical locomotion speed a bifurcation (or non-equilibrium phase transition) from walk to trot occurred when increasing locomotion speed. Speed was the bifurcation parameter in this experiment.

As mentioned already above, the four examples of systems shown in Fig. 3.5 belong to different classes of systems. Table 3.1 characterizes the systems addressed in Fig. 3.5. To begin with the ice-to-water phase transition, as mentioned above, in the literature there is not a general consensus about the definition of self-organization. Water is typically not considered as a self-organizing system. Therefore, in Fig. 3.2 it is acknowledged that there are systems that are not self-organizing systems that exhibit equilibrium phase transitions. Having said that, the minimal definition of a self-organizing system given above in fact applies to water in its solid and liquid aggregate states because these states are characterized by mutual interactions between the water molecules. In Table 3.1 the question whether or not water is a self-organizing system is left open for debate. Irrespective of the debate whether water qualifies as self-organizing system or not, water is a system that exhibits phase transitions between different aggregate states. Water is an equilibrium system. That is, for a given value of the bifurcation parameter, which is the temperature of the environment, the system is supposed to have the same temperature as the environment. Accordingly, the ice-to-water phase transition is an equilibrium phase transition. The density graph of water as a function of the temperature exhibits a jump at the phase transition point. Water belongs to the inanimate world.

Table 3.1 Characterization of the systems shown in Fig. 3.5 showing phase transitions and bifurcations (the star "*" indicates that this issue is open for debate)

	Water	Ferromagnetic material	Heated fluid layers	Horses in motion
Self-organizing system	*	Yes	Yes	Yes
Phase transition or bifurcation	Yes	Yes	Yes	Yes
Equilibrium/non-equilibrium	Equil.	Equil.	Non-equil.	Non-equil.
Jump or kink	Jump	Kink	Kink	Kink
Inanimate/animate	Inanimate	Inanimate	Inanimate	Animate

Ferromagnetic materials are considered to be self-organizing systems. They are composed of so-called elementary magnets (i.e., magnets on the atomic or molecular level). These elementary magnets exhibit mutual interactions that give rise to the magnetic state. Ferromagnetic materials exhibit phase transitions between magnetic and non-magnetic phases. Ferromagnetic materials are equilibrium systems just as water. Accordingly, the phase transitions of ferromagnetic materials are equilibrium phase transitions. The strength of the magnetic field of ferromagnetic materials as a function of the temperature exhibits a kink at the critical temperature. Ferromagnetic materials belong to the inanimate world.

Fluid systems that are heated from below and exhibit under appropriate conditions fluid streams and roll patterns are self-organizing systems. The fluid cells at different locations interact with each other. The up- or down-streaming dynamics of a particular fluid cell cannot be explained when considering that fluid cell out of the context of all other fluid cells. Under appropriate conditions, fluid systems heated from below exhibit transitions between two categories of states. The first category of states is characterized by heat transport due to diffusion and the absence of fluid streams and dynamic rolls. The second category of states is characterized by heat transport due to convection and the presence of fluid streams and rotating rolls. Fluid systems heated from below are pumped systems. There is an energy flow through the fluid systems. Consequently, they are non-equilibrium systems (in this context note also that heated systems exhibit a temperature gradient and we cannot define a unique temperature of the environment, which implies again that these systems cannot be considered as equilibrium systems). The transition from the so-called homogeneous state that exhibits no rotating rolls to the state that exhibits rotating rolls is a non-equilibrium phase transition. Bifurcation theory applies to this kind of systems. Therefore, the non-equilibrium phase transition is also a bifurcation. The systems are pattern formation systems. The fluid stream velocity as function of the heat source temperature exhibits a kink at the critical heat source temperature. The fluid system belongs to the inanimate world just as the two previous examples.

From the pattern formation perspective taken in this book, horses just like humans are self-organizing systems. A horse in motion is characterized by a particular pattern on brain activity and body movements. The brain activity pattern emerges due to the interaction of various neuronal components of the animal brain such that the emergence of the pattern and the dynamics of the brain activity in general cannot be understood when the components are studied separately. In addition, the produced muscle forces result in the changes of external forces acting back on the brain level. The produced forces and the activity in the components of the sensory system are relevant for the locomotion of horses. Therefore, the self-organization system at hand involves the brain and body level. Horses exhibit gait transitions from walk to trot and vice versa from trot to walk. These gait transitions are non-equilibrium transitions because horses (and animals in general) are non-equilibrium systems. The oxygen consumption as a function of the locomotion speed exhibits a kink at the critical locomotion speed. Horses belong to the animate world. It is assumed that bifurcation theory applies such that the non-equilibrium transitions can also be understood as bifurcations. Horses in locomotion are pattern formation systems.

3.2.4 Not a Bifurcation

Phase transitions and bifurcations connect qualitatively different states with each other. In this context, the questions arises how to determine that something is qualitatively different from something else. A possibility to define qualitatively different state is to require that they are connected by phase transitions or bifurcations. That is, the jump-like or non-smooth change of a physical quantity may be used as a measure for states "being qualitatively different". If so, we end up with a truism. By definition, qualitative changes involve phase transitions and bifurcations. If there is no bifurcation or no phase transition, then the two things being compared are not qualitatively different.

In contrast, taking a conservative point of view, we may assume that there are other ways to determine that two states are qualitatively different. Following this more conservative point of view, changes between qualitatively different states do not necessarily be related to phase transitions and bifurcations. When structure parameters change physical quantities may change in a dramatic but smooth fashion. In such situations, from a conservative point of view, we may be attempted to conclude that the systems under consideration change qualitatively without featuring phase transitions or bifurcations.

Figure 3.6 provides an example in this regard. Figure 3.6 shows the activity of a particular DNA as a function of a chemical driving force measured in terms of the concentration of a particular chemical substance, called an activator [251]. For relatively low chemical driving forces the DNA activity is relatively low. In contrast, for sufficiently large chemical driving forces the DNA activity is relatively high. Importantly, the DNA activity reaches a plateau value. The low DNA activity is considered to be the baseline activity or the "off" state of the DNA. In contrast, the activity that the DNA exhibits in the plateau domain of large driving forces

Fig. 3.6 Switch of DNA activity from an "off" state to an "on" state. The DNA activity is shown as a function of activator concentration. Data taken from [251]

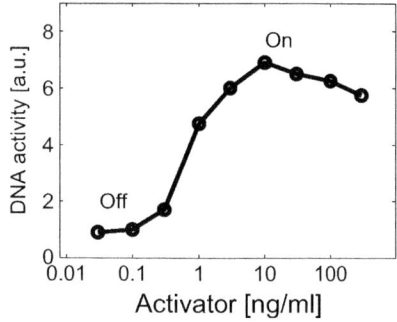

is considered to be the "on" state of the DNA. Therefore, when increasing the strength of the chemical driving force acting on the DNA, then the DNA switches from the "off" state to the "on" state. This switch is not related to a bifurcation. Taking a conservative point of view, the "off" and "on" states might be considered as qualitatively different states.

Let us discuss a technical example of an electric circuit that switches an electric fan on. Let us imagine an assembly of the following items: a sandclock, a light source, a light sensor, an electric circuit connected to an electric fan. The light source points to the upper part of the sandclock. If the sandclock is completely filled the light is reflected by the sand. If the sandclock is emptied by say 1 min, then the light can pass through the glass. Behind the class the light sensor is mounted. The light sensor is part of the circuit that connects to the fan. If the sensor does not receive light from the light source, then the sensor shuts down the circuit. That is, the circuit is open. The electric fan does not work. In contrast, if the sensor receives light from the light source, then the sensor closes the circuit. The electric fan is switched on and blows air into the room. Accordingly, if we fill the sandclock, in the beginning the fan is switched off. As soon as the sandclock is run down by 1 min, the light can pass through the glass. The light sensor closes the circuit and the fan is switched on. We may consider the on and off states of the fan as two categorical states (just like the on and off states of the DNA). The transition from the off to the on state does not involve a bifurcation.

Let us dwell on the example of the electric fan. Let us replace the light sensor in the circuit by a power "controller" with an turning knob. If the knob is in the most left position, the circuit is shut down. If the knob is in the most right position, the circuit receives full power. If the knob is anywhere between the left and right position, the circuit receive only a certain percentage of the full power. Accordingly, if we turn gradually the knob from left to right, then the electric fan will initially work with low power (e.g., the fan will rotate slowly) and finally work with full power (e.g., the fan will rotate with maximal speed). In this context, the two aforementioned states, the on and off states, are just two extreme states on a spectrum of possible states. That is, the change from on to off is not a between-category-transition but a state change within a category.

Analogous considerations can be made for the DNA example above. The two states illustrated in Fig. 3.6 may not be interpreted as two qualitatively different states. They may be considered as two extreme states of a spectrum of states that belong to the same category.

3.2.5 Strong and Weak Categories: Physics Perspective

The aforementioned discussion suggests that the theory of pattern formation, indeed, can be used to define categories. Accordingly, two states A and B belong to different categories if they are connected by a bifurcation. That is, states A and B are qualitatively different, if state B can be obtained from state A by changing a bifurcation parameter beyond a critical value at which a bifurcation or phase transition occurs. In contrast, two states A and B belong to the same category of states if state B can be obtained from state A by changing a parameter such that the system does not exhibit a bifurcation.

Having said that the situation at hand might be more complicated. In material physics there are systems that can be taken from a state A to a state B in two different ways. One way involves a phase transition. The other way does not involve a phase transition. This is the case when substances exhibit a so-called critical point in their phase diagrams. For example water exhibits a critical point that connects the liquid phase with the gas phase in such as way that it is possible to go from a liquid state to a gaseous state without a phase transition.

Likewise, in dynamical systems theory, there a dynamical system in which a state B can be obtained from a state A by changing parameter such that a bifurcation occurs. However, it is also possible to obtain the state B from state A changing the parameters in a different way such that there is no bifurcation. The so-called cusp model (related to the cusp catastrophe) is an example of such a dynamical system [148].

Therefore, the definition of categories by means of the theory of pattern formation may be revised by introducing weak and strong categories. For weak categories there must be at least one way to obtain state B from state A that involves a bifurcation or phase transition in order to state that B and A belong to different categories. The liquid and gaseous phases of water satisfy this definition and can be considered as weak categories. In contrast, for strong categories all states that belong to a given category A can be obtained from all states that belong to another category B only if bifurcation parameters are changed beyond critical values such that bifurcations or phase transitions occur. For example, the emerging rolls in the Benard experiment form a strong category. The rolls emerge from the spatially homogeneous state that does not exhibit rolls by means of a bifurcation.

3.2.6   How to Decide? Critical Phenomena

The notion of bifurcations and phase transitions is closely connected with the observation of jumps and kinks. While jumps in observable values can easily be defined, the question arises whether or not in experimental research non-smooth changes of an observable (i.e., kinks) can be uncovered. From a mathematical point of view, a smooth function and a non-smooth function are clearly different concepts. Therefore, in order to determine whether or not a change observed in a system is related to a phase transition or bifurcation we "just" need to make sufficiently precise measurements. However, such precise measurements might not be possible for practical reason.

In such cases, there are other ways to test whether or not a qualitative change is related to a phase transition or bifurcation. At bifurcation points, several so-called critical phenomena occur [4, 14, 25, 78, 160, 252, 253, 303, 305, 349]. In fact, such

critical phenomena that have been studied originally on non-living systems have
recently also been observed in living systems such as cells [295, 340]. In particular,
attempts have been made to uncover such critical phenomena in studies on human
"perception", "cognition", and "behavior" [122, 158, 182, 281, 285, 339].

3.3 State and Time

3.3.1 State and State Space

Let us consider two types of systems: discrete systems and spatially extended
systems. Discrete systems are systems that are composed of a number of discrete
components (or parts). Spatially extended systems are systems that continuously
extend over a certain space.

Let us consider discrete systems with the help of the collection of their components. Without loss of generality, it is assumed that each component is described
by a single, real-valued variable. For sake of simplicity, only systems with a finite
number of components will be considered. In this case, the real-valued variables
of all components constitute a vector, which will be called state vector. When we
imagine the components as axes of a high-dimensional space, then the space thus
obtained is the state space. The state vector lives in the state space and specifies a
point in that space. The point in the state space of a system that is defined by the
state vector is the state of the system. Figure 3.7 illustrates state spaces for systems
involving a single variable (panel a), two variables (panel b), and three variables
(panel c). The variables are denoted by x, y, and z.

The definition of state can be generalized to spatially extended systems. If the
system of interest is a spatially extended system, it is assumed that the system
can be described by a function over the space of interest. That function is called

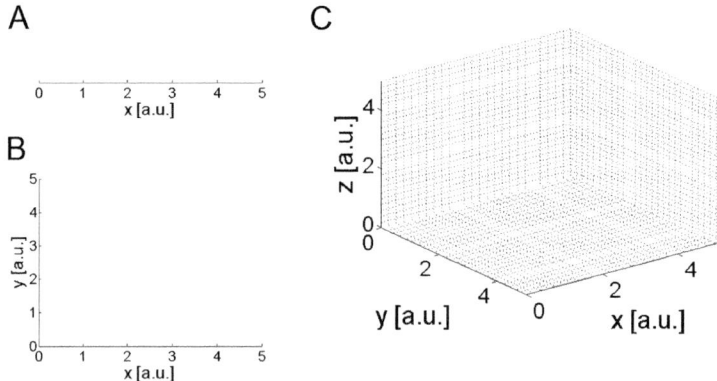

Fig. 3.7 Illustration of one-dimensional (panel **a**), two-dimensional (panel **b**) and three-
dimensional (panel **c**) state spaces

a field variable or simply a field. Again, without loss of generality, it is assumed that the field is a real-valued function. The function values of a field over all spatial positions describe the state of the spatially extended system. The state of a spatially extended system can be mapped to a state of a discrete system with infinitely many components. To this end, a basis defined by so-called eigenfunctions can be used. We return briefly to this issue in Chap. 4 and mathematical details in this regard can be found in Sect. 4.7.2. Therefore, a state vector and a state space may be defined for spatially extended system (despite the fact that the state vector and the state space are infinite dimensional).

Phase Space

The phrase phase space [252] is sometimes used as a synonym to state space. In particular, the term phase space is used in the context of Hamiltonian systems, which are a special class of dynamical systems.

3.3.2   Time and Dynamical Systems

Self-organization is a process that unfolds over time. Therefore, the state of self-organizing systems depends on time. In general, systems that evolve in time are called dynamical systems, which implies that self-organizing systems belong to the class of dynamical systems and elements of dynamical systems theory are part of the theory of self-organization (see Fig. 1.3). The state as a function of time corresponds to a graph in the state space. The graph is called trajectory. If the state is described by a single variable, the trajectory can conveniently be plotted. However, as argued above, self-organization requires interacting components. Therefore, the state should be described by two or more than two variables or a field variable. For discrete systems characterized by $n > 1$ variables, the state space is a n-dimensional space. The n variables as functions of time may be plotted in n different panels. These panels are projections of the trajectory on one of the axes of the state space. The trajectory in the n-dimensional state space can then only be imagined on the basis of those projections.

3.3.3   Phase Portrait

In line with our considerations on the physics perspective and physics-based determinism, the evolution of a system, to which physics-based determinism applies, depends on the conditions of the system at a given reference point in time. In other words, trajectories of states depend on initial states (e.g., see Fig. 1.5d). For discrete systems involving two components (or variables) it is useful to plot the

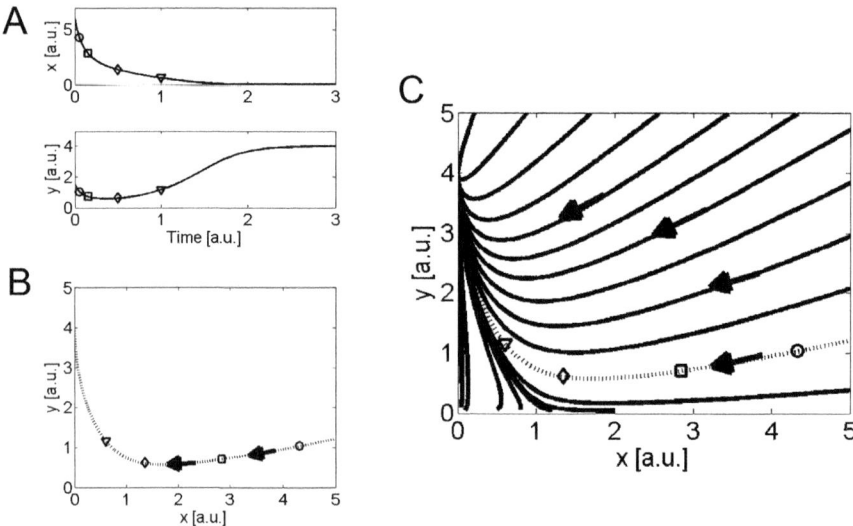

Fig. 3.8 Construction of a phase portrait (here: for a two species system in which only one species survives). Panel (**a**): Trajectory for a given initial state plotted in separate panels for the two species. Panel (**b**): Same trajectory but plotted in the two-dimensional state space. Panel (**c**): Phase portrait constructed from trajectories drawn for various initial conditions

trajectories for all possible initial states (in a certain domain) in a two-dimensional plane. In doing so, a comprehensive picture is obtained about how the system under consideration evolves in time depending on its initial state.

Figure 3.8 provides the phase portrait of a system of two competing species (e.g., two sub-species of owls that hunt for the same kind of pray). The underlying model is the Lotka-Volterra model that will be discussed in later chapters. At this stage, let us only note that each species is described by its number of animals such that the state of the entire two-species system is given by a two-component vector and the state space is two-dimensional. Panel (a) of Fig. 3.8 shows a trajectory of the two-species system. To this end, the animal population size measures denoted by x and y are plotted as functions of time $x(t)$ and $y(t)$, respectively, where t denotes time, in two separate panels. The two panels taken together describe the trajectory in the two-dimensional state space. This trajectory is shown in panel (b) when ignoring time. That is, panel (b) shows the states that were visited during the whole duration of the simulation (here from 0 to 3 time units). Arrows have been added in order to indicate in which direction the state evolves along the trajectory. In addition, markers have been added to panels (a) and (b) in order to demonstrates with pairs of species numbers shown in panel (a) correspond to which states shown in panel (b). The trajectory shown in panel (a) describes the case in which the first species initially outnumbers the second species. Nevertheless, during the course of time, the first species dies out and the population number of the second species converges to a saturation value. In fact, Fig. 3.8 describes a two species system that is monostable in the sense that the second species dies out due to the competition with the first species for any initial condition. Panel (c) displays the phase portrait of the two species system. The trajectory shown in panel (b) is shown as a special case. In addition, trajectories for all kind of initial conditions are shown. Arrows indicate in which direction the state evolves along the trajectories. The phase portrait illustrates that for any initial condition, the dynamical systems converges to the state in which the first species is extinct (i.e., x converges to $x = 0$ units) and the second species assumes a particular saturation value (here: $y = 4$ units).

3.4 Attractors and Repellors: A First Look

Attractors and repellors are particular domains or subspaces in the state spaces of dynamical systems that come with certain stability properties. More precisely, attractors and repellors are single points or sets of points in the state space of a dynamical system for which we can say something about how the state evolves in time close to those attractor and repellor points. There are different types of attractors and repellors [149, 160, 252]. For the sake of simplicity, only two types of attractors and repellors will be considered: fixed point attractors and repellors and limit cycle attractors and repellors. Figure 3.9 summarizes the different types of fixed point and limit cycle attractors and repellors. Table 3.2 points out that attractors and repellors differ with respect to their stability property.

Fixed points are locations in the state space of a dynamical system at which the state of the dynamical system does not change with time [149, 160, 252]. Fixed point attractors are fixed points such that when the state is perturbed out of the fixed point then the perturbation will decay and the state will return to the fixed point [149, 160, 252]. This property holds at least for sufficiently small perturbations. It does not necessarily hold for large perturbations. In this context, note that in general a perturbation of a state means that the state is shifted by a small amount. For fixed point attractors a system returns back to the attractor for arbitrary perturbations as long as they are sufficiently small. In this sense the fixed point attracts all states

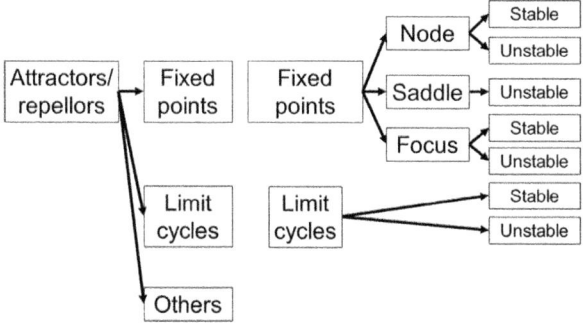

Fig. 3.9 Overview over different types of attractors and repellors

Table 3.2 Classification and notations of fixed points in one-dimensional and two-dimensional state spaces

Attractor/repellor	Dimension of state space	Synonym	Subtype
Fixed point attractor	1	Stable fixed point	
Fixed point repellor	1	Unstable fixed point	
Fixed point attractor	2	Stable fixed point	
			Stable node
			Stable focus
Fixed point repellor	2	Unstable fixed point	
			Unstable node
			Saddle
			Unstable focus
Limit cycle attractor	2	Stable limit cycle	
Limit cycle repellor	2	Unstable limit cycle	

in its neighborhood. Fixed point repellors are fixed points for which at least one perturbation exist such that the state does not return to the fixed point even if the perturbation is made arbitrarily small. In this sense, repellors repel states under certain conditions. Fixed point attractors and repellors are special cases of fixed points, see Fig. 3.9.

Fixed point attractors are also called stable fixed points. Likewise, fixed point repellors are also called unstable fixed points, see Table 3.2.

3.4.1   Single Variable Case

For dynamical systems described by a single variable Fig. 3.10 describes schematically stable and unstable fixed points (panels b and d) in terms of their phase portraits [252] and gives examples of trajectories (panels a and c) for both. In line with the general definition of stable and unstable fixed points, in the case of a stable fixed point any perturbation decays as a function of time as exemplified in panel (a). Moreover, any state close to the fixed point is attracted to the fixed point as illustrated in the phase portrait depicted in panel (b). In contrast, in the case of an unstable fixed point perturbations out of the fixed point increase in magnitude as exemplified by the trajectory shown in panel (c). In fact, for single variable dynamical systems, any state close to an unstable node is pushed away from the fixed point as shown in the phase portrait of panel (d). Note that in Fig. 3.10 the fixed points under consideration are located at the origin. In general, fixed points can be located at any point in the state space of a system.

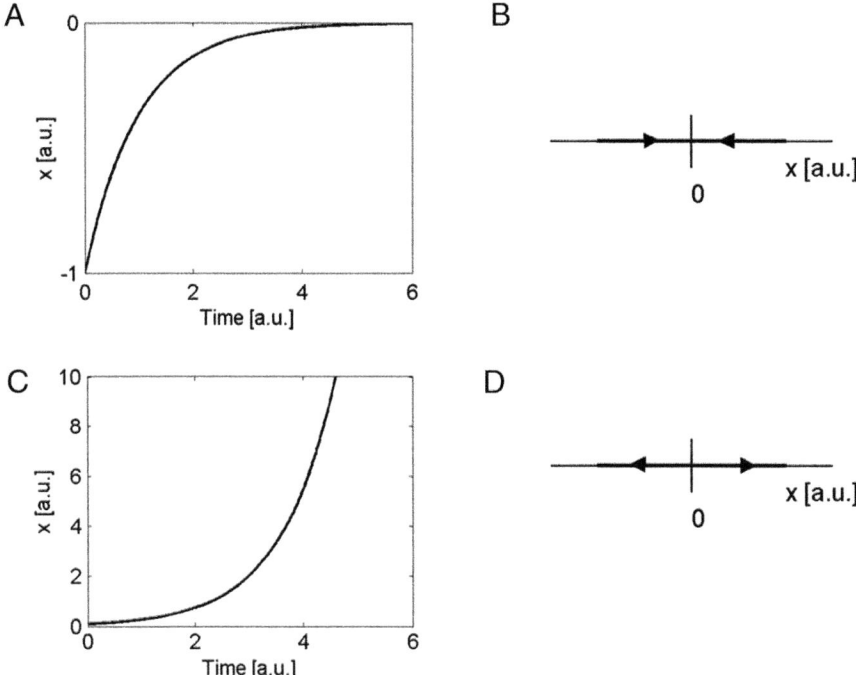

Fig. 3.10 Characterization of stable and unstable fixed points of single variable systems. Panels (**a**) and (**b**) describe a stable fixed point, while panels (**c**) and (**d**) describe an unstable fixed point

3.4.2   Two Variables Case

For self-organizing systems and dynamical systems described by two variables, we can distinguish between three types of fixed points: nodes, saddles, and foci, see Fig. 3.9 and Table 3.2. Nodes and foci can be found as fixed point attractors or repellors. That is, they can be stable or unstable. Saddles are fixed point repellors in any case. That is, they are unstable fixed points. Figures 3.11, 3.12, and 3.13 describe schematically these fixed point attractors and repellors. Figure 3.11 describes stable and unstable nodes. Panel (a) shows a trajectory of a system describe by a state vector with two components x and y. The system exhibits a stable node at the origin. Accordingly, the perturbation starting at $x = -1$ and $y = 2$ decays to zero. The trajectory depicted in panel (a) in Fig. 3.11 is shown in the state space in panel (b). Panel (c) shows the phase portrait composed of trajectories like the one shown in panel (b) but for various initial conditions [160, 252]. From the phase portrait it follows that for arbitrary initial conditions that are in a neighborhood of the fixed point, the state converges back to the fixed point. Panel (d) shows the phase portrait of an unstable node at the origin. States are repelled by the origin or driven out of

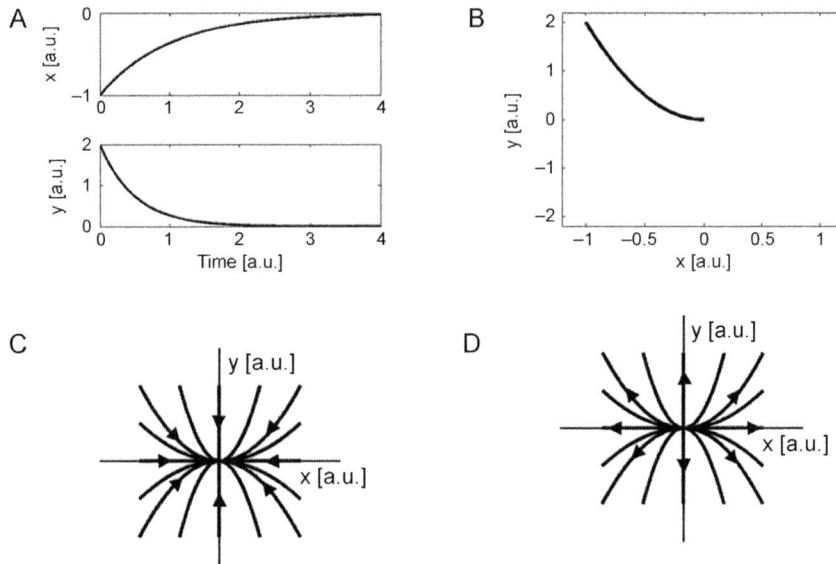

Fig. 3.11 Characteristic features of nodes. Panel (**a**): Example of a trajectory of a stable node. Panel (**b**): Trajectory depicted in the state space. Panel (**c**): Phase portrait of a stable node. Panel (**d**): Phase portrait of an unstable node

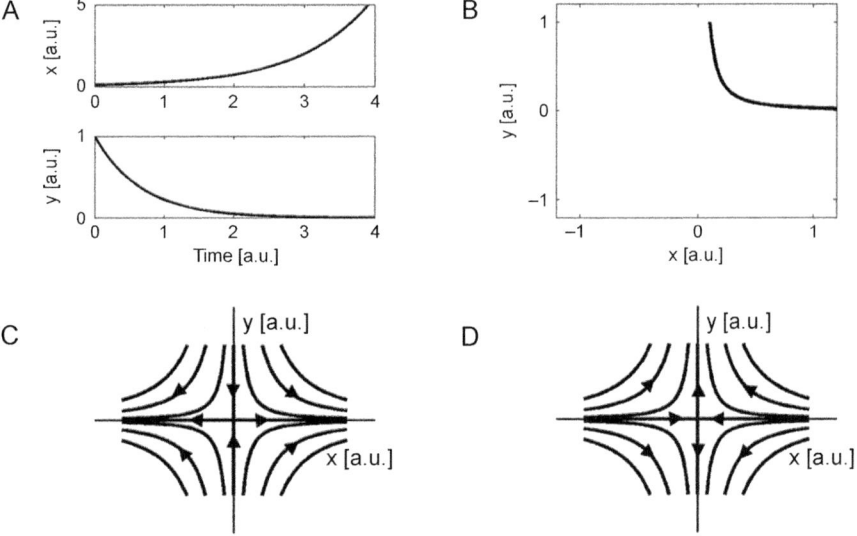

Fig. 3.12 Characteristic features of a saddle point. Panels (**a**) and (**b**) show an exemplary trajectory as function of time (panel **a**) and in state space (panel **b**). Panels (**c**) and (**d**) show phase portraits of saddle points exhibiting different stable and unstable directions

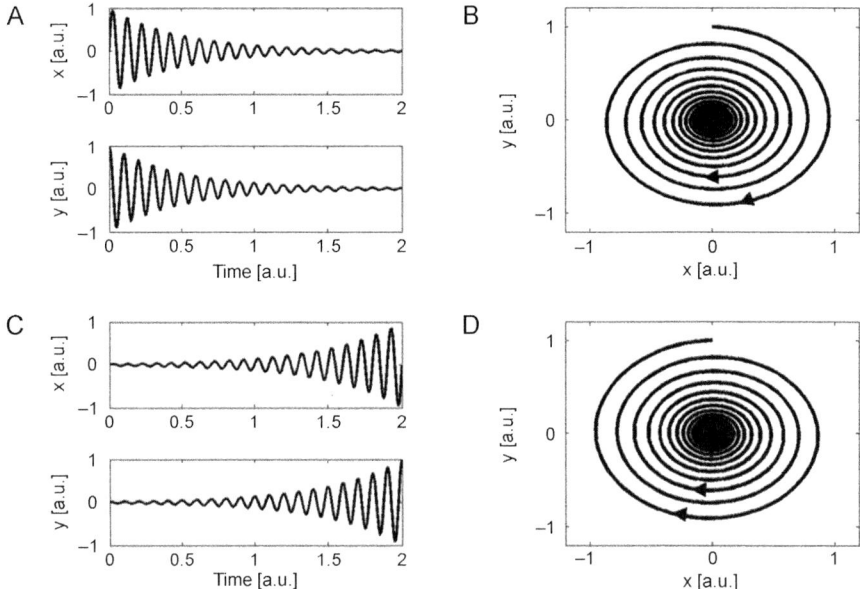

Fig. 3.13 Characteristic trajectories (panels **a** and **c**) and phase portraits (panel **b** and **d**) of foci. A stable focus (panels **a** and **b**) and an unstable focus (panels **c** and **d**) are shown

the origin. Consequently, any perturbation increases in magnitude over time even if the perturbation is initially very small.

Figure 3.12 describes the characteristic features of a saddle point in a two-dimensional state space. Panel (a) of Fig. 3.12 shows a trajectory close to the saddle for a particular initial condition. In the y direction the state approaches zero. In contrast, in the x direction the state increases exponentially. Panel (b) shows the trajectory in the state space. Panel (c) presents the phase portrait of the saddle [160, 252]. The saddle is said to exhibit a stable direction in the y direction (because in this direction perturbations decay) and an unstable direction in the x direction (because in this direction perturbations increase in magnitude. Panel (d) shows a saddle with a stable x direction and an unstable y direction. In general, a saddle exhibits a stable and an unstable direction in the two-dimensional state space. These two directions can point in any angle and do not necessarily correspond to the two axes of the state space.

Finally, Fig. 3.13 describes key properties of foci, see Table 3.2 again. Panels (a) and (b) of Fig. 3.13 describe a stable focus at the origin in terms of a trajectory (panel a) and the corresponding phase portrait (panel b) [160, 252]. The characteristic feature of a stable focus is that a perturbation decays in a oscillatory fashion. Consequently, the phase portrait displays a spiral on which the state evolves inwards (i.e., towards the fixed point of the focus, which happens to be the origin in Fig. 3.13). In order to obtain a phase portrait it is typically sufficient to plot a trajectory for a single initial condition. Due to the oscillatory dynamics of the state, the state will visit more or less the entire neighborhood around the fixed point. Therefore, it is not necessary to plot additional trajectories into the state space with different initial conditions. Moreover, any point on the spiral shown in panel (b) can be considered as an initial condition. Panels (c) and (d) of Fig. 3.13 present the characteristic features of an unstable focus. Panel (c) presents a trajectory, while panel (d) displays the phase portrait obtained by plotting the trajectory of panel (c) into the two-dimensional state space. In the case of an unstable focus, any perturbation increases in magnitude in an oscillatory fashion. This leads to a spiral-like phase portrait again. In contrast to the stable focus shown in panel (b), in the case of the unstable focus shown in panel (d) the state moves on the spiral outwards.

The fixed points shown in Figs. 3.11, 3.12, and 3.13 are all located at the origin of the two-dimensional state space spanned by the x and y variables. In general, a fixed point can correspond to any point of a state space.

3.4.3 Limit Cycle Attractors

As mentioned above, fixed point attractors of two-component systems (i.e., stable nodes and foci) are stable fixed points in the sense that perturbations decay of time. The system returns back to the attractor. To re-iterate, this stability property needs to be satisfied only for arbitrarily small perturbations. That is, if, for example, the state of a dynamical system is shifted out of a fixed point by a relatively large amount and, subsequently, does not return to the fixed point, then this does not necessarily violate the definition of a fixed point attractor. In contrast, if one can find arbitrarily small perturbations out of a fixed point such that the system does not return to the fixed point, then the fixed point under consideration cannot be a fixed point attractor.

Limit cycles attractors exhibit a similar stability property. The difference between fixed point attractors and limit cycle attractors is that an attractor fixed point is a point or location in the state space, whereas a limit cycle attractor is a closed line (or close curve or closed orbit [252]) in the state space. If the system is in any state on that closed line (i.e., on the limit cycle attractor) then the state will stay on that close line forever [160, 252]. The state will evolve along the line. Since the line is closed a state labeled A will return to that state A after a certain period. Consequently, limit cycle attractors describe oscillations. If a systems is initially close to a limit cycle attractor but not on the attractor, then it will be attracted to the limit cycle attractor (i.e., it will approach the attractor). Figure 3.14 illustrates a limit cycle attractor and the approach towards the attractor. Figure 3.14 presents a limit cycle attractor that surrounds the origin of the state space. Panel (a) shows a trajectory for an initial condition close to the origin of the state space. Panel (b) displays the trajectory in the state space. The state spirals out and eventually converges to the limit cycle. Note that the origin correspond to an unstable focus (compare panel (b) of Fig. 3.14 with panel (d) of Fig. 3.13). Panel (c) of Fig. 3.14 shows a trajectory for an initial state that is located outside of the limit cycle. The oscillation decays in magnitude

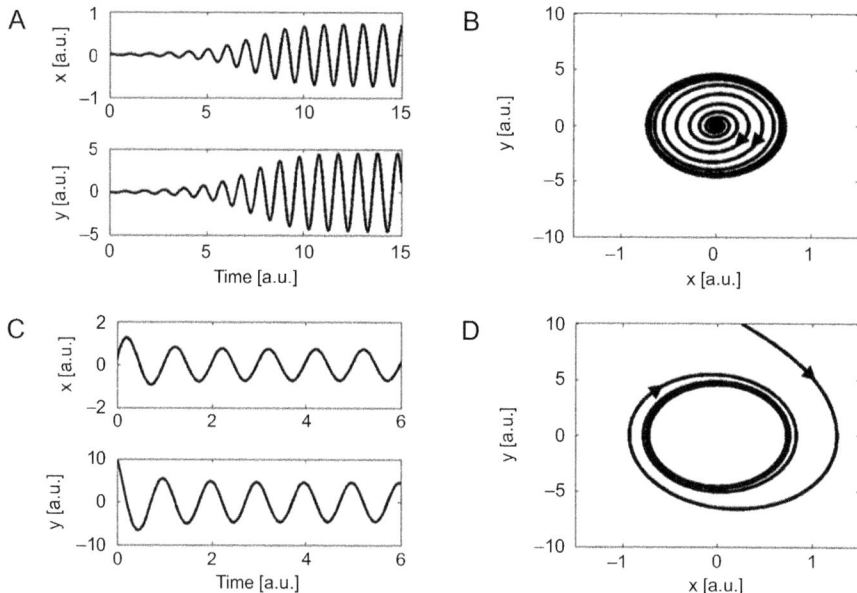

Fig. 3.14 Characteristic trajectories and phase portraits of a limit cycle attractor. Panels (**a**) and (**c**) show trajectories as functions of time starting inside and outside, respectively, the limit cycle attractor under consideration. Panels (**b**) and (**d**) show the corresponding phase portraits

and approaches the limit cycle. Panel (d) illustrates this approach in the state space. The full phase portrait of the limit cycle attractor would be given by overlapping panels (b) and (d) [160, 252].

While Fig. 3.14 shows a limit cycle attractor centered around the origin of a two-dimensional state space, in general, a limit cycle attractor can be located anywhere in a given (two-dimensional) state space. That is, a limit cycle attractor does not necessarily describe oscillations around the origin. For example, a limit cycle attractor may describe oscillations that only assume positive values and, consequently, are centered around a positive baseline level.

3.4.4 General Attractors and State Spaces

In general, attractors are sets of points in a state space such that sufficiently small perturbations out of the attractors return to those points in the state space. However, the perturbed states never reach the attractors in finite time. In other words, perturbed states converge to the domains or subspaces described by attractors and come arbitrarily close to those subspaces in finite amount of time but will never be exactly on those spaces. This is a general feature of any attractor in any n-dimensional state space. This feature is nicely illustrated by the fixed point attractor of a single variable dynamical system addressed in Fig. 3.10. Panel (a) of Fig. 3.10 displays a trajectory approaching the fixed point attractor located at the origin. The state variable decays monotonically towards zero. The state will come arbitrarily close to zero if we wait for a sufficiently long period. However, the state will never reach a value of exactly being zero in a finite amount of time.

3.4.5 Globally Stable Systems

With the definition of attractors and repellors at hand, we are in the position to defined globally stable dynamical systems. For our purposes, globally stable dynamical systems are dynamical systems that exhibit at least one attractor and have the property that their states converge to an attractor over time for arbitrary initial conditions. That is, when a globally stable system is prepared in an arbitrary state of its state space, then it will evolve in any case such that the system converges to an attractor. If there is only one attractor, it will converge to that unique attractor. If there are several attractors, it will converge to one of them dependent on the initial state. Importantly, the state does not increase beyond any boundary.

In the subsequent chapters, we will make frequent use of the notion of globally stable systems because it simplifies the presentation and argumentation. In particular, it will be assumed that the pattern formation systems under consideration are globally stable systems. However, most of what will be presented in the following chapters can be generalized for systems that do not correspond to globally stable systems or applies to systems that cannot be classified in this regard (e.g., because a proof whether or not they are globally stable cannot be given).

3.5 Bifurcations and Pattern Formation Systems: A Second Look

3.5.1 Definition of Bifurcations

Let us define bifurcations from an applied perspective as follows:

> A bifurcation is a jump-like or non-smooch change of an appropriately defined state variable induced by a change of a bifurcation parameter and related to the appearance or disappearance of attractors and repellors or to the change of stability of an attractor to a repellor and vice versa.

That is, like equilibrium and non-equilibrium phase transitions, a bifurcation is characterized by a jump or kink in the relation between an observable and a structure parameter that acts as bifurcation parameter. In the case of a bifurcation the jump or kink is related to a qualitative change of the underlying dynamical system. The qualitative change of the underlying dynamical system is in turn described in terms of the appearance and disappearance of attractors and repellors or as the change of stability from an attractor to a repellor or vice versa.

As far as the appearance of attractors is concerned, for example, it can be shown that in the Benard convection experiment mentioned above in the state space of the roll velocity a fixed point attractor emergence at a non-zero velocity when the temperature is increased beyond the critical temperature value. Therefore, non-equilibrium phase transitions in Benard experiments correspond also to bifurcations. As far as the change of stability from an attractor to a repellor is concerned, note that this requirement may be dropped because it can be regarded as a special case of a situation in which attractors and repellors appear and disappear. For example, when a stable fixed point becomes unstable, we may say that a stable fixed point disappears and at the same time an unstable fixed point appears.

3.5.2   Two Types of Bifurcation Diagrams

Bifurcation diagrams present attractors and repellors of systems as functions of bifurcation parameters. We may distinguish between bifurcation diagrams that involve only fixed point attractors and repellors and bifurcation diagrams that involve systems exhibiting other attractors and repellor such as limit cycles. For fixed point attractors and repellors, bifurcation diagrams present the fixed points of the attractors and repellors as functions of the relevant bifurcation parameters. In many cases we are primarily interested in the stationary stable states of the systems (i.e., the states to which the systems converge after transient periods). If so, since the stationary stable states of the systems under consideration correspond to the fixed point attractors, it follows that bifurcation diagrams reveal the jump-like or non-smooth changes of those states of interest.

Panel (a) of Fig. 3.15 presents the bifurcation diagram of a pattern formation system that involves only fixed points. The bifurcation parameter is denoted by α and goes from negative to positive values. The critical bifurcation parameter value equals zero. For negative values of the bifurcation parameter the system exhibits only a stable fixed point at zero. In contrast, for positive values of the bifurcation parameter, there are three fixed points. Two stable fixed points that assume non-zero values that are symmetric around zero and an unstable fixed point at zero. At the bifurcation point the stable fixed point at zero turns into an unstable fixed point (i.e., there is a change in stability). At the same time, two stable fixed points appear. This bifurcation is called a pitchfork bifurcation (because it looks somewhat like a pitchfork, see insert) [149, 160, 252].

Panel (b) presents the non-equilibrium phase transition diagram of the Benard experiment shown previously in Fig. 3.5c. Comparing the two diagrams in panels (a) and (b) of Fig. 3.15, we see that in the bifurcation diagram in panel (a) the unstable fixed point at zero is shown that appears for a bifurcation parameter larger than zero, whereas in panel (b) this fixed point is ignored. In addition, in panel (a) both stable branches are shown, whereas in panel (b) only one branch is shown. In fact, as far

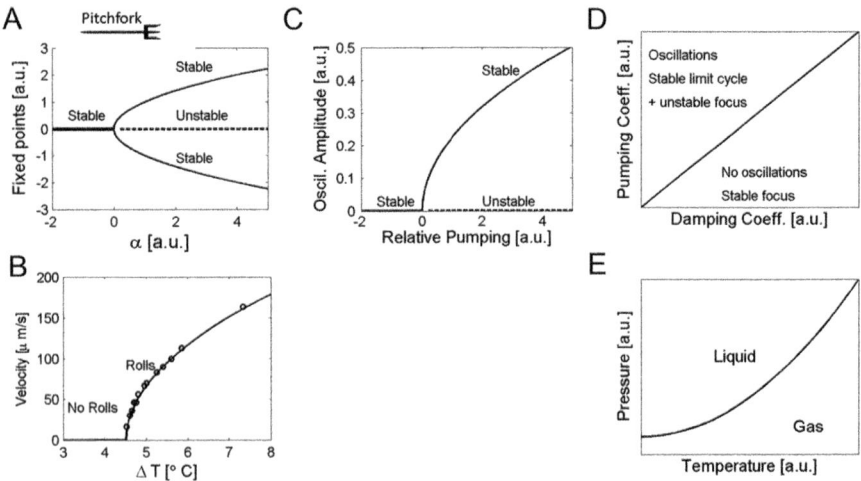

Fig. 3.15 Two types of bifurcation diagrams. Panels (**a**)–(**c**): Locations of fixed point attractors and repellors are plotted versus bifurcation parameters. Panels (**d**) and (**e**): Categorical states (panel **d**) and aggregate phases (panel **e**) are plotted in appropriately defined parameter spaces

as the Benard experiment is concerned, from theoretical reasonings it follows that the bifurcation of the Benard experiment corresponds to a pitchfork bifurcation as shown in panel (a). For temperature differences larger than the critical value the spatially homogeneous state with no rolls and zero roll velocity exists as stationary unstable state. Due to the fact that the state is unstable, the state is difficult to observe experimentally. As far as the second stable branch for temperature differences larger than the critical value is concerned, note that roll patterns typically can rotate in both ways. Therefore, in experimental studies—with appropriate experimental designs—both branches could be observed (not only the one shown Fig. 3.15b).

Note that by convention, for bifurcation diagrams of the first kind as shown in panels (a)–(c), stable fixed points (i.e., fixed point attractors) are indicated by solid lines, whereas unstable fixed points (i.e., fixed point repellors) are indicated by dashed lines.

If bifurcations involve attractors and repellors different from fixed points, the presentation of the attractors as functions of the bifurcation parameter becomes more challenging. In the case of a limit cycle attractor the oscillation on the limit cycle may be characterized by means of an oscillation amplitude. If so, the oscillation amplitude as function of the bifurcation parameter can be presented. An example in this regard is given in panel (c) of Fig. 3.15. Figure 3.15c presents the oscillation amplitude of a limit cycle oscillator as a function of the bifurcation parameter given by the relative pumping parameter of the oscillator. For negative values of the relative pumping parameter, the oscillator exhibits a stable focus but not a limit cycles attractor. Accordingly, the oscillation amplitude equals zero. For positive relative pumping parameters, the oscillator exhibits an unstable focus and a limit cycle attractor. The "diameter" of the closed orbit of the attractor (as shown in panels (b) and (d) of Fig. 3.14) increases monotonically as function of the relative pumping parameter. This implies that the oscillation amplitude increases. A detailed mathematical analysis shows that the amplitude increases in the same manner as the branches of the pitchfork bifurcation (compare panels (a) and (c) of Fig. 3.15; both graphs correspond to square-root functions of the respective bifurcation parameters). The bifurcation diagram of the limit cycle oscillator presented in panel (c) of Fig. 3.15 presents also the unstable focus for positive relative pumping parameters (see the dashed line).

In general, bifurcation diagrams can be presented in a similar way as phase diagrams showing aggregate states of materials. This is illustrated in panels (d) and (e). The relative pumping parameter of the limit cycle oscillator discussed earlier is composed of two terms: a damping and a pumping term [93]. If the damping term exceeds the pumping term the relative pumping parameter is negative. The oscillator does not oscillate. In contrast, if the pumping term exceeds the damping term, the relative pumping parameter is positive, the stable focus turns into an unstable focus and a limit cycle attractor appears. The oscillator begins to oscillate. Therefore, in the two-dimensional plane spanned by the damping parameter (horizontal axis) and the pumping parameter (vertical axis) the diagonal separates two parameter regions in which the oscillator exhibits qualitatively different states. For all parameters that fall in the parameter space below the diagonal, the oscillator exhibits a stable focus and does not oscillate. In contrast, for all parameters that fall in the parameter space above the diagonal, the oscillator exhibits the aforementioned unstable focus and the limit cycle attractor. For any initial condition (except the one that correspond to the fixed point given by the unstable focus) the oscillator dynamics converges to the limit cycle attractor and therefore, eventually, the oscillator settles down in an oscillation with a constant amplitude (see, e.g., panel (a) of Fig. 3.14 for times larger than 10 time units). The bifurcation diagram shown in panel (d) of Fig. 3.15 should be seen in analogy to phase diagrams that present aggregate states of matter as functions of thermodynamic parameters such as temperature, volume, and pressure. A detail of a typical phase diagram is shown in panel (e). Panel (e) shows a detail of a pressure-temperature phase diagram. Typically, for not too low and not too high temperatures the two-dimensional parameter space spanned by temperature (on the horizontal axis) and pressure (on the vertical axis) is separated by a line in two parameter regions. In the example shown in panel (e) the line is a monotonically increasing accelerating function of the temperature. For temperature and pressure values that fall in the parameter space above the line the matter under consideration is in the liquid state. In contrast, for temperature and pressure values that fall below the line, the matter under consideration is in its gas state.

In Fig. 3.15 panels (a)–(c), on the one hand, and panel (d), on the other hand, present two types of bifurcation diagrams typically used to describe bifurcations. Both diagrams refer to attractors and repellors. In the first type of bifurcation diagrams, fixed points of fixed point attractors and repellors are presented. Bifurcation diagrams of this kind reveal the jump-like and non-smooth changes at bifurcation points. In contrast, bifurcation diagrams of the second type present a map of parameter domains that feature different kind of attractors and repellors and/or different types of characteristic dynamical patterns [93, 149]. At the boundaries between the domains bifurcations take place. Bifurcation diagrams of the second type do not reveal the jump-like or non-smooth changes that occur at the bifurcation points (at the boundary lines). In order to see such changes, appropriate measures need to be defined. An example is given in panel (c) that presents the bifurcation that occurs on the diagonal of panel (d) in terms of a bifurcation diagram for the oscillation amplitude. Panel (c) illustrates that at every point of the diagonal in panel (d) a kink-like change of the oscillator amplitude takes place when the line is crossed from one side to the other.

3.5.3   Definition of Pattern Formation Systems Revisited

In Sect. 3.1.3 pattern formation systems were defined as follows:

> Pattern formation systems are a special case of non-equilibrium self-organizing systems that exhibit bifurcations due to the self-organization of their components and are globally stable.

Moreover, in Sect. 3.5.1 bifurcations were defined like:

> A bifurcation is a jump-like or non-smooch change of an appropriately defined state variable induced by a change of a bifurcation parameter and related to the appearance or disappearance of attractors and repellors or to the change of stability of an attractor to a repellor and vice versa.

Taking these two definitions together and taking into account the rather vague definition for self-organizing systems given earlier, pattern formation systems are systems that feature all of the following properties:

1. They exhibit mutually interacting components.
2. They exhibit bifurcations due to their mutually interacting components. This implies that

   o The systems feature bifurcation parameters.
   o At critical values of the bifurcation parameters some system properties change in a jump-like or non-smooth fashion.
   o The systems exhibit attractors and repellors.
   o At the aforementioned critical values of the bifurcation parameters attractors and repellors appear or disappear and/or switch stability from attractors to repellors and vice versa.

3. They are globally stable

In this definition the notion of a pattern itself does not show up. In fact, in the following chapter (Chap. 4) the pattern concept will be discussed in detail. Among other things, it will be shown that in some applications the patterns at hand are rather abstract objects, while in other applications the patterns correspond to geometric objects like stripes or rolls (e.g., the rolls shown in Fig. 3.5c).

3.6   Mathematical Notes

3.6.1   Definition of Fixed Points, Attractors, and Repellors

Fixed point attractors and repellors as well as limit cycle attractors can conveniently be defined on state spaces involving a set of real-valued variables. In this case, the states X under consideration corresponds to vectors $\mathbf{X}$ describing points in the state space. Attractors and repellors are subsets of points in such a state space. Let $\mathbb{R}^n$ denote a n-dimensional state space. Then fixed points, attractors, and repellors are given by sets

$$W \subset \mathbb{R}^n . \tag{3.1}$$

If the problem at hand requires the use of field variables, the field variables may be decomposed into basis patterns and amplitudes (see Chap. 4). Under certain conditions, the amplitudes can be considered as coordinates of a state space. Attractors and repellors can then be defined on those amplitude spaces.

Fixed Points and Stationary States

Let $\mathbf{X}$ a n-dimensional state vector that evolves like

$$\frac{\mathrm{d}}{\mathrm{d}t}\mathbf{X} = \mathbf{N}(\mathbf{X}) , \tag{3.2}$$

see Eq. (1.6) in Sect. 1.11.1. A fixed point $\mathbf{X}_{st}$ is defined by [149, 160, 252]

$$\mathbf{N}(\mathbf{X}_{st}) = 0 . \tag{3.3}$$

That is, the state does not evolve in time at a fixed point. Therefore, the fixed point also corresponds to a stationary state of the system. That is, in the context of Eq. (3.2) the phrases fixed point and stationary state are interchangeable. In particular, the subindex "st" refers to the word stationary.

Fixed Point Attractor

A fixed point attractor is a fixed point $\mathbf{X}_{st}$ that possesses the property [160, 252]

$$\forall \mathbf{X}(t_0) : \{\mathbf{X}(t_0) \neq \mathbf{X}_{st} \wedge |\mathbf{X}(t_0) - \mathbf{X}_{st}| < \epsilon\} \Rightarrow \mathbf{X}(t \to \infty) = \mathbf{X}_{st} , \qquad (3.4)$$

where ϵ is arbitrarily small and $|\mathbf{y}|$ is the length (or Euclidean norm) of the vector $\mathbf{y}$. According to Eq. (3.4), any state in a neighborhood of the fixed point that differs from the fixed point by an arbitrarily small amount converges back to the fixed point. Strictly speaking, the fixed point is asymptotically stable. In the main text fixed point attractors are called stable fixed points and the phrase asymptotically has been dropped. For the purpose of this book the phrase "stable fixed point" actually means "asymptotically stable fixed point". That is, while as such a distinction between "stable" and "asymptotically stable" should be made, we will not dwell on this issue unless it is necessary.

A fixed point attractor is composed of a set that contains a single point: the fixed point. That is, we get

$$W = \{\mathbf{X}_{st}\} . \qquad (3.5)$$

Fixed Point Repellors

A fixed point repellor is a fixed point $\mathbf{X}_{st}$ that possesses the property

$$\exists \mathbf{X}(t) : \{\mathbf{X}(t_0) \neq \mathbf{X}_{st} \wedge |\mathbf{X}(t_0) - \mathbf{X}_{st}| < \epsilon \wedge \frac{d}{dt}|\mathbf{X}(t) - \mathbf{X}_{st}| > 0 \text{ for } t \in [t_0, t_0 + \delta]\} , \qquad (3.6)$$

where $\delta > 0$ denotes a period that can be arbitrarily short but does not have to be short. According to Eq. (3.6), there exist states in arbitrarily small neighborhoods of the fixed point such that the states do not converge back to the fixed point. Rather, the distance (as measured by $|\mathbf{X}(t) - \mathbf{X}_{st}|$) between those states and the fixed point increases with time. Fixed point repellors are said to be unstable fixed points. That is, the phrases "unstable fixed point" and "fixed point repellors" are synonyms, see Table 3.2. Equation (3.5) also holds for fixed point repellors. The only difference is that the fixed point that constitutes the set W corresponds to an unstable fixed point.

Perturbations

Note that the states $\mathbf{X}(t_0)$ that are considered above and that are located in neighborhoods of a fixed point can be considered as perturbed (fixed point) states. That is, the state of a system initially corresponds to a fixed point. Subsequently, the state is perturbed out of the fixed point such that at time t_0 it ends up somewhere close to the fixed point. The perturbation at time t_0 is defined by [160, 252]

$$\mathbf{u}(t_0) = \mathbf{X}(t_0) - \mathbf{X}_{st} . \tag{3.7}$$

Equation (3.4) states that a fixed point attractor is a fixed point for which all perturbations (that belong for sufficiently small neighborhoods) decay to zero in time (i.e., $|\mathbf{u}| \to 0$) such that all perturbed states return to the unperturbed state (i.e., the original fixed point). In contrast, Eq. (3.6) states that a fixed point repellor is a fixed point for which there exist perturbations that increase in magnitude in time.

Limit Cycle Attractors

A limit cycle is a subset W of the state space under consideration defined by

$$W = \text{closed orbit} \wedge \mathbf{X}(t_0) \in W \Rightarrow \{\mathbf{X}(t \geq t_0) \in W \wedge \mathbf{X}(t_0 + T) = \mathbf{X}(t_0)\} . \tag{3.8}$$

That is, the limit cycle is a closed orbit (i.e., a closed line) such that if the state at an arbitrary reference time point t_0 is on the orbit, then it stays on the orbit and it satisfies the periodicity condition $\mathbf{X}(t_0 + T) = \mathbf{X}(t_0)$, where $T > 0$ is the period. A limit cycle attractor is a limit cycle that satisfies the property

$$\forall \mathbf{X}(t_0) \in \{\mathbf{X}(t_0) \notin W \wedge \mathbf{X}' \in W \wedge |\mathbf{X}(t_0) - \mathbf{X}'| < \epsilon\} \Rightarrow \mathbf{X}(t \to \infty) \in W , \tag{3.9}$$

where ϵ is a arbitrarily small parameter, again. That is, any state $\mathbf{X}(t_0)$ that is at time t_0 not on the limit cycle but in a small neighborhood of the limit cycle converges to the limit cycle. A more sophisticated definition of a limit cycle attractor can be given in terms of so-called Poincare maps and Floquet exponents [149] or in terms of eigenvalues (that will be discussed below) [28, 333].

3.6.2 Globally Stable Systems: Definition and Example

A globally stable system is a dynamical system described by a state X that evolves in time t such that

○ A trajectory $X(t)$ exists for $t \to \infty$ for any initial condition.
○ The system exhibits at least one attractor (i.e., the system may exhibit more than one).

○ Let W_k, $k = 1, 2, \ldots$ denote the subsets of the attractors and repellors of the
system and $W_{tot} = W_1 \cup W_2 \cup \ldots$. Then any trajectory $X(t)$ satisfies $X(t \to \infty) \in W_{tot}$.

That is, for globally stable systems any trajectory (or any solution of our fundamental model equation $dX/dt = N(X)$) converges to an attractor or remains on a
repellor if it initially is placed on a repellor.

An example of a globally stable dynamical system is the bistable system of the
state variable x defined on the one-dimensional state space of real numbers with

$$\frac{d}{dt}x = -\frac{d}{dx}V \, , \; V(x) = -\frac{ax^2}{2} + \frac{x^4}{4} \; \Rightarrow \; \frac{d}{dt}x = ax - x^3 \qquad (3.10)$$

with $a > 0$. Here, V is a so-called bistable potential function that exhibits two
minima at $x_{st} = \pm\sqrt{a}$ and a maximum at $x_{st} = 0$. These locations x_{st} correspond to
fixed points. The locations of the minima correspond to fixed point attractors, while
the location of the maximum corresponds to a fixed point repellor. Accordingly, we
obtain

$$W_1 = \{-\sqrt{a}\} \, , \; W_2 = \{\sqrt{a}\} \, , \; W_3 = \{0\} \, , \; W_{tot} = \{-\sqrt{a}, 0, \sqrt{a}\} \, . \qquad (3.11)$$

For any initial condition $x(t_0) \in \mathbb{R}$ with $x(t_0) \neq 0$ we have $x(t \to \infty) = \pm\sqrt{a}$.
That is, the solution converges to one of the two minima. If the state is located
initially at time t_0 at $x = 0$ such that $x(t_0) = 0$, then it remains there forever:
$x(t \geq t_0) = 0$. In any case, we have $x(t \to \infty) \in W_{tot}$.

Note that the model (3.10) describes the pitchfork bifurcation shown in Fig. 3.15a
if the parameter a is considered as bifurcation parameter and is allowed to assume
both positive and negative values. For $a \leq 0$ there is only one stable fixed point
(fixed point attractor) at $x_{st} = 0$. At $a = 0$ the stable fixed point turns into an
unstable fixed point (i.e., the fixed point attractor turns into a repellor). That is,
for $a > 0$ the two aforementioned stable fixed points (fixed point attractors) at
$x_{st} = \pm\sqrt{a}$ appear and $x_{st} = 0$ corresponds to an unstable fixed point.

3.6.3 Limit Cycle Attractor Example:
The Canonical-Dissipative Oscillator

Two famous limit cycle oscillators are the Rayleigh oscillator [267] and the van der
Pol oscillator (for a review see Jenkins [174]). In the field of human movements
sciences the two oscillators have been merged to obtain a so-called hybrid limit
cycle oscillator [19, 180]. In what follows, the canonical-dissipative oscillator will
be considered. On the one hand, the canonical-dissipative oscillator is a special
case of a hybrid limit cycle oscillator [70]. On the other hand, it is a benchmark
example [77, 92, 274] of so-called canonical-dissipative systems [77, 88, 151]. Let

$\mathbf{X} = (x_1, x_2)$ denote the two-components state vector of the oscillator, where x_1 denotes position and x_2 velocity. Then, the oscillator equations read [77, 92, 274]

$$\frac{d}{dt}x_1 = x_2 , \quad \frac{d}{dt}x_2 = -\omega_0^2 x_1 - \frac{\gamma}{m}x_2(H - B) , \tag{3.12}$$

where $\omega_0 > 0$ is the angular oscillation frequency, $\gamma > 0$ the damping constant, $m > 0$ the oscillator mass, and B the effective pumping coefficient. The function H is the Hamiltonian energy of the conservative oscillator for $\gamma = 0$ and is given by $H = m(x_2^2 + \omega_0^2 x_1^2)/2$. For $B \leq 0$ the oscillator describes a stable focus at the origin $(x_1, x_2) = (0, 0)$. In contrast, for $B > 0$ the origin is an unstable focus and there exists a limit cycle attractor around the origin. The evolution of the position x_1 can approximately be described by means of the slowly evolving amplitude $A(t)$ and the phase $\phi(t)$ like $x_1(t) = A(t)\cos(\omega_0 t + \phi(t))$. A detailed calculation yields the evolution equation of the amplitude [239]

$$\frac{d}{dt}A = \lambda A - C A^3 , \tag{3.13}$$

with $\lambda = \gamma B/(2m)$ and $C = \gamma\omega_0^2$. For $B \leq 0$ there is a stable fixed point at $A_{st} = 0$. In contrast, for $B > 0$ there is an unstable fixed point at $A_{st} = 0$ and two stable fixed points at $A_{st} = \pm\sqrt{\lambda/C}$. The amplitude equation (3.13) satisfies the pitchfork bifurcation described in Sect. 3.5.2 (compare also Eq. (3.13) with Eq. (3.10)). Moreover, Eq. (3.13) corresponds to the Lotka-Volterra-Haken model for a single amplitude, see Eq. (4.37) for $m = 1$ in Chap. 4.

3.6.4 Miscellaneous

The Lotka-Volterra model used in Sect. 3.3.3 to produce the phase portrait shown in Fig. 3.8 will be discussed in Sects. 4.5, 4.6, and 4.7. Mathematical details will be given there.

Chapter 4
Pattern Formation

This chapter consists of two parts. In the first part, the fundamental mechanism of pattern formation will be explained. In this context, the notion of a pattern will be defined explicitly. In fact, it will be shown that pattern formation involves two different but related types of patterns: attractor and repellor patterns, on the one hand, and basis patterns, on the other. Attractor and repellor patterns can come, for example, as fixed point patterns. In this case, we are dealing with fixed point attractor and fixed point repellor patterns. Attractor patterns correspond to the patterns that we humans form with our bodies (e.g., when walking) or to brain activity patterns in the human brain. Attractor patterns corresponds to the patterns that are usually observed in experiments in the first place. Attractor patterns can turn into repellor patterns at bifurcation points, and in this context are like two sides of the same coin. Attractor patterns are composed of elementary units: the basis patterns. In special cases, an attractor pattern may be composed of a single basis pattern. In such cases, an attractor pattern is identical to a basis pattern. In the context of the basis patterns the abstract concept of eigenvalues will be introduced. By definition, these eigenvalues determine how quickly basis patterns emerge and disappear. However, taking all eigenvalues together as a set, the set of eigenvalues constitutes an eigenvalue spectrum. This spectrum is the key for understanding pattern formation in general, and transitions between attractor patterns, the emergence of attractor patterns, and bifurcations, in particular. That is, eigenvalues play a key role in the theory of pattern formation and synergetics. Finally, in the context of basis patterns also the concept of pattern amplitudes will be introduced. As will be explained below, pattern amplitudes describe how much basis patterns contribute to attractor patterns. Pattern amplitudes, in general, and the so-called reduced amplitude space, in particular, allow for a convenient description of the formation of patterns and transitions between attractor patterns. The second part of this chapter is devoted to the Lotka-Volterra-Haken amplitude equations. These equations describe pattern formation in the aforementioned reduced amplitude space. The Lotka-Volterra-Haken amplitude equations correspond to a general class

© Springer Nature Switzerland AG 2019
T. Frank, *Determinism and Self-Organization of Human Perception and Performance*, Springer Series in Synergetics,
https://doi.org/10.1007/978-3-030-28821-1_4

of amplitude equations in the theory of pattern formation. They will be used in all applications of this book to describe the formation of brain and body activity patterns in humans and animals (i.e., "perception", "cognition", and "behavior").

Note that in the literature amplitude equations different from the Lotka-Volterra-Haken amplitude equations can be found. They typically exhibit similar properties as the Lotka-Volterra-Haken amplitude equations. Therefore, the reader should feel free to replace the Lotka-Volterra-Haken amplitude equations in the applications shown in the following chapters with other kind of amplitude equations. In principle, given a pattern formation phenomenon at hand, the underlying amplitude equations can be derived from an appropriate bottom-up (i.e., microscopic) approach. However, in this book, a top-down modeling approach will be used. In view of the generality of the Lotka-Volterra-Haken amplitude equations, it will be assumed that they capture the essential aspects of the phenomena that will be discussed in subsequent chapters. The amplitude equations will serve as a tool to demonstrate explicitly how the formation of brain and body activity patterns in humans and animals (i.e., "perception", "cognition", and "behavior") can be understood from a pattern formation perspective and within the framework of synergetics.

4.1   Fixed Point, Attractor, and Repellor Patterns

Patterns are typically defined based on geometric shape. Accordingly, there are dots patterns, stripes, spirals, and roll patterns. The V-formation of a flock of birds is another example of a geometric pattern. In the context of the theory of pattern formation, patterns may exhibit a particular geometric shape. However, the shape is not the defining feature. Rather, in what follows, the concept of a pattern will be defined in an abstract way on the basis of the mechanism that underlies in self-organizing systems the formation of all kinds of patterns including the aforementioned geometric patterns as special cases.

Let us define fixed point patterns, attractor patterns, and repellor patterns as follows:

A fixed point pattern is a state of a system on a fixed point. An attractor or repellor pattern is the evolution of a state on an attractor or repellor.

In other words, the trajectory of a state in its state space on an attractor or repellor is considered to be a pattern: an attractor pattern in the case of an attractor and a repellor pattern in the case of a repellor. Likewise, fixed point patterns are defined by the fixed points in state spaces of pattern formation systems. If the attractor or repellor corresponds to a fixed point, then the definition for fixed point patterns and the definition for attractor and repellor patterns lead to the same result. Attractor and repellor patterns of fixed point attractors and repellors are given by the states that describe the locations of the fixed point attractors and repellors. That is, the patterns correspond to the respective fixed points. If the attractor or repellor is a limit cycle, then the pattern is a temporal pattern that corresponds to the time-dependent state evolving along the limit cycle.

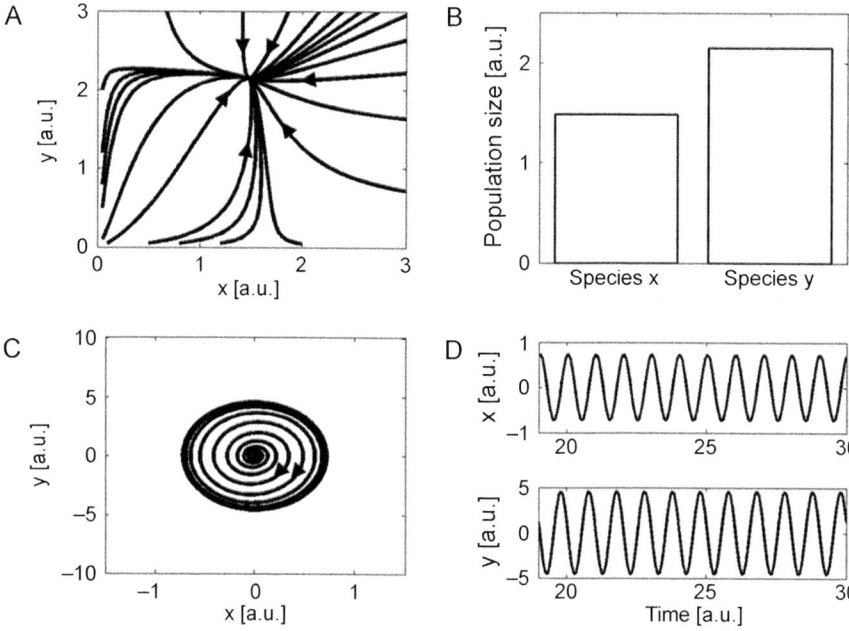

Fig. 4.1 Attractors versus attractor patterns. Panels (**a**) and (**b**) illustrate an attractor pattern (panel **b**) and its fixed point attractor (panel **a**). The pattern is identical to the fixed point of the attractor. Panel (**c**) and (**d**) illustrate an attractor pattern (panel **d**) of a limit cycle attractor (panel **c**). The pattern is a time-dependent quantity, whereas the attractor is a set of points. Therefore, the attractor pattern is not identical to its attractor

Recall that in general an attractor or repellor is a subset of the state space. In the case of a limit cycle attractor, the attractor is a closed line or orbit in the state space. In contrast, the attractor pattern of a limit cycle attractor is a function of time assuming all the points on the limit cycle at different time points. That is, for limit cycle attractors and repellors, attractor and repellor patterns are not identical to their corresponding attractors or repellors. Figure 4.1 provides two examples of attractors and their corresponding attractor patterns. In panel (a) the phase portrait of a Lotka-Volterra model describing two competing species is shown for the case in which the species can co-exist. The number of species of each type are described by the variables x and y. Panel (a) demonstrates that there is a fixed point attractor at about $x = 1.5$ units and $y = 2.2$ units. Panel (b) shows the corresponding attractor pattern. The pattern is a vector with two components and corresponds to the fixed point. That is the attractor pattern is identical to the fixed point of the attractor. Panel c show a limit cycle attractor as discussed in Sect. 3.4.3. Note that the limit cycle attractor in panel (c) corresponds only to the thick, black circle and does not correspond to the spiral that converges to the thick black circle. Panel (d) shows the evolution of the state described by the variables x and y on the limit cycle. In doing so, panel (d) shows the attractor pattern of the limit cycle attractor shown in panel (c). The attractor pattern given in terms of the trajectories $x(t)$ and $y(t)$ shown in panel (d) is not identical to the attractor given in terms of the set of points shown in panel (c).

Attractor and repellor patterns can often be described more conveniently in reduced state spaces. In particular, for spatially extended systems featuring geometric patterns as mentioned above it is useful to considered reduced state spaces rather than the original state spaces. Attractor and repellor patterns in the original state spaces of spatially extended systems may correspond to relatively simple attractors and repellors in reduced state spaces. For example, stripe patterns such as zebra stripes are described in state spaces defined by functions. Frequently, the class of all smooth functions that satisfy certain boundary conditions is used. These functions can be described in terms of Fourier functions and their weights. If Fourier functions with very short periods can be neglected, a finite set of Fourier functions and their corresponding weights is sufficient to describe the original state space. The Fourier weights can then conveniently be used as coordinates of a reduced state space. A stripe pattern in the original state space corresponds then to a fixed point attractor in the reduced space of Fourier weights.

In Sect. 3.5.2 it has been argued that certain limit cycle oscillators can be described in terms of their oscillation amplitudes (see e.g., the bifurcation diagram in panel (c) of Fig. 3.15). In doing so, the oscillator is described in a reduced state space that is described by a single variable only. The limit cycle attractor in the original state space (spanned by the coordinates x and y) corresponds then to a fixed point attractor in the reduced single-variable state space of the oscillation amplitude. Likewise, the oscillatory attractor pattern shown in Fig. 4.1d corresponds to a fixed point attractor in the reduced state space of the oscillation amplitude. Therefore, an oscillatory pattern can be described in terms of a fixed point attractor.

Finally, in Chap. 3, the bifurcation diagram shown in Fig. 3.5c (or Fig. 3.15b) of the Benard experiment was introduced, in which roll patterns emerge in fluid or gas systems. The bifurcation diagram describes the fluid or gas systems in a reduced state space given by the roll velocity. In the state space of the roll velocity, roll patterns correspond to fixed point attractors. The rotating rolls form a spatio-temporal pattern. Therefore, in the Benard experiment a spatio-temporal attractor pattern in its original state space correspond to a fixed point attractor in a suitably defined reduced state space.

Attractor and repellor patterns typically form strong categories (see Sect. 3.2.5). Given a particular bifurcation parameter, all attractor and repellor patterns that emerge when varying the bifurcation parameter such that it does not reach a bifurcation point belong to the same category. In contrast, two attractor or repellor patterns belong to different categories if they are connected by a bifurcation. For example, when horses walk at different locomotion speeds, then they from quantitatively different but not qualitatively different locomotion patterns. In contrast, if the locomotion speed is increased beyond the critical value for walk-trot transitions (which is about 1.6 m/s for miniature horses, see Fig. 3.5d) then horses transition from walk to trot. The locomotion patterns of the trot are qualitatively different from those of the walk. Walking and trotting patterns form different attractor pattern categories.

4.2   Basis Patterns, Eigenvalues, and Pattern Amplitudes

In order to describe the formation of attractor patterns, some key concepts need to be introduced: basis patterns, eigenvalues, and pattern amplitudes. They will be introduced for fixed point attractors and repellors. That is, basis patterns of fixed point attractors and repellors will be discussed and in this context eigenvalues and pattern amplitudes will be defined.

4.2.1   Discrete Pattern Formation Systems

Let us first consider discrete pattern formation systems described by a set of real-valued variables. Let us consider the evolution of states close to a fixed point. It can be shown that the evolution of states close to fixed points is determined by two types of special vectors called eigenvectors.

A type 1 eigenvector is given by a direction in the state space of a dynamical system or a pattern formation system that has the following two properties. First, the direction is defined by a straight line. Second, if the state is initially located on that line, then the system will evolve approximately along that line as long as it is close to the fixed point. Mathematically speaking, the direction is defined by the eigenvector. Note that the phrase "direction" is somewhat misleading because what matters is the line or the axis defined by the vector. If we let point an eigenvector in just the opposite direction then we have the same line or axis. In fact, the vector that points in the opposite direction is an eigenvector again. Mathematically speaking, the sign of the vector does not matter. Likewise, the length or magnitude of the vector is of no concern. However, it is convenient to consider eigenvectors that have a magnitude (i.e., a length) of one unit. Panel (a) of Fig. 4.2 shows an eigenvector of a fixed point in a two-dimensional state space spanned by the variables x and y. The fixed point is located at $(x_{st}, y_{st}) = (2, 2)$ units. Panel (b) shows the evolution of the state for the initial state indicated by the diamond in panel (a) assuming that the fixed point is a stable node. The state evolves along the axis defined by the eigenvector and converges to the fixed point $(2, 2)$. Importantly, the axis can be used to introduce a variable A that measures the distance of a point (x, y) from the fixed point along the axis defined by the eigenvector. The variable A has a sign as well. The reason for this is that the axis is separated by the fixed point into two half-lines. Using the sign of A, we can distinguish between points (x, y) whose projections fall on either of the two half-lines of the axis. In the example shown in panels (a) and (b) the variable A is defined in such a way that it is positive for the initial position indicated by the diamond. The top subpanel of Fig. 4.2b shows the evolution of $A(t)$ along the axis of the eigenvector. A becomes smaller in the amount as a function of time because the fixed point is stable. The middle and bottom panels in Fig. 4.2b show the evolution of x and y. Those variables decay as well as functions of time. In contrast, A increases in the amount as function of time in panel (c) because in panel (c) it is

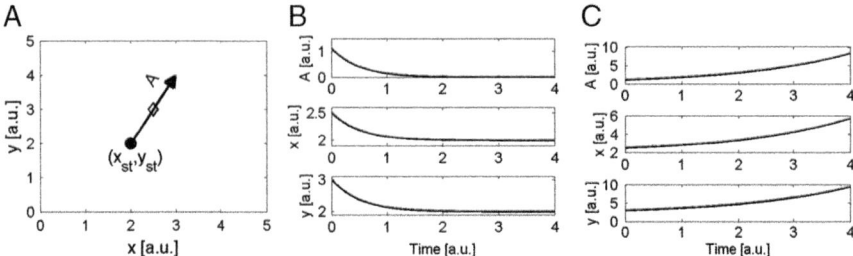

Fig. 4.2 Eigenvector of a fixed point and evolution of the system close to the fixed point described by the pattern amplitude variable A. Panel (**a**): Fixed point and eigenvector. Variable A measures the distance of a point (x, y) to the fixed point (x_{st}, y_{st}) along the eigenvector. Panel (**b**): Evolution of A (top) and state (x, y) (middle and bottom) in the case in which the fixed point is a stable node. Panel (**c**): A and (x, y) as functions of time in the case of an unstable node

assumed that the fixed point is unstable. The component variables x and y increase in magnitude consistent with the increase of the amplitude A.

The components of an eigenvector define a basis pattern. This holds for state spaces of any dimension. In the example shown in Fig. 4.2, the relevant state space is two-dimensional and the eigenvector is described by two components. Therefore, the basis pattern given by the eigenvector exhibits two components. For a fixed point in a three dimensional state space, type 1 eigenvectors have three components and, consequently, the corresponding basis patterns have three components. Let us point out an analogy between the definitions of attractor patterns and basis patterns. Fixed point attractor patterns are defined by the components of fixed points. Similarly, basis patterns of type 1 eigenvectors are defined by the components of eigenvectors.

Having defined (type 1) basis patterns, we can re-interpret the meaning of the variable A. In synergetics and the theory of pattern formation, the amplitude variable A associated with a type 1 eigenvector is considered as the pattern amplitude of the basis pattern related to that eigenvector.

The functions $A(t)$, $x(t)$, and $y(t)$ shown in panels (b) and (c) are exponential functions of time. The three exponential functions in panel (b) have the same exponent, which is a negative number. This exponent is the eigenvalue of the eigenvector. Likewise, the three exponential functions in panel (c) have the same positive exponent. Again, this exponent is the eigenvalue of the eigenvector in the case of an unstable node. The character of a fixed point (stable/unstable node, saddle, stable/unstable focus) defines characteristic features of the eigenvalues of the fixed point and vice versa. We will return to this issue in a moment.

Type 1 eigenvectors have the property that they are orthogonal to each other. Therefore, if we take two type 1 eigenvectors together, we can describe all points in a two dimensional state space. The eigenvectors constitute a basis. If they are used as a basis to describe the state, then they are referred to as basis vectors. In order to describe states of a two-dimensional state space (containing a fixed point with two type 1 eigenvectors) the two eigenvectors as basis vectors and the two pattern amplitudes as coordinates can be used. This construction of a two-dimensional state space spanned by means of two eigenvectors (or two basis patterns) can be generalized to arbitrary dimensions. That is, if a fixed point only features type 1 eigenvectors then for a n-dimensional state space there are n eigenvectors that defined n new axes (i.e., basis vectors). The eigenvectors are interpreted as basis patterns and the values along the axes as pattern amplitudes. With the help of the n pattern amplitudes (and the n basis patterns) any point in the original state space can be described. Importantly, close to the fixed point the evolution along the new axes is determined by exponential functions. The exponents of those exponential functions are the eigenvalues. Using (1) eigenvectors (basis patterns) and their (2) amplitudes and (3) eigenvalues, we obtain a comprehensive picture of the evolution of the state from the perspective of the fixed point under consideration [160, 252].

Let us return to two-variables systems (i.e., systems with two-dimensional state spaces). Let us consider a more general perspective. Let us consider planes that either correspond to two-dimensional state spaces of two-variables systems or two-dimensional subspaces of n-dimensional state spaces of n variable systems. There exist pattern formation systems with such planes that can not be described in terms of two type 1 eigenvectors. This leads to the definition of type 2 eigenvectors. Type 2 eigenvectors are eigenvectors that occur in pairs. These pairs span a plane on which states that are initially located on the plane remain to a good approximation as long as they stay close to the fixed point under consideration. In the planes defined by type 2 eigenvectors states do not evolve along straight lines. In contrast, they evolve in spirals either away or towards the fixed point. Again, the sign and the length (magnitude) of type 2 eigenvectors is arbitrary. To ease mathematical calculations they have a magnitude of one unit. Moreover, they are orthogonal to each other. Figure 4.3 illustrates various aspects of type 2 eigenvectors. In Fig. 4.3 a two-variables system is considered involving the variables x and y and exhibiting a fixed point at $(x_{st}, y_{st}) = (2, 2)$ units (just as in the previous example). The plane spanned by the type 2 eigenvectors is indicated in Fig. 4.3a by the hatched area. However, in fact, that plane extends to infinity in both x and y directions. That is, for the system with a two-dimensional state space, the two-dimensional plane of two type 2 eigenvectors is identical to the original state space. For higher dimensional systems (e.g., systems with a three-dimensional state space) this is not the case. Panels (b) and (c) demonstrate the phase portrait (panel b) and the evolution of the state (panel c) in the case of a stable fixed point. The initial state is indicated by the diamond in panels (a) and (b). The state converges to the fixed point in an oscillatory fashion. The fixed point is a stable focus. Importantly, the oscillatory trajectories for x and y shown in panel (c) decay in amplitude in an exponential manner. That is, they exhibit envelopes given in terms of exponentially decaying functions. For both trajectories the exponential functions exhibit the same exponent. This exponent is a negative number and is the real part of the eigenvalue of the two type 2 eigenvectors. Panels (d) and (e) show the phase portrait (panel d) and the trajectories for x and y (panel e) in the case of an unstable fixed point. The state spirals away from the fixed point. The x and y trajectories show oscillations with exponentially increasing oscillation amplitudes. The oscillation amplitudes increase according to the same exponential law. The exponent of that law is the real part of the eigenvalue of the two type 2

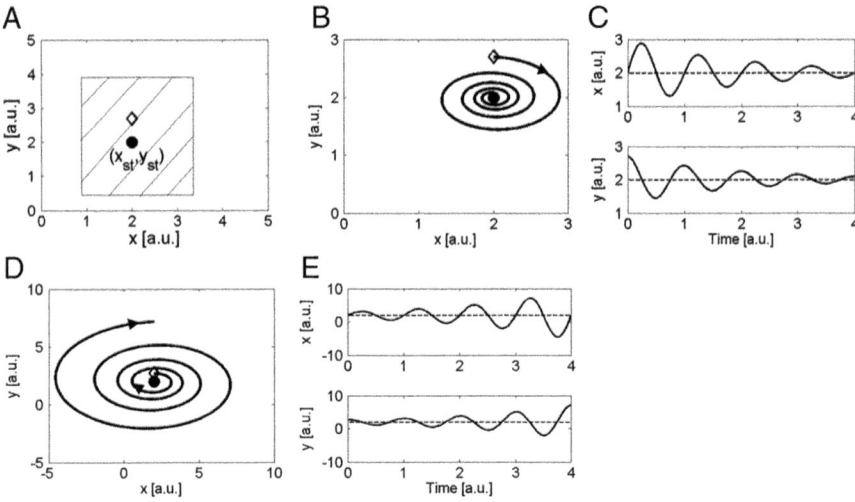

Fig. 4.3 Illustration of type 2 eigenvectors and their complex-valued eigenvalues. Panel (**a**): Plane spanned by a pair of type 2 eigenvectors. Panels (**b**) and (**c**): Evolution of a perturbation in the case of a stable fixed point. The exponential decay rate is determined by the real value of the eigenvalues, whereas the oscillation frequency corresponds to the imaginary value of the eigenvalues. Panels (**d**) and (**e**): As for (**b**) and (**c**) but for an unstable fixed point

eigenvectors. Panels (c) and (e) show oscillatory functions with particular oscillation frequencies. Type 2 eigenvectors exhibit eigenvalues that are composed of real and imaginary parts. As mentioned above, the real parts correspond to exponents that describe the exponential increase or decay of oscillation amplitudes. The imaginary parts are related to the oscillation frequencies. Since the states of type 1 eigenvectors do not exhibit an oscillatory evolution, the eigenvalues of the type 1 eigenvectors do not exhibit any imaginary parts. They are real-valued [160, 252].

As mentioned above type 1 eigenvectors are orthogonal to each other. Likewise, two type 2 eigenvectors are orthogonal to each other. In addition, type 1 eigenvectors are orthogonal to type 2 eigenvectors. Therefore, taken all eigenvectors together they form a vector basis that allows to describe any point in the original state space. Just as for pattern formation systems that feature only type 1 eigenvectors, in the general case that involves both type 1 and 2 eigenvectors, all eigenvectors constitute a basis in terms of basis vectors. The basis vectors are considered as basis patterns. The coordinates along the directions given by the basis vectors (eigenvectors/basis patterns) are considered as pattern amplitudes.

The pattern amplitude of a basis pattern tells us how much of the basis pattern is contained in a particular state. In this context, two key phrases are important. First, if a state is expressed in terms of a set of basis patterns (eigenvectors) then we say that we have an expansion of the state into eigenvectors. Second, if we add basis patterns (eigenvectors) together and weight the individual basis patterns by their respective pattern amplitudes, then we arrive at a superposition of basis patterns.

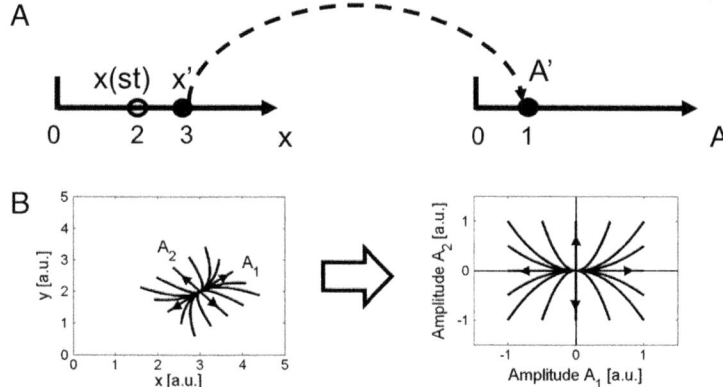

Fig. 4.4 State spaces and amplitude spaces. Panels (**a**) and (**b**) illustrate the change in perspective from state spaces to amplitude spaces for a single variable system (panel **a**) and a two-variables system (panel **b**)

Using that terminology, we can say that states in state spaces can be expressed by superpositions of basis patterns of fixed point attractors and repellors. In fact, the two phrases describe in some sense the same issue. A superposition of eigenvectors that yields a particular state is the same as the expansion of the state into those eigenvectors.

Pattern formation from the perspective of given fixed point attractors and repellors can be described with the concepts that we have introduced so far. Figure 4.4 illustrates this point. Figure 4.4a refers to a one-variable dynamical system for the variable x that exhibits a fixed point repellor at $x = 2$ units. In this one-dimensional state space the point x' at 3 units is marked. From the perspective of the fixed point repellor the point x' is 1 unit to the right. For one-dimensional state spaces there is only one eigenvector, which is the vector pointing in the direction of the space. Therefore, the eigenvector is frequently not explicitly mentioned. Let us define the pattern amplitude A (or more precisely, the orientation of the eigenvector) such that every state to the right of the fixed point repellor is described by a positive amplitude and every state to the left is described by a negative amplitude. Then the point $x' = 3$ in the state space corresponds to a point of $A = 1$ unit in the one-dimensional space of the amplitude A centered around the fixed point repellor.

Figure 4.4b illustrates this change of perspective from the original state space to a space that is centered around a fixed point for a two-dimensional state space spanned by the variables x and y containing a fixed point repellor in form of an unstable node. The node is located at $(x, y) = (3, 2)$. The two eigenvectors of the fixed point are indicated by straight lines. The two eigenvectors describe the two basis patterns of the fixed point. Distances along the axes defined by those basis patterns are measured in terms of the pattern amplitudes A_1 and A_2 as indicated. On the right hand side of panel (b), the situation is shown from the perspective of the

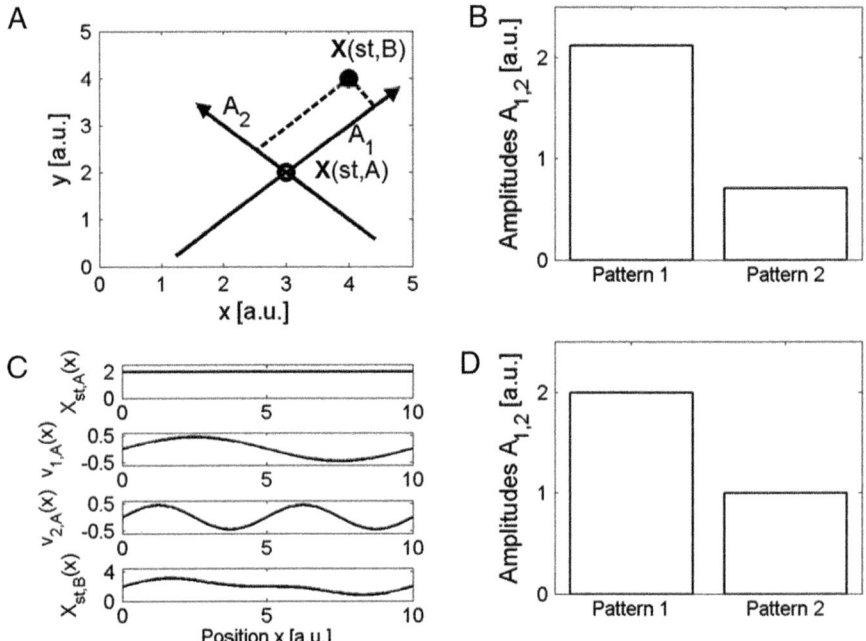

Fig. 4.5 Expansion of states into basis patterns of a reference point. Panels (**a**) and (**b**) refer to a discrete system, while panels (**c**) and (**d**) refer to a spatially extended system. Panels (**a**) and (**c**) show the reference points that are given in term of the fixed point $X(st, A)$ (panel **a**) and a function $X_{st,A}(x)$ (panel **c**, top) and the states of interest given in terms of the fixed point $X(st, B)$ (panel **a**) and the function $X_{st,B}(x)$ (panel **c**, bottom). The states of interests are expanded into eigenvectors (panel **a**) or eigenfunctions (panel **c**, second and third subpanels). Eigenvectors and eigenfunctions are both considered as basis patterns. The amplitudes of the basis patterns (eigenvectors/eigenfunctions) are shown in panels (**b**) and (**d**)

fixed point, that is, from the perspective of the space spanned by the amplitudes A_1 and A_2.

Figure 4.5 builds on Fig. 4.4b and illustrates again the expansion of a state into basis patterns. Panels (a) and (b) refer to the case of discrete systems that has been discussed so far and, in particular, generalize the example shown in Fig. 4.4b. In panels (a) and (b), a two-dimensional state space spanned by the variables x and y is considered. As shown in panel (a), at $(x, y) = (3, 2)$ there is a fixed point repellor, while at $(x, y) = (4, 4)$ the pattern formation system exhibits a fixed point attractor. The fixed point repellor exhibits two type 1 eigenvectors as indicated that are considered as basis patterns. Along the corresponding axes the pattern amplitudes A_1 and A_2 are measured. The location of the fixed point attractor measured in terms of A_1 and A_2 is at $A_1 = 3/\sqrt{2} \approx 2.1$ units and $A_2 = 1/\sqrt{2} \approx 0.7$ units. The fixed point attractor at $(x, y) = (4, 4)$ corresponds to an attractor pattern. This attractor pattern can be expressed in terms of the basis patterns of the repellor as a superposition of basis vectors with weights $A_1 = 2.1$ and $A_2 = 0.7$ units. Panel

(b) shows graphically the contributions of the basis patterns to the attractor pattern. Panels (c) and (d) will be addressed in Sect. 4.2.3 below in the context of spatially extended systems.

4.2.2 Amplitude Space

The space spanned by the pattern amplitudes of basis patterns is referred to as amplitude space Ω_A. From the discussion above it follows that any point in the original state space Ω_X can be described by means of the pattern amplitudes of basis patterns. Therefore, in terms of the states that are described by the state space the amplitude space is identical to the state space $\Omega_X = \Omega_A$. The difference between the state space and the amplitude space is the perspective. In the original state space Ω_X the reference point is the origin. In contrast, in the amplitude space Ω_A the reference point is a fixed point attractor or repellor. The state space Ω_X of discrete pattern formation systems is spanned by the original variables x, y, z, etc. that form a coordinate system. The amplitude state Ω_A of a discrete pattern formation system is spanned by the eigenvectors that form a vector basis and are interpreted as basis patterns. The eigenvectors define axes in the original state space. Distances along those axes are described in terms of the pattern amplitudes A_1, A_2, A_3, etc. Figure 4.6 illustrates the step from state space to amplitude space. Figure 4.6 also indicates that there is a reduced amplitude space. We will return to this issue below.

4.2.3 Spatially Extended Pattern Formation Systems

The discussion so far has been focused on discrete pattern formation systems described by n real-valued variables and n-dimensional state spaces. A similar

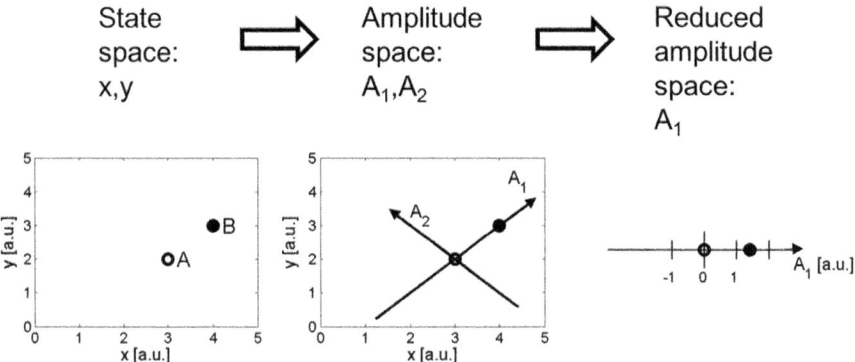

Fig. 4.6 State space, amplitude space, and reduced amplitude space

discussion can be conducted for spatially extended systems described by field variables. Typically, spatially extended systems exhibit state spaces formed by functions that satisfy certain properties (e.g., smoothness and boundary conditions). Fixed points correspond to special functions. Panels (c) and (d) of Fig. 4.5 illustrate the concepts of basis patterns and pattern amplitudes for spatially extended systems. Let us consider a pattern formation system that exhibits a fixed point repellor labeled A described by a constant function $X_{st,A}(x) = $ const. as shown in the top subpanel. In addition, it is assumed that the system exhibits a fixed point attractor labeled B described by the function $X_{st,B}(x)$ shown in the bottom subpanel. In the state space of functions we can define at the repellor fixed point A special directions (just as in the case of discrete systems). These directions correspond to functions. The second and third subpanels show two functions that are assumed to correspond to these special directions. They have been labeled as $v_{1,A}$ and $v_{2,A}$, respectively. The subindex A indicates that they are related to the fixed point repellor A (not to the fixed point attractor B). These special functions have the property that if the state of the system initially deviates from the fixed point repellor (i.e., deviates from the constant function) by means of an arbitrary but small multiple of those functions, then the state of the system evolves in an exponentially growing or decaying fashion. In this context, exponentially growing means that the initial deviation from the constant function given in terms of a multiple of one of the special functions $v_{1,A}$ and $v_{2,A}$ (e.g., $0.001v_{1,A}$) increases exponentially in magnitude. Likewise, exponentially decaying means that the deviations decay in magnitude exponentially. The special functions $v_{1,A}$ and $v_{2,A}$ are called eigenfunctions. These eigenfunctions are considered to be the basis patterns of the fixed point repellor A. In general, the eigenfunctions of fixed point repellors and attractors of spatially extended pattern formation systems are considered as basis patterns just as the eigenvectors of fixed point repellors and attractors of discrete pattern formation systems. The exponents determining the exponential increase or decay correspond to the eigenvalues of the basis patterns (or eigenfunctions).

Above, only the case of basis patterns that grow or decay exponentially (i.e., in a monotonic fashion) has been addressed. In fact, by analogy to the type 2 eigenvectors discussed in Sect. 4.2.1, it is also possible that the evolution of the basis patterns shows an oscillatory character, where the oscillations exhibit exponentially increasing or decreasing envelop functions. In this case, the eigenvalues of the eigenfunctions (basis patterns) are composed of a real and an imaginary part. The imaginary part determines the oscillation frequency (just as for type 2 eigenvectors, see Fig. 4.3). The real part determines the properties of the envelop functions (i.e., the real part is positive in the case of exponential increase and negative in the case of exponential decay).

Importantly, the scaling factors of basis patterns (eigenfunctions) are regarded as pattern amplitudes. That is, if a state corresponds to a certain multiple of a particular basis pattern (e.g., the state correspond to the constant pattern $X_{st,A}$ plus five time the basis pattern $v_{1,A}$), then the multiplication factor is regarded as the pattern amplitude of that basis pattern. In this context, the amplitude does not need to be a small number. In the example shown in panel (c), the fixed point attractor B described by the function $X_{st,B}(x)$ deviates from the constant function $X_{st,A}$ by two times the basis pattern $v_{1,A}$ and one time the basis pattern $v_{2,A}$. That is, the attractor pattern $X_{st,B}(x)$ can be expressed in terms of the basis patterns $v_{1,A}(x)$ and $v_{2,B}(x)$ by the amplitudes $A_1 = 2$ and $A_2 = 1$. This is illustrated in panel (d) of Fig. 4.5.

Comparing panels (c) and (d) with panels (a) and (b) of Fig. 4.5, we see that eigenfunctions play the same role as eigenvectors. They correspond to basis patterns. Moreover, the scaling factors of eigenfunctions play the same role as the coordinates along the axes defined by eigenvectors. They correspond to the amplitudes of the aforementioned basis patterns and are simply referred to as pattern amplitudes. Basis patterns (eigenvectors as well as eigenfunctions) exhibit eigenvalues that can be real-valued or complex-valued. They describe the evolution of states close to the fixed points for which the basis patterns have been defined. Both for discrete and spatially extended systems any attractor pattern can be expressed in terms of the basis patterns and the corresponding pattern amplitudes.

4.2.4 The Human Body and the Scale of Patterns

Self-organization is a phenomenon or mechanism that takes place on various spatial scales relevant for understanding human and animal brain and body activity. As pointed out in the introduction, see Table 1.6 in Chap. 1, self-organization occurs on the cellular level, neuronal level, level of human and animal bodies, and social level. Therefore, the question arises what is the relationship between the scale of the emerging patterns and scales of the physical bodies of humans or animals. In general, all thinkable relations are possible: (1) pattern formation can take place in parts of human and animal bodies, (2) human and animal bodies form patterns, or (3) human and animals are parts of emerging patterns. That is, patterns are smaller in scale than the human and animal bodies, they are on the scale of the bodies, or patterns emerge on scales that are larger than the bodies of human individuals and animals. For example, from a pattern formation perspective, brain activity of the human isolated system (e.g., "thinking" or "dreaming") corresponds to patterns that form in certain areas of the human brain. BA patterns of this kind are patterns that form in human bodies. They are smaller in scale than the bodies. In contrast, when humans or animals move, then the corresponding body movement patterns are on the scale of the human and animal bodies. Finally, imagine a parent with his or her child on a children's playground that has a child swing. The child sits on the swing. The parent stands behind the moving swing and gives the child a push every time the child swings back. In doing so, the parent forms an oscillatory movement pattern. That pattern, however, is part of a larger pattern that involves the moving swing and even the body movements of the child.

We will discuss a somewhat more comprehensive classification scheme of patterns and pattern formation systems in the context of the human pattern formation reaction model in Sect. 5.2 of Chap. 5. That scheme addresses to some extent also the issue about the scale of patterns.

4.2.5   Interim Summary and a Second Look at Attractors and Repellors

Table 4.1 summarizes some aspects that have been discussed above. Both discrete and spatially extended self-organizing pattern formation system exhibit attractor and repellor patterns which correspond to states evolving on attractors and repellors, respectively. Fixed point attractors and repellors exhibit a set of basis patterns. For discrete systems these patterns correspond to eigenvectors. For spatially extended systems they correspond to eigenfunctions. A pattern amplitude is assigned to each basis pattern. The pattern amplitude describes how much the pattern contributes to a given state in the state space. The reference point in this case is the fixed point under consideration (not the origin of the state space). All amplitudes span the amplitude space. Close to fixed points states move away or approach fixed points according to exponential laws. That is, within the framework of the amplitude space, pattern amplitudes (describing states of interest) increase or decrease as exponential functions of time close to their respective fixed points. The exponents of the exponential functions are called eigenvalues and can be real-valued or assume complex numbers.

Although attractor patterns and basis patterns have been introduced separately, these two types of patterns are related to each other. Since attractor patterns are (evolving) states and states can be expressed in terms of superpositions of basis patterns it follows that attractor patterns can be expressed in terms of the amplitudes of the basis patterns. To re-iterate, as such attractor patterns are described by the variables of interest of the pattern formation systems of under consideration. However, the basis patterns of fixed points define another basis for the description in terms of pattern amplitudes. Therefore, attractor patterns can be expressed in terms of pattern amplitudes of basis patterns.

If an attractor pattern is expressed in terms of basis patterns, then in general it is composed of several basis patterns that are weighted with their respective pattern amplitudes. In this case, several basis patterns emerge at the same time. They co-exist. In general, these co-existing basis patterns make different contributions to attractor patterns. That is, in general, the amplitudes of the basis patterns differ from each other as exemplified in panels (b) and (d) of Fig. 4.5. In other scenarios that will

Table 4.1 Summary and comparisons of key concepts of pattern formation for discrete and spatially extended systems

Discrete or spatially extended systems	Discrete systems	Spatially extended systems
Attractor patterns		
Basis patterns	Eigenvectors	Eigenfunctions
Pattern amplitudes		
Amplitude space		
Eigenvalues		

Table 4.2 Description of fixed point attractors and repellors in terms of eigenvalues

Dimension	Attractor/repellor	Subtype	Eigenvalue(s) λ
1	Attractor	Stable fixed point	λ negative
	Repellor	Unstable fixed point	λ positive
2	Attractor	Stable node	λ_1, λ_2 real and negative
		Stable focus	λ_1, λ_2 complex, real parts negative
	Repellor	Unstable node	λ_1, λ_2 real and positive
		Saddle	λ_1, λ_2 real, one positive, one negative
		Unstable focus	λ_1, λ_2 complex, real parts same and positive

be addressed below, there are mutually exclusive basis patterns. In such scenarios, only one basis pattern can emerge at a time. Basis patterns cannot co-exist with each other. Consequently, for systems with mutually exclusive basis patterns, there is a one-to-one mapping between basis patterns and attractor patterns. Basis patterns form attractor patterns. Vice versa, attractor patterns are given by basis patterns.

Table 4.2 illustrates the connection between eigenvalues and the type of a fixed point. That is, the eigenvalues of a fixed point define whether the fixed point is an attractor or repellor and the subtype of attractor or repellor. Vice versa, the type of fixed point attractors and repellors defines the characteristic features of the fixed point eigenvalues.

4.3 Amplitude Equations

4.3.1 Bifurcations of Two Variable Systems Involving a Single Positive Eigenvalue

How does an attractor pattern emerge? Using the concept of basis patterns and pattern amplitudes, this question can now be answered. Attractor patterns emerge at bifurcation points. As mentioned in Sect. 3.5.1, bifurcations are characterized by a change in stability of attractors and repellors and/or the appearance and disappearance of attractors and repellors. For the purposes of this book it is sufficient to consider the fundamental case in which a stable fixed point becomes unstable (change in stability) such that the state converges to another fixed point attractor. We leave it open whether the second fixed point attractor exists or does not exist before the first attractor becomes unstable. In the latter case, the second attractor would appear at the bifurcation point.

Figure 4.7 summarizes the emergence of an attractor pattern at such a bifurcation point. As shown in panel (a), we consider the transitions between two attractor

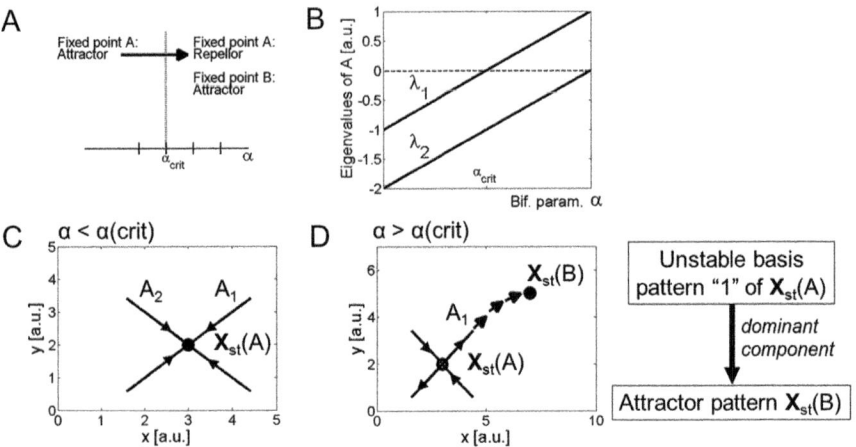

Fig. 4.7 Emergence of an attractor pattern B at a bifurcation point at which a reference point (the attractor pattern A) becomes unstable. See text for details

patterns, say A and B, that correspond to two fixed points. The fixed points will be labeled A and B as well. The transition is assumed to be induced by a bifurcation parameter α and occurs when the bifurcation parameter exceeds a critical value $\alpha(crit)$. The fundamental mechanism can be described as follows. Initially, the bifurcation parameter α is smaller than the critical value. It is assumed that for those subcritical values the fixed point A corresponds to a stable node. The state X at the fixed point A corresponds to the attractor pattern A. For sake of simplicity, a two-dimensional state space is considered spanned by the variables x and y. In our example, the fixed point A is at $(x, y) = (3, 2)$, see panel (c). That is, the attractor pattern is given by the state $(3, 2)$ as well. The fixed point has eigenvalues λ_1 and λ_2. As mentioned above, it is assumed that for subcritical values of α the fixed point A corresponds to a stable node. That implies that the eigenvalues λ_1 and λ_2 of fixed point A are real and assume negative values. Panel (b) shows the two eigenvalues of A as functions of the bifurcation parameter α. For $\alpha < \alpha(crit)$ both eigenvalues assume negative numbers. Finally it is assumed that initially (i.e., when α smaller than $\alpha(crit)$ holds) the state of the pattern formation system is located sufficiently close to the fixed point attractor A. As a result, the state converges to the fixed point such that initially the system exhibits the attractor pattern A. In summary, initially the pattern formation system is characterized by a bifurcation parameter $\alpha < \alpha(crit)$, an attractor pattern A, and a fixed point attractor given by a stable node that exhibits two basis patterns with two negative, real eigenvalues.

As shown in panel (c), the basis vectors/eigenvectors of A (i.e., basis patterns of A) define two axes. Distances along those axes are measured in terms of the pattern amplitudes A_1 and A_2, respectively. Recall that the amplitudes tell us how much the two basis patterns contribute to any state in the state space that is different from the attractor pattern A. Having discussed the initial conditions, let us assume that a force is applied to the pattern formation system such that the bifurcation parameter α increases gradually. For sake of simplicity, as shown in panel (b), the case is considered in which the eigenvalues of the fixed point attractor A assume different values. This case will be referred to as the case of a completely inhomogeneous eigenvalue spectrum (see below). In this case, when the bifurcation parameter α is increased towards the critical value $\alpha(crit)$, one of the eigenvalues becomes zero at the bifurcation point and assumes a positive value for bifurcation parameter values slightly above the critical value. At that stage, when the bifurcation parameter assumes a value slightly above the critical value, only this particular eigenvalue is positive, whereas the other eigenvalue is negative. That means the fixed point A slightly above the bifurcation point is characterized by two qualitatively different basis patterns: one basis pattern (axis, direction, or eigenvector) exhibits a positive eigenvalue, the other basis pattern (axis, direction, or eigenvector) exhibits a negative eigenvalue. In the example shown in panel (b), the eigenvalue λ_1 of basis pattern 1 becomes positive, while the eigenvalue λ_2 of basis pattern 2 remains negative (see panel (b) for α slightly larger than $\alpha(crit)$). Due to the fact that there is one positive eigenvalue, the fixed point attractor A becomes unstable. It becomes a repellor of the type of a saddle. Panel (d) illustrates the fixed point A for bifurcation parameters slightly larger than the critical value. Note that in panel (d) the fixed point A is shown at the same location as in panel (c). In general, the force acting on the pattern formation system and changing the bifurcation parameter will also shift the location of the fixed point attractor A. For sake of simplicity, this effect has not been taken into account in panel (d).

Let us return to the two basis patterns of the saddle point A. The basis pattern 1 with the positive eigenvalue is referred to as the unstable basis pattern. Any state close to the repellor that is located on the axis defined by the basis pattern will evolve away from the fixed point. The amplitude A_1 of that state will increase exponentially as function of time. The basis pattern 2 with the negative eigenvalue is referred to as stable basis pattern. States that exhibit a non-zero amplitude A_2 evolve close to the saddle such that the amplitude 2 decays to zero. Consequently, states evolve towards the axis defined by the basis pattern 1. Let us combine these two effects. First, states evolve such that they are pushed to the axis of basis pattern 1. Second, if they are on that axis (of close to it) they are pushed away from the fixed point in the direction of the axis. Therefore, the system evolves along the trajectory indicate in panel (d) by the arrow. This implies that the amplitude A_1 of basis pattern 1 increases in magnitude. Consequently, the attractor pattern A disappears and a new attractor pattern emerges. Typically, the second fixed point attractor B that is assumed to exist for $\alpha > \alpha(crit)$ exhibits a location somewhere in the direction of the unstable basis pattern (because the unstable basis pattern is the pattern whose amplitude increases in magnitude). The stable basis pattern may make a small contribution or does not make any contribution to the emerging attractor pattern. In general, due to the mechanism just described, the unstable basis pattern makes the dominant contribution to the emerging attractor pattern. For example, in panel (d) the fixed point attractor is located at $(x, y) = (7, 5)$. Accordingly, in the amplitude space of fixed point A, the attractor B exhibits the amplitudes $A_1 = 4.2$ and $A_2 = 1.4$ units.

For sake of simplicity, in the description above, it was assumed that pattern amplitudes increase over time. In fact, amplitudes given in terms of real-valued numbers may assume initially negative values. In this case, they may become "more negative" over the course of time. That is, in general, pattern amplitudes of unstable basis patterns increase in magnitude. Moreover, it is important to realize that—within the framework of the amplitude space—the transition from an attractor pattern A to an attractor pattern B is seen from the perspective of the attractor pattern A. In contrast, when looking at the bifurcation from the perspective of the state space, the increase in magnitude of the amplitude A_1 of the unstable basis vector implies that the attractor pattern A disappears. The formation of attractor pattern B and the disappearance of the attractor pattern A are not two separate processes but just two aspects of the same process.

4.3.2 Dominance of Unstable Patterns Over Stable Patterns and Order Parameter Concept

The discussion above suggests that unstable basis vectors at bifurcation points have a special role. While initially the amplitude of an unstable basis pattern increases in magnitude, the amplitudes of all stable basis patterns decay in magnitude. Therefore, the unstable basis pattern typically dominates the emerging attractor pattern. In synergetics, unstable basis patterns are seen as counterparts to macroscopic quantities that have been introduced to describe the order of self-organizing systems (e.g., magnetic fields of ferromagnetic materials). These macroscopic quantities are called order parameters. For this reason, unstable basis patterns have also been referred to as order parameters [160]. The amplitudes of unstable basis patterns have been referred to as unstable amplitudes or order parameter amplitudes. The amplitudes of stable basis patterns are called stable amplitudes. Finally note that frequently basis patterns can be regarded as modes. An overview of the terminology and synonyms in this regards is given in Table 4.3.

Table 4.3 Terminology used to distinguish between basis patterns with positive and negative eigenvalues and to distinguish between the corresponding pattern amplitudes

Physical quantity	Eigenvalue positive or positive real part	Eigenvalue negative or negative real part
Pattern	Unstable basis pattern	Stable basis pattern
	Unstable mode	Stable mode
	Order parameter	
Amplitude	Unstable pattern amplitude	Stable pattern amplitude
	Unstable amplitude	Stable amplitude
	Order parameter amplitude	

4.3.3   Derivation of Amplitude Equations

The transition from an attractor pattern to another attractor pattern or the emergence of an attractor pattern has been discussed qualitatively in Sect. 4.3.1. For the scenario in Sect. 4.3.1 involving an unstable and a stable basis pattern, it has been argued in Sect. 4.3.1 that the amplitude of the unstable basis pattern increases dramatically in magnitude and dominates the emerging attractor pattern. In contrast, the amplitude of the stable basis pattern increases in magnitude only slightly or not at all such that the stable basis pattern makes only a minor contribution to the emerging attractor pattern. Eventually, both amplitudes approach fixed point values. That is, the emerging attractor pattern can be expressed in terms of the amplitudes of the basis patterns at hand. In view of this situation it would be useful to be able to describe quantitatively the evolution of pattern amplitudes. In fact, the evolution of pattern amplitudes can be described by means of amplitude equations. Just as the amplitude space is a different perspective of the state space, amplitude equations are a different perspective of the original evolution equations of the pattern formation systems under consideration. That is, while the general evolution equation $dX/dt = N(X)$ introduced in Chap. 1 (see Sect. 1.11.1 and Figs. 1.5 and 1.11) describes pattern formation with respect to the original state variable X, an amplitude equation describes the formation of the same patterns from the perspective of pattern amplitudes. In order to derive an amplitude equation from a general evolution equation of the form $dX/dt = N(X)$, the state X of the pattern formation system under consideration is expressed in terms of a superposition of basis patterns of a fixed point with weights given by the relevant pattern amplitudes. This superposition also involves the fixed point under consideration as reference point. The superposition is substituted into the original evolution equation for X (i.e., the equation $dX/dt = N(X)$) and solved for the amplitudes. In doing so, a set of coupled first order differential equations for the amplitudes is obtained. These are the amplitude equations of interest [160] (see also the mathematical notes in Sect. 4.7 below).

4.3.4   Reduced Amplitude Space

From the previous discussion it follows that the eigenvalue of an unstable basis pattern changes its sign and increases from a negative value to a small positive value for bifurcation parameter values that are slightly above the critical value (see also Fig. 4.7b). In other words, the eigenvalue must pass through a value of zero at the bifurcation point. In contrast, all other eigenvalues assume negative values (at least in the inhomogeneous case that we consider at the moment). Since they do not pass through zero when the bifurcation parameter exceeds the critical value, from a mathematical point of view, it follows that the eigenvalues of the stable basis patterns are much larger in magnitude than the eigenvalue of the unstable basis pattern. As mentioned above, the eigenvalues determine the exponential increase and decay of their respective amplitudes. Consequently, the amplitude of the unstable basis pattern evolves much slower in time than the amplitudes of the stable basis patterns (i.e., the remaining basis patterns). Due to this separation of the time scales, from the perspective of the slowly evolving unstable basis pattern, all other amplitudes are either "immediately" constant or they follow "immediately" any changes of that slowly evolving amplitude [160].

The first case applies when the amplitudes of the stable basis patterns are independent of the amplitude of the unstable basis pattern. The second case applies when the evolution of the amplitudes of the stable basis patterns are coupled to the evolution of the amplitude of the unstable basis pattern. In particular, in the second case, if we could stop the evolution of the amplitude of the unstable basis pattern at an arbitrary time point, then "immediately" all other amplitudes would stop evolving as well. In this context, "being immediately constant" and "following immediately" means that the degree of lagging behind (i.e., the violation of the ordinary meaning of the word immediately) can be made arbitrarily small if we consider a bifurcation parameter that is larger than the critical value but arbitrarily close to that critical value. Figure 4.8 points out that in general the question arises how does the amplitude of a stable basis pattern depend on the amplitude of the (dominating) unstable basis pattern. If the stable amplitude is independent of the unstable amplitude, then we are dealing with the first case. If the stable amplitude depends on the unstable amplitude, we are dealing with the second case.

Irrespective whether or not the amplitude of a stable basis pattern depends on the amplitude of an unstable basis pattern, from the discussion above it follows that attractor patterns appear and disappear on the time scales that are defined by the dynamics of unstable basis patterns. Therefore, unstable basis patterns do not only make the major contribution to emerging attractor patterns but they also determine how quickly the formation of attractor patterns takes place. In addition to these two outstanding properties, unstable basis patterns come with another useful feature. A detailed mathematical analysis shows that the amplitude of a stable basis pattern can be described either by a constant (first case) or by a function of the amplitude of the unstable basis pattern (second case). This implies that it is sufficient to consider the evolution of the amplitude of the unstable basis pattern at hand in order to describe the formation of an attractor pattern of interest. This leads to the introduction of a reduced amplitude space, see panel (c) of Fig. 4.6. The pattern formation problem at hand can be described by considering only the amplitude of the relevant unstable

Fig. 4.8 For the formation of patterns it can be important whether or not stable amplitudes depend on unstable amplitudes

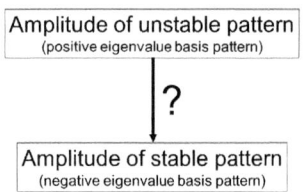

pattern. In some sense, this means that the evolution of the amplitudes of stable basis patterns can be ignored or that we can ignore stable basis patterns entirely.

4.3.5   Dominance of Unstable Patterns Over Stable Patterns: A Second Look

Let us continue the discussion of the previous section but focus on the amplitude of a stable basis pattern at an attractor fixed point. If the amplitude of a stable basis pattern does not depend on the amplitude of the unstable basis pattern, then the amplitude typically decays exponentially to zero in line with the fact that the eigenvalue of the stable basis pattern is negative. Therefore, the amplitude does not make a contribution to the emerging attractor pattern. In contrast, if the amplitude of the stable basis pattern depends on the amplitude of the unstable basis pattern, then the unstable basis pattern may produce a pumping force that causes the stable amplitude to increase. However, in view of the relative large (in the amount) negative eigenvalue, the pumping force must act against the damping of the basis pattern that is simply due to its nature of being a stable basis pattern (and not an unstable one). As a result of the interplay between exponential damping as characterized by the negative eigenvalue and pumping via the increasing amplitude of the unstable basis pattern, the amplitude of the stable basis pattern becomes stationary, on the one hand (and on a relatively fast time scale as discussed above), and approaches a relatively small fixed point value, on the other hand. Therefore, even if the amplitude of a stable basis pattern is pumped by the amplitude of an unstable basis pattern, the amplitude typically assumes a relative small value, when the pattern formation system becomes stationary. Therefore, stable basis patterns typically make only small contributions to emerging attractor patterns.

4.3.6   Bifurcations for Multi-Variable Systems Involving a Single Positive Eigenvalue

The scenario described in detail in Sect. 4.3.1 focused on pattern formation systems with two-dimensional state spaces and fixed points exhibiting two basis patterns. At the bifurcation point one pattern becomes unstable while the other pattern remains stable. The scenario can be generalized to discrete pattern formation systems with n-dimensional state spaces that involve fixed points exhibiting n basis patterns and to spatially extended pattern formation systems exhibiting and infinitely large number of basis patterns. In general, the eigenvalues of the relevant basis patterns depend on the bifurcation parameter. In the case of an inhomogeneous spectrum of eigenvalues it is assumed that all eigenvalues differ from each other. Accordingly, for bifurcation parameters smaller than the critical value all eigenvalues are negative. The fixed

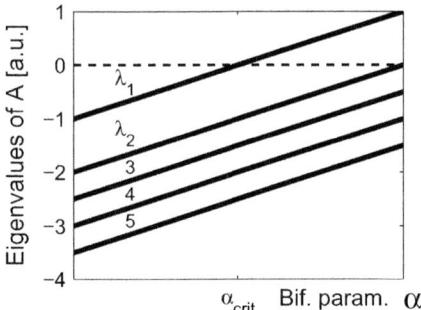

Fig. 4.9 Bifurcation from a reference point A given in terms of a stable node fixed point attractor A that becomes unstable in the case of a discrete system with an arbitrary dimension n or a spatially extended system exhibiting infinitely many basis patterns. Only the first five eigenvalues of the fixed point A are shown schematically. The figure can be generalized by adding an arbitrary number of additional eigenvalues

point under consideration corresponds to an attractor given by a stable node. At the bifurcation point exactly one of the eigenvalues crosses the zero line. This is shown in Fig. 4.9. Without loss of generality, in Fig. 4.9 it is assumed that the eigenvalue λ_1 of basis pattern 1 crosses the zero line at the critical bifurcation parameter value $\alpha(crit)$, while all other eigenvalues remain negative. Figure 4.9 generalizes Fig. 4.7b. Note that eigenvalues are not necessarily linear functions of bifurcation parameters (as shown in Figs. 4.7b and 4.9). Eigenvalues can depend in a nonlinear way on bifurcation parameters. However, the explicit dependency of eigenvalues on bifurcation parameters is not the crucial issue. At issue is that in the case of inhomogeneous eigenvalue spectrums there is only a single eigenvalue that becomes positive at the bifurcation point. The basis pattern of that eigenvalue is the unstable basis pattern that drives the transition from the "old" (i.e., disappearing) to the "new" (i.e., emerging) attractor pattern. All other basis patterns are stable basis patterns. By analogy to the discussion above, the unstable basis pattern contributes most to the emerging attractor pattern, it determines the time scale on which the "new" attractor pattern emerges, and its amplitude can be used to defined a reduced one-dimensional amplitude space that is sufficient to describe the dynamics of the pattern formation system.

4.3.7   Outstanding Properties of Unstable Basis Patterns and Order Parameters

The discussion above generalizes the case of two-variables systems with two-dimensional state spaces to discrete systems involving an arbitrary number of variables, on the one hand, and spatially extended systems, on the other hand.

However, only basis patterns of type 1 featuring real-valued eigenvalues have been considered. In fact, it can be shown that similar considerations can be made for basis patterns of type 2 that exhibit complex-valued eigenvalues [160]. Therefore, in general, unstable basis patterns (of type 1 or type 2) come with several outstanding properties.

First, unstable basis patterns dominate the emerging attractor patterns in the sense that they either correspond to the attractor pattern (as seen from the reference point of the unstable fixed point under consideration) or they make the major contribution to the attractor pattern. Second, as mentioned above, the dynamics of the amplitudes of the unstable basis patterns determine the time scale on which pattern formation takes place. Third, pattern formation can be understood by means of the dynamics of a single variable: the unstable amplitude (order parameter amplitude). That is, a reduced amplitude space can be introduced that exhibits a single dimension and is spanned by the unstable amplitude at hand. Pattern formation can be described as a process taking place in this reduced, one-dimensional amplitude space.

The third property leads to an enormous simplification of the problem at hand. Figure 4.6 illustrates the steps towards the reduced amplitude space for a pattern formation system described by two variables x and y. However, a similar picture can be drawn for any discrete pattern formation system involving an arbitrary number of variables x_1, x_2, x_3, etc. Let us consider a n-dimensional system with a n-dimensional state space spanned by $x_1, \ldots, x_n$. In a first step a new basis is constructed given in terms of a n-dimensional amplitude space spanned by the pattern amplitudes $A_1, \ldots, A_n$ of n basis patterns related to a reference fixed point. In the case of the inhomogeneous spectrum of eigenvalues mentioned above, only one of these basis patterns becomes unstable at the bifurcation point. Without loss of generality, the basis patterns can be labeled such that basis pattern 1 correspond to the unstable basis pattern. Then the n-dimensional amplitude space with $A_1, \ldots, A_n$ reduces to a one-dimensional reduced amplitude space given by A_1. This one-dimensional space is sufficient to describe the pattern formation system at the bifurcation under consideration. Importantly, the reduce amplitude space can also be introduced for spatially extended system. In this case, the state is described by a field variable $X(x, t)$, where x denotes a coordinate and t is time. The field variable at a fixed point is expressed in terms of a set of eigenfunctions and their weights. As discussed above, these eigenfunctions are considered as basis patterns and the weights correspond to the pattern amplitudes. In general, the set of eigenfunctions is infinitely large. Therefore, we end up with infinitely many amplitudes. For practical purposes (e.g., simulations) we may truncate the expansion into eigenfunctions (e.g., if the eigenfunctions are Fourier modes, then we may neglect all Fourier modes with very short periods). In any case, the amplitude space is either infinitely dimensional or it is relatively large. However, in the case of an inhomogeneous eigenvalue spectrum, only one eigenvalue becomes zero at the bifurcation point. That is, there is only one basis pattern (eigenfunction) that becomes unstable. It is then sufficient to focus on the dynamics of the amplitude A of that pattern (i.e., to study how the weight of that particular eigenfunction evolves in time). In doing so, the problem at hand that started with a field variable $X(x, t)$ and was turned into a problem

Fig. 4.10 For the formation
of patterns it is also important
whether or not unstable
amplitudes depend on stable
amplitudes. Compare with
Fig. 4.8

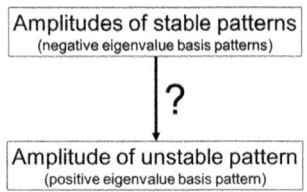

involving a infinitely large set of coupled amplitude equations turns into a relatively simple problem given in terms of a first order dynamics evolution equation for a single amplitude variable $A(t)$.

4.3.8   Dynamics of Unstable Amplitudes and Order Parameter Amplitudes

As mentioned above, the amplitude equations in the amplitude space correspond to a set of coupled first order differential equations. That is, in general the dynamics of a particular pattern amplitude depends on the dynamics of all remaining pattern amplitudes. In Sect. 4.3.4 it was discussed how stable amplitudes may depend on unstable amplitudes. The two somewhat trivial possibilities were address: stable amplitudes are either independent of unstable amplitudes or they depend in some way on unstable amplitudes. In this section, the remaining question is asked, see Fig. 4.10. How do unstable amplitudes depend on stable amplitudes? Again, only the two somewhat trivial answers are considered. Unstable amplitudes either do or do not depend on stable amplitudes.

Decoupled Case

If unstable amplitudes do not depend on stable amplitudes then the evolution equations for the unstable amplitudes are decoupled from the evolution equations of the stable amplitudes. That is, they do not contain the amplitude variables of the stable basis patterns. However, a key feature of the unstable basis patterns is that their amplitudes increase in magnitude exponentially at least as long as the state is close to the unstable fixed point involved in the bifurcation at hand. This exponential increase does not continue forever. The state converges to the fixed point of another attractor. Therefore, pattern formation systems for which unstable amplitudes are decoupled from stable amplitudes must feature a damping mechanism. When an unstable pattern amplitude increases in magnitude a damping force occurs that acts against a further increase of the amplitude such that eventually the amplitude approaches it stationary value. That stationary value describes the emerging attractor pattern.

Coupling Case and Circular Causality

If an unstable amplitude depends on stable amplitudes then the evolution equation of the unstable amplitude is coupled to the evolution equations of the stable amplitudes.

Again, as pointed out in the previous paragraph, amplitudes of unstable basis patterns do not increase in magnitude beyond any boundary (in fact this follows from the definition of pattern formation systems as globally stable systems). Amplitudes of unstable basis patterns saturate, that is, approach fixed point values. One mechanism that leads to saturation due to coupling between unstable and stable basis patterns is shown in Fig. 4.11a. When the amplitude of an unstable basis pattern (labeled here by $k = 1$) increases in time the amplitude result in pumping forces that lead to the increase of the amplitudes of stable basis patterns (labeled here by $k = 2, 3, \ldots$). When the amplitudes of those stable basis patterns increase in magnitude, then they produce damping forces that slow down the increase of the unstable basis pattern. Eventually, there is a balance between the pumping and damping forces acting on the amplitude of the unstable basis pattern. The situation at which this balance is established correspond to the newly emerged attractor pattern.

The key parts of the circular causality loops are highlighted in panel (b) of Fig. 4.11. A positive eigenvalue (unstable) basis pattern causes the pumping of negative eigenvalue (stable) basis patterns. Those pumped negative eigenvalue basis patterns in turn result in a damping of the positive eigenvalue basis pattern.

While panel (b) gives the key aspects of the circular causality loop, panel (a) also points out that unstable basis patterns increase in magnitude (as measured in terms of their corresponding pattern amplitudes) close to their unstable fixed points. This indicates that they are subjected to an built-in pumping mechanism (that might be called "self-excitation"). In contrast, stable basis patterns decay in magnitude close to the reference fixed point of interest. Consequently, pattern formation systems feature built-in damping mechanisms for stable basis patterns that lead to the exponential decay of those patterns.

Although in the scenario under consideration the dynamics of the unstable amplitude of a pattern formation system depends on the amplitudes of all stable basis patterns, the pattern formation system can be described by means of the dynamics of the unstable amplitude alone. That is, a reduced, one-dimensional amplitude space can be introduced. The reason for this is that due the fact that the stable and unstable amplitudes evolve on different characteristic time scales (fast versus slow), the stable amplitudes can be expressed as functions of the unstable amplitude. Therefore, from a mathematical point of view, they can be eliminated in the evolution equation for the unstable amplitude. In the end, an effective evolution equation can be obtained that only involves the amplitude of the unstable basis pattern. The circular causality involved in the scheme shown in Fig. 4.11 and the fact that nevertheless a simplified description for the dynamics of the unstable amplitude can be obtain that is closed and sufficient to describe the pattern formation under consideration has also been called the slaving principle [160]. The unstable amplitude acts like a "master" and determines the values of the stable amplitudes. In this context, however, it should be noted that this "master" is in fact not an

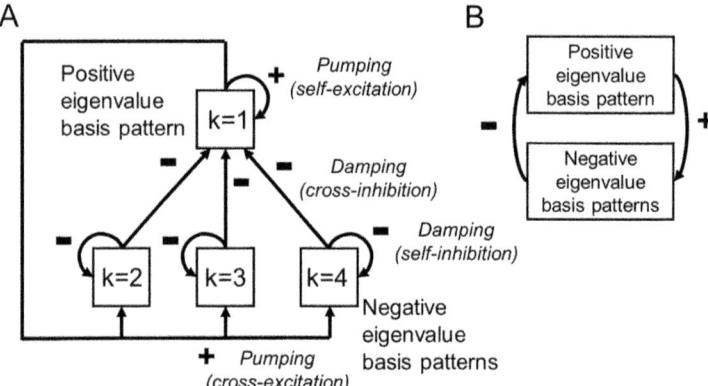

Fig. 4.11 Circular causality in pattern formation systems for which the amplitude of the dominating, unstable basis pattern depends on the amplitudes of the remaining stable basis patterns. Panel (**a**) provides details, whereas panel (**b**) only shows the key elements of the circular causality loop

independent entity. The amplitude dynamics of the unstable basis pattern depends on the evolution of all other amplitudes. A mathematical description for the unstable amplitude ("master" amplitude) can be obtain that does not involve the other amplitudes. Although the stable amplitudes do not show up in this description, the terms in this mathematical description reflect the impacts of the stable amplitudes.

4.4   Pattern Formation in Reduced Amplitude Spaces with Several Amplitudes

So far, we have discussed pattern formation that involves a single unstable basis pattern. In this case the reduced amplitude space is one-dimensional and is spanned by the amplitude of the unstable basis pattern. This scenario holds in the case of pattern formation systems featuring inhomogeneous eigenvalue spectrums at their bifurcation points. However, due to certain symmetry properties of the pattern formation systems under consideration, it is possible that the eigenvalues of several basis patterns assume the same value and that the eigenvalues of several basis patterns vary in the same way with the bifurcation parameter. Those basis patterns form a group of basis patterns that exhibits a homogeneous sub-spectrum of eigenvalues. At the bifurcation point, all eigenvalues becomes positive. In this case, reduced amplitude spaces are composed of several unstable amplitudes and have dimensions larger than one.

4.4.1   Symmetric Systems with Homogeneous Groups of Positive Eigenvalues

Let us illustrate the implication of symmetry properties on pattern formation systems and the eigenvalue spectrum of such systems at bifurcation points, see Fig. 4.12. First, let us consider a chemical reaction in a fluid layer that leads to the emergence of stripe patterns, see panel (a). The reaction is assumed to take place in a squared-shaped dish. Due to this geometric feature of the pattern formation system it follows that if a stripe pattern in one direction, say the x direction, can emerge, then the stripe pattern in the other direction, say y direction, can emerge as well. Let us assume that the stripe patterns emerge from a state in which all chemical substances under consideration are homogeneously distributed over the area of the squared dish. This state is referred to as spatially homogenous state. Let us assume that there is a bifurcation parameter with a critical value. For values of the bifurcation parameter smaller than the critical value the spatially homogeneous state is an attractor pattern of a certain fixed point attractor. If the bifurcation parameter is increased slightly above the critical value, then the fixed point of the homogenous state becomes unstable. Given the symmetry of the problem at hand, there must be at least two unstable basis patterns. One basis pattern that describes x-stripes and another basis pattern that describe y-stripes. Therefore, slightly above the bifurcation point there is not a single positive eigenvalue but there are at least two of them. Moreover, assuming perfect symmetry, these two eigenvalues assume the same value. For bifurcation parameters slightly above the critical value, the spatially

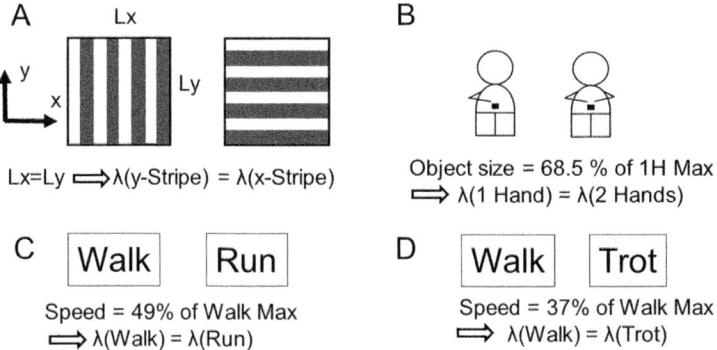

Fig. 4.12 Examples of symmetric pattern formation systems of the inanimate and animate worlds that are assumed to exhibit multiple (here: two) unstable basis patterns at their respective bifurcation points. Panel (**a**): Chemical system in a squared dish producing stripe patterns either oriented in x or y direction. Panel (**b**): Human individuals grasping medium sized objects producing either a one-hand grasping or two-hands grasping pattern. Panel (**c**): Hypothetical experiment with humans stepping on a moving treadmill that moves at a medium speed producing either a walk or a running pattern. Panel (**d**): Hypothetical experiment as in panel (**c**) but for horses stepping on a moving treadmill that moves at a medium speed producing either walk or trot gait movement patterns

homogeneous state becomes unstable and either the x-stripe basis pattern or the y-stripe basis pattern grows in magnitude. Typically, during a transient period the amplitudes of both patterns increase and compete with each other. Which pattern eventually wins the competition and emerges as attractor pattern or as dominant component of an attractor pattern depends on the initial conditions. For example, if the chemical substances initially are not perfectly homogeneously distributed, then there is an initial bias towards either the x-stripe or y-stripe attractor pattern. Due to this initial bias one of the two basis patterns wins the competition and the corresponding attractor pattern emerges.

In general, symmetries of pattern formation systems lead to groups of eigenvalues that all assume the same value and therefore become all positive when the bifurcation parameter exceeds the critical value. If the system is considered sufficiently close to the bifurcation point, then there is only one such group of eigenvalues. This implies that there is a group of unstable basis patterns. Moreover, the spectrum of eigenvalues constitutes a subspectrum that is composed of a set of eigenvalues that assume the same value. This set is considered as a homogeneous subspectrum of the eigenvalue spectrum.

The outstanding properties of unstable basis patterns discussed in Sect. 4.3.7 also apply in the case of groups of unstable basis patterns. First, the unstable basis patterns typically make the major contributions to the emerging attractor patterns. Second, the amplitudes of the unstable basis patterns evolve relatively slowly as compared to the amplitudes of the stable basis patterns. Therefore, the dynamics of the unstable basis patterns determines how fast or slow the overall pattern formation is completed. Third, the time scale separation also implies that the dynamics of the amplitudes of the stable basis patterns can be eliminated and a closed description for the evolution of the amplitudes of the unstable basis patterns can be obtained. That closed description involves only the amplitudes of the unstable basis patterns. This closed description is a reduced description because it does not involve the amplitudes of the stable basis patterns.

Panels (b)–(d) of Fig. 4.12 provide examples a human and animal pattern formation systems that for theoretical reasonings are assumed to exhibit certain symmetry properties. As mentioned in Sect. 2.5.1, humans grasp medium sized objects both with one hand or with two hands. The two types of grasping movements and their corresponding brain activity dynamics are considered as two BBA attractor patterns. Experimental and theoretical work using the human pattern formation reaction model has shown that a human individual acts under a symmetric grasping situation when the object exerting forces on that individual (e.g. via the visual system) has the size of 68.5% of the hand space of the individual [112]. In other words, a model-based evaluation of experimental data [112] suggests that the eigenvalues of the two unstable basis patterns related to one-hand grasping and two-handed grasping assume the same value provided that a person grasps an object that measures about 68.5% of the person's hand span, see panel (b). The reference point in the case of grasping movements is the condition in which the individual does not perform a grasp. That is, just as the spatially homogeneous state of the chemical reaction system addressed in panel (a) becomes unstable and exhibits at its bifurcation point two unstable basis patterns related to x-stripes and y-stripes, it is hypothesized that for human individuals under the forces exerted by medium sized objects with 68.5% relative size the pattern of not grasping becomes unstable and that the corresponding bifurcation point exhibits two unstable basis patterns (rather than a single one) related to grasping with one hand or grasping with two hands. Note that according to the theoretical model suggested in Ref. [112] the two possible grasping patterns are assumed to be mutually exclusive. For this reason, it is assumed that they compete with each other.

Panel (c) illustrates a similar example for human gaits. In general, humans prefer walking at low locomotion speeds and running at high locomotion speeds. Using the data from Diedrich and Warren [64] as presented in Frank [112] in the context of the human pattern formation reaction model, we can conclude that humans are in a symmetric condition when the locomotion speed is about 49% of the maximal possible walking speed. In a hypothetical experiment in which an individual would need to step on a moving treadmill that moves with 49% of the maximal walking speed of that individual, the not walking pattern would correspond to an unstable fixed point featuring two unstable basis patterns with identical eigenvalues. One basis pattern would describe walking. The other basis pattern would describe running. Again, both patterns would compete with each other. Again, depending on the initial conditions one of the two patterns would win the competition and the corresponding pattern amplitude would increase in magnitude. This pattern would make the major contribution to or entirely determine the emerging walking or running attractor pattern (see Sects. 4.3.2, 4.3.5 and 4.3.7). Finally note this is a hypothetical experiment. For safety reasons, participants should not be asked to step on a moving treadmill (i.e., the experiment should not be conducted in this way).

Panel (d) provides an another example from the field of gait transitions, here, gait transitions in horses rather than humans. As mentioned briefly in Sect. 3.2.3, see Fig. 3.5d, horses make a gait transition from walk to trot when locomotion speed is increased. Walking and trotting is considered as two BBA attractor patterns. Standing still is considered as reference point that becomes unstable under appropriate conditions. From the data reported by Griffin et al. [143] we may determine the situation under which the animal pattern formation system exhibits symmetry in the sense that there are two unstable basis patterns (one for walking and one for trot) at the unstable reference fixed point such that the basis patterns exhibit the same eigenvalues. Although the study by Griffin et al. was not tailored to address the human pattern formation reaction model, under certain assumptions (and following the argument in [112]) we can conclude that for miniature horses the symmetric case occurs when they are confronted with a situation to move with 37% of their maximal walking speed.

The examples shown in Fig. 4.12 illustrate pattern formation at bifurcation points that exhibit two basis pattern with positive eigenvalues due to the symmetry features of the systems under consideration. In general, the symmetry group of unstable basis patterns can involve an arbitrary number of unstable basis patterns. In the extreme case, there are infinitely many unstable basis patterns. An example is shown in Fig. 4.13 and has been discussed in detail by Bestehorn and Haken

Fig. 4.13 Stripe pattern emerging in a system with rotational symmetry. The symmetry implies that at the bifurcation point there is an infinitely large set of unstable basis patterns. For bifurcation parameters slightly above the critical value those basis patterns exhibit all the same positive eigenvalue

[23]. In Fig. 4.13 we consider a system that can exhibit periodic patterns (e.g. roll patterns as in the Benard experiment or stripe patterns as in chemical reaction). The system is considered under a circular boundary. Consistent with the boundary, it is assumed that the system has identical properties if we rotate it by an arbitrary angle. Consequently, if the conditions are such that a particular periodic pattern A with a certain orientation Z can emerge, then under the same conditions any periodic pattern A' can occur that is identical to A expect for its orientation angle Z'. Therefore, there are infinitely many attractor patterns. For example, the periodic pattern shown in Fig. 4.13 could be rotated by an arbitrary angle to give rise to another periodic pattern. At the bifurcation point (which involves in this example again a spatially homogeneous state as unstable reference fixed point) there is an infinitely large set of unstable basis patterns. Each basis pattern describes a periodic pattern with a particular orientation. Each basis pattern with orientation Z gives rise to an attractor pattern with orientation Z. The attractor pattern Z either is identical to the basis pattern Z or the attractor pattern has as major contribution the unstable basis pattern with orientation Z and includes in addition some stable basis patterns (i.e., basis patterns with negative eigenvalues), see Sects. 4.3.2 and 4.3.5.

In general, if there is a symmetry group that involves a continuous parameter (such as the orientation angle) then there is an infinitely large set of attractor patterns and likewise an infinitely large set of unstable basis patterns. The reduced amplitude space becomes infinite dimensional. Nevertheless, the reduced amplitude space is still a useful framework to discuss pattern formation in such situations because it does not contain all the stable basis patterns. In our example depicted in Fig. 4.13 such stable basis patterns could be periodic patterns with arbitrary orientation again but periods different from the period of the pattern shown in Fig. 4.13.

4.4.2   Asymmetric Systems with Inhomogeneous Groups of Positive Eigenvalues

So far, we have discussed pattern formation involving a single unstable basis pattern that by definition exhibits a positive eigenvalue or a group of unstable basis patterns that all exhibit the same positive eigenvalue. Let us turn to the case of a group

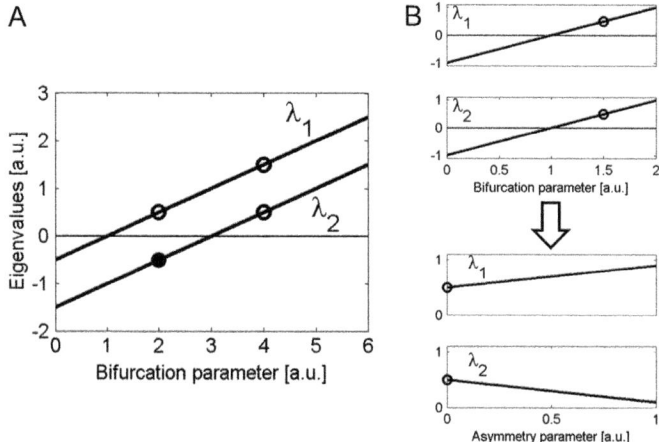

Fig. 4.14 Two scenarios (panels (**a**) and (**b**)) leading to a pattern formation system exhibiting multiple unstable basis patterns with different eigenvalues. Panel (**a**): A system that is shifted to a relatively large amount beyond its bifurcation point. Panel (**b**): A symmetric system that is brought into an instability by increasing the bifurcation parameter beyond the critical value and whose symmetry is broken by a (external) force

of unstable basis patterns that exhibit positive but different eigenvalues. That is, the sub-spectrum of eigenvalues of this group is inhomogeneous. Let us consider two scenarios that can lead to this situation. The first scenario involves forces that affect bifurcation parameters. The second scenario involves forces that affect the symmetry properties of pattern formation systems.

The first scenarios is illustrated in panel (a) of Fig. 4.14 for a discrete system described by two variables x and y (which gives us two basis patterns at a bifurcation point). Figure 4.14a shows the eigenvalues λ_1 and λ_2 of two basis patterns 1 and 2 as functions of the bifurcation parameter α. The bifurcation parameter has a critical value $\alpha(crit) = 1$. For parameter values $\alpha < 1$ both eigenvalues are negative. Accordingly, the system exhibits a fixed point attractor in terms of a stable node. At the critical value, eigenvalue λ_1 becomes positive, the stable node becomes unstable and turns into a saddle. In particular, for a bifurcation parameter $\alpha = 1.5$ we have $\lambda_1 = 0.5$ and $\lambda_2 = -0.5$ as indicated in panel (a). Up to this stage, there is an inhomogeneous eigenvalue spectrum with one positive eigenvalue only—as discussed in Sect. 4.3.1. However, let us consider next the case in which a force acts on the system such that the bifurcation parameter is shifted even further away from the critical value. For $\alpha = 3$ the eigenvalue λ_2 becomes positive as well. In particular, for $\alpha = 4$ both eigenvalues are positive. We have $\lambda_1 = 1.5$ and $\lambda_2 = 0.5$. In doing so, the pattern formation system exhibits two positive but different eigenvalues. In general, systems with inhomogeneous eigenvalue spectrums may exhibit several positive eigenvalues with different values when bifurcation parameters exceed their critical values by relatively large amounts.

Panel (b) demonstrates the second scenario leading to multiple positive but different eigenvalues. The departure point is a pattern formation system exhibiting a particular symmetry. In panel (b) a two-variables system is considered featuring a fixed point attractor in terms of a stable node when the bifurcation parameter α is smaller than the critical value $\alpha(crit) = 1$. Accordingly, for $\alpha < 1$ both eigenvalues λ_1 and λ_2 are negative. Due to the assumed symmetry property, the eigenvalues assume the same value (see top subpanel). They become both positive at $\alpha(crit) = 1$. In particular, for $\alpha = 1.5$ we have $\lambda_1 = \lambda_2 = 0.5$. Let us assume that the bifurcation parameter is fixed at $\alpha = 1.5$. Subsequently, a force acts on the system that affects the symmetry of the system. In order to describe the impact of the force an asymmetry parameter is introduced. If the asymmetry parameter equals zero, the system exhibits the symmetry property. For values different from zero the symmetry property is violated. It is assumed that the violation of the symmetry affects differently the eigenvalues λ_1 and λ_2. The bottom subpanel shows λ_1 and λ_2 as functions of the asymmetry parameter for a representative case. In this case, increasing the asymmetry increases λ_1 but decreases λ_2. Consequently, for asymmetry parameters different from zero, the pattern formation system exhibits two positive but different eigenvalues.

The second scenario applies to the examples shown in panels (b)–(d) of Fig. 4.12. As far as the example of grasping transitions is concerned (see panel b), recall that humans prefer to grasp small objects with one hand and larger objects with two hands. Therefore, it is plausible to assume that for objects that have a relative size smaller than 68.5% the eigenvalue of the one-hand grasping basis pattern becomes larger than the eigenvalue of the two-handed grasping basis pattern: $\lambda(1 \text{ hand}) > \lambda(2 \text{ hands})$. Vice versa for objects with relative size larger than 68.5% it is plausible to assume that we have $\lambda(2 \text{ hands}) > \lambda(1 \text{ hand})$. In short, the forces exerted by the to-be-grasped objects break the symmetry of the human pattern formation system if objects differ in relative size from 68.5%. The grasping transition model (a special case of the human pattern formation reaction model) proposed in Frank et al. [120] and Lopresti-Goodman et al. [215] describes this symmetry breaking explicitly as will be discussed in Sect. 6.5.4. Similar considerations can be made for the formation of walk-run patterns (panel c) and walk-trot patterns (panel d). In the case of the formation of walk-run patterns the human pattern formation is assumed to exhibit a certain symmetry property for locomotion speeds of 49% of the maximal walking speed. For lower speeds the symmetry is broken such that $\lambda(\text{walk}) > \lambda(\text{run})$ consistent with the preference of humans to walk at low speeds. Likewise, for higher speeds it is plausible to assume that $\lambda(\text{run}) > \lambda(\text{walk})$. Given the analogy between transitions between one-hand and two-handed grasping and walk-run transitions it does not come as a surprise that it has been suggested to apply the grasping transition model also to walk-run gait transitions [105, 110]. The symmetry breaking mechanism is explicitly modeled in those studies. Finally, in the case of the walk-trot example it is plausible to assume that $\lambda(\text{walk}) > \lambda(\text{trot})$ holds for relative locomotion speeds below 37%, while $\lambda(\text{trot}) > \lambda(\text{walk})$ holds for relative locomotion speeds above 37%.

4.5 Lotka-Volterra-Haken Amplitude Equations

As mentioned above, due to the time-scale separation between unstable and stable basis patterns, the formation of patterns can approximately be described in the reduced amplitude space by means of a closed set of amplitude equations for unstable basis patterns. The Lotka-Volterra-Haken amplitude equations are a set of amplitude equations of this kind [107, 109, 114, 186, 187]. They are very general in the sense that they include various models as special cases. In particular, they include the Lotka-Volterra model of population dynamics used in ecology [20, 134, 137, 140, 217, 236] and the amplitude equations of Haken's neuronal network for pattern formation, "perception", and "associative memory" [23, 36, 37, 61, 95, 97, 125, 127, 157, 242, 282, 299, 358]. They also correspond to amplitude equations of various pattern formation system in physics. In fact, Haken's neuronal network model was motivated by such amplitude equations that have been studied in classical areas of physics.

First of all, the amplitude equations describe pattern formation due to the Benard instability in fluid and gas layers heated from below [24, 57, 78, 247, 252, 292, 293]. Furthermore, the amplitude equations have been discussed in the context of pattern formation systems exhibiting so-called Turing instabilities [74, 76, 260] and satisfying the Stuart-Landau equation [22, 128, 129]. Amplitude equations of the Lotka-Volterra-Haken type have been used to describe how patterns emerge in certain chemical systems [354], in epidemic outbreaks [341], during sputtering processes [259], and during certain cell signaling processes [266, 308]. Moreover, the emergence of phase synchronization [30] has been described by means of amplitude equations that are of a similar type as the Lotka-Volterra-Haken amplitude equations. The formation of concentration waves carrying self-moving oil droplets satisfies the amplitude equations as well [314]. Moreover, amplitude equations of the Lotka-Volterra-Haken type have been discussed in the context of theoretical models for the formation of roll patterns and other type of patterns within the human skin [58]. The Lotka-Volterra-Haken model for a single amplitude also describes the amplitude dynamics of the canonical-dissipative oscillator presented in Sect. 3.6.3. In fact, the so-called van der Pol oscillator satisfies the same approximative amplitude equation [157]. In particular, for an arbitrary number of amplitudes the Lotka-Volterra-Haken amplitude equations involving cubic nonlinearities as suggested by Haken can be derived from a set of coupled neuronal self-oscillators similar to the van der Pol oscillators [157]. The Lotka-Volterra-Haken amplitude equations have also been studied from the perspective of neuronal network theory [82, 165].

We will refer to the Lotka-Volterra-Haken amplitude equations as LVH amplitude equations or the Lotka-Volterra-Haken (LVH) model. The LVH model is about the evolution of unstable pattern amplitudes or order parameter amplitudes. Frequently, to improve readability, the unstable pattern amplitudes of the LVH model will be referred to as pattern amplitudes or simply as amplitudes. The reader "should keep in mind" that in those cases we are talking about unstable pattern amplitudes.

4.5.1 Model Components

The LVH model involves three mechanisms illustrated in Fig. 4.15 for three unstable basis patterns 1, 2, and 3 with pattern amplitudes.

First of all, as shown in panel (a), there is the pumping mechanism of unstable basis patterns related to the fact that all patterns exhibit positive eigenvalues (see also Sect. 4.4). The pumping mechanism describes the dynamics close to the reference point that is given by the origin $A_1 = A_2 = A_3 = 0$ of the reduced amplitude space. Mathematically speaking, this domain is described by linearized evolution equations. Consequently, the pumping terms are linear with respect to the amplitudes. As shown in panel (b) the pumping mechanism leads to an unbounded increase of all amplitudes. However, the LVH model also accounts for damping of the amplitudes as discussed in Sect. 4.3.8. The LVH model does not specify the nature of the damping (ordinary self-damping as in the decoupled case or

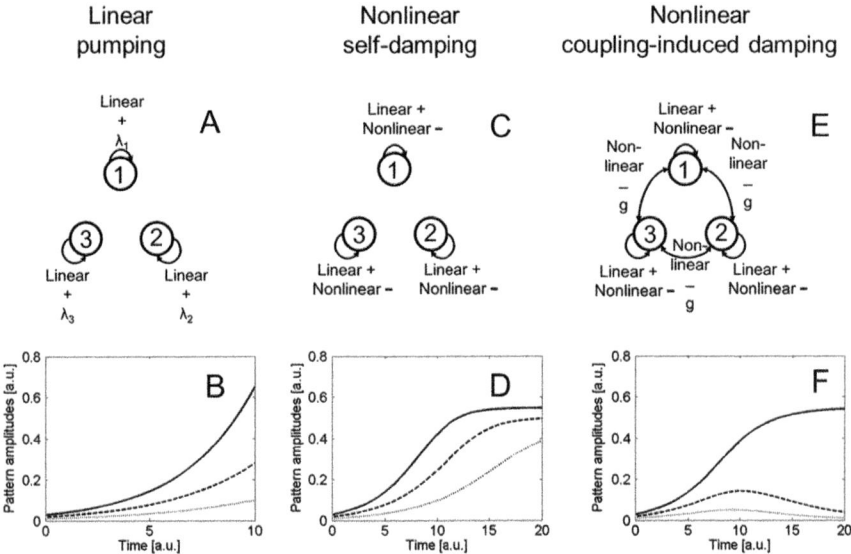

Fig. 4.15 Three components of the LVH model illustrated for a model with $m = 3$ unstable basis pattern amplitudes A_1, A_2, and A_3. Panel (**a**): Eigenvalues λ_1, λ_2, and λ_3 describe a linear pumping mechanism that results in an exponential increase of the amplitudes. Panel (**b**) shows A_1 (dotted), A_2 (dashed), and A_3 (solid) as functions of time for $\lambda_3 > \lambda_2 > \lambda_1$ and $A_3(0) > A_2(0) > A_1(0)$ computed from the LVH model defined by Eq. (4.38) when only the linear pumping is taken into account. Panel (**c**): The standard LVH model exhibits a nonlinear self-damping term normalized to 1. Panel (**d**) shows the evolution of the amplitudes when linear pumping and nonlinear self-damping are taken into account. Panel (**e**): Competitive interaction between basis patterns leads to competitive coupling between amplitudes and coupling-induced damping as measured by the coupling constant g. Panel (**f**) shows the evolution of the amplitudes when all three model components are taken into account. The LVH model is a winner-takes-all model: only one amplitude survives

self-damping via stable basis patterns as in the case of circular causality). In any case, the damping involves nonlinear terms, see panel (c). Due to the interplay between the linear pumping and the nonlinear damping the amplitudes approach fixed point values, as shown in panel (d). Without any coupling between the pattern amplitudes all pattern amplitudes would approach non-zero fixed point values. In general, pattern amplitudes interact with each other. That is, the amplitude evolution equations are coupled. The LVH model has originally been introduced for competing basis patterns and pattern formation systems exhibiting mutually exclusive attractor patterns [157]. Examples of such mutually exclusive patterns are given in panels (b)–(d) of Fig. 4.12. Objects can be either grasped with one hand or with two hands. Humans can either walk or run. Horses can either walk or trot. In the case of competitive coupling between pattern amplitudes, only the amplitude of one unstable basis pattern approaches a finite fixed point value. Panel (e) illustrates the coupling between amplitudes. From a mathematical point of view, due to the coupling between amplitudes, the amplitudes interact via nonlinear damping terms. Panel (f) illustrates a typically amplitude dynamics of the LVH model involving all three components: linear pumping, nonlinear self-damping, and nonlinear coupling-induced damping. In the example, the amplitude A_1 approaches a non-zero stationary value. The other amplitudes A_2 and A_3 decay to zero. Although the LVH model has originally been introduced to describe pattern formation of mutually exclusive patterns, it can also be used to describe co-existence of patterns. We will return to this issue in Sect. 4.6.2.

As far as the eigenvalue spectrum is concerned, the Lotka-Volterra-Haken model applies to all three cases discussed above in Sects. 4.3.6, 4.4.1, and 4.4.2. That is, the model can describe pattern formation in the case of a single unstable basis pattern (i.e., a system with a bifurcation point that exhibits only one positive eigenvalue). The model can describe pattern formation of a system characterized by some symmetry such that the formation of the pattern involves multiple unstable basis patterns whose eigenvalues all assume the same value. Finally, the model can address the formation of patterns when there are multiple unstable basis patterns with eigenvalues that differ in magnitude, that is, there is an inhomogeneous sub-spectrum of positive eigenvalues. In the very first case, the LVH model corresponds to a single first order differential equation of the amplitude of the unstable basis pattern at hand. In the case of several unstable basis patterns with a homogeneous or inhomogeneous spectrum of eigenvalues the LVH model consists of several coupled amplitude equations. Table 4.4 summarizes these three cases. Note that in Table 4.4 the sub-spectrum only refers to the eigenvalues of the unstable basis patterns. The eigenvalues of the stable basis patterns are by definition negative. These eigenvalues do not show up explicitly in the LVH model because the LVH model is a reduced model that does not explicit address the dynamics of the amplitudes of the stable basis patterns.

Table 4.4 Three cases of pattern formation systems that differ with respect to their eigenvalue spectrums

Case	Number of amplitudes and eigenvalues	Eigenvalue (sub)spectrum
I	Single amplitude, one eigenvalue	One positive eigenvalue
II	Several amplitudes and eigenvalues	Homogeneous (sub)spectrum: identical positive eigenvalues
III	Several amplitudes and eigenvalues	Inhomogeneous (sub)spectrum: eigenvalues all positive but different

4.5.2 Model Parameters

The LVH amplitude equations involve as parameters the eigenvalues of the subspectrum of the unstable basis patterns. From our previous discussion in Sect. 4.5.1 and from Fig. 4.15 and Table 4.4 it follows that they are part of the linear pumping terms. Throughout this book, eigenvalues are denoted by the variable λ (e.g., if there are m amplitudes then the model exhibits m parameters $\lambda_1, \ldots, \lambda_m$). In the most parsimony version of the LVH model, the model exhibits only two more additional parameters. The first additional parameter describes the strength of the coupling between amplitudes. The parameter will be referred to as amplitude-amplitude coupling parameter or simply as coupling parameter and will be denoted by g. The second additional parameter describes the type of nonlinearity of the nonlinear terms occurring in the LVH model. Recall that the LVH model involves two types of nonlinear damping terms, see Fig. 4.15. The nonlinearity parameter will be called nonlinearity exponent and denoted by q. The LVH model parameters (i.e., the eigenvalues $\lambda_1, \ldots, \lambda_m$, coupling parameter g, and nonlinearity exponent q) describe the structure of the human or animal under consideration.

4.5.3 Lotka-Volterra-Haken Model and Human Pattern Formation Reaction Model

If pattern formation does not involve changes of the structure, then the LVH model is assumed to describe the reduced description of the general human pattern formation reaction model $dX/dt = N(X)$ (see Chap. 1 and the mathematical notes in Sect. 4.7 below). If pattern formation involves changes of the structure, then the human pattern formation reaction model describes those changes by additional evolution equations (see Chap. 1 again and Sect. 1.11.1 in particular). In the context of the LVH model this implies that the model parameters (e.g., $\lambda_1, \ldots, \lambda_m, g, q$) become time-dependent variables. In the applications that will be discussed in the following chapters, in order to capture structural changes, the LVH model will be supplemented with evolution equations for the model parameters, in general, and the eigenvalues, in particular.

4.5.4 Lotka-Volterra-Haken Model and Attractor Patterns

The model describes emergence of attractor patterns on the basis of the amplitudes of unstable basis patterns. As discussed in Sect. 4.2, attractor patterns can be described in terms of the pattern amplitudes of unstable and stable basis amplitudes. Typically, the unstable basis patterns contribute primarily to attractor patterns (see Sects. 4.3.2, 4.3.5 and 4.3.7). The LVH model does not account for the evolution of the stable basis patterns. Therefore, there are two possible scenarios. First, the stable basis patterns do not contribute to the emerging attractor patterns under consideration or their contributions are negligible. In this case, the unstable basis patterns described by the LVH model in terms of their amplitudes correspond to the emerging attractor patterns. Second, the attractor patterns under consideration are composed of stable and unstable basis patterns. The amplitudes of the stable basis patterns are relatively small but sufficiently large such that they should not be neglected. This scenario may apply in particular for pattern formation systems exhibiting the circular causality loop discussed in Sect. 4.3.8 (see Fig. 4.11). In this case, the unstable basis patterns addressed by the LVH model approximately describe the emerging attractor patterns. Finally note that the LVH model does not specify the basis patterns themselves. That is, in the case of discrete pattern formation systems the LVH model does not provide a description of unstable eigenvectors. Likewise, for spatially extended pattern formation systems the LVH model does not provide a description of the relevant eigenfunctions with positive eigenvalues. The LVH model describes how amplitudes of eigenvectors and eigenfunctions (i.e., basis patterns) with positive eigenvalue evolve in time and, in doing so, how attractor patterns appear and disappear.

4.5.5 Two Examples

Let us give two examples of the LVH amplitude equations. A benchmark example of the LVH model involving cubic nonlinearities (i.e., $q = 3$) is the amplitude equation for the formation of convection rolls in the Benard experiment introduced in Sect. 3.2.3 (see Fig. 3.5c) that leads to the pitchfork bifurcation discussed in Sect. 3.5.2 (see Fig. 3.15a). The amplitude equation describes the evolution of a single pattern amplitude. The pattern amplitude in turn describes the rotation velocity of the rolls (or the velocity of the upwards or downwards streaming fluid or gas particles in the streams of the rolls). As such the velocity can be positive or negative, which indicates clockwise or counterclockwise rotations. Since there is only a single pattern amplitude, the state space of the amplitude dynamics is a

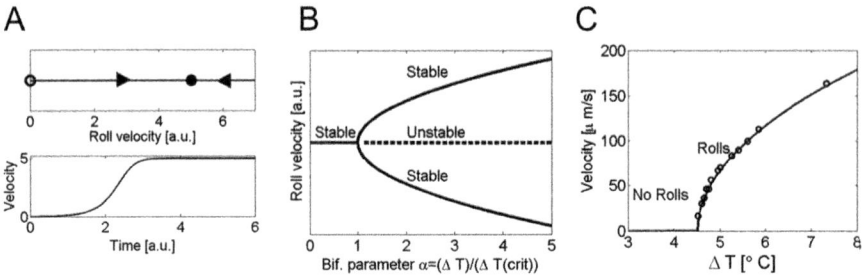

Fig. 4.16 The cubic LVH model for the special case of a single amplitude A as defined by Eq. (4.39) when applied to describe the emergence of roll patterns in the Benard experiment [349]. In this case, the amplitude A corresponds to the roll velocity. Panel (**a**): One-dimensional state space (top subpanel) with unstable (open circle) and stable (full circle) fixed points (only the positive axis is shown) and evolution towards the stable fixed point (bottom subpanel). Panel (**b**): Pitchfork bifurcation diagram described by the LVH model. Panel (**c**): Application to experimental data reported by Wesfreid et al. [349]

one-dimensional space given by a line as shown in panel (a) of Fig 4.16 (only the positive axis is shown). Panel (a) illustrates the situation when the temperature of the heat source is sufficiently large such that a roll pattern emerges. In this case, the spatially homogeneous pattern characterized by a zero roll velocity is a fixed point of the pattern formation system but it is an unstable fixed point. In the reduced, one-dimensional amplitude space, the spatially homogeneous pattern with a pattern amplitude of zero corresponds to an unstable fixed point (as indicated in panel (a) by an empty circle). In contrast, the roll pattern is a stable pattern. The roll pattern corresponds to an attractor pattern of a fixed point attractor (as indicated by the full circle in panel (a)). The arrows in panel (a) point in the direction in which states of the reduced amplitude space evolve in time. For any positive initial velocity different from zero, the roll velocity (i.e., pattern amplitude) as a function of time converges to the value given by the stable fixed point. Panel (a) also presents an example of a trajectory as computed from the LVH model.

Panel (b) of Fig. 4.16 displays the pitchfork bifurcation diagram of the single variable LVH model with cubic nonlinearities. The pitchfork bifurcation was discussed in Sect. 3.5.2, in general, and, in particular, with respect to its application to the Benard experiment. Here, let us briefly re-iterate that depending on the bifurcation parameter, the model exhibits either a single or three fixed points. For bifurcation parameter values smaller than the critical value, there is only one stable fixed point that is characterized by a zero roll velocity (and related to the spatially homogeneous attractor pattern). For bifurcation parameters larger than the critical value, the situation corresponds to the state space description shown in the top subpanel of panel (a). The system exhibits three fixed points. The first fixed point is at zero roll velocity for all values of the bifurcation parameter ΔT. It describes the location of the unstable fixed point describing a spatially homogeneous fixed point pattern (that corresponds to an attractor pattern for $\Delta T < \Delta T(crit)$ and to an repellor pattern for $\Delta T > \Delta T(crit)$). The second fixed point increases monotonically as a function of the bifurcation parameter ΔT. It describes clockwise rotating roll patterns of the pattern formation system for temperature differences above the critical temperature differences. The third fixed point assumes negative values, is symmetric to the second fixed point, and describes counter-clockwise rotating rolls. Finally, panel (c) presents experimental results shown earlier in Fig. 3.5c taken from the study by Wesfreid et al. [349]. The solid line is a fit to the data using the bifurcation diagram of the LVH model. Note that in the experiment, for temperature values above the critical value, only the stable branch with positive velocities was observed.

The Lotka-Volterra model is a special case of the LVH model for quadratic nonlinearities (i.e., $q = 2$). The Lotka-Volterra model has its origin in ecology as mentioned above. The model describes how species grow and decay in size due to interactions between species and various other factors such as the availability of food [20, 134, 137, 140, 217, 236]. Consequently, the origin of the model has little to do with pattern formation. Nevertheless, the model is of interest for the study of pattern formation and in particular for understanding human and animal "perception", "cognition", and "behavior" for two reasons. First, from a mathematical perspective the model exhibits a structure that is similar to the structure of amplitude equations of pattern formation systems. Therefore, what is known about the solutions of the Lotka-Volterra model can be used to improve our understanding of the dynamics of amplitude equations of pattern formation systems [98, 186], see also Sects. 4.6.6 and 4.7.10 below. Second, it has been suggested that the Lotka-Volterra model can be applied to describe patterns of brain activity in humans [265]. The variables that originally are interpreted as the size of populations are re-interpreted in this context as amplitudes of patterns of brain activity. In our context, the variables of the Lotka-Volterra model are re-interpreted as patterns amplitudes of unstable basis patterns.

Let us describe a Lotka-Volterra model in its original interpretation for the competition of two species. In what follows, the case is considered in which two species compete for survival such that only one of the two species can survive. Moreover, the species under consideration are assumed to live in the same environment (i.e., habitat) and do not differ with respect to their ability to survive. Since the Lotka-Volterra model is a special case of the LVH model, the Lotka-Volterra model equations exhibit two eigenvalues. For two species with the same degree of "fitness", the model features a symmetric structure and, consequently, the two eigenvalues assume the same value. Panel (a) of Fig. 4.17 shows the phase portrait for a representative set of model parameters. The horizontal and vertical axes describe the sizes of the species 1 and 2, respectively. In the two-dimensional plane spanned by those two axes there are four fixed points. The origin is an unstable fixed point. This implies that all trajectories evolve in time away from the origin. There are two fixed point attractors given in terms of stable nodes. They are located on the axes and describe the situation when species 1 has survived the competition process and species 2 is extinct or vice versa when species 2 has survived the competition process and species 1 is extinct. The solutions of the Lotka-Volterra model converge to either of those two stable fixed points. More precisely, the diagonal separates

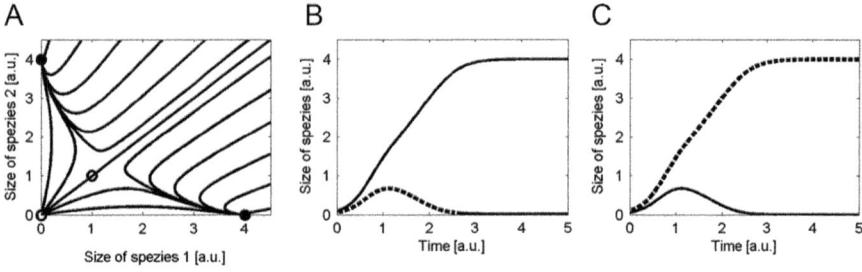

Fig. 4.17 Illustration of a Lotka-Volterra model for population dynamics. Panel (**a**): Phase portrait of two species competing for food with equal "fitness". Panels (**b**) and (**c**): State dynamics (i.e., population dynamics) for two different initial conditions. Solid and dashed lines describe the population sizes of species 1 and 2, respectively

the state space in two domains. The right-bottom half of the state space is the basin of attraction of the fixed point attractor that describes survival of species 1. Likewise, the left-top half of the state space is the basin of attraction of the fixed point attractor that describes survival of species 2. On the diagonal there is a saddle point. The saddle point and the origin are indicated by open circles (because they are both unstable fixed points). In contrast, the locations of the fixed point attractors are shown as full circles.

Panels (b) and (c) of Fig. 4.17 exemplify the state dynamics (i.e., population dynamics) described by the Lotka-Volterra model. Panel (b) shows the size variables of the two species as functions of time when species 1 initially outnumbers species 2. That is, initially the model is prepared in the right-bottom half. Consequently, the dynamics approaches the fixed point on the horizontal axis indicating that species 1 survives and species 2 dies out. Panel (c) illustrates the case when species 2 initially outnumbers species 1. In this case, the dynamics converges towards the fixed point on the vertical axis indicating that species 2 survives and species 1 dies out.

4.6   General Properties of the Lotka-Volterra-Haken Model

Let us discuss some general properties of the LVH amplitude equations.

4.6.1   Determinism and the Shoe Example Revisited

The LVH amplitude equations describe pattern formation as a deterministic process. For a given initial state of the pattern formation system, the state of the system evolves according to the LVH model in exactly one possible way. For example, the phase portrait shown in panel a of Fig. 4.17 illustrates that for every state of the state space there is a unique trajectory that describes how that state evolves in time.

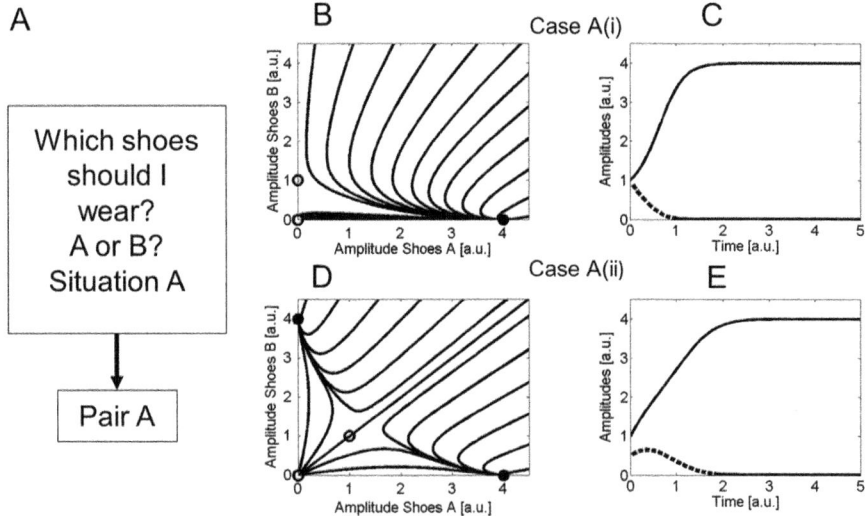

Fig. 4.18 Determinism is an in-built feature of pattern formation systems and their corresponding amplitude equations such as the LVH amplitude equations. Illustration of determinism for the shoe example as described in terms of a LVH model with $m = 2$ basis patterns for putting shoes A or B on. Panel (**a**): Question and condition as cause and the corresponding effect. Panels (**b**)–(**e**): Phase portraits and exemplary trajectories for the monostable case (panels (**b**) and (**c**)) and the bistable case (panels (**d**) and (**e**))

In general, the solutions of the LVH amplitude equations do not feature a special point in the state space that could be interpreted to reflect "free will", "choice" or "decision making" of the human under consideration. Figures 4.18 and 4.19 illustrate in the context of the shoe example introduced in Sect. 2.5.2 that the concept of determinism is in-built in the LVH model [112].

It is assumed that putting shoes A or shoes B on are two BBA attractor patterns that involve two mutually exclusive unstable basis patterns. The amplitudes of the unstable basis patterns are then described by the LVH amplitude equations. Figure 4.18 illustrates the case in which the person under consideration eventually put shoes A on. Accordingly, as illustrated in panel (a), the question which shoes to wear is asked under case A circumstances, which are circumstances that make that the parts of the person organize themselves into the BBA attractor pattern of putting shoes A on. There are two subcases of case A, labeled A(i) and A(ii). Case A(i) is shown in panels (b) (phase portrait) and (c) (example of trajectory). The pattern formation system exhibits only one fixed point attractor which corresponds to the attractor pattern of putting shoes A on. Consequently, irrespective of the initial pattern amplitudes, the amplitudes approach this unique attractor and the person under consideration puts shoes A on. Since the initial amplitudes reflect the initial state of the pattern formation system, if case A(i) holds then the initial state of the pattern formation system does not play a role. In contrast, in case A(ii) the initial state does play a crucial role. Case A(ii) is illustrated in panels (d) (phase portrait)

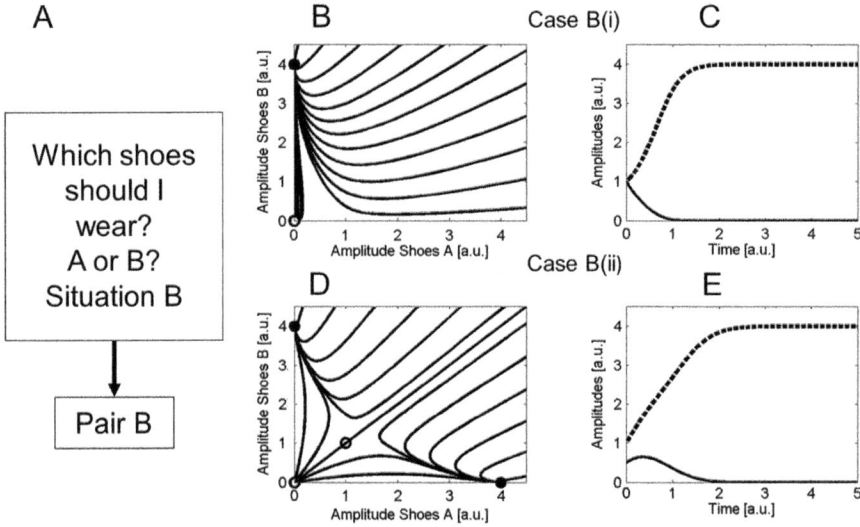

Fig. 4.19 As in Fig. 4.18 but for situations (or conditions) that cause the individual under consideration to put shoes B on

and (e) (example of trajectory). Accordingly, the pattern formation system exhibits two fixed point attractors (as in the example of the Lotka-Volterra system discussed in the previous section). They will be labeled as A and B, where A and B have fixed points location on the horizontal and vertical axes, respectively, and corresponds to the two possible BBA attractor patterns of putting shoes A or B on. In case A(ii) the initial state of the pattern formation system is such that the initial amplitudes of the two unstable patterns fall into the basin of attraction of attractor A (i.e., in the right-bottom half). Comparing cases A(i) and A(ii) we see that in both cases the future "behavior" is determined at the beginning of the pattern formation. While in case A(i) the initial state has no qualitative impact on the formation of the pattern, in case A(ii) the initial state plays a crucial role. Figure 4.19 describes the case in which the person takes on shoes B. Panels (a)–(e) are the counterparts to panels (a)–(e) of Fig. 4.18.

4.6.2   Haken's Principle for Winner-Takes-All Dynamics

In general, an attractor pattern is composed of several stable and unstable basis patterns (see Sect. 4.2). In this context, an important special case occurs when unstable basis patterns are mutually exclusive (see specifically Sect. 4.2.5). Only one unstable basis pattern can emerge and exhibit a finite amplitude. All others unstable basis patterns disappear (i.e., have zero amplitudes). In this case, any attractor pattern can maximally involve one unstable basis pattern. In other words, an attractor pattern is composed of one unstable basis pattern and a number of stable basis patterns that make some minor contributions (see Sects. 4.3.2, 4.3.5 and 4.3.7) and may or may not be pumped by the unstable basis pattern (see Sect. 4.3.8). This case is referred to as a winner-takes-all case. Therefore, the fact that unstable basis patterns are mutually exclusive implies that the corresponding attractor patterns are mutually exclusive as well. In summary, the winner-takes-all case addresses pattern formation system exhibiting mutually exclusive attractor patterns.

Examples of mutually exclusive BBA patterns are gaits like walk and run of humans and walk and trot of horse. In particular, the examples given in panels (b)–(d) of Fig. 4.12 on symmetric systems exhibiting homogeneous spectrums of positive eigenvalues are assumed to correspond to systems with mutually exclusive BBA patterns. Reactions to foreground-background aspects of figures (i.e., "foreground-background perception") may involve mutually exclusive attractor patterns. For example, the picture shown in panel (b) of Fig. 1.7 of Chap. 1 is assumed to induce one of two mutually exclusive brain activity patterns of the visual system [187]. One BBA pattern (the visual world cross BBA pattern) is similar to BBA patterns that emerge when crosses are shown (i.e., when forces exerted by figures with crosses act on the human pattern formation system). The other BBA pattern (the visual world squares BBA pattern) is similar to BBA patterns that emerge when squares are shown.

Haken suggested a principle that leads to mutually exclusive unstable basis patterns [157]. The principle is motivated by the theoretical analysis of pattern formation systems in physics such as the fluid and gas layers showing roll patterns in the Benard experiment (see Sect. 3.2.3). The principle focuses on the two types of damping mechanisms involved in pattern formation in systems with multiple unstable basis patterns: the self-damping and the damping via other unstable basis patterns (see Sect. 4.5.1).

Recall that the self-damping mechanism describes that unstable basis patterns when they increase in magnitude are subjected to a damping force that acts against a further increase. The damping force increases monotonically with the amplitude of an unstable basis pattern and only depends on the amplitude of the pattern that is affected by the damping force (therefore this mechanism has been referred to in Sect. 4.5.1 as self-damping and has also been referred to in the literature as "self-inhibition"). In contrast, the coupling-induced damping mechanism describes that the increase of an unstable basis pattern is damped due to the coupling with other unstable basis patterns. That is, the coupling-induced damping force acting on a particular basis pattern does not primarily depend on the amplitude of that pattern but on the amplitudes of the remaining unstable patterns (which is the reason why this damping mechanism has also been referred to as "cross-inhibition").

Haken's principle is about the strength of the self-damping and coupling-induced damping mechanisms. Haken's principle states that if the coupling-induced damping is stronger than the self-damping, then the pattern formation system exhibits mutually exclusive unstable basis patterns [157]. In order to measure the strength of the two damping mechanisms, we may assign to each mechanism a parameter. In the LVH model used in this book the strength of the damping mechanisms will be characterized in another way. The coupling parameter g that describes the strength of the coupling between amplitudes describes the strength of the coupling-induced damping relative to the strength of the self-damping. If g is larger than one, then coupling-induced damping is stronger than self-damping. If g is smaller than one, then the opposite is true.

If Haken's principle apply, that is, if g is larger than one, then the LVH model describes a winner-takes-all dynamics. In this case, the LVH model exhibits fixed point attractors or repellors on the axes of the reduced amplitude state. By definition, on those axes only one amplitude is finite, whereas all other amplitudes are zero. These fixed points are called winner-takes-all fixed points. Fixed points other than winner-takes-all fixed points may exist (i.e., fixed points for which more than one amplitude is finite). If they exit, then they correspond to repellors. The pattern formation system will not approach those fixed points (unless it is prepared in some very exceptional initial conditions, e.g., on the stable "direction" of a saddle). In contrast, if g is smaller than one, the LVH model may exhibit fixed point attractors for which more than one amplitude is finite [104]. That is, attractor patterns may emerge that are composed of more than one unstable basis pattern. In this sense, unstable basis patterns may co-exist. The corresponding fixed points are called co-existence points.

Haken's principle can be understood heuristically without the need to do explicit mathematical calculations. Accordingly, if the coupling-induced damping (i.e., "cross-inhibition") is stronger than the self-damping (i.e., "self-inhibition"), then the co-existence of unstable basis pattern would create relatively strong damping forces. In contrast, a winner-takes-all attractor pattern (i.e., a attractor pattern that involves only one unstable basis pattern) involves only the relatively weak self-damping. As a result, in the reduced amplitude space fixed point attractors that describe the co-existence of basis patterns cannot exist. In contrast, if coupling-induced damping is relatively weak, then there is no "penalty" for a co-existence pattern. In the extreme case of the absence of any coupling between the amplitudes (i.e., when the coupling parameter g is put to zero), we re-obtain the case discussed in panel (c) of Fig. 4.15. All pattern amplitudes evolve independently of each other. Since their dynamics is characterized by positive eigenvalues, they all increase to certain fixed point values that are determined by the magnitude of their eigenvalues, on the one hand, and the strength of the self-damping. In this extreme scenario all unstable basis patterns co-exist with each other.

This book addresses primarily pattern formation systems exhibiting mutually exclusive attractor and basis patterns. In this context, the LVH model will be used for coupling parameters g larger than one as a dynamical system featuring a winner-takes-all dynamics. This does not mean that the case of co-existing unstable basis patterns is irrelevant for our understanding of human and animal brain and body activity. However, pattern formation systems with winner-takes-all dynamics should be discussed first, before pattern formation systems with co-existence fixed point should be addressed. The reason for this is that pattern formation systems of the latter kind are more challenging conceptually and from a mathematical point of view.

Finally note that the correspondence between Lotka-Volterra population dynamical models, on the one hand, and amplitude equation models of pattern formation systems, on the other hand, suggests that Haken's principle also plays a role in the field of ecology. In fact, it has been proposed to see Haken's principle as counterpart to the competitive exclusion principle in ecology [104]. The competitive exclusion principle states under which conditions species can co-exist with each other [140, Chap. 8]. It has been argued that the relationship of the competition within species and the competition between species determines whether or not species can co-exist [20, Chap. 7]. Competition within species and competition between species may be interpreted as self-damping and coupling-induced damping. Consequently, according to the LVH model, co-existence is not possible if the competition between species is stronger than the competition within species (i.e., $g > 1$).

4.6.3 Stability Band

The LVH model describes pattern formation in reduced amplitude spaces spanned by the amplitudes of unstable basis patterns. Focusing on the winner-takes-all dynamics mentioned above, the LVH model exhibits fixed point attractors on the axes of reduced amplitude spaces. Those attractors describe that one of the amplitudes is finite, whereas all other amplitudes are at zero. However, not every amplitude or axis of the amplitude space necessarily possesses such a fixed point attractor.

Panel (b) of Fig. 4.20 shows an example of a phase portrait of a LVH model for two amplitudes A_1 and A_2. In the example, there is only one fixed point attractor located on the A_1 axis. The A_2 axis displays a fixed point repellor. In this example, for any initial state (except for the state given by the repellors at the origin and at $A_1 = 0, A_2 = 1$) the dynamics converges to the fixed point attractor at $A_1 = 4, A_2 = 0$. Therefore, eventually the pattern formation system exhibits an attractor pattern that involves basis pattern 1 as major component or is even identical to basis pattern 1. In any case, the attractor pattern does not contain basis pattern 2. In view of the phase portrait shown in panel (b), the question arises which winner-takes-all fixed point attractors exist and which not. More precisely, the objective is to derive a rule that tells us for which basis patterns fixed point attractors exist at non-zero pattern amplitudes. Which unstable basis patterns can emerge as dominant parts of attractor patterns, which not?

In this section, only the case will be considered in which a single coupling parameter is sufficient to describe interactions between amplitudes leading to nonlinear coupling-induced damping forces. This case hold when all amplitudes interact with each other in approximately the same way. Following Sect. 4.5.2, the parameter will be referred to as coupling parameter g. Moreover, we assume that $g > 1$ holds such that the LVH model describes a winner-takes-all dynamics. Under these conditions, fixed points of the LVH models with an arbitrary number of

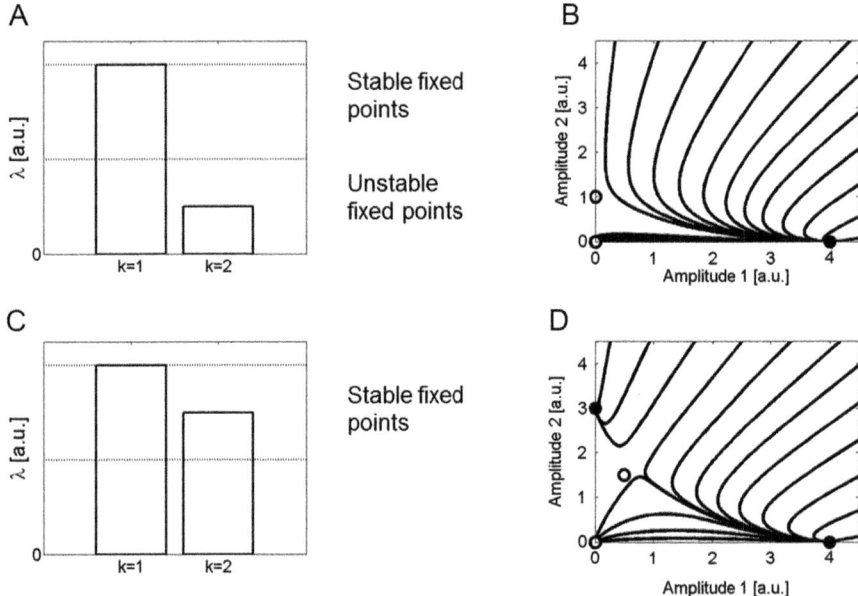

Fig. 4.20 Construction of the stability band for the special case of $m = 2$ unstable basis patterns. The LVH model is assumed to exhibit a coupling parameter $g = 2$. Panels (**a**) and (**c**) show two examples of eigenvalue spectrums. The top of the stability band is given by the largest eigenvalue λ_{max}. The lower boundary is given by λ_{max}/g, which for $g = 2$ is $\lambda_{max}/2$. The stability band is the band between the upper and lower boundaries (see dotted lines). In panel (**a**) only λ_1 falls into the stability band. The system is monostable. Panel (**b**) shows the corresponding phase portrait. In panel (**c**) both eigenvalues fall into the stability band. The system is bistable. Panel (**d**) shows the corresponding phase portrait

amplitudes (i.e., reduced amplitude spaces of arbitrary dimensions) can be discussed in a general way as follows.

On each axis of the reduced state space there is exactly one winner-takes-all fixed point at a non-zero amplitude value. That is, for each unstable basis pattern there is a fixed point such that the amplitude of the basis pattern assumes a finite value at that fixed point. However, that does not mean that the fixed point is necessarily a stable one. The fixed point can either correspond to a fixed point attractor in terms of a stable node or a repellor. The stability is determined by the eigenvalue spectrum of the LVH model and the magnitude of the coupling parameter. In order to determine the stability of a given fixed point, first the maximal eigenvalue of the spectrum is determined. Subsequently, the maximal eigenvalue is divided by the coupling parameter. This ratio of the maximal eigenvalue over the coupling parameter yields a lower bound value. Fixed points are attractors if their corresponding eigenvalues are larger than that lower bound value. Fixed point are repellors if they are smaller than the lower bound value [90, 98, 104, 120]. The aforementioned rule to determine the stability of fixed points can be visualized. In doing so, the stability of winner-takes-all fixed point can be determined graphically without any calculations. To this end, the concept of a stability band has been introduced [90, 96–98, 104, 106, 112, 121, 215]. First, all eigenvalues are plotted next to each other in a diagram. Subsequently, the maximal eigenvalue is divided by the coupling parameter (as mentioned above) and the value thus obtained, the lower boundary, is drawn as a horizontal line. In addition, a horizontal line is drawn at the maximal eigenvalue, which represents the upper boundary. The stability band is the region between the upper and lower boundary. All unstable basis pattern that have eigenvalues that fall into that stability band can emerge as attractor patterns (or as parts of attractor patterns in combination with several stable basis patterns). Roughly speaking, unstable basis patterns with eigenvalues in the stability band "exhibit" fixed point attractors. In contrast, amplitudes of unstable basis patterns with eigenvalues that fall below the lower boundary only exhibit fixed point repellors at finite stationary values. Therefore, basis patterns with eigenvalues that do not fall in the stability band do not emerge. Panel (a) of Fig. 4.20 shows the stability band for the LVH model from which the phase portrait in panel (b) is drawn. Only the eigenvalue λ_1 of the basis pattern 1 with amplitude A_1 is in the stability band. The other eigenvalue λ_2 does not fall in the stability band. Therefore, in the phase portrait shown in panel (b) there is only one fixed point attractor.

Let us assume there is a force that increases the eigenvalue λ_2 such that it falls into the stability band. Panel (c) shows a possible result of such a force-induced eigenvalue shift. Both eigenvalues fall into the stability band. Panel (d) displays the phase portrait of the LVH model for the eigenvalue spectrum shown in panel (c). There are two winner-takes-all fixed point attractors: one at $A_1 = 4, A_2 = 0$ and the other at $A_1 = 0, A_2 = 3$. Therefore, after a transient period, the pattern formation system is located either close to the first attractor at $A_1 = 4, A_2 = 0$ or close to the second attractor at $A_1 = 0, A_2 = 3$ units. At this stage, the pattern formation system exhibits an attractor pattern that is based either on the basis pattern 1 or on the basis pattern 2. Both attractor patterns are stable in the sense that if one of the two states (or attractor patterns) is perturbed then the perturbation decays and the pattern formation systems returns to the unperturbed state. Since the pattern formation system exhibits two stable attractor patterns it is called bistable. If the pattern formation system is initially close to the origin at $A_1 = 0, A_2 = 0$, then eventually one of the attractor patterns will emerge. Which pattern emerges depends on the precise initial state given in terms of the amplitudes A_1 and A_2. The phase portrait shown in panel (d) allows to identify roughly the basins of attraction of the two fixed point attractors and, in doing so, allows to classify initial states that will lead to the emergence of the attractor patterns with $A_1 = 4, A_2 = 0$ or $A_1 = 0, A_2 = 3$. For any initial state the amplitude equations of the LVH model can be solved numerically in order to determine which attractor pattern emerges for that given initial state.

4.6.4   Degree of Multistability and Eigenvalues as Pumping Parameters

From the graphical construction of the stability band it follows that the width of the stability band is determined by the magnitude of the coupling parameter. The width increases with the magnitude of the coupling parameter. Since the width is a rough measure for the degree of multistability of a pattern formation system, it follows that the degree of multistability increases when the strength of the coupling increases.

This observation might be counter-intuitive. However, it can be explained as follows. First of all, the eigenvalues of unstable basis patterns are positive and describe the exponential increase of amplitudes when they are initially relatively small. For this reason, they may be regarded as pumping parameters (see also Sect. 4.5.1). In fact, deriving amplitude equations from microscopic descriptions, eigenvalues can sometimes explicitly related to pumping sources (for lasers see e.g. Ref. [156]). Given two unstable basis patterns with different eigenvalues, where one eigenvalue is small and the other eigenvalue is large, then due to the coupling between the pattern amplitudes, and the fact that the system is pumped strongly in the direction of one basis pattern but not in the direction of the other basis pattern, typically (i.e., for not too larger coupling parameters g), the smaller eigenvalue drops out of the stability band such that corresponding unstable basis pattern exhibits an unstable fixed point and the associated attractor pattern cannot emerge. A mathematical analysis shows [120] that the key requirement for stability in this regard is that for the basis pattern with the smaller eigenvalue fluctuations that increase in magnitude the amplitude of the basis pattern with the larger eigenvalue must be sufficiently damped. That is, if the "weaker" basis pattern (i.e., pattern with the weaker pumping/smaller eigenvalue) can indeed damp the increase of the amplitude of the "stronger" basis pattern (i.e., pattern with stronger pumping/larger eigenvalue), then the "weaker" basis pattern can exhibit a stable fixed point. In this context, the coupling parameter g becomes important. If the coupling is strong, then even a "weak" pattern can induced a relatively large damping force. Therefore, for sufficiently large coupling parameters g it follows that a basis pattern 1 with small eigenvalue (i.e., weak pumping) can produce sufficiently large coupling-induced damping forces on the amplitudes of other basis patterns 2, 3, 4,... with large eigenvalues (i.e., strong pumping), which implies that pattern 1 has a stable fixed point and the eigenvalue λ_1 of the pattern 1 is the stability band. Therefore, the larger the parameter g is, the larger is the width of the stability band, and the larger is the degree of multistability.

4.6.5   Stationary Amplitudes and Eigenvalues

An unstable basis pattern with a small eigenvalue (i.e., weak pumping) exhibits an amplitude that increases initially in magnitude relatively slowly, whereas an

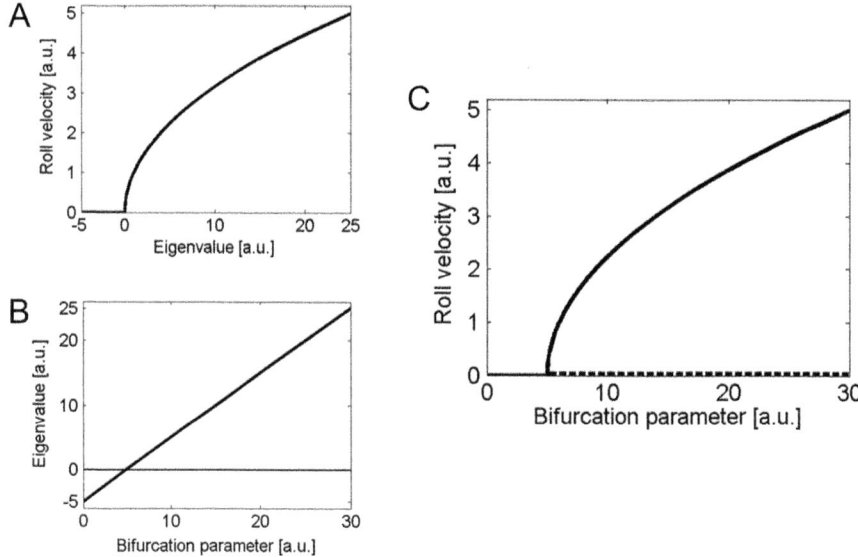

Fig. 4.21 Dependency of pattern amplitudes on eigenvalues and implication for bifurcation diagrams illustrated for the Benard experiment. Panel (**a**): Roll velocity (pattern amplitude) as function of the eigenvalue as computed form the LVH amplitude equations. Panel (**b**): Example of a relationship between bifurcation parameter and eigenvalue (here: a linear relationship is shown which in fact holds for the Benard instability). Panel (**c**): Bifurcation diagram obtained from merging panels (**a**) and (**b**)

unstable basis pattern with a large eigenvalue (i.e., strong pumping) exhibits an amplitude that increases initially in magnitude relatively fast. In other words, the eigenvalue of a basis pattern determines the degree of pumping specific to that basis pattern (see previous section).

In line with this observation, it does not come as a surprise that the eigenvalue (degree of pumping) of an unstable basis pattern also determines the magnitude of the amplitude in the stationary case, that is, at the fixed point attractor or repellor. If only positive-valued amplitudes are considered, then the stationary amplitude of a basis pattern is a monotonically increasing function of the eigenvalue. Figure 4.21 illustrates this monotonous relationship for the single-amplitude LVH model introduced in Sect. 4.5.5 that describes the rotation velocity of convection rolls in the Benard experiment (Sect. 3.2.3). Panel (a) shows the stationary amplitude given in terms of the roll velocity as function of the eigenvalue. Figure 4.21 also illustrates how this relationship between stationary amplitude and eigenvalue can be used to construct a bifurcation diagram. To this end, panel (b) exemplifies a possible relationship between the eigenvalue and the bifurcation parameter. In this example there is a linear relationship. Importantly, the eigenvalue becomes positive at a bifurcation parameter of 5 units. Only if the eigenvalue of a basis pattern is positive, then the amplitude of the basis pattern can exhibit a finite (i.e., non-zero) stable fixed point. For negative values of the eigenvalue the amplitude exhibits a stable fixed point at zero only. Consequently, taking panels (a) and (b) together, the stationary amplitude (i.e., the roll velocity) can be expressed as a function of the bifurcation parameter. This procedure leads to the bifurcation diagram shown in panel (c), which has been shown earlier in Fig. 3.5c, see Sect. 3.2.3.

4.6.6   Stationary Amplitudes and Nonlinearity Exponent

The LVH model involves two nonlinear terms: a self-damping term and a coupling-induced damping term. These terms are nonlinear functions of the amplitudes. The original Lotka-Volterra model as proposed in ecology features quadratic nonlinearities. In contrast, Haken's model for "perception" and other benchmark models for pattern formation exhibit cubic nonlinearities [22–24, 57, 58, 74, 76, 128, 129, 247, 252, 259, 260, 266, 292, 293, 308, 341, 354]. The LVH model exhibits a nonlinearity exponent that describes the type of nonlinearity. The nonlinearity parameter is a positive real-valued number q that can assume any value larger than 1. For $q = 2$ the LVH model recovers the quadratic nonlinearities as used in ecology. For $q = 3$ the LVH model features cubic nonlinearities as in the studies on pattern formation mentioned above. In general, the nonlinearities of the LVH amplitude equations are power law functions with exponent q. Importantly, the nonlinearity exponent q does not affect qualitatively the layout of the fixed points in the reduced amplitude space of a LVH model. That is, if for a given parameter set $\lambda_1, \ldots, \lambda_n, g, q$ including a particular nonlinearity exponent q the LVH model exhibits m fixed point attractors, then the nonlinearity exponent q can be varied in the whole range from unity to infinity and the LVH models thus obtained all exhibit m fixed point attractors as well. That is, all LVH models that differ only with respect to the nonlinearity exponent q are topological equivalent [98, 186], see also Sect. 4.7.10 below. The reason for this property is that the amplitudes of a LVH model with a particular exponent q can be mapped to the amplitudes of a LVH model with a different exponent q' by a power law function with exponent larger than zero. Such functions are monotonically increasing functions and describe one-to-one mappings. Therefore, solutions (given in terms of trajectories) of a LVH model for a given nonlinearity exponent can be used to construct the solutions of all other LVH models with different nonlinearity exponents.

While the exponent q does not have a qualitative impact on the LVH model and its solutions, the exponent does have a quantitative impact on the solutions. In particular, the magnitude of a stationary amplitude does not only depend on the respective eigenvalue as described in the previous section (Sect. 4.6.5) but also on the nonlinearity exponent. Figure 4.22 presents the stationary amplitude of a LVH model as function of its eigenvalue for several nonlinearity exponents. Note

Fig. 4.22 Stationary
amplitudes of LVH models
with different nonlinearity
exponents q as functions of
the eigenvalue λ. The cases
$q = 2, 3, 5$ are shown

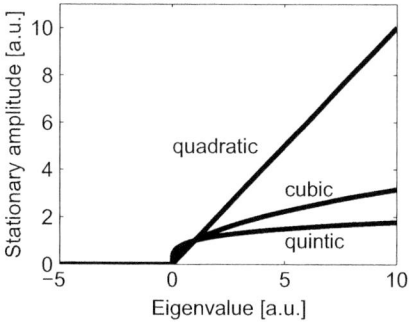

that Fig. 4.22 contains the graph shown in panel a of Fig. 4.21 as the special case
for a nonlinearity exponent $q = 3$. We will exploit these power law relationships
that are in-built into the LVH amplitude equations in Sect. 8.4, when discussing
psychophysical power laws.

4.6.7 Model Parameters Revisited

Table 4.5 summarizes the model parameters of the LVH model that have been
discussed so far and provides a short description of how they impact the formation
of patterns according to the LVH model.

Table 4.5 Overview of LVH
model parameters and their
roles/functions

Parameter	Function
Eigenvalues	Stability of fixed points
	Existence of fixed point attractors/repellors
	Magnitude of stationary amplitudes
Coupling parameter	Type of pattern formation system (winner-takes-all versus co-existence)
	Width of stability band[a]
Nonlinearity exponent	Model type (e.g., Lotka-Volterra versus Haken)
	Magnitude of stationary amplitudes

[a] Applies only to the winner-takes-all case

4.7   Mathematical Notes

4.7.1   Fixed Point, Attractor, and Repellor Patterns

In Sect. 3.6.1 attractors and repellors have been defined as sets W of states (points) that exhibit special features with respect to neighboring states. Fixed point, attractor, and repellor patterns as defined in Sect. 4.1 are given by trajectories on those attractors. Accordingly, fixed point, attractor, and repellor patterns $X(t)$ are defined by

$$X(t) = \Phi_t(X_0) \in W , \forall t \geq t_0 , \tag{4.1}$$

where $X_0 \in W$ is an initial state at time t_0 on the fixed point, attractor or repellor under consideration and Φ_t is the mapping of that initial state to the later state $X(t)$ due to the evolution equation $dX/dt = N(X)$ at hand. As such the initial state X_0 is equal to the special value $X(t_0)$ of the trajectory X: $X(t_0) = X_0$. Let us briefly discuss implications of the fact that the definition given by Eq. 4.1 involves this initial state $X(t_0)$. For fixed points, fixed point attractors, fixed point repellors, limit cycle attractors, and limit cycle repellors the patterns actually do not depend on that initial state or only depend on the initial state in terms of a phase shift (which will be shown next). For other attractors and repellors (e.g., chaotic attractors) this issue should be considered in more detail.

For fixed points, fixed point attractors, and fixed point repellors $W = \{X_{st}\}$ from Eq. (4.1) it follows that the corresponding patterns are defined by the locations of the attractors and repellors and fixed points

Fixed point, fixed point attractor, fixed point repellor $W = \{X_{st}\}$

$$\Rightarrow \text{ pattern } X(t) = X_{st}. \tag{4.2}$$

The initial state $X(t_0)$ also equals X_{st}. In other words, the location of the attractor, repellor, or fixed point under consideration defines the initial state. Therefore, the patterns does not depend on the initial state. For limit cycle attractors and repellors we have $X(t + T) = X(t)$, where T is the period of the limit cycle. Therefore, after one period T the state $X(t)$ has visited all points in W. It is sufficient to consider one period such that the pattern is described by

Limit cycle attractor, limit cycle repellor W

$$\Rightarrow \text{ pattern } X(t) = \Phi_t(X(t_0)) \in W , t \in [t_0, t_0 + T] . \tag{4.3}$$

Depending on the initial state $X(t_0) \in W$ the patterns $X(t)$ given in terms of functions of time look different. However, all patterns can be mapped onto each other by shifting the time appropriately. That is, limit cycle attractor and repellor patterns are uniquely defined except for an arbitrary time shift. That time shift is typically considered as the "phase" of the pattern.

4.7.2 Eigenvalues, Eigenvectors, Eigenfunctions, and Amplitude Space

Let $X(t)$ denote a vector or a field variable that satisfies the fundamental "symbolic" equation $dX/dt = N(X)$, see Eq. (1.1), which reads explicitly $d\mathbf{X}/dt = \mathbf{N}(\mathbf{X})$ for discrete systems and $\partial X/\partial t = N(\nabla, X(\mathbf{x}, t), \mathbf{x})$ for spatially extended systems, see Sect. 1.11.1. Fixed points (i.e., stationary states) X_{st} are defined by

$$N(X_{st}) = 0, \tag{4.4}$$

see also Sect. 3.6.1 for the special case of discrete systems. Let $u = X - X_{st}$ denote the relative state as seen from the fixed point X_{st}. If u is small in the amount (i.e., corresponds to a small perturbation), then $N(X)$ can be linearized at X_{st} and from $dX/dt = N(X)$ it follows that

$$\frac{d}{dt}u = L_N u \tag{4.5}$$

holds, where L_N is the linear operator derived from N. Equation (4.5) is a symbolic equation and reads explicitly $d\mathbf{u}/dt = \mathbf{L}_N\mathbf{u}$ for discrete systems and $\partial u/\partial t = L_N(\nabla, \mathbf{x})u(\mathbf{x}, t)$ for spatially extended systems. In any case, the right hand side of Eq. (4.5) is linear with respect to u. The operator L_N has basis patterns v_k that satisfy

$$L_N v_k = \lambda_k v_k \tag{4.6}$$

with $k = 1, 2, \ldots$, where λ_k are the eigenvalues of the basis patterns. For discrete systems involving n variables Eq. (4.6) reads $L_N\mathbf{v}_k = \lambda_k\mathbf{v}_k$ with $k = 1, \ldots, n$, where $\mathbf{v}_k$ are n-dimensional eigenvectors. For spatially extended systems Eq. (4.6) reads $L_N v_k(\mathbf{x}) = \lambda_k v_k(\mathbf{x})$, where $v_k(\mathbf{x})$ are eigenfunctions satisfying certain boundary conditions. Typically, there is an infinite but countable set of eigenfunctions such that $k = 1, 2, \ldots$.

As discussed in Sect. 4.2, eigenvectors and eigenfunctions are regarded as the elementary patterns or basis patterns of the human pattern formation system under consideration. Type I and II basis patterns are defined by

$$\text{Type I basis pattern } v_k : \lambda_k \in \mathbb{R} \implies \text{Imag}(\lambda_k) = 0, \tag{4.7}$$

$$\text{Type II basis patterns } v_k, v_j :$$

$$\text{Re}(\lambda_k) = \text{Re}(\lambda_j) \wedge \text{Imag}(\lambda_k) = -\text{Imag}(\lambda_j) \neq 0. \tag{4.8}$$

Type II basis patterns only occur in pairs.

Any relative state u can be expressed in terms of basis patterns by means of the superposition

$$u = \sum_k A_k v_k \ , \tag{4.9}$$

where A_k denote the pattern amplitudes (or basis pattern amplitudes). Likewise, any state X of the original state space can be expressed like

$$X = X_{st} + \sum_k A_k v_k \ . \tag{4.10}$$

The amplitude space is spanned by the pattern amplitudes A_k. For discrete systems involving n variables, there are n pattern amplitudes $A_1, \ldots, A_n$. For spatially extended systems there are typically infinitely many pattern amplitudes. For practical purposes (e.g., numerical simulations) only a finite number of basis patterns may be considered. For example, if the basis patterns correspond to Fourier functions, then Fourier functions with small periods may be neglected. Furthermore, frequently, amplitudes of eigenfunctions with negative eigenvalues that are large in the amount are neglected. If only a finite set of amplitudes is considered, then the amplitude space is spanned by the amplitudes $A_1, \ldots, A_{n(trunc)}$, where $n(trunc)$ is the index at which the infinitely large set of amplitudes is truncated.

Close to the fixed point X_{st} under consideration, Eq. (4.5) holds and from Eqs. (4.5), (4.6), (4.9) and (4.10) it follows that

$$u(t) = \sum_k A_k(t_0) v_k \exp\{\lambda_k (t - t_0)\} \ ,$$

$$X(t) = X_{st} + \sum_k A_k(t_0) v_k \exp\{\lambda_k (t - t_0)\} \tag{4.11}$$

holds, where $A_k(t_0)$ are the initial amplitudes satisfying

$$X(t_0) = X_{st} + \sum_k A_k(t_0) v_k \ . \tag{4.12}$$

4.7.3   Eigenvalue Characterization of Fixed Point Attractors and Repellors in Arbitrary Dimensions

From Eq. (4.11) and the definitions of fixed point attractors and repellors in Sect. 3.6.1 we can derive the classification scheme reported in Table 4.2. Moreover, in arbitrary dimensions fixed point attractors, fixed point repellors, and stable nodes can be characterized in terms of eigenvalues as shown in Table 4.6. Table 4.2 in

Table 4.6 Mathematical description of fixed point attractors and repellors in terms of eigenvalues for arbitrary dimensions n

Dimension	Attractor/repellor	Subtype	Eigenvalue(s)
n	Attractor		$\forall k \; : \; \{\lambda_k \in \mathbb{R} \wedge \lambda_k < 0\}$ or $\mathrm{Re}(\lambda_k) < 0$
n	Attractor	Stable node	$\forall k \; : \; \lambda_k \in \mathbb{R} \wedge \lambda_k < 0$
n	Repellor		$\exists k \; : \; \{\lambda_k \in \mathbb{R} \wedge \lambda_k > 0\}$ or $\mathrm{Re}(\lambda_k) > 0$

Sect. 4.2 is the verbal re-formulation of Table 4.6 for the special cases of $n = 1$ and $n = 2$.

4.7.4 Formal Derivation of Amplitude Equations

From Eq. (4.10) it follows that any trajectory $X(t)$ can be expressed in the amplitude space by means of time-dependent amplitudes $A_k(t)$ like

$$X(t) = X_{st,A} + \sum_k A_k(t)v_k , \qquad (4.13)$$

where $X_{st,A}$ denotes the location of a fixed point attractor or repellor A. Let us consider an initial state $X(t_0)$ close to $X_{st,A}$ and let us further assume that the attractor turns into a repellor, see Sect. 4.3. In this case, we have the initial amplitudes $A_k(t_0)$, see Eq. (4.12). Moreover, close to repellor A (i.e., for amplitudes A_k that are all small in the amount) Eq. (4.11) holds, which implies

$$A_k(t) = A_k(t_0) \exp\{\lambda_k(t - t_0)\} . \qquad (4.14)$$

At least one of the eigenvalues is positive or has a positive real value such that there is at least one amplitude that increases in magnitude exponentially. Let us assume that there is a fixed point attractor B at $X_{st,B}$ and that the corresponding attractor pattern B emerges. That is, the state $X(t)$ evolves to $X_{st}(B)$ for $t \rightarrow \infty$, which implies that

$$A_k(t) \rightarrow A_{k,st,B} \; : \; X_{st,B} = X_{st,A} + \sum_k A_{k,st,B}v_k . \qquad (4.15)$$

In summary, the dynamics of the amplitudes is characterized by the initial values $A_k(t_0)$ defined by Eq. (4.12) that are small in the amount, the exponential dynamics (4.14) that holds for amplitudes that are small in the amount, and the final amplitude values $A_{k,st,B}$ defined by Eq. (4.15).

The amplitude dynamics (4.14) cannot explain the approach of the amplitudes towards the stationary values $A_{k,st,B}$. Therefore, at issue is to derive evolution equations that describe the dynamics of $A_k(t)$ for $t \in [t_0, \infty)$. In order to derive these evolution equations, following Ref. [160], the superposition (4.13) is substituted into $dX/dt = N(X)$, see Eq. (1.1). For discrete, n-dimensional pattern formation systems this means that $\mathbf{X}(t) = \mathbf{X}_{st,A} + \sum_{k=1}^{n} A_k(t)\mathbf{v}_k$ is substituted into $d\mathbf{X}/dt = \mathbf{N}(\mathbf{X})$. For spatially extended pattern formation systems this means that $X(\mathbf{x},t) = X_{st,A}(\mathbf{x}) + \sum_{k=1}^{\infty} A_k(t)v_k(\mathbf{x})$ is substituted into $\partial X(\mathbf{x},t)/\partial t = N(\nabla, \mathbf{x}, X)$. In both cases, a set of coupled first-order differential equations for the amplitudes A_k is obtained that reads

$$\frac{d}{dt} A_k = N_{A,k}(A_1, A_2, \dots) , \qquad (4.16)$$

where $N_{A,k}$ are nonlinear functions of the amplitudes. For discrete, n-dimensional systems the functions $N_{A,k}$ depend on $A_1, \dots, A_n$. For spatially extended system the functions $N_{A,k}$ depend on an infinitely large set of amplitudes $A_1, A_2, \dots$. That set may be truncated at $n(trunc)$, see above. The nonlinear functions $N_{A,k}$ exhibit linear parts. Decomposing $N_{A,k}$ into their linear and "purely" nonlinear parts, Eq. (4.16) becomes [160]

$$\frac{d}{dt} A_k = \lambda_k A_k + G_k(A_1, A_2, \dots) , \qquad (4.17)$$

where λ_k denote the eigenvalues of the respective basis patterns and G_k are functions that do not exhibit any terms linear in any of the amplitudes $A_1, A_2, \dots$. The functions G_k may feature terms like A_1^2, A_1^3, $A_1 A_2$, $A_1 A_2 A_3$, etc. If we put the functions G_k for a moment equal to zero, then Eq. (4.17) describes the exponentially increasing or decreasing dynamics of amplitudes as given by Eq. (4.14). In other words, the linear terms (involving the eigenvalues λ_k) of the amplitude equation (4.17) determine the exponential increase and decrease of (small) amplitudes for states close to A. In contrast, the functions G_k determine the approach towards the attractor B, that is, the convergence of $A_k(t)$ to $A_{k,st,B}$. The coupled set of differential equations (4.17) describes the amplitude equations of pattern formation systems, in general, and human pattern formation systems, in particular. The amplitude equations describe the formation of patterns from reference points given in terms of fixed points (in our example, the fixed point A). The functions G_k as well as the eigenvalues λ_k depend on the particular system at hand. As we have shown these functions can in principle be derived explicitly from bottom-up (microscopic) approaches that start with evolution equations of the form $dX/dt = N(X)$, see Eq. (1.1).

4.7.5   Reduced Amplitude Equations: Cases I, II, III

Let us consider a bifurcation parameter α that becomes critical at α_c. Let us consider case I discussed in Sect. 4.3.1 of a single real-valued eigenvalue that becomes positive at the bifurcation point $\alpha = \alpha_c$. Without loss of generality, let us assume

λ_1 becomes zero at $\alpha = \alpha_c$ and switches from $\lambda_1 < 0$ to $\lambda_1 > 0$ if α is increased beyond α_c. For λ_j with $j = 2, 3, \ldots$ we have $\lambda_j < 0$. Eq. (4.15) holds in general. Since for $\alpha = \alpha_c + \epsilon$ with $\epsilon > 0$ and ϵ small we have $\lambda_1 > 0$ and $\lambda_j < 0$ for $j = 2, 3, \ldots$ only A_1 will increase in magnitude for small amplitudes. Therefore, as mentioned in the main text, the new fixed point attractor B will be in the direction of the basis pattern v_1 such that

$$|A_{1,st,B}| \gg |A_{k,st,B}| \; \forall \, k = 2, 3, \ldots \tag{4.18}$$

Equation (4.18) describes in mathematical terms what has been argued in the main text, namely, that the amplitude of the unstable basis pattern makes the major contribution to the emerging attractor pattern B (see, e.g., Sects. 4.3.2, 4.3.5 and 4.3.7). Second for $\alpha = \alpha_c + \epsilon$, where $\epsilon > 0$ is a small number, we have

$$\lambda_1 \ll |\lambda_k| \; \forall \, k = 2, 3, \ldots \tag{4.19}$$

because λ_1 passes zero at $\alpha = \alpha_c$, whereas all other eigenvalues are negative and finite at $\alpha = \alpha_c$ (for an illustration involving only two eigenvalues λ_1 and λ_2 see Fig. 4.7b, for several eigenvalues see Fig. 4.9). From Eqs. (4.17) and (4.19) it follows that the stable amplitudes A_k with $k = 2, 3, \ldots$ evolve relatively fast as compared to A_1. Using the so-called adiabatic elimination technique [107, 160], the left-hand-side of Eq. (4.17) can be put equal to zero such that

$$k = 2, 3, \ldots \; : \; A_k = \frac{G_k(A_1, A_2, \ldots)}{|\lambda_k|} \; . \tag{4.20}$$

That is, the stable amplitudes A_k can be expressed in terms of mappings from the unstable amplitude A_1 and all unstable amplitudes A_j with $j = 2, 3, \ldots$. Note that in Eq. (4.20) we restrict ourselves to consider real-valued eigenvalues λ_k only. Moreover, note that the mappings in Eq. (4.20) are implicit definitions because A_k appears on the left and right hand sides of the equals sign. Nevertheless, the mappings can frequently be solved for A_k and, in doing so, the amplitudes A_k can be expressed as functions f_k of the only "independent" variable A_1 such that

$$k = 2, 3, \ldots \; : \; A_k = f_k(A_1) \; . \tag{4.21}$$

The step from Eqs. (4.20) to (4.21) may become mathematically difficult, when dealing with an infinitely large set of amplitudes. As mentioned above, the set may be truncated (e.g., for eigenvalues λ_k that are large in magnitude). In particular, Eq. (4.20) suggests that if $|\lambda_k|$ is large for a given index k, then the value A_k on the left hand side is small in magnitude irrespective of the function G_k. Substituting Eq. (4.21) into Eq. (4.17) for $k = 1$, we obtain

$$\frac{\mathrm{d}}{\mathrm{d}t} A_1 = \lambda_1 A_1 + G_{1,eff}(A_1) \tag{4.22}$$

with the effective "purely" nonlinear function

$$G_{1,eff}(A_1) = G_1(A_1, f_2(A_1), f_3(A_1), \ldots) . \tag{4.23}$$

The function $G_{1,eff}(A_1)$ does not involve a term linear in A_1. It involves terms like A_1^2, A_1^3, A_1^4, etc. Equation (4.22) is a closed description for the unstable amplitude or order parameter amplitude A_1. The pattern formation phenomenon of interest can be described by solving Eq. (4.22) and the mappings (4.21). Since the mappings (4.21) do not involve time (by construction) the dynamical system of interest becomes one-dimensional. It is sufficient to study the emergence of the pattern of interest in a one-dimensional space spanned by A_1. This is the reduced amplitude space introduced in Sect. 4.3.4.

Cases II and III addressed in Sects. 4.4.1 and 4.4.2, respectively, can be treated by analogy. Case II addresses a set of eigenvalues that assume the same value for any bifurcation parameter α. In particular, they all become zero at $\alpha = \alpha_c$ and positive when α exceeds the critical value α_c. Without loss of generality, the indices of the basis patterns and eigenvalues can be arranged in such a way that the indices $k = 1, \ldots, m$ refer to the unstable basis patterns with $\lambda_k = \lambda(\alpha) \ \forall \ k = 1, \ldots, m$, while $k = m + 1, m + 2, \ldots$ denote the stable basis patterns with $\lambda_k < 0$ at $\alpha \leq \alpha_c + \epsilon$, where $\epsilon > 0$ is a finite value that might be small. That is, the eigenvalues of the stable basis patterns are all negative at the bifurcation point and remain negative when the bifurcation parameter is increased slightly above the critical value. In case II, Eq. (4.20) holds for $k = m + 1, m + 2, \ldots$ such that Eq. (4.21) becomes

$$k = m + 1, m + 2, \ldots \ : \ A_k = f_k(A_1, \ldots, A_m) . \tag{4.24}$$

Likewise, Eq. (4.22) becomes a set of m coupled first-order differential equations for the unstable amplitudes $A_1, \ldots, A_m$. Explicitly, we obtain

$$k = 1, \ldots, m \ : \ \frac{\mathrm{d}}{\mathrm{d}t} A_k = \lambda A_k + G_{k,eff}(A_1, \ldots, A_m) . \tag{4.25}$$

Equation (4.25) involves m effective "purely" nonlinear functions $G_{k,eff}$. These functions do not contain linear terms in $A_1, \ldots, A_m$. They may involve terms like $A_1^2, A_1 A_2$, etc. They do not involve stable amplitude A_k with $k > m$. As indicated in Eq. (4.25) all eigenvalues of the unstable basis patterns $k = 1, \ldots, m$ are equal and assume the value λ. The functions $G_{k,eff}$ may feature a similar symmetry property (for an example, see the Lotka-Volterra-Haken model below). Equation (4.25) provides a closed description for the unstable amplitudes (order parameter amplitudes) $A_1, \ldots, A_m$. The pattern formation phenomenon of interest can be described by solving Eq. (4.25) and the mappings (4.24). Only the m-dimensional set of amplitude equations accounts for the temporal evolution of the phenomenon. Therefore, it is sufficient to study the formation of the pattern under consideration in the reduced, m-dimensional amplitude space spanned by $A_1, \ldots, A_m$. Note that just as in case I the amplitudes of the stable basis patterns are assumed to make only minor contributions to the emerging attractor pattern. That is, Eq. (4.18) generalized for case II to

$$|A_{j/max,st,B}| \gg |A_{k,st,B}| \; \forall \, k > m \; , \tag{4.26}$$

where $A_{j/max,st,B}$ is the amplitude that is largest in the amount of all unstable amplitudes.

Let us illustrate the meaning of $A_{j/max,st,B}$. For example, if the unstable basis patterns are mutually exclusive, we may have $A_{1,st,B1} > 0$ and $A_{k,st,B1} = 0$ for $k = 2, \ldots, m$ for an attractor pattern B1 and $A_{2,st,B2} > 0$ and $A_{k,st,B1} = 0$ for $k = 1, 3, \ldots, m$ for an attractor pattern B2. Then, in the case of the attractor pattern B1 we have $A_{j/max,st,B} = A_{1,st,B1}$, whereas in the case of the attractor pattern B2 we have $A_{j/max,st,B} = A_{2,st,B2}$.

Finally, case III differs with respect to case II only in the fact that the eigenvalues of the unstable basis patterns assume different values. Therefore, in case III the reduced amplitude equations (4.25) are replaced by

$$k = 1, \ldots, m \; : \; \frac{d}{dt} A_k = \lambda_k A_k + G_{k,eff}(A_1, \ldots, A_m) \; . \tag{4.27}$$

Case III includes cases I and II as special cases. If we put $\lambda_k = \lambda$ in Eq. (4.27) then we re-obtain Eq. (4.25), that is the model of case II. If we put $m = 1$ then there is a single unstable amplitude. In this case both cases II and III reduce to case I.

4.7.6 Dependencies of Stable and Unstable Amplitudes

In Sect. 4.3 the dependency of stable amplitudes on unstable amplitudes and, vice versa the dependency of unstable amplitudes on stable amplitudes was addressed. Let us address these issues in the context of the mathematical framework developed in the previous section. To this end, it is sufficient to consider the most general case III because case III includes cases I and II as special cases.

Let us first address the two qualitatively different scenarios for the dependency of unstable amplitudes on stable amplitudes: unstable amplitudes either do or do not depend on stable amplitudes. If unstable amplitudes are not coupled to (i.e., do not depend on) stable amplitudes, then Eq. (4.17) reads

$$\frac{d}{dt} A_k = \lambda_k A_k + G_k(A_1, \ldots, A_m) \tag{4.28}$$

for $k = 1, \ldots, m$. Comparing Eq. (4.28) with the case III Eq. (4.27) of the reduced amplitude space, we see that

$$G_k = G_{k,eff} \; . \tag{4.29}$$

That is, the "purely" nonlinear functions G_k introduced in Eq. (4.17) in the context of the full (i.e., not the reduced) amplitude space correspond already to the effective "purely" nonlinear functions $G_{k,eff}$. Consequently, the damping forces that lead to a saturation of the unstable amplitudes can be only due to a self-damping like $-A_k^2$ or $-A_k^3$ or due to a coupling-induced damping involving other unstable amplitudes like $-A_k A_j$ with $j \neq k$ and $j \in \{1, \ldots, m\}$.

In contrast, if the unstable amplitudes are coupled to the stable amplitudes amplitudes then we need to follow the procedure outlined above and substitute the amplitudes A_k with $k > m$ in $G_1, \ldots, G_m$ by the functions $f_k(A_1, \ldots, A_m)$ (i.e., we replace in $G_1, \ldots, G_m$ the variables $A_{m+1}, A_{m+2}, \ldots$ by $f_{m+1}, f_{m+2}, \ldots$). As a result, we obtain the effective "purely" nonlinear functions $G_{1,eff}, \ldots, G_{m,eff}$, which are not identical to the functions $G_1, \ldots, G_m$. In this case, damping of the unstable amplitudes can be due to the two aforementioned mechanisms (coupling-induced damping involving unstable amplitudes and self-damping). In addition, unstable amplitudes may be damped due to coupling with the stable amplitudes. For example, a function G_k for $k = 1, \ldots, m$ may involve a term like $-A_k A_j$ with $j > m$. This term would become a term $-A_k f_j(A_1, \ldots, A_m)$ in the effective function $G_{k,eff}$. This example also illustrates that on the level of the reduced amplitude equations (4.27), it cannot be seen whether or not damping is due to a coupling between unstable and stable amplitudes. For example, for $m = 2$ unstable amplitudes terms like $-A_1^2$ and $-A_1 A_2$ that occur on the right hand side of the evolution equation for A_1 might describe self-damping $(-A_1^2)$ or coupling-induced damping by means of another unstable amplitude $(-A_1 A_2)$. However, it is also possible that they reflect damping due to the coupling of A_1 with stable amplitudes $A_3, A_4, \ldots$ that in turn depend on A_1 and A_2.

Next, let us address the two qualitatively different scenarios for the dependency of stable amplitudes on unstable amplitudes: stable amplitudes either do or do not depend on unstable amplitudes. If stable amplitudes are not coupled to unstable amplitudes and in that sense do not depend on unstable amplitudes, then they converge to zero for any initial state close to the reference fixed point. The reason for this is that they all feature negative eigenvalues or complex-values eigenvalues with negative real parts. Let us demonstrate this issue explicitly. If stable amplitudes are not coupled to unstable amplitudes then from Eq. (4.17) it follows that

$$\forall\, k > m \; : \; \frac{\mathrm{d}}{\mathrm{d}t} A_k = \lambda_k A_k + G_k(A_{m+1}, A_{m_2}, \ldots) \tag{4.30}$$

with $\lambda_k \in \mathbb{R} \; \wedge \; \lambda_k < 0$ or $\mathrm{Re}(\lambda_k) < 0$. This is closed description for an attractor in the (infinitely large) subspace spanned by A_k with $k > m$. For any small perturbation out of the "origin" $(A_{m+1}, A_{m+2}, \ldots) = (0, 0, \ldots)$ the state will return to that origin such that $A_k = 0$ for all stable amplitudes $k > m$. Consequently, in the context of pattern formation at bifurcation points as described above, stable amplitudes that are not affected by unstable amplitudes do not contribute to the formation of patterns.

In contrast, if stable patterns depend on unstable patterns, then they may contribute with finite (i.e., non-zero) values to emerging attractor patterns. The values of those stationary stable amplitudes are given by Eq. (4.24) for stationary values of the unstable amplitudes. For example, if the fixed point attractor pattern B with location $X_{st,B}$ and unstable amplitudes $A_{1,st,B}, \ldots, A_{m,st,B}$ emerges then the stable amplitudes are given by

$$k = m+1, m+2, \ldots : \quad A_{k,st,B} = f_k(A_{1,st,B}, \ldots, A_{m,st,B}) \,. \tag{4.31}$$

The case of circular causality addressed in Sect. 4.3.8 (i.e., the slaving principle) applies if both the unstable amplitudes are coupled to the stable amplitudes and the stable amplitudes are coupled to the unstable amplitudes. In this case, the stable amplitudes can increase due to pumping by means of the unstable amplitudes. If so, they assume finite values as described by Eqs. (4.24) and (4.31). At the same time, the stable amplitudes may be part of coupling-induced damping forces that stop the increase of the unstable amplitudes (see Sect. 4.3.8 and Fig. 4.11). The coupling-induced damping terms may correspond, for example, to expressions like $-A_k A_j$, $-A_k^2 A_j$, $-A_k A_j^2$ with $k \in \{1, \ldots, m\}$ and $j > m$. A worked-out example in this regard is the Haken-Zwanzig model [160] with $n = 2$, $m = 1$, $\lambda_1 > 0$, $\lambda_2 < 0$ that reads

$$\frac{d}{dt} A_1 = \lambda_1 A_1 - A_1 A_2 \,, \quad \frac{d}{dt} A_2 = \lambda_2 A_2 + A_1^2 \,. \tag{4.32}$$

For the Haken-Zwanzig model, we obtain

$$G_2 = A_1^2 \Rightarrow A_2 = f_2(A_1) = G_2/|\lambda_2| = A_1^2/|\lambda_2| \tag{4.33}$$

and

$$G_1 = -A_1 A_2 \Rightarrow G_{1,eff} = -A_1 f_2(A_1) = -A_1^3/|\lambda_2| \tag{4.34}$$

such that the reduced amplitude space description $dA_1/dt = \lambda_1 A_1 + G_{1,eff}$ is given by

$$\frac{d}{dt} A_1 = \lambda_1 A_1 - \frac{1}{|\lambda_2|} A_1^3 \,. \tag{4.35}$$

The amplitude equation for A_1 describes a pitchfork bifurcation (see Sect. 3.6.2 and Eq. (3.10)).

4.7.7 Lotka-Volterra-Haken Amplitude Equations

Definition of the Model Equations

The Lotka-Volterra-Haken (LVH) amplitude equations are a special case of the case III amplitude equations (4.27) describing pattern formation in m-dimensional reduced amplitude spaces. They are defined by Eq. (4.27) for

$$G_{k,eff} = -BA_k \sum_{j=1, j\neq k}^{m} A_j^{q-1} - CA_k^q \tag{4.36}$$

with $B, C > 0$. As indicated in Eq. (4.36), the sum in the B-term of $G_{k,eff}$ runs over all m indices $j = 1, \ldots, m$ except for the index k. For example, for $k = 1$ we have $-BA_1 \sum_{j=2,\ldots,m} A_j^{q-1}$ and for $k = 2$ we have $-BA_2(A_1^{q-1} + \sum_{j=3,\ldots,m} A_j^{q-1})$. Substituting Eq. (4.36) into Eq. (4.27) the LVH model reads explicitly [107, 109, 114, 186, 187]

$$\frac{d}{dt} A_k = \lambda_k A_k - BA_k \sum_{j=1, j\neq k}^{m} A_j^{q-1} - CA_k^q \tag{4.37}$$

for $k = 1, \ldots, m$. The parameter $q > 0$ is the nonlinearity exponent. B and C describe the strength of the coupling-induced damping (B) and the self-damping (C). Note that the phrases coupling-induced damping and self-damping only refer to the level of the reduced amplitude equations. As shown in the previous section, the B- and C-terms may reflect interactions involving stable amplitudes that do not occur in the reduced amplitude equations description.

To reduce the number of parameters it has been suggested to rescale the amplitude equations (4.37) by dividing left and right hand sides with the parameter C and rescaling time [215]. In doing so, Eq. (4.37) becomes $dA_k/dt' = \lambda'_k A_k - gA_k \sum_{j=1, j\neq k}^{m} A_j^{q-1} - A_k^q$ with $g = B/C$, $t' = tC$, $\lambda'_k = \lambda_k/C$. To improve readability, in what follows the prime will be dropped. Accordingly, the LVH amplitude equations in the rescaled form read

$$\frac{d}{dt} A_k = \lambda_k A_k - gA_k \sum_{j=1, j\neq k}^{m} A_j^{q-1} - A_k^q \tag{4.38}$$

for $k = 1, \ldots, m$. For the special case of $m = 1$ and $q = 3$, Eq. (4.38) reduces to

$$\frac{d}{dt} A = \lambda A - A^3 \tag{4.39}$$

and describes a pitchfork bifurcation.

Mathematics of Winner-Takes-All Fixed Points: Qualitative Aspects

It can be shown that the LVH model exhibits only fixed points attractors and repellors [90, 104, 157]. That is, Eqs. (4.38) do not exhibits limit cycles or chaotic attractors. Fixed point attractors and repellors can either be co-existence fixed points or winner-takes-all fixed points. Winner-takes-all fixed points are defined by

$$\text{Winner-takes-all fixed point } k \; : $$
$$A_{k,st} \neq 0 \; \wedge \; A_{j,st} = 0 \; \forall j \in \{1, \dots, m\} \neq k \; . \tag{4.40}$$

That is, only one amplitude is finite. All remaining amplitudes equal zero. Winner-takes-all fixed points are nodes or saddle points. That is, the LVH model does not exhibit winner-takes-all fixed points with complex-valued eigenvalues at those fixed points [120]. Winner-takes-all fixed points can be stable nodes, unstable nodes, and saddle points. If they correspond to stable nodes, then they describe fixed point attractors. In the remaining cases, they describe fixed point repellors. The stability of winner-takes-all fixed points can be determined by means of the stability band.

Haken's Principle for Winner-Takes-All Dynamics and the Coupling Parameter

The parameter g denotes the coupling parameter introduced in the main text. The coupling parameter measures the strength of coupling-induced damping in units of self-damping. From Haken's principle (Sect. 4.6.2) for a winner-takes-all dynamics it follows that for $B > C$ (or $g > 1$), the LVH model does not exhibit stable co-existence fixed points. If co-existence fixed points exist, then they correspond to repellors. Vice versa, if attractors exist then they involve only a single finite amplitude. In other words, the LVH model only exhibits attractors in form of winner-takes-all fixed point attractors.

4.7.8 Mathematics of the Stability Band

In general, winner-takes-all fixed points can be attractors or repellors. As discussed in Sect. 4.6.3, the stability of winner-takes-all fixed points can be determined by means of the stability band. Let us provide some mathematical details in this regard. The stability band is defined by an upper bound λ_{max} and a lower bound, the critical value λ_c. The upper bound λ_{max} corresponds to the largest eigenvalue

$$\lambda_{max} = \max\{\lambda_1, \dots, \lambda_m\} \; . \tag{4.41}$$

The lower bound is defined by

$$\lambda_c = \frac{\lambda_{\max}}{g} \; . \tag{4.42}$$

The lower bound is the critical bound (whence the subindex "c"). Note that the stability band applies to the LVH model as a winner-takes-all dynamics. That is, we have $g > 1$, which implies

$$\lambda_c < \lambda_{\max} \; . \tag{4.43}$$

Winner-takes-all fixed points k are stable if $\lambda_k > \lambda_c$ and unstable if $\lambda_k < \lambda_c$ [90, 98, 104, 120]. Since in applications of the LVH model this relationship is frequently used, we present it in form of the following equations

$$\lambda_k > \lambda_c = \frac{\lambda_{\max}}{g} \Rightarrow \text{winner-takes-all fixed point } k \text{ stable} \; ,$$

$$\lambda_k < \lambda_c = \frac{\lambda_{\max}}{g} \Rightarrow \text{winner-takes-all fixed point } k \text{ unstable} \; . \tag{4.44}$$

Eigenvalues with $\lambda_c < \lambda_k \leq \lambda_{\max}$ are located within the stability band. The corresponding winner-takes-all fixed points are attractors (in form of stable nodes). Eigenvalues with $\lambda_k < \lambda_c$ do not belong to the stability band. The corresponding winner-takes-all fixed points are repellors (in form of saddle points and unstable nodes).

4.7.9 Mathematics of Attractor Patterns

Above and in the main text the notations $X_{st,A}$ and $X_{st,B}$ have been used to refer to fixed points A and B or fixed point attractors or repellors. Let $X_{st,k}$ denote fixed points in the original state space Ω_X (which is $\mathbb{R}^n$ for discrete systems and a suitably defined space of functions for spatially extended systems) of a pattern formation system. Then the question arises how can a fixed point $X_{st,k}$ expressed in the context of a pattern formation system be described in the reduced amplitude space by a LVH model with m unstable amplitudes $A_1, \ldots, A_m$. The amplitudes are the pattern amplitudes of the unstable basis patterns v_k. In line with the LVH model as a winner-takes-all model, the basis patterns are assumed to be mutually exclusive. As discussed in Sects. 4.3.2, 4.3.5 and 4.3.7, a given attractor pattern k can either be identical to its corresponding unstable basis pattern v_k or involve the unstable basis pattern v_k and a number of stable basis patterns v_j with $j > m$. In the first case, the attractor pattern is completely described by means of the amplitude $A_{k,st}$ of the stable winner-takes-all fixed point k of the LVH model. We have

$$X_{st,k} = X_{st,A} + A_{k,st} v_k . \tag{4.45}$$

In the second case, the attractor pattern is given by

$$X_{st,k} = X_{st,A} + A_{k,st} v_k + \sum_{j \geq m+1} A_{j,st} v_j . \tag{4.46}$$

The sum in Eq. (4.46) runs from $j = m + 1$ to $j = n$ in the case of discrete pattern formation systems and from $j = m + 1$ to $j = n(trunc)$ or goes to infinity in the case of spatially extended systems. Using Eq. (4.24) in the stationary case (for an example see Eq. (4.31)), we obtain

$$X_{st,k} = X_{st,A} + A_{k,st} v_k + \sum_{j \geq m+1} f_j(A_{k,st}) v_j . \tag{4.47}$$

Note that Eqs. (4.46) and (4.47) include Eq. (4.45) as special case if the summation operator $\sum (\cdot)$ does not run over any elements.

4.7.10   Nonlinearity Exponent and Quantitative Aspects of Winner-Takes-All Fixed Points

Equivalence of LVH Models with Different Nonlinearity Exponents

The LVH model (4.38) for m amplitudes $A_1, \ldots, A_m$ and quadratic nonlinearities (i.e., $q = 2$) reads

$$\frac{d}{dt} A_k = \lambda_k A_k - g A_k \sum_{j=1, j \neq k}^{m} A_j - A_k^2 . \tag{4.48}$$

Let

$$\frac{d}{d\tau} B_k = \tilde{\lambda}_k B_k - g B_k \sum_{j=1, j \neq k}^{m} B_j^{q-1} - B_k^q \tag{4.49}$$

describe another LVH model for m amplitudes $B_1, \ldots, B_m$ and eigenvalues $\tilde{\lambda}_k$. The amplitudes B_k evolve in time, where time is denoted by τ. The second LVH model (4.49) involves the nonlinearity exponent $q > 1$ but $q \neq 2$. Both models are assumed to exhibit the same coupling parameter g. Importantly, the second model describes the evolution of the amplitudes B_k by means of the time variable τ. In general, the time variable τ is not identical to the time variable t. It can be shown that any solution $\mathbf{B}(\tau) = (B_1(\tau), \ldots, B_m(\tau))$ corresponds to a solution $\mathbf{A}(t) = (A_1(t), \ldots, A_m(t))$ if we put

$$A_k(t) = \frac{1}{q-1}[B_k(\tau)]^{(q-1)} , \ \tau = \frac{t}{(q-1)^2} , \ \tilde{\lambda}_k = (q-1)\lambda_k . \tag{4.50}$$

That is, any LVH model with an arbitrary nonlinearity exponent $q > 1$ can be mapped to the quadratic model by means of a variable transformation and a rescaling of the eigenvalues and the time variable. That also means that two models with exponents $q > 1$ and $q' > 1$ and $q' \neq q$ can be mapped into each other (e.g., we may have $q = 3$ and $q' = 4$). The rescaling of the eigenvalues does not affect the relationship between the eigenvalues (i.e., the ratios of eigenvalues are not affected by the rescaling procedure). However, the ratios of eigenvalues determine the stability of winner-takes-all fixed points (e.g., if $\lambda_k/\lambda_{max} > 1/g$ holds then the eigenvalue λ_k is in the stability band and the corresponding fixed point describes and attractor, whereas if $\lambda_k/\lambda_{max} < 1/g$ holds, then we deal with a repellor, see Eq. (4.44)). Consequently, if a model with exponent q exhibits a winner-takes-all fixed point attractor k, then the corresponding model with exponent q' exhibits a winner-takes-all fixed point attractor k as well. In general, all models exhibit qualitatively the same phase portraits in their m-dimensional reduced amplitude spaces (i.e., the models are topologically equivalent) [98, 186].

Stationary Amplitudes of Winner-Takes-All Fixed Points

Substituting a winner-takes-all fixed point k defined by Eq. (4.40) into the LVH model (4.38) yields the stationary amplitudes

$$A_{k,st} = \lambda_k^{1/(q-1)} \tag{4.51}$$

and $A_{j,st} = 0$ for $j \neq k$. In Eq. (4.51) we restrict ourselves to positive-valued amplitudes $A_k \geq 0$. That is, the reduced amplitude space is spanned only by the positive half axes $A_1 \geq 0, A_2 \geq 0, \ldots, A_m \geq 0$. Equation (4.51) states that the finite amplitudes of winner-takes-all fixed points are monotonically increasing functions of their eigenvalues. The functions are power-law functions with exponents $z = 1/(q - 1)$. For $q > 2$, $q = 2$ (quadratic nonlinearities, Lotka-Volterra case), and $q \in (1, 2)$ we have

$$q > 2 \ : \ A_{k,st} = \lambda_k^z , \ z = 1/(q - 1) < 1 ,$$
$$q = 2 \ : \ A_{k,st} = \lambda_k ,$$
$$q \in (1, 2) \ : \ A_{k,st} = \lambda_k^z , \ z = 1/(q - 1) > 1 . \tag{4.52}$$

That is, for $q > 2$ we have de-accelerating monotonically increasing functions (like the square-root function). For $q = 2$ the amplitude increases linearly with the eigenvalue. For $q \in (1, 2)$ we have accelerating monotonically increasing functions. In many applications (see main text) the cubic case $q = 3$ is of interest. In this case, we have

$$q = 3 \; : \; A_{k,st} = \sqrt{\lambda_k} \; . \tag{4.53}$$

That is, the amplitude is given by a square root function of its eigenvalue. The curves shown in Fig. 4.22 have been computed from Eq. (4.51) for $q = 2$ (quadratic case), $q = 3$ (cubic case), and $q = 5$ (quintic case).

For particular nonlinearity exponents q, we may need to consider stationary solutions in addition to those described by Eq. (4.51). For example, for $q = 3$ the m-dimensional reduced amplitude space can be taken as $\mathbb{R}^m$. In this case winner-takes-all fixed points are located on the amplitude axes at $A_{k,st} = \pm\sqrt{\lambda_k}$ and $A_{j,st} = 0$ for $j \neq k$. That is, in this case we may take negative values for stationary amplitudes into account as well.

Chapter 5
Human Reactions

In this chapter the first and second laws of classical mechanics will be reviewed. Subsequently, in analogy to those laws, the first and second laws of self-organizing, pattern formation systems will be introduced. These laws suggest to conduct the study of human brain and body activity in a particular order. Accordingly, first, the brain activity dynamics of the idealized isolated human pattern formation system should be considered. Subsequently, the human pattern formation system in contact with its environment, that is, under the impact of external forces, should be examined. In this context and within the framework of the human pattern formation reaction model introduced in Sect. 1.6, a systematic classification of human pattern formation systems into classes and subclasses will be presented. In doing so, the first part of this chapter is about the laws, forces, and components involved in the pattern formation of the human system.

The second part of this chapter addresses a question that has not yet been addressed to some extent. The question arises how attractor patterns should be labeled at all. While it seems obvious to talk about a walking pattern when an individual is walking, the question about the labeling of patterns becomes more difficult in the context of the isolated human system. For example, how should verbal thoughts or the imagination of objects be addressed in terms of brain activity (BA) attractor patterns? In order to answer this question another fundamental property of attractors will be introduced: attractors have basins of attraction. Using the symmetry property of pattern formation systems addressed in Chap. 4 and the concept of basins of attraction, attractors can be labeled in a systematic way in terms of external forces. In this context, it will be pointed out that attractor patterns that emerge under the impact of external forces (and are labeled accordingly) are likely to emerge as well (as presumably modified variants) in the absence of such forces, which allows us to label BA patterns that emerge in the absence of forces.

Finally, note that the chapter title, human reactions, is somewhat misleading. This chapter title has the advantage of being short. However, this chapter is about humans as well as animals. In addition, this chapter is not only about human reactions to

© Springer Nature Switzerland AG 2019

T. Frank, *Determinism and Self-Organization of Human Perception and Performance*, Springer Series in Synergetics, https://doi.org/10.1007/978-3-030-28821-1_5

external forces but also about the brain activity dynamics of the isolated human and
animal system.

5.1   First and Second Laws

5.1.1   First and Second Laws of Classical Mechanics

The first Newtonian law of classical mechanics addresses system that are not
subjected to any forces. In our context, it is not only interesting to point out what
the first Newtonian law states explicitly. Rather, it is important to see what the
law is about. To re-iterate the first law is about systems in the absence of forces.
For sake of simplicity, in what follows, a single point-like particle with a certain
mass will be considered as the system under consideration. The particle is located
at every time point at some location in a three-dimensional space spanned by the
coordinates x, y, z. The first law addresses the situation when the particle is isolated
from everything else in the sense that there are no forces that act on the particle.
Note that this notion of being isolated is an idealization. For example, on our planet
Earth there is no such place at which a particle with mass would be free of forces.
At any place on our planet, a particle is subjected to gravitational forces exerted by
the planet. The same holds for our solar system and the milky way galaxy. At any
place in our milky way galaxy, a massive particle is subjected to at least some weak
gravitational forces. Therefore, in physics, the hypothetical case is considered in
which a mass particle is located infinitely far away from all other mass particles. In
this hypothetical case, the mass particle under consideration would not be subjected
to gravitational forces. Let us return to the explicit statement of the first law.

The first law states that a force-free particle remains in its state of movement. If
it does not move with respect to the origin of the aforementioned x, y, z coordinate
system, then in the force-free condition it continues not to move. That is, it remains
at the location at which it was located initially. If it moves with a particular velocity
in a particular direction, then it continuous to move with the same velocity in the
same direction.

Figure 5.1a illustrates the evolution of the state of a force-free particle as stated
by the first law. Figure 5.1a shows the particle location in a single dimension, say
the x-dimension. In the example, the particle is located at $x = 2$m at time $t = 0$
and it moves along the x-direction with a velocity of 1 m/s. Consequently, within
every second the x-position of the particle increases by one meter. This is shown in
Fig. 5.1a.

Let us summarize three key aspects of the first law. First, it is about force-
free systems. Second, it involves to some extent an idealized point of view. Third,
force-free systems remain in their states of movements. These three aspects will be
taken below from particle systems of classical mechanics to self-organizing, pattern
formation systems.

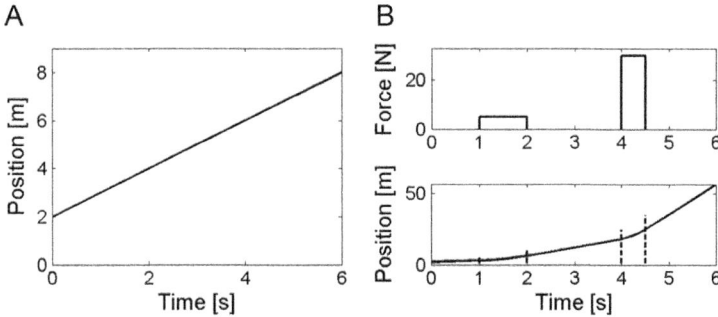

Fig. 5.1 Demonstrations of the first (panel **a**) and second (panel **b**) laws of classical mechanics

The second law states how a particle evolves under the impact of forces. There are several possible formulations of the second law. For our purposes, it is sufficient to formulate the second law as follows. A particle with a certain mass evolves under the impact of a force such that it is accelerated and that the acceleration times the mass equals the applied force. Note that here acceleration also includes de-acceleration. Figure 5.1b illustrates the evolution of a particle with a mass of 1 kg under the impact of certain forces. Between 1 and 2 s and between 4 and 5 s forces are applied to the particle. The forces differ in magnitude as can be seen in the top subpanel. Due to the forces the particle is accelerated such that the particle velocity at $t = 2$ s exceeds the velocity at $t = 1$ s and, likewise, the particle velocity at $t = 5$ s exceeds the velocity at $t = 4$ s. During the force-free periods, the particle satisfies the first law and moves with constant velocity, as can be seen in the bottom panel of Fig. 5.1b. Taking a simplified perspective, Newton's second law states that a force applied to a particle changes the state of movement of the particle.

5.1.2   First and Second Laws of (Human and Animal) Pattern Formation

As we have seen above, in classical mechanics, the first and second laws make statements about the movement of particles that are not subjected to forces (first law) or that are subjected to forces (second law). It would be desirable to have similar laws for pattern formation systems. In this context, recall the discussion above about the first law of classical mechanics. The notion of an isolated massive particle in terms of a force-free particle refers to an idealized situation. Likewise, the notion of an isolated self-organizing, pattern formation system is an idealization. For example, the human brain as a living organ cannot exist in isolation. However, a hypothetical situation can be considered in which the human brain as a pattern formation system is set up in particular way including all its life-sustaining mechanisms. All forces acting on the brain exerted by the body sensory system would be shut down. In this sense, the human brain would be isolated or free of the impact of external forces. In general, an isolated pattern formation system is a pattern formation system that is set up in particular way and characterized by a particular phase portrait in its state space. Importantly, the phase portrait does not vary with time. That is, in the context of pattern formation systems the phrases "free of external forces" or "isolated" should be understood as exhibiting a time-invariant (i.e., constant) phase portrait.

The first laws states that an isolated pattern formation system as defined above remains in its dynamic state. This dynamic state is characterized by the state variable of the system converging towards an attractor. In other words, the first law states that the characteristic feature of isolated pattern formation systems is the emergence of attractor patterns. Just as in classical mechanics a particle continues to move in a certain direction (or stands still) as long as there are no forces acting on the particle, within the field of pattern formation we have that the state of an isolated pattern formation system continues to converge to one of the system's attractors (or to the unique attractor if there is only one) as long as there are no forces that act on the system and affect the system's phase portrait. Consequently, isolated systems build up attractor patterns. Figure 4.1 in Chap. 4 illustrates fixed point and limit cycle attractor patterns. The first law states that the approach to such attractor patterns and the emergence of such attractor patterns is the characteristic feature of isolated pattern formation systems.

The second law in the field of pattern formation states that an external force applied to a pattern formation system results in a change of the system with respect to its ordinary state or structure. That is, the external force either shifts the ordinary state of the system or changes the structure of the system or both. Importantly, structural changes necessarily imply changes in the evolution of the ordinary state. Therefore, the second law states that external forces either directly or indirectly result in a change of the evolution of the ordinary state, that is, in a change of the phase portrait. This implies further that external forces changes the evolution of systems as defined by the first law.

The second law of the theory of pattern formation is formulated in analogy to the second law of classical mechanics. In classical mechanics, a force applied to a massive particle changes the state of movement of the particle. By analogy, in the context of pattern formation systems, an external force affects quantitatively or qualitatively how the state variable of such a system converges towards one of its attractors (i.e., the force affects the phase portrait).

Figure 5.2 illustrates the second law. Panels (a) and (b) illustrates a force leading to a state change of a pattern formation system. The variable α denotes the external force. The time course of the force α is shown in the top subpanel of panel (a). The pattern formation system itself is described in its reduced amplitude space that is assumed to be two-dimensional. Furthermore, the system is assumed to be bistable. As shown in panel (b) there are two fixed point attractors located at $(A_1, A_2) = (1, 0)$ and $(A_1, A_2) = (0, 1)$, where A_1 and A_2 denote the unstable pattern amplitudes of the system under consideration.

Let us turn back to panel (a) of Fig. 5.2. The pattern formation system initially exhibits the attractor pattern with $A_1 = 1$, $A_2 = 0$ based on the first unstable basis

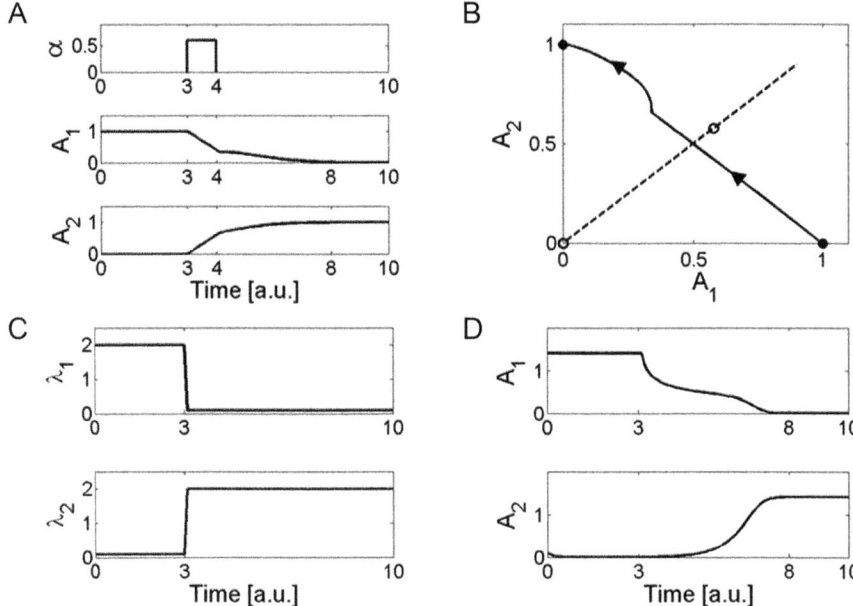

Fig. 5.2 Illustration of the second law for two special cases. Panels (**a**) and (**b**) demonstrate an external forces leading to a shift of the system's state (e.g., brain state or ordinary state). Panels (**c**) and (**d**) demonstrate an external force leading to a change of structure (here: eigenvalues) of a pattern formation system

pattern. At $t = 3$ time units, the force α starts to shifts the state of the pattern formation system such that the amplitude A_1 decays and the amplitude A_2 increases, see middle and bottom subpanels of panel (a). With respect to the reduced amplitude space shown in panel (b) the force shifts the state beyond the diagonal (indicated by the dashed line). It is assumed that the pattern formation system exhibits a symmetry property such that all states in the amplitude space "above" the diagonal are attracted by the $A_1 = 0$, $A_2 = 1$ attractor, whereas all states "below" the diagonal are attracted by the $A_1 = 1$, $A_2 = 0$ attractor. At $t = 4$ time units the force α is switched off. At that time point, the system is located "above" the diagonal. Consequently, the state converges to the attractor at $A_1 = 0$, $A_2 = 1$. The attractor pattern with $A_1 = 0$, $A_2 = 1$ based on the second unstable basis pattern emerges. In this example, the external force α leads to a state change that in turn results in a transition from the attractor pattern with $A_1 = 1$, $A_2 = 0$ to the attractor pattern with $A_1 = 0$, $A_2 = 1$. As far as the phase portrait is concerned, during the time interval from $t = 3$ to $t = 4$ units the original phase portrait (i.e., the phase portrait for t smaller than $t = 3$) is changed by the external force. The straight line shown in Panel (b) starting at $A_1 = 1$, $A_2 = 0$ describes the evolution under the system under the external force and differs qualitatively from the "natural" evolution of the system (i.e., the attraction of all states towards either $A_1 = 1$, $A_2 = 0$ or $A_1 = 0$, $A_2 = 1$).

Panels (c) and (d) illustrate the second law for a force causing a structural change in a pattern formation system. The force is not shown, only its impact is shown. The pattern formation system is described again in terms of a reduced amplitude space with dimension 2 and is assumed to exhibit two fixed points with $A_1 > 0$, $A_2 = 0$, on the one hand, and $A_1 = 0$, $A_2 > 0$, on the other hand. Here A_1 and A_2 refer to the unstable pattern amplitudes again. The two unstable basis patterns exhibit eigenvalues λ_1 and λ_2 that describe the structure of the system. Panel (c) describes the impact of the force on the structure in terms of changes of those eigenvalues. At $t = 3$ time units the force acts on the structure and flips the eigenvalues such that λ_1 drops from 2 to 0.1 units. In contrast, λ_2 increases from 0.1 to 2 units. The pattern formation systems is assumed to satisfy the LVH model with a coupling parameter such that the system is monostable for structures with $\lambda_1 = 2$, $\lambda_2 = 0.1$, on the one hand, and $\lambda_1 = 0.1$, $\lambda_2 = 2$, on the other hand. Consequently, initially, the pattern formation system exhibits the attractor pattern with $A_1 > 0$, $A_2 = 0$ based on the first unstable basis pattern. At $t = 3$ time units a bifurcation takes place due to the change of the eigenvalues (i.e., the structure of the system). The fixed point with $A_1 > 0$, $A_2 = 0$ becomes unstable. In contrast, the fixed point with $A_1 = 0$, $A_2 > 0$ becomes stable. As shown in panel (d), after the force has been applied, the state of the pattern formation system (depicted in the reduced amplitude space) converges to the attractor at $A_1 = 0$, $A_2 > 0$ based on the second unstable basis pattern. The attractor pattern with $A_1 = 0$, $A_2 > 0$ emerges. As far as the phase portrait is concerned (not shown in Fig. 5.2), for $t < 3$ the phase portrait is given by attraction of all states (except for the unstable fixed points) towards $A_1 = 1$, $A_2 = 0$, whereas for $t > 3$ the phase portrait is characterized by all trajectories (except for those starting at the unstable fixed points) converging to $A_1 = 0$, $A_2 = 1$. In summary, the phase portrait is changed by the external force.

In both examples, the external force induces a transition between two attractor patterns. In the first example, we deal with a bistable system, an external force is applied to the system, and the force results in a shift of the state variable out of an attractor towards the so-called basin of attraction (see below) of another attractor. In the second example, we deal with a system again that has two fixed points A and B. However, the system is initially monostable with an attractor A and a repellor B. A force is applied and changes the structure of the system such that the fixed point A becomes unstable and B becomes stable. In doing so, the force acts as a bifurcation parameter and induces a bifurcation.

Let us dwell on certain aspects of the first and second laws of pattern formation systems and classical mechanics. The first law of pattern formation systems is a truism. In Chap. 4 we have defined pattern formation systems in such a way that their states converge to attractors in the long time limit. The fact that the first law is a truism does not mean that it is useless. In fact, the first law of classical mechanics is a truism as well. The reason for this is that in classical mechanics the concept of a force can be defined as everything that changes the state of movement of a particle. If we define a force in this way, then by definition the state of movement of particle does not change in the absence of forces. The first laws in classical mechanics and in the theory of pattern formation (as introduced here) are important departure points because in some sense they describe the characteristic dynamics (i.e., some sort of "baseline" dynamics or "default" dynamics) that can be observed in systems of classical mechanics, on the one hand, and pattern formation systems, on the other hand, when those systems are set up once and then are "left alone".

In contrast, the second laws of classical mechanics and the theory of pattern formation address situations when the characteristic dynamics of systems is disturbed. In particular, for pattern formation systems the second law addresses the situation when a pattern formation system is set up (such that it exhibits a particular phase portrait) and then an additional, external force is imposed on the system. The external force can result in a change of the ordinary state (i.e., a state shift) or a structural change. The change of the structure can, in turn, affect (and typically will do so) the evolution of the ordinary state (see the example in Fig. 5.2d). When the external force is switched off, the second law does no longer apply. Rather, the first law applies and the pattern formation system under consideration exhibits a characteristic dynamics again. If the force did not involve a structural change, then the characteristic dynamics (i.e., phase portrait) before the force occurred is identical to the characteristic dynamics (phase portrait) after the force ended. If the force did involve a structural change, then the system exhibits a different characteristic dynamics after the force has terminated. In any case, once the force is switched off, the first law applies and the state variable converges to one of the attractors of the pattern formation system under consideration.

The second law involving structural changes (not state changes) is important for our understanding of pattern formation as a phenomenon that involves phase transitions and bifurcations. The important scenario in this context (see panels (c) and (d)) is that an external force changes the structure of a system and, in doing so, the force corresponds to or affects a bifurcation parameter. Due to the impact of the force, the bifurcation parameter is shifted beyond a critical value and a bifurcation takes place. The bifurcation implies that the state of the pattern formation system undergoes a qualitative change. The attractor pattern of the system switches from one category to another category (see Chap. 4). For example, when a horse is set up on a treadmill that runs with a low speed such that the horse is walking and, subsequently, an additional force changes the structure of the system by increasing the treadmill speed beyond the critical speed for walk-trot transitions, then the horse makes a gait transition from walk to trot.

The first and second laws in the field of pattern formation are about external forces and apply to all kind of pattern formation systems. That is, the laws apply to pattern formation systems (as defined in Chap. 4) of the animate and inanimate worlds. For example, if a laser is set up with a low pumping current such that the laser operates below the laser threshold and produces ordinary light and, subsequently, the pumping current is increased (and in this sense an additional force is applied that changes the structure parameter that describes the flow of energy through the laser, i.e., the degree to which the laser deviates from equilibrium) beyond the critical value, then a bifurcation occurs and the laser switches from the production of ordinary light to the production of laser light [156, 160].

5.1.3 First and Second Laws of Dynamical Systems

The first and second laws described above are special cases of the first and second laws of dynamical systems. The first law of dynamical systems states that a dynamical system free of external forces has a constant phase portrait. In fact, this statement can be regarded as a definition of force or as a definition of force-free or isolated dynamical systems. In dynamical systems theory such force-free or isolated dynamical systems are also called autonomous systems. Special cases of the first law of dynamical systems are Newton's first law and the first law of in the theory of pattern formation. The second law states that a force either results in a state shift (in addition to the shift determined by the phase portrait in the absence of forces) or a change of system parameters. In both cases, the phase portrait of the isolated (i.e., force-free) case is changed. Newton's second law and the second law of the theory of pattern formation systems are special applications of this second law of dynamical systems.

5.2 Classes of Human Pattern Formation Systems

5.2.1 Classes and Subclasses Ax to Ex

Let us define classes of pattern formation systems on the basis of the components and mechanisms involved in the pattern formation. The component level of the human pattern formation reaction model has been introduced in Sects. 1.6.1 and 1.11.1. The component model is shown in panel (a) of Fig. 5.3 again. In order to arrive at a more succinct presentation, in what follows the notation shown in panel

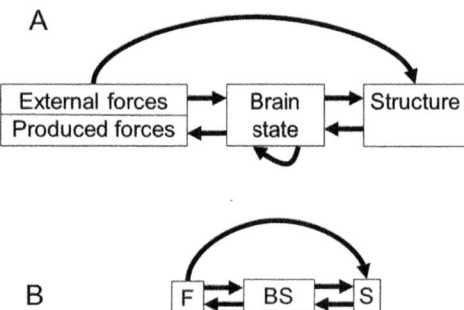

Fig. 5.3 Two steps in simplifying the presentation of the components of the human pattern formation reaction model introduced in Figs. 1.5 and 1.11. Panel (**a**): Produced and external forces are taken together to one component F. Arrows indicate whether we deal with a produced or external force or both. Panel (**b**): Symbols F, BS, S are used to denote the force, brain state, and structure components

(b) will be used. In particular, the box labeled F will stand for both produced forces and external forces. If an arrow points from the brain state component (BS) to the force component (F), then this means that the brain state triggers the production of muscle forces. That is, produced forces are involved in the phenomenon of interest. If an arrow points away from the force component (F) either to the brain state component or the structure component, then the phenomenon of interest involves external forces. The arrow from F to BS stands for a state change induced by the external force as mentioned in the context of the second law. The arrow from the force component (F) to the structure component (S) stands for a structural change induced by the external forces as mentioned again in the context of the second law. Finally, note that for sake of simplicity, we drop the curved arrow pointing from the brain state to the brain state. Nevertheless, the box labeled brain state describes brain activity as a dynamical system such that the ongoing brain activity at a particular reference time point affects among other factors the brain activity at the very next point in time (see also the consideration on circular causality in Sect. 1.6.3).

Before we address the definition of classes, it is worthwhile to discuss the three schemes shown in Fig. 5.4. Panel (a) show the most fundamental case of pattern formation in the human system. Accordingly, brain activity forms a pattern. This process is determined by the structure (here: brain structure) of the human system under consideration. Panels (b) and (c) illustrate how the structure (i.e., internal structure describing brain structure, structure of the sensory system and muscular system, see Sect. 1.6.1) can be changed. Panel (b) demonstrates the case in which brain activity triggers internal forces that result in structural changes. Panel (c) refers to the case in which, in line with the second law, external force result in structural changes.

Finally, let us turn to Fig. 5.5 that provides an overview of possible classes and subclasses of human pattern formation system. There are five main classes A,B,C,D,E. Each main class has several subclasses. Class A describes the human isolated system. In the case of the human isolated system there are neither external forces acting on the individual via the body sensory system nor produced forces. Only two components are relevant: the brain state and the structure. There are

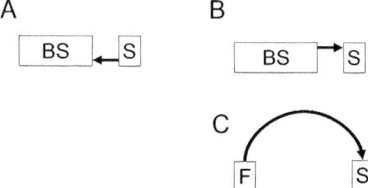

Fig. 5.4 Three fundamental relations between the components of human pattern formation reaction systems: Panel (**a**): Structure determines how patterns form in the brain state. Panel (**b**): Brain activity produces internal forces that result in structural changes. Panel (**c**): External forces exerted by the current circumstances of the human system under consideration lead to structural changes

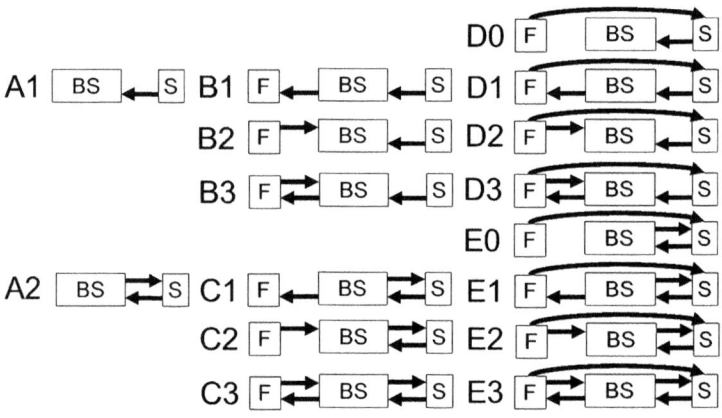

Fig. 5.5 Overview of classes and subclasses of human pattern formation systems

two subclasses A1 and A2. A1 systems correspond to the most fundamental case discussed above in the context of Fig. 5.4a. A2 systems describe human pattern formation systems in which structural changes take place that are induced by the current brain activity. That is, A2 systems describe cases in which emerging BA attractor patterns act back on the conditions (described by the structure) under which they emerged in the first place.

The remaining pattern formation system classes B,C,D, and E involve either external forces or produced forces or both. The system classes B and C describe systems whose structure does not change due to the direct impact of external forces. In contrast, the classes D and E describe systems under external forces that cause structural changes that are relevant for understanding the pattern formation phenomena of interest. The system class B describes pattern formation systems whose structure does not change over the time period of interest, whereas class C refers to pattern formation systems featuring structural changes due to the formation of patterns in the brain. The subclasses B1,B2,B3 just as the subclasses C1,C2,C3 differ with respect to the interactions between the brain state, produced forces, and external forces. The subclasses D1,D2, and D3 correspond to the system subclasses B1,B2, and B3 except for the fact that systems of type D involve external forces that change the structure of the systems. Likewise, the systems E1,E2, and E3 correspond to systems C1,C2, and C3 with the additional feature that E1,E2,E3 systems are restructured due to the impact of external force. Finally, D0 and E0 systems are special cases of human pattern formation systems for which the brain state is neither affected directly by external forces (i.e., is shifted due to external forces, see second law) nor the brain state is involved in force production.

5.2.2   Brain Activity Patterns and Brain and Body Activity Patterns

In the previous section, human pattern formation systems have been classified according to the mechanisms and basic components involved in pattern formation. However, in various scientific disciplines patterns have been classified according to the systems in which they occur or based on the explicit mechanisms that are involved in the formation of the patterns. Some patterns classified in this way are listed in Table 5.1.

Table 5.1 Terminology that denotes patterns according to the systems, locations, or mechanisms involved in the formation of those pattern (BA and BBA stands for brain activity and brain and body activity, respectively)

X- or Y-patterns	X = system or location	Y = mechanism
Ant trails	Ants	
Billow clouds	Clouds	
Chemical patterns	Chemical systems	Chemical reactions
Cloud patterns	Clouds	
Convection patterns	Fluid, gas	Convection
Flock patterns	Flocks of birds	
Gas discharge patterns	Gas	Gas discharge
Hurricane spirals	Hurricanes	
Lake effect snow	Snow at lake shore	
Mantle convection cells	Earth mantle	Convection
Optical patterns	Optical systems	
Polar cells	Polar region	
Reaction diffusion patterns		Reaction diffusion processes
Swarm patterns	Swarms of fish	
Tornado funnels	Tornadoes	
Water cycle	Involves water in various forms at several locations	
Zebra stripes	Zebras	
Isolated BA patterns	Isolated brain	
Sensory body activity and BA patterns	Sensory body system and brain	
Muscular body activity and BA patterns	Muscular body system and brain	
Brain and body activity patterns	Body and brain	
BA patterns	Brain but may involve other parts of the body as well	
Xx systems attractor patterns	Xx systems	

Let us briefly address the examples listed in Table 5.1. Ant trials are patterns formed by ants. Billow cloud patterns are clouds that look like waves (i.e., like billows). The name refers to the system and the shape of the patterns. Chemical patterns are patterns in chemical systems that are produced by chemical reactions. Cloud patterns are patterns formed by clouds. Here, the name refers to the system in which the pattern can be found. In particular, stratocumulus clouds form large size roll patterns and can be considered as a form of convection patterns. Convection patterns refer to patterns that emerge due to the mechanism of convection. In this context, convection means mass transport. Convection patterns typically occur in fluids and gases heated from below. Streams pointing upwards and downwards emerge that transport gas particles and fluid particles and, in doing so, transport heat energy from hot to cold regions. The upwards and downwards streaming particles eventually create rotating rolls. Convection patterns have extensively been studied in Benard experiments that have briefly described in Sect. 3.2.3. Flock and swarm patterns are observed in flocks of birds and swarms of fish, respectively. If a gas is put between two plates that are charged such that there is a relatively high voltage gap between the plates then various discharge patterns emerge [263]. The benchmark pattern is a filament or light column [263] similar to lightning in a storm. The hurricane spiral and the hurricane rain bands are patterns again named after the system in which they occur. Roughly speaking, hurricanes are a special case of convection patterns. The lake effect snow is part of a water cycle that takes place over large lakes and involves snow rather than rainfall. Mantle convection cells are large-scale rotating cells in the earth mantle. These are convection cells similar to polar cells in the atmosphere (see below) or similar to convection rolls produced in laboratories in Benard experiments. Optical systems exhibit patterns like filaments and hexagon patterns [130, 304, 316, 335]. Polar cells are large-scale, atmospheric, convection cells that exist over the regions close to the north and south poles. That is, these cells are named after the geographical region, where they can be found. Polar cells are one of the three global circulation cells in the earth atmosphere. The other two circulation cells are called Ferrel and Hadley cells. Like chemical patterns, reaction diffusion patterns are patterns in reaction diffusion systems produced by diffusion processes and chemical reactions. Swarm patterns are observed in swarms of fish. Tornado funnels are rapidly rotating columns of air that form in tornados. The columns are typically small in diameter over the ground and then increase in size with increasing altitude. That is, spatio-temporal patterns of this kind look like funnels. The water cycle is a general pattern that describes, roughly speaking, how water evaporates over the oceans of the earth and rains down over landsides. The water cycle describes spatio-temporal patterns that involve various water-related components such as the oceans, evaporating water, clouds, rain, river water, and groundwater. Zebra stripes are patterns on the skin of zebras. That is, the name refers to a system or location (the zebra) and refers to a geometric shape (stripes).

This approach to denote patterns can be used to labeled patterns emerging in various components of human pattern formation systems (see the entries at the end of Table 5.1). Let us dwell on those entries in Table 5.1. To this end, let us relate the pattern notations in Table 5.1 more explicitly to the components of human pattern

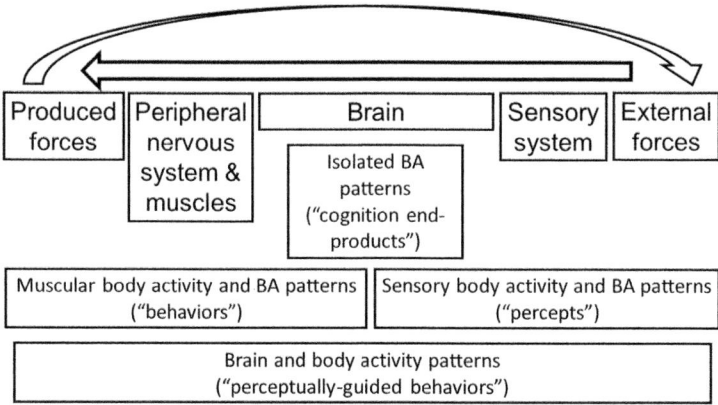

Fig. 5.6   Four classes of patterns (relevant for understanding human reactions, movements, and the dynamics of human brain activity) mapped to the components (F,BS,S) of human pattern formation systems and the system classes A,B,C,D,E

formation system discussed in Figs. 5.3 and 5.5, see Fig. 5.6. The first class of patterns listed in at the end of Table 5.1 and addressed in Fig. 5.6 are those patterns that occur in the isolated human system (i.e., isolated human brain) as defined in Sect. 5.2.1. The formation of those attractor patterns is not affected by external forces, on the one hand, and is not involved in the production of muscle forces, on the other hand. The patterns occur in the brain and exclusively reflect brain activity for A1 systems, whereas they involve brain activity and dynamic changes of the brain structure in the case of A2 systems, see Fig. 5.5. Attractor patterns of this kind are special cases of brain activity (BA) attractor patterns.

Let us return to Fig. 5.6. The second class of patterns listed at the end of Table 5.1 and illustrated in Fig. 5.6 on the right-hand-side are patterns that emerge under the impact of external forces. The patterns involve sensory body activity as well as brain activity. Accordingly, the patterns will be referred to as sensory body activity and BA attractor patterns. The third class of patterns listed in Table 5.1 and illustrated on the left-hand-side of Fig. 5.6 involves the production of muscle forces that, for example, lead to limb movements or are posture related. Consequently, we are dealing with muscle activity patterns as well as brain activity patterns. The patterns will be referred to as muscular body activity and BA attractor patterns. Pattern formation that is both affected by external forces and results in the production of muscle forces may be considered as a case in which we are dealing with two classes at the same time: the second and the third. Taking a slightly different perspective, such patterns may be considered as a class of its own, that is, as a forth class of patterns. In Fig. 5.6 and Table 5.1 they are referred to as brain and body activity attractors patterns. Table 5.2 assigns the patterns emerging in the system classes Ax to Ex to one of the four pattern classes discussed so far. Let us return to the last two entries in Table 5.1. Accordingly, brain activity (BA) patterns, in general, are patterns that form in the human brain, see Table 5.1. They either are identical

Table 5.2 System classes and type of patterns

Patterns	Classes/subclasses
BA patterns of isolated human systems	A1,A2
Sensory body and brain activity patterns	B2,C2,D0,D2,E0,E2
Muscular activity and brain activity patterns	B1,C1
Brain and body activity patterns (involving both activity of the body sensory system and muscular activity)	B3,C3,D1,D3,E1,E3

to isolated BA patterns or are part of patterns that extend beyond the human brain system. That is, they can be part of sensory body and BA patterns, muscular activity and BA patterns, or brain and body activity (BBA) patterns. Attractor patterns can emerge in all system classes Xx displayed in Fig. 5.5. Here, the capital letter X in Xx stands for A,B,C,D,E and the lower case letter x in Xx stands for 0,1,2,3. If a pattern emerges in a certain system class Xx, then the pattern can be denoted accordingly as Xx systems attractor pattern. For example, from the pattern formation perspective, $X2$ systems attractor patterns describe (where X can be B,C,D,E) humans reacting to external circumstances without producing forces. The formation of $X2$ systems attractor patterns in humans is consistent with what has been called in the literature "making judgments", see Sect. 6.5.2. That is, as will be argued in Sect. 6.5.2, the phrase "to judge something" means to form a $X2$ systems attractor pattern as a reaction to something (or to react to something by forming a $X2$ systems attractor pattern).

Note that in A1 and A2 systems only BA patterns can emerge (e.g., BBA patterns cannot emerge in A1 and A2 systems).

To ease the presentation, throughout this book at several places a simplified terminology will be used.

o Attractor patterns will frequently just referred to as patterns. For example, in Fig. 5.6 the phrase attractor has been dropped in the pattern descriptions.
o Brain and body activity (attractor) patterns will be abbreviated as BBA (attractor) patterns.
o Frequently, no explicit distinction between sensory body activity and muscular body activity will be made. If pattern formation involves either forces produced by the human body or forces acting on the human body or both, we will simply refer to the emerging or disappearing patterns as BBA (attractor) patterns.

The approach outlined in Table 5.1 to denote patterns according to the systems, locations, or mechanisms that are involved in the formation of the patterns puts us in the position to relate the concepts of "perception", "cognition", and "behavior" frequently used in the literature to concepts that exist in the physics perspective of human beings. As indicated in Fig. 5.6, the *formation* of isolated BA attractor patterns corresponds to what has often been called "cognition" or "cognitive activity". Accordingly, isolated BA attractor patterns, themselves, may be interpreted as the "end-products of cognitive activities". The *formation* of sensory body and BA patterns seems to corresponds more or less to what has been often called "perception". In line with this terminology, the patterns themselves would correspond to "percepts". Muscular body activity and BA patterns may describe human and animal "behavior". The *formation* of those patterns corresponds to the build-up of "behavior". Finally, brain and body activity patterns may in general reflect something that has been called "perceptually-guided behaviors".

At this stage, the author would like to remind the reader that the physics perspective, in general, and the pattern formation perspective, in particular, of human and animal activities treats systems of the animate and inanimate worlds on an equal footing. Mystification of humans and animals is not supported. The concepts of "perception", "cognition", and "behavior" are unnecessary concepts within the physics perspective. However, since most readers will be familiar with those unnecessary terms, the author of this book will still use them from time to time to make it easier for the reader to understand (i.e., to react in an appropriate way to) this text.

5.2.3   Neuroanatomy and Pattern Formation

In the previous section, it has been suggested to introduce classes of pattern of brain and body activity based on body components that are involved in the pattern formation. This kind of classification does to a certain extent take neuroanatomical aspects of humans and animals into account. In general, the pattern formation approach is consistent with neuroscientific approaches that take the anatomy of the brain into account. Let us dwell on this issue. Panel (a) of Fig. 5.7 illustrates that the human central nervous system can be divided into various subsystems. The central nervous system consists of the brain and spinal cord. The brain itself involves the

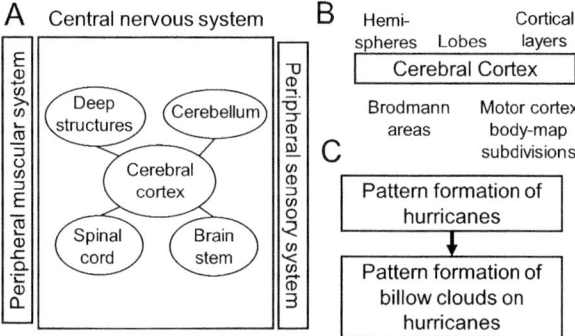

Fig. 5.7 Substructures of the central nervous system (panel **a**) and the cerebral cortex (panel **b**). Substructures may interact with each other and, in doing so, pattern formation in one structure may induce the formation of a pattern in another structure. Panel (**c**): Billow clouds on hurricanes illustrate how an emerging pattern (hurricane) can induce the emergence of another pattern (billow clouds)

cerebral cortex, the cerebellum, the brain stem, and the deep structures such as the thalamus and the hypothalamus. Pattern formation may restricted to any of these subsystems. For example, we may consider all BA patterns that are formed in the cerebellum as a class of patterns of its own. This is similar to studying the lake snow effect as a special case of the water cycle that occurs specifically at the seashore of large lakes. Panel (b) illustrates that the cerebral cortex may be decomposed in various different ways into different kinds of subsystems. Pattern formation may take place as a local phenomenon in one of those subsystems of the cerebral cortex. Importantly, as indicated in panel (a) by the connecting lines, there is typically cross-talk between the various subsystems of the central nervous system (and the same holds for the subsystems of the human and animal body in general). From the pattern formation perspective this implies that pattern formation may spread over several subsystems or pattern formation takes place in different subsystems simultaneously such that the localized processes interact with each other. The formation of a pattern in a subsystem A could act as a force (see the second law above) on the formation of a pattern in a subsystem B.

In fact, pattern formation can be nested in the sense that the formation of a pattern in one system on a particular spatial scale can affect the formation of a pattern in another system on a different spatial scale. Two examples from meteorology is the formation of billow clouds on top of hurricanes and the formation of tornados, when hurricanes make landfall. Hurricanes are rotating large-scale patterns. Due to their rotational movements, hurricanes can create conditions such that billow clouds emerge on their rotating top surface [1]. If so, the attractor pattern of a hurricane affects the bifurcation parameter relevant for the emergence of billow attractor patterns and shifts it beyond its critical value. Panel (c) illustrates schematically this example of two pattern formation systems that are connected with each other. Just as the emergence of billow clouds can be induced by hurricanes, hurricanes frequently induce the emergence of tornadoes, when hurricanes make landfall. Again, hurricanes change the structure of the atmospheric system such that an atmospheric instability (i.e., a bifurcation) occurs at which tornadoes are created.

5.3 Basins of Attraction

In Sects. 1.6 and 1.11.1 the role of initial conditions was pointed out. Which attractor pattern emerges in a multistable pattern formation system depends on the initial state of the system, that is, on the state variable of the system at an appropriately defined initial time point. In this context, it is useful to define the basin of attraction of attractors.

The basin of attraction of a fixed point attractor or limit cycle attractor in a given state space are the domains in the state space that describe all states that converge to the attractor. That is, every point in a given state space is considered and it is determined to which attractor the point evolves in time. All points (states) that

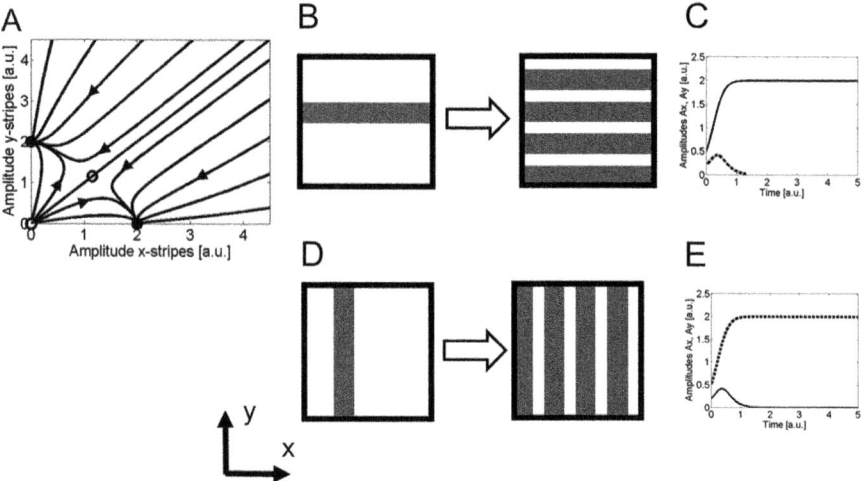

Fig. 5.8 Basins of attraction of the attractors of a chemical system producing stripe pattern that exhibits symmetry with respect to the x-direction and y-direction of the system. See text for details

converge to the same attractor belong to the basin of attraction of that attractor. For systems with two-dimensional state spaces, the phase portrait tells us the attractor basins of the attractors at hand. Let us discuss some examples.

Figure 5.8 refers to the pattern formation system exhibiting stripe patterns introduced in Sect. 4.4.1. The system exhibits a symmetry property such that if stripe patterns in the x-direction can emerge, then stripe patterns in the y-direction can emerge as well and vice versa. Panel (a) shows the phase portrait in the reduced amplitude space of the amplitudes of the two unstable basis patterns related to the x-stripe and y-stripe attractor patterns. It is assumed that the pattern formation system satisfies the LVH amplitude equations. Due to the symmetry property, all initial states for which the amplitude of the x-stripe basis pattern exceeds the amplitude of the y-stripe basis pattern converge to the x-stripe attractor at (0,2). Vice versa, all initial states for which the amplitude of the y-stripe basis pattern exceeds the amplitude of the x-stripe basis pattern converge to the y-stripe attractor at (2,0). Accordingly, the basin of attraction of the x-stripe attractor is the half space of the two-dimensional reduced amplitude space "below" the diagonal, whereas the basin of attraction of the y-stripe attractor is the half space "above" the diagonal. Panels (b) and (c) exemplifies the evolution of the pattern formation system for an initial state for which the y-stripe amplitude is larger than the x-stripe amplitude. Panels (d) and (e) illustrate the opposite case, in which the x-stripe amplitude is initially larger than the y-stripe amplitude.

The basin of attraction for a monostable system is illustrated in Fig. 5.9. In this example human locomotion is considered from a pattern formation perspective. Walking and running are considered as attractor patterns. The symmetric case was

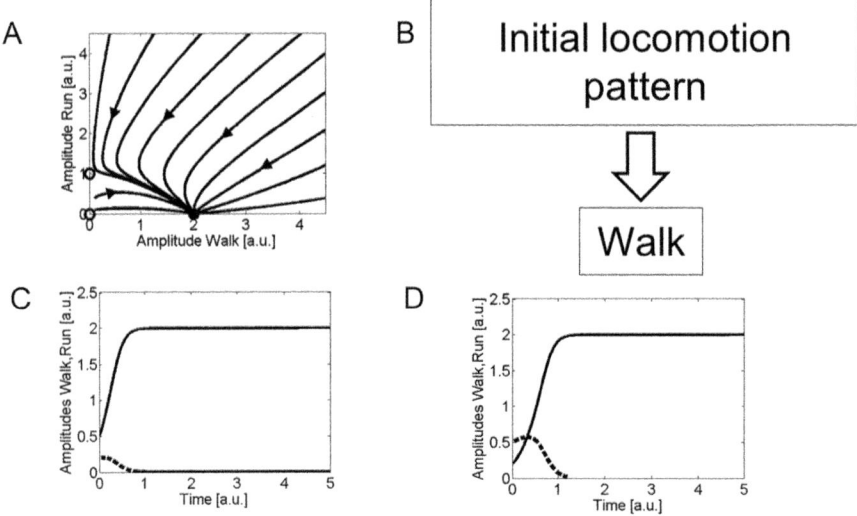

Fig. 5.9 Basin of attraction of the attractor of a monostable system here illustrated for the human pattern formation system for locomotion (in walk and run) in the case of sufficiently small locomotion speeds. Panel (**a**): Phase portrait in reduced amplitude space spanned by the amplitudes of the unstable basis patterns for walking (horizontal axis) and running (vertical axis). Panel (**b**): Monostability means that for any initial state the same BBA pattern emerges (here: the walking pattern). Panels (**c**) and (**d**): Evolution of brain and body activity as described in terms of pattern amplitudes for the monostable case for different initial conditions (solid line: walking amplitude, dashed line: running amplitude)

discussed in Sect. 4.4.1 (see Fig. 4.12). Figure 5.9 refers to the asymmetric case that is assumed to hold for low locomotion speeds. Panel (a) shows the phase portrait in the reduced two-dimensional amplitude space of the amplitudes of the walking and running basis patterns. Again, it is assumed that the LVH amplitude equations can capture the amplitude dynamics. For low locomotion speeds the eigenvalue of the walking basis pattern is assumed to be much larger than the eigenvalue of the running basis pattern such that the system is monostable. In this case, there is an unstable fixed point (repellor) on the vertical axis corresponding to the unstable running pattern. On the horizontal axis there is a stable fixed point reflecting the stable walking pattern. This fixed point attractor of the BBA walking pattern has a basin of attraction that corresponds to the whole two-dimensional amplitude space excluding the unstable fixed point related to a "perfect" running attractor pattern and excluding the origin. The latter may be interpreted as the individual of interest not moving at all. As pointed out in panel (b), for any initial state of locomotion (except for the "perfect" running fixed point pattern and the "perfect" not-moving-at-all fixed point pattern) the brain and body activity converges to the walking BBA attractor pattern. Panels (c) and (d) illustrate the convergence of the human pattern formation system to this unique attractor and, in doing so, the emergence of the walking BBA pattern for two different initial conditions.

In what follows let us assume that the amplitudes shown in Fig. 5.9 reflect primarily brain activity and to lesser extent muscular activity. If so, then in panel (c) the walking BA basis pattern makes a relatively strong contribution to the initial locomotion-related brain activity (e.g., activity in the motor cortex), whereas the running BA basis pattern makes only a small contribution. Over the course of time, the contribution of the running BA basis pattern "dies out" (see dashed line). In contrast, the amplitude of the walking BA basis pattern increases over time and eventually reaches a finite, stationary value (see solid line). Panel (d) illustrates what happens when the initial scenario is just opposite to what is shown in panel (c). The running basis pattern makes a large contribution to the initial locomotion-related brain activity, whereas the walking basis pattern only makes a small contribution. Nevertheless, due to the fact that the system is monostable, the component of the brain activity that "looks like" the running basis pattern "dies out" again. The brain state converges to the walking attractor and the BA walking attractor pattern emerges. In both cases illustrated in panels (c) and (d), it is assumed that the emergence of the walking BA attractor pattern goes along with the emergence of the complete walking attractor pattern involving muscular activity such that in the end for both initial conditions the walking BBA (i.e., brain and body activity) attractor pattern emerges.

Before we turn to our final example let us highlight the importance of the distinction between state and structure in the context of the basins of attraction of attractors. The structure defines the basins of attraction of the attractors of a pattern formation system. If the structure does not change in a given time internal, then the basins of attraction are fixed and do not change. If the structure varies, then at every time point, we deal with different basins of attraction. The concept of basins of attraction of a system is a useful concept only if the structure does not vary during the formation of a particular pattern or the structure changes only slowly (relative to the ordinary state dynamics/amplitude dynamics) such that the basins vary only slowly and/or to a small extent. This notion of structure as those system properties that are constant during pattern formation or as those system properties that evolve slowly as compared to the formation of patterns is in line with the description of structure given in Sect. 1.6.1, see also Table 1.2.

However, if there are structural changes and the structural changes are part of the formation of a pattern that involves the ordinary state and structure of a human individual, then a grand state description of the individual can be taken. That is, if structural changes occur due to the mechanisms discussed in Sect. 5.2.1 (see panels (b) and (c) of Fig. 5.4) and these changes can be considered as part of an emerging spatiotemporal BBA or BA pattern, then the relevant state for understanding the phenomenon at hand is the grand state. In this case, the state space of interest is spanned by all system components described by the ordinary state variable and in addition by all system components that refer to the structure of the system. Attractors and their basins of attraction can then be defined with respect to the grand state of interest.

Figure 5.10 illustrations basins of attraction for the human pattern formation systems affected by forces from the visual world (panel a) and the acoustic world

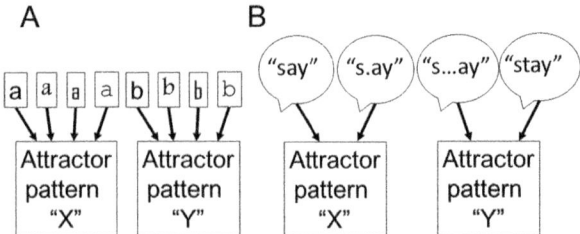

Fig. 5.10 Forces exerted by "similar" objects and sounds are assumed to put brain states into basins of attraction of the same attractors. Panel (**a**): Variants of the letter "a" and "b" are assumed to move the brain state into the basins of attraction of two attractors X and Y, respectively. Panel (**b**): Variants of the sound "s..ay" involving short gaps, on the one hand, and variants of the sound "s..ay" involving relatively large gaps, on the other hand, are assumed to put the brain state in the basins of attraction of two attractors X and Y, respectively

(panel b). As illustrated in panel (a), the letter "a" can be written in various different ways. Letters "a" written in these different ways exert different forces on the human brain (via electromagnetic forces of the light, chemical driving forces in the eye receptors, and electrical forces of the axonal pulses and dendritic currents connecting the eye receptors to various brain areas). Let us assume that according to the second law this leads to different shifts of the brain state (i.e., changes in the brain activity as measured in terms of firing rates, membrane potentials, etc.). In other words, the different variants of the letters "a" are assumed to set different initial brain states. Although these states differ from each other, they are assumed to belong to the basin of attraction of the same attractor that is labeled in Fig. 5.10a as attractor X [61, 95, 97, 125, 127, 157, 299]. Consequently, the human brain reacts to those letters such that an attractor pattern X emerges. Likewise, letters "b" written in various different ways are assumed to exert external forces that lead to different initial brain states. Despite differences between those initial brain states, those initial brain states are assumed to belong to the basin of attraction of the same attractor Y. As a result, the various different types of letters "b" of the visual world lead to the emergence of the same attractor pattern Y. The attractors X and Y are different attractors describing different attractor patterns. In particular, the basis patterns underlying the attractors (or pointing in the directions of those attractors) are assumed to be mutually exclusive basis patterns such that either the attractor pattern X or the attractor pattern Y can emerge but they cannot co-exist [61, 95, 97, 125, 127, 157, 299].

By analogy to panel (a), panel (b) illustrates that pronouncing the word "say" in different ways leads to forces acting on the human brain such that an attractor pattern X emerges. The underlying attractor X has as basin of attraction all initial brain states that are caused by various pronunciations of the word "say". In particular, the notation "s..ay" refers to a speaker that pronounces the word "say" with a short gap between the letter "s" and the letter group "ay". In line with experimental research [45, 325], it can be assumed that as long as this gap is sufficiently short, the initial state caused by the forces exerted by gap-equipped sound "s..ay" will be in the same basin of attraction as the gap-free sound "say". In contrast, again in line with experimental research [45, 325], it is plausible to assume that various types of pronunciation of the word "stay" as well as sounds "s..ay" that feature relative long gaps exert forces that lead to initial conditions that are located in the basin of attraction of a different attractor, the attractor Y.

5.4   Human Reactions to Forces

The concept of attractor basins of attraction will help us to label attractors. In the previous examples shown in Figs. 5.8 and 5.9 we labeled attractors as x-stripe or y-stripe attractors and as walking or running attractors. With respect to the two examples in Fig. 5.10, we took a conservative point of view and referred to the attractors as X and Y using the different labels just to indicate that we were dealing with two different attractors. Labeling patterns is important because as researchers we would like to put them on a (categorical or nominal) scale such that we can talk about them. Moreover, as will be speculated in Sect. 5.4.5, patterns that can be labeled may occur in BBA pattern sequences that describe that humans talk about their experiences, whereas patterns that cannot be labeled cannot be part of (i.e., integrated into) such pattern sequences.

5.4.1   Attractor Patterns as Reactions to Forces, Objects, and Events

Attractors and the corresponding attractor patterns can be labeled with the help of external forces. Figure 5.11 illustrates this labeling method. As illustrated in panel (a), spatio-temporal events that are external to an individual of interest can exert forces on the individual. Here, spatio-temporal events are understood as events that enfold over certain periods such as a person moving in the environment of the individual of interest from A to B. Spatio-temporal events also include the special case of objects, other individuals, or animals situated somewhere in the environment of the individual and not moving at all. In general, any aspect of the environment of an individual may exert a force on the individual. In line with the second law, external forces can determine the initial conditions of pattern formation in the brain state of the individual. Assuming for a moment that the brain as a pattern formation system exhibits multistability, the force-induced initial conditions determine which attractor pattern emerges. More precisely, the brain activity will converge to the attractor whose basin of attraction includes the brain state induced by the external force. Therefore, as pointed out in panel (b), attractors can be assigned to spatio-temporal external events and forces. Likewise, the corresponding attractor patterns can be regarded as reactions to certain external events and forces.

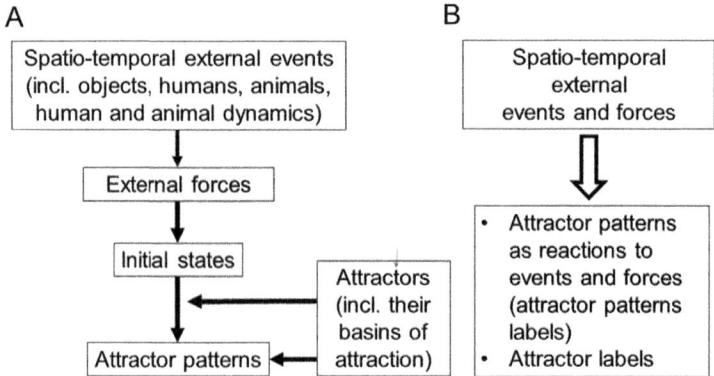

Fig. 5.11 An attractor and its corresponding attractor pattern can be labeled according to all external forces that put the brain state into the basin of attraction of that attractor. Panel (**a**): Detailed scheme leading to the emergence of BBA attractor patterns as reaction to external forces exerted by events. Panel (**b**): Reduction of panel (**a**) to a mapping that allows to label patterns and attractors by means of external forces and events

For example, if a parent sees his or her child, then a particular BBA pattern emerges. That pattern may be referred to as my-child pattern. The pattern corresponds to a certain attractor that can be labeled as my-child attractor. Irrespective whether or not the child smiles or looks bored, whether or not it is young or old, the forces exerted by the child on the parent put the brain state of the parent in the basin of attraction of the same attractor, which is the my-child attractor. Likewise, if an individual sees (i.e., is exposed to the visual world forces exerted by) a table or a chair, BBA patterns emerge that can be labeled as table and chair BBA patterns, respectively. The corresponding attractors are the table and chair attractors, respectively. Likewise, forces exerted by buildings such as houses, towers, churches, mosques, etc. act on the human brain (via the electromagnetic forces of the light that is reflected by those buildings and acts on the retina of human individuals) and are assumed to shift the brain activity into the basin of attraction of a particular attractor. This attractor may be referred to as building attractor.

As mentioned above, this labeling method requires multistability. From a mathematical point of view, we need to consider the hypothetical case of symmetry with respect to all external events or external forces of interest. For example, consider the reactions of an individual to single letters of the Latin alphabet. Let us assume that in the beginning the individual is in a reference brain state that refers to the situation in which no letter is presented. Subsequently, one of the 26 letters is presented to the individual. By presenting any letter, it is assumed that the aforementioned reference brain state becomes unstable. At the bifurcation point, the relevant brain areas (when considered as a pattern formation system) feature an infinitely large set of basis patterns. We consider next the case in which the brain of the individual has been structured (during various childhood encounters with external forces when the individual grew up and became "familiar" with the alphabet) such that there are 26 unstable basis patterns (i.e., patterns with positive eigenvalues) and all remaining basis patterns are stable (i.e., exhibit negative eigenvalues). Modeling the human pattern formation system explicitly as a multistable winner-takes-all system as described by the LVH model, we are dealing with a 26 dimensional reduced amplitude space that exhibits 26 attractors located on its 26 axes. Assuming symmetry among the unstable basis patterns, all eigenvalues assume the same value. At this stage, a detailed description of the attractor basins of attraction is not necessary. From the assumed symmetry, it follows that the brain state is attracted to the attractor with the largest initial amplitude. Therefore, all letters that come with forces that make a particular amplitude largest, say amplitude number 17, result in the emergence of the same attractor pattern and belong from the perspective of the individual to the same letter category.

Let us consider the four examples of "a" letters shown in panel (a) of Fig. 5.10. For most individuals "familiar with" (i.e., appropriately restructured with regard) the Latin alphabet all four letters will make one particular amplitude largest such that for all four letters one particular BBA attractor pattern will emerge. The pattern can be regarded as reaction to "a" letters. The attractor can be labeled as letter "a" attractor and the pattern as letter "a" pattern. Importantly, as will be argued below, in the absence of external forces the letter "a" pattern may emerge (or a part of it or a pattern similar to that pattern). If so, if the individual is awake, then he or she experiences a verbal or graphical thought about the letter "a". If the individual is in a dream stage, then the individual dreams about the letter "a".

Let us return to the general case of labeling attractors and the corresponding attractor patterns by means of external events and their forces. An attractor can be labeled on the basis of all forces that shift brain states into the basin of attraction of that attractor. This situation is in analogy to the impact of forces in classical mechanics on the trajectory of massive particles. Let us consider a particle with a particular mass that moves in a two-dimensional plane described by x and y coordinates. Let us assume that the particle for the duration of 1 sec is exposed to a force in this two-dimensional space. The force is assumed to be constant in magnitude during that 1 sec period. Panel (a) in Fig. 5.12 shows five forces that all have the same magnitude but point in different directions. The particle movement under each of those forces is shown in panel (b). From time t_0 to t_1 the particle is assumed to move force-free in x-direction with a particular initial velocity. At time point t_1 one of the five forces is applied for 1 sec. After the period of 1 sec the force is switched off. The resulting trajectories under the impact of the forces and after the forces are switched off are shown in panel (b). From the trajectories shown in panel (b), it is clear that the particle reacts to the different forces in different ways. The particle trajectories can be labeled 1,2,3,4,5 according to the forces 1,2,3,4,5. Coming back to attractor patterns of the human system, the patterns (and their underlying attractors) can be labeled with respect to external forces in a similar way as the trajectories of a particle can be labeled with respect to forces acting on the particle.

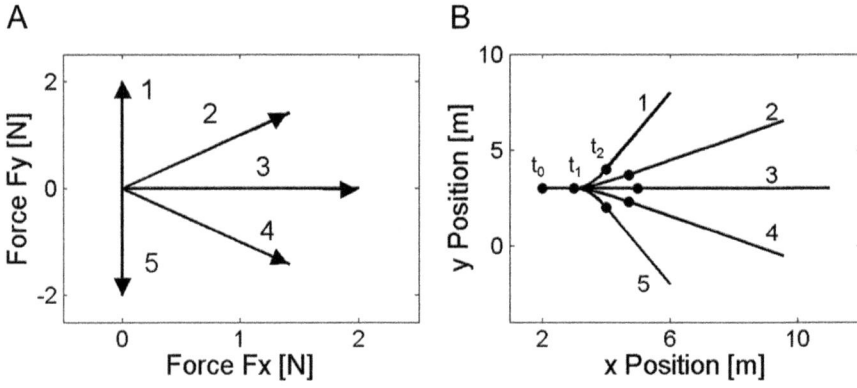

Fig. 5.12 Labeling trajectories of a massive particle according to the forces that have been applied to the particle. Panel (**a**): Forces. Panel (**b**): Trajectories

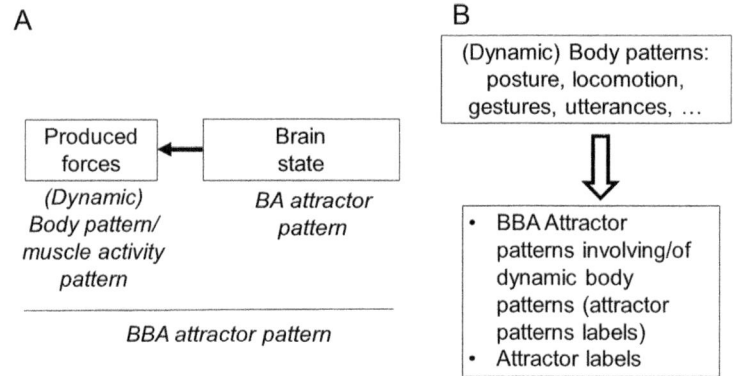

Fig. 5.13 An attractor pattern and its corresponding attractor can be labeled according to produced forces specific to the pattern. See text for details

5.4.2   Attractor Patterns Specific to Produced Forces

Attractor patterns that involve body activity, in general, and muscle activity, in particular, may be labeled with respect to the activity described by the patterns, as illustrated in Fig. 5.13. Systems that exhibit produced forces are minimally of the class B1 (see Sect. 5.2.1 and Fig. 5.5). Accordingly, a pattern of brain activity triggers the build-up of a muscle activity pattern and the production of forces. This in turn leads to a body pattern related to posture and movement such as sitting and standing, sitting up, rolling, crawling, walking, running, jumping, reaching, grasping, finger tapping, smiling, talking, etc. The pattern including the body activity constitutes a BBA attractor pattern, see panel (a). As suggested in panel (b), this BBA attractor pattern may be labeled by means of the activity described by the body. Likewise, the attractor may be labeled accordingly. For example, according to the pattern formation perspective, a person who stands upright forms a particular BBA attractor pattern. That pattern could be referred to as the standing upright pattern and the underlying attractor could be called the standing upright attractor.

5.4.3 Skinner's Perspective and Patterns Specific to Produced Forces

Using produced forces related to body movements and posture rather than external forces in order to label attractor patterns and attractors can be more convenient. Importantly, using produced forces is to some extent consistent with the labeling method based on external forces discussed in Sect. 5.4.1. Any attractor by definition exhibits a basin of attraction. The attractor pattern of an attractor can only emerge if at a certain time point the state of the human system is located in that basin of attraction. When talking about a particular body pattern attractor (e.g., the standing upright attractor), then we refer implicitly to forces that put an appropriately defined state variable into the basin of attraction of that body pattern attractor. In line with Skinner's perspective of human beings and animal as entities determined by the circumstances of their environments, these forces would be external forces.

5.4.4 Attractor Patterns Specific to Localized Brain Activity

As discussed in Sect. 5.2.3 pattern formation may take place in specific areas of the central nervous system (see Fig. 5.7a) and/or in specific areas of the cerebral cortex (see Fig. 5.7b). For example, brain activity patterns associated with finger tapping are typically located in the motor cortex [47, 88, 117, 131, 132, 197]. Patterns and attractor patterns can then be described in terms of their locations in the central nervous system or more specifically in the cerebral cortex. Table 5.1 and Fig. 5.6 can be revised in this regard in order to account for more specific locations of attractor patterns. For example, a BA pattern that emerges in the motor cortex (and is part of or leads to the emergence of a muscular body activity and BA pattern), can be referred to as motor cortex activity attractor pattern. Using simplified terminology (see Sect. 5.2.2) such a pattern may simply called a motor cortex activity pattern. The location of BBA attractor patterns in certain areas of the human brain is also consistent with the dipole hypothesis of human brain activity. According to the dipole hypothesis there are localized centers of brain activity that produce electromagnetic fields similar to the fields produced by electronic dipoles [59, 86, 88, 113, 117, 119, 132]. These electromagnetic fields can be measured, for example, using electroencephalography (EEG) [254].

5.4.5   Neuroanatomically Defined But Not Further Specified Attractor Patterns

According to the pattern formation perspective of human "perception", "cognition", and "behavior", the human system operates close to bifurcation points. At the bifurcation points, the system organizes into a set of stable and unstable basis patterns. The unstable basis patterns under certain conditions (see, e.g., stability band) give rise to attractor patterns. Some of the attractor patterns can be labeled as shown above, for example, as letter "a" attractor pattern or walking pattern. In general, not every attractor pattern can be associated with an external or produced force. That is, in general, at bifurcation points there are basis patterns leading to attractor patterns that can mathematically be described but otherwise are not further specified in terms of external or produced forces. The mathematical descriptions describe brain activity on a neuroanatomic level (see e.g. [38, 39, 116, 175, 176]). The fact that those patterns cannot be associated to forces does not mean that they are irrelevant. Any attractor pattern can be part of a chain of pattern formation events. As indicated above and will be described in detail in later chapters, attractor patterns may change the internal structure of a human (brain structure or structure of the peripheral sensory or muscular system) or the external structure (i.e., external forces acting on an individual). Such changes may trigger secondary bifurcations. Likewise, an emerging attractor pattern in a particular brain area may produce internal forces or lead to external forces that shift brain states in different brain areas from one basin of attraction to another basin of attraction. If so, a transition between two attractor patterns takes place. In summary, not further specified attractor patterns may exist that can only be described in mathematical terms but, nevertheless, determine how a human individual evolves in time as a pattern formation system. We will return to the issue of not further specified attractor patterns in Sect. 6.1.1.

5.4.6   Failure of One-to-One Mappings Between Patterns and Forces

In Sect. 5.4.1 a method was discussed to assign attractors to external forces. Let us briefly discuss how this method applies to pattern formation systems of the classes A,B,C,D,E introduced in Sect. 5.2.1. Subsequently, let us show that there is not a one-to-one mapping between patterns and forces.

The method outlined in Sects. 5.4.1 and 5.4.2 to relate patterns to forces involves the basin of attraction of attractors. Therefore, it is based on a hypothetical situation in which the human pattern formation system of interest exhibits multistability and symmetry at the same time. Moreover, forces act on the brain state only and do not affect structure. Therefore, the method does not apply to the system classes D and E introduced above. Since the structure should exhibit symmetry, it would be desirable to consider pattern formation under a fixed structure, which rules out the system class C. As far the class B is concerned only B2 and B3 refer to systems in which external forces shift brain states. B3 would not be an ideal system because the brain state produces forces that may affect the external forces that we want to use for labeling purposes. In short, the labeling method works well for multistable, symmetric systems of the type B2. In this case, one-to-one mappings between forces and attractors can be established.

The idea is to consider a two step approach. In the first step, the human pattern formation system of interest is modified and reduced such that it belongs to the system class B2 and exhibits multistability and symmetry. In this first step, attractors can be labeled. Subsequently, the original system is reestablished. Due to the second step, some of the attractors may become repellors or disappear entirely. Moreover, the original system may exhibit other attractors in addition to those that could be labeled in the first step. The two step method is useful if (most of) the attractors labeled in the first step are still present after the second step has been completed. If there is a one-to-one mapping then we deal with a situation as exemplified in panel (a) of Fig. 5.14 (and has been shown earlier in Fig. 5.10). Forces of letters of type "a" produce certain initial conditions that all fall into the basin of attraction of a particular attractor that we call the letter "a" attractor. By analogy, letters of type "b" can be used to label another attractor that we denote as letter "b" attractor. In summary, the human brain reacts to force exerted by letters of type "a" and "b", respectively, by producing the letter "a" and letter "b" attractor patterns.

However, in general, according to the pattern formation perspective to human reactions there is not a one-to-one mapping between human reactions and forces. Let us address in what follows some of the reasons why such a one-to-one mapping in general does not exists. First, the human pattern formation phenomenon of interest may show a strong asymmetry in the reduced amplitude space. Due to such an

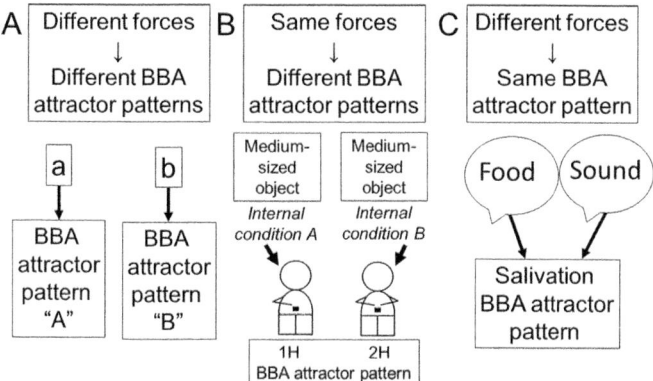

Fig. 5.14 Possible relationship between categorically different external forces and different attractor patterns. Panel (**a**): One-to-one mapping. Panel (**b**): Due to the impact of the brain state, the same force may be mapped to different patterns. Panel (**c**): Different forces may be mapped to the same pattern

asymmetry, some unstable basis pattern may exhibit eigenvalues that are not in the stability band. Such unstable basis patterns would have repellors rather than attractors. B2 systems with strong asymmetry exemplify this case. A force that would shift the brain state in the direction of an unstable basis pattern with a repellor would not induce the emergence of the attractor pattern related to the unstable basis pattern under consideration.

Furthermore, forces may not affect at all the brain state but only structure. Pattern formation systems of classes D0,D1, E0, E1 exemplify such cases. In this case, the current or previous brain state acts as initial condition for a given pattern formation phenomenon. Consequently, different attractor patterns can emerge for the same external forces. The shoe example discussed throughout this book is an example in this regard. Consider the symmetric cases A(ii) and B(ii) addressed in panels (d) of Figs. 4.18 and 4.19 in Sect. 4.6.1. Depending on the initial brain activity, a person will put on shoes A or B, when confronted with the question, which of the two pairs of shoes he or she should put on. In Panel (b) of Fig. 5.14 the analogous example from grasping transition experiments is illustrated. For medium sized objects it is assumed that the pattern formation system exhibits symmetry or is at least sufficiently close to being symmetry (see our explicit calculations in Sect. 4.4.1). A person grasps an object with one hand if he/she is in a certain brain state that is in the basin of attraction of the one-hand grasping attractor and grasps the object with two hands if he/she is in a brain state that is located in the basin of attraction of the two-hands grasping attractor. The external force exerted by a given medium sized object is the same in both cases.

Furthermore, as pointed out in panel (c), categorically different forces may be mapped to the same BBA attractor patterns. For example, in the experiments by Pavlov on dogs discussed in Sect. 2.1 dogs were restructured that they showed salivation production not only when food was presented but also when the experimenter rang a bell. The chemical driving forces of food, on the one hand, and the mechanical forces produced by a ringing bell acting on the receptors on the ear, on the other hand, can be regarded as qualitatively different forces. Nevertheless, after restructuring the animal brains appropriately, those categorically different forces lead to the emergence of the same attractor pattern that is related with the production of saliva.[1] We will return to this issue in more detail in Sect. 9.1 when discussing restructuring in animals in Pavlov's experiments.

A situation that is even more extreme as the situation shown in panel (c) is when a particular BBA pattern emerges under a certain external force Z but may also emerge in the absence of any external force. For example, a certain brain activity pattern is assumed to emerge when an person stands in front of a building and is viewing the building. In this case the building exerts forces that lead to the emergence of the building BA attractor pattern. However, a similar BA pattern may emerge in the

---

[1] Having said that taking a different perspective we may say that after restructuring the attractor has changed to a food-and-ringing bell attractor. I.e., the basin of attraction includes a larger repertoire of forces.

absence of any building around the person if the person is awake (when the person is "thinking" about a building) or if the person is in a dream stage. We will return to this issue in Sect. 5.5 below.

5.4.7   Physical Systems React to Forces But Do Not Have "Representations" of Forces: Why Humans Have as Much "Representations" as Billiard Balls

As discussed above, some attractor patterns can be related to external forces. For example, as far as human reactions to objects of the visual world are concerned, the pattern formation perspective allows us to defined the letter "a" attractor pattern as the reaction pattern to external forces exerted by letters of type "a". In this context, the question arises whether or not attractor patterns or the corresponding attractors should be interpreted as "representations" of the forces or the objects or events causing the forces under consideration. For example, as questioned in panel (a) of Fig. 5.15, if a parent sees his or her daughter and the corresponding my-daughter attractor pattern emerges in the parent, does this pattern "represent" the daughter? According to the theory of pattern formation a pattern emerges at a bifurcation point. In the case of a multistable system the initial conditions determine which pattern emerges. The bifurcation might be induced by external forces that change the structure by shifting relevant bifurcation parameters and eigenvalues (see second law). The initial conditions may be set by external forces by shifting brain states to appropriate locations (see second law again). The emerging pattern is the effect or end-product of the process of forming a pattern under the forces acting on the system. A pattern (or the corresponding attractor) is not interpreted as an entity that "represents" the forces (or the objects and events causing the forces) that set the initial conditions of that process.

Let us illustrate this point by comparing humans exposed to letters "a" and "b" with so-called bistable laser devices that operate at low and high pumping currents. Certain laser devices are bistable in the sense that for the same pumping current they can produce two different types of laser light that we will denote here by type 1 and type 2 light. For relatively low pumping currents (but still sufficiently large currents such that laser light is produced at all) type 1 laser light is produced. In contrast, for relatively high currents type 2 laser light is produced. For medium pumping currents both types of light can emerge depending on the initial conditions [16, 48, 244]. For somewhat more details about this topic see Sect. 6.3.4. As discussed previously, forces from letters "a" are assumed to put the state of the multistable human brain into the basin of a letter "a" attractor, which leads to the emergence of a letter "a" BA pattern. In a similar way, forces from letters "b" are assumed to result in the emergence of a letter "b" BA pattern. This should be seen in analogy to low pumping currents that put the state of a bistable laser in the basin of attraction of the fixed point attractor for producing type 1 laser light such that type 1 laser light is produced. High pumping currents put the state of a bistable laser in the basin of attraction of the type 2 laser light attractor such that the state converges to that attractor and the device produces type 2 laser light. In the case of the laser, laser engineers have not designed the laser such that it has any "representations" about pumping currents. In particular, neither the type 1 laser light attractor of the laser not the emerging type 1 laser light are "representations" of low pumping currents. Likewise, neither the type 2 laser light attractor of a laser nor the corresponding state (i.e., the type 2 light) are "representations" of high pumping currents. By analogy, neither the letter "a" or letter "b" attractors of a human being described above within the pattern formation perspective are representations of the letters "a" and "b". Nor the corresponding BA activity patterns are "representations". If humans would have "representations" (which they do not have according to the physics perspective presented in this book), then bistable lasers, bistable chemical systems (see Fig. 5.8), bistable atmospheric systems (that can produce clockwise or counterclockwise rotating cloud roll patterns, see insert in panel (c) of Fig. 3.5), and many more multistable physical systems would all exhibit "representations". For example, bistable equilibrium systems such as ferromagnets would have "representations" (see panel (b) of Fig. 8.5). The ferromagnetic state, in which the elementary magnets of a rod shaped magnet point mainly in the one direction, would "represent" something. The ferromagnetic state, in which the elementary magnets point mainly just in the opposite direction, would "represent" something else.

Just as a pattern or its attractor should not be interpreted (and mystified) to "represent" something (e.g. forces) that puts the system at hand into the basin of attraction of the pattern's attractor, a pattern or its attractor should not be interpreted (and mystified) to "represent" forces (or the objects and events causing the forces) that shift bifurcation parameters beyond critical values. For example, cloud patterns may emerge in the atmosphere when temperature gradients across altitude differences are sufficiently large. However, if this happens, then those cloud patterns do not "represent" temperature gradients. Ordinary laser devices produce laser light when they are pumped by sufficiently high pumping currents. However, just as in the previous example about bistable lasers, laser light does not "represent" pumping currents.

In general, an attractor is a structural property of a pattern formation system that has a basin of attraction such that states shifted by external forces such that they end up in that basin converge to the attractor. The emerging attractor pattern is the reaction of the system or human being of interest to the forces acting on the system or the human individual.

An attractor pattern may be seen as in analogy to the trajectory of a billiard ball moving on a billiard table under the impact of forces exerted by the billiard player on the ball. As illustrated in panel (b) of Fig. 5.15 we could ask whether the particular trajectory or movement direction of the billiard ball is a "representation" of the force that acts on the ball. Typically, in physics, a trajectory of a particle like a billiard ball is not considered as a "representation" of the forces acting on that particle. Consequently, if an advocate of the "representation" concept argues

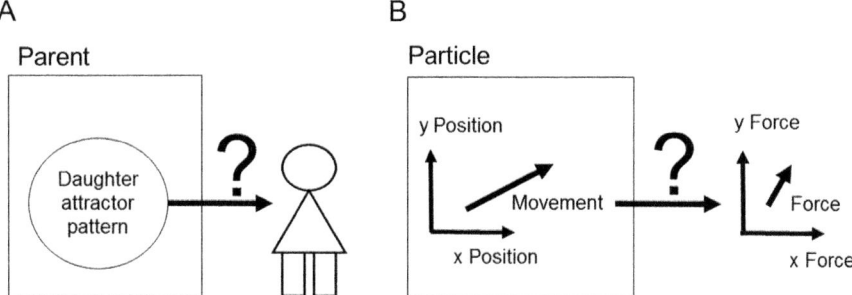

Fig. 5.15 Attractor patterns are reactions to forces just as particle trajectories are reactions to forces. Panel (**a**): My-daughter attractor pattern emerging in a parent as a reaction to forces exerted by the daughter on the parent. Panel (**b**): Particle trajectory as a reaction to the forces acting on the particle

that humans have "representations" in terms of attractor patterns, then one could argue by analogy that particles like billiard balls have "representations" in terms their movement trajectories.

5.5   Brain Activity Dynamics of Human Isolated Systems

From a modeling point of view, attractor patterns that emerge under the impact of forces can also emerge in the same way or as variants in the human isolated brain, that is, in the absence of the relevant forces. The reason for this is that there are two fundamental mechanisms leading to the emergence of an attractor pattern. First, the human pattern formation system may be multistable and a force shifts the system from an attractor A into the basin of attraction of another attractor XYZ (see second law). As a result, the attractor pattern XYZ emerges. Second, the structure of the system changes (see second law again) such that the current attractor pattern A becomes unstable, while at the same time the system of interest features an attractor XYZ. If the state (in the case of a fixed point attractor) or states (e.g., in the case of a limit cycle attractor) that describe the former attractor A are in the basin of attraction of attractor XYZ then the system switches from attractor pattern A to the pattern XYZ. Panels (a) and (b) of Fig. 5.16 exemplify these two mechanisms leading to the emergence of an attractor pattern for systems of classes B2 and D0. In the B2 system an external force shifts the brain state out of an attractor into the basin of the attractor XYZ. Consequently, the XYZ BA attractor pattern emerges. In the D0 system an external force leads to a structural change that makes the reference state at A unstable such that the unstable state is in the basin of attraction of the XYZ attractor. Consequently, the XYZ attractor pattern emerges. A special case of this scenario is when the external force makes the D0 system monostable with the XYZ attractor being the only attractor.

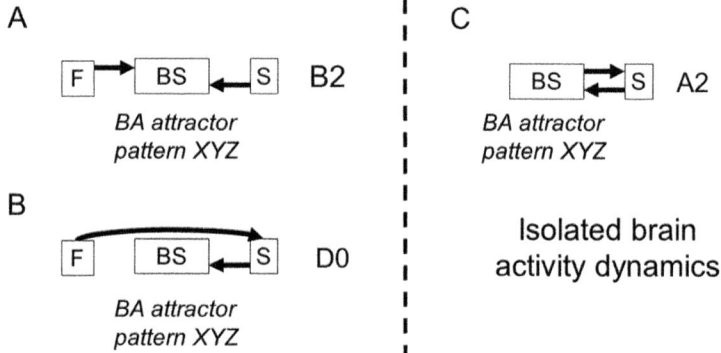

Fig. 5.16 Three mechanisms leading to the emergence of attractor pattern XYZ. While the mechanisms in panels (**a**) and (**b**) involve external forces, the mechanism in panel (**c**) does not require the presence of an external force. Consequently, patterns XYZ that emerge under certain external forces can emerge without the presence of those forces

However, as discussed in Sect. 5.2.1 structural changes are not only due to external forces. BA attractor patterns may result in changes of the structures that supported the emergence of those patterns in the first place. This kind of circular causality can occur in the isolated brain and is formally described by systems of class A2, see panel (c) of Fig. 5.16. Accordingly, an attractor pattern A may change the structure that supported its emergence such that the pattern becomes unstable. As a result, the brain activity forms a new attractor pattern (assuming as we do throughout this book that the system is globally stable). In particular, the brain state may converge to the attractor XYZ as if there would be external forces leading to the emergence of an attractor pattern XYZ.

For example, consider a teen driver, who has problems with parking. While parking the car in the garage at his parents' house, the car hits the garage wall. The teen driver gets out of the car, checks the car, and sees a dent at the front of the car. That is, when the driver gets out of the car, forces exerted by the dent put the brain state of the driver into the basin of attraction of a certain attractor that we may call the dent attractor or more specifically the car-dent attractor. Therefore, the corresponding car-dent BBA attractor pattern emerges (and may trigger a sequence of emerging and disappearing follow-up BBA or BA patterns like being shocked or producing "verbal complaining thoughts" about the garage that is just too small). If the driver goes to sleep, in the dream stage, on-going brain activity may lead to a change of the brain structure such that the eigenvalue of the unstable basis pattern of the car-dent basis pattern becomes positive and large. Consequently, the basin of attraction of the car-dent attractor increases (the system may become even monostable with the car-dent attractor being the only attractor). Let us assume that the ongoing brain activity happens to be located in this "inflated" basin of attraction of the car-dent attractor. Consequently, the corresponding attractor pattern emerges. The teen driver dreams about the car dent.

The possibility that BA patterns related to external forces can also emerge without the presence of such forces demonstrates that in general there is not a one-to-one mapping between forces and patterns, see Sect. 5.4.6.

5.6 Mathematical Notes

5.6.1 Single Particle Movement: Under the Impact of External Forces and Isolated Case

Let us consider a particle with mass m that moves in a one-dimensional space described by the coordinate $x(t)$. In the absence of forces, the evolution equation of x reads

$$m \frac{d^2}{dt^2} x = 0 . \tag{5.1}$$

This implies that the particle velocity $v(t)$ is given by

$$v(t) = v_0 , \tag{5.2}$$

where v_0 is the initial velocity at a reference time point t_0. Equation (5.2) describes Newton's first law in mathematical terms for a particle moving along a single dimension. Finally, the trajectory is given by

$$x(t) = x_0 + v_0(t - t_0) , \tag{5.3}$$

where x_0 is the particle position at time t_0. Figure 5.1 illustrates a particle trajectory (5.3) for $t_0 = 0$, $x_0 = 2$m, and $v_0 = 1$m/s. Newton's second law (for a particle moving in a single dimension) reads

$$m \frac{d^2}{dt^2} x = F(t) , \tag{5.4}$$

where $F(t)$ is the force acting on the particle. Equation (5.4) considers the so-called non-relativistic case. Accordingly, the force F accelerates the particle. Let $a = d^2x/dt^2$ denote the acceleration. Then $a(t) = F(t)/m$. For a pulse-like forces that is constant during an interval $[t_a, t_b]$ with $t_b > t_a$ the velocity is given by

$$v(t) = v_a + \frac{F}{m}(t - t_a) \tag{5.5}$$

with $v_a = v(t_a)$. The trajectory is given by

$$x(t) = x_a + v_a(t - t_a) + \frac{F}{m}(t - t_a)^2 \tag{5.6}$$

with $x_a = x(t_a)$. The trajectory shown in Fig. 5.1b has been computed using Eq. (5.6) for $m = 1$ kg and $F = 5$ N in the interval $[t_a, t_b]$ with $t_a = 1$ s, $t_b = 2$ s and $F = 30$ N in the interval $[t_a, t_b]$ with $t_a = 4.0$ s, $t_b = 4.5$ s.

5.6.2   First Law of Human Isolated Systems

Isolated pattern formation systems satisfy $dX/dt = N(X)$, where X denotes the state variable under consideration, see Eq. (1.1). The systems under considerations are assumed to be globally stable systems. Therefore, at least one attractor exists. Let $M \geq 1$ denote the number of attractors and W_k with $k = 1, \ldots, M$ denote the attractor sets. Let us assume that there are $0 \leq S \leq M$ fixed point attractors and $M - S$ limit cycle attractors with $k = 1, \ldots, S$ denoting the indices of the fixed point attractors and $k = S + 1, \ldots, M$ denoting the indices of the limit cycle attractors. Then, by definition of globally stable systems, we have

$$t \to \infty \ : \ X(t) \to X' \in W_{i,FP} \ \vee \ \min_{X' \in W_{j,LC}} \{X(t) - X'\} \to 0 , \tag{5.7}$$

where $W_{i,FP}$ is one of the fixed point attractors (if there is a fixed point attractor at all) with $i \in [1, S]$ and $W_{j,LC}$ is one of the limit cycle attractors (if there is a limit cycle attractor at all) with $j \in [S + 1, M]$. That is, the state X either converges to a fixed point or limit cycle attractor (note again that throughout this book other types of attractors are ignored). Equation (5.7) describes the first law of pattern formation systems in mathematical terms.

5.6.3   Second Law: Pattern Formation Systems Under External Forces

The second law states that an external force either changes the state of the human pattern formation system under consideration or its structure. Let α denote an external force. A force-induced state change is described in the state space by

$$X(t) = \Phi_t^X(X(t_0), \alpha) \tag{5.8}$$

and in the reduced amplitude space by

$$A_j(t) = \Phi_t^{A,j}(A(t_0), \alpha) , \tag{5.9}$$

where Φ_t^X and Φ_t^A are appropriately defined functions and A_j denote amplitudes spanning the reduced amplitude space. In the simulation shown in Fig. 5.2, it has been assumed that Eq. (5.9) is given by

$$A_1(t) = A_1(t_a) - \alpha(t - t_a) , \quad A_2(t) = A_2(t_a) + \alpha(t - t_a). \tag{5.10}$$

Finally, note that a force-induced structural change is described by $d\alpha/dt = R_\alpha(q, N_q, \alpha)$ for the external structure and $dN_q/dt = R_{N/q}(q, N_q, \alpha)$ for the internal structure, see Eq. (1.5).

5.6.4 The Fundamental Systems Class A1

The fundamental systems class A1, depicted in Fig. 5.4a, is described by the brain state q_b. The brain state evolves like $dq_b/dt = N_{q/b}(q_b)$, as discussed earlier, see Eq. (1.3). As discussed in Chap. 4, the formation of patterns in the brain state can be expressed by means of amplitude equations in the reduced amplitude space. In terms of the LVH model, those amplitude equations read $dA_k/dt = \lambda_k A_k - g A_k \sum_{j=1, j \neq k}^{m} A_j^{q-1} - A_k^q$ assuming that there are m unstable basis patterns $k = 1, \ldots, m$, see Eq. (4.38).

5.6.5 Two Fundamental Mechanisms Leading to Structural Changes

Figure 5.4b expresses that the brain state affects its own structure. The brain state produces internal forces that act on the brain structure. Mathematically speaking, this phenomenon is captured by $dN_q/dt = R_{N/q}(q, N_q)$, see Eq. (1.5). A special case that will be used in later chapters is the impact of the brain state on eigenvalues of unstable basis patterns. For example,

$$\frac{d}{dt}\lambda_j = f_j(\lambda_j, A_1, \ldots, A_m) \tag{5.11}$$

with $j \leq m$ describes how the eigenvalue of the jth unstable basis pattern evolves in time under the impact of the brain state described in terms of amplitudes A_k, where f_j are arbitrary functions. In a similar vain,

$$\lambda_j = h_j(A_1, \ldots, A_m) \tag{5.12}$$

describes how the eigenvalue λ_j depends on the brain activity as expressed in terms the amplitudes $A_1, \ldots, A_m$, where h_j is an appropriately defined function.

In contrast, Fig. 5.4c expresses that structure changes can be caused by external forces. Mathematically speaking, this effect is described by $dN_q/dt = R_{N/q}(q, N_q, \alpha)$, see Eq. (1.5), where $R_{N/q}(q, N_q, \alpha)$ depends on the external force α, as indicated. For example, we may have

$$\frac{d}{dt}\lambda_j = f_j(\lambda_j, \alpha) \tag{5.13}$$

or

$$\lambda_j = h_j(\alpha) \tag{5.14}$$

with $j \leq m$, where f_j and h_j are appropriately defined functions (that are not identical with those functions occurring in Eqs. (5.11) and (5.12).

Chapter 6
Pattern Formation of Ordinary States

This chapter considers human pattern formation systems for which pattern formation takes place in their ordinary states. Recall that the ordinary state is given by the brain state and variables describing force production as well as position and movement of the body and limbs. Pattern formation of ordinary states implies that the formation of patterns does not affect the structure of those systems. With respect to the system classes introduced in Chap. 5 (see Fig. 5.5) this further implies that we are dealing with A1 systems and all B systems. However, in fact, while the pattern emerges in an appropriately defined ordinary state and, consequently, the pattern does not involve structure components, the structure is not necessarily fixed. In particular, external forces may affect the structure of the systems under consideration in order to induce bifurcations that lead to the formation of BA and BBA patterns. Therefore, not only A1 systems and B systems are considered but also D0 and D2 systems. D1 and D3 systems are considered when considering single events of pattern formation (which will be explained in more detail in the sections below).

In other words, in this chapter systems are not considered for which emerging brain activity patterns result in structural changes as in A2 systems and all C and E systems. Moreover, D1 and D3 systems are not considered in this chapter in the context of pattern sequences because the formation of patterns in such systems is assumed to change external circumstances that lead to structural changes. Consequently, as far as the emergence of pattern sequences in such systems is concerned, pattern formation is not restricted to the ordinary state. In summary, A2 systems, C and E systems, D1 and D3 systems exhibit pattern formation that comes with structural changes. Patterns of this kind emerge in the grand state (not in the ordinary state). Pattern formation of grand states will be discussed in Chap. 7.

Table 6.1 summarizes the systems that will be addressed in this chapter and the systems that will be addressed in the subsequent chapter.

© Springer Nature Switzerland AG 2019
T. Frank, *Determinism and Self-Organization of Human Perception and Performance*, Springer Series in Synergetics,
https://doi.org/10.1007/978-3-030-28821-1_6

Table 6.1 Systems showing pattern formation of ordinary and grand states and chapters in which they will be discussed

Chapter	State space of pattern formation	Systems
Chapter 6	Ordinary state	A1, Bx, D0, D2, D1[a], D3[a]
Chapter 7	Grand state	A2, Cx, D1, D3, Ex

[a]Single events of pattern formation

Note that the distinction between the system classes D0 versus D2 (and D1 versus D3) depends on the scale of our perspective, which will be briefly discussed at the end of Sect. 6.5.2.

6.1   Dynamics of the Isolated Human and Animal Brain: A1 Systems

A1 pattern formation systems correspond to isolated human and animal brains that form attractor patterns that leave brain structures unchanged. Recall that isolated systems are defined as systems that are not affected by external forces and are not involved in force production. In what follows, the emergence of attractor patterns in A1 systems will be discussed. Subsequently, sequences of emerging attractor patterns and pattern formation cycles will be considered.

6.1.1   Emergence of Single Attractor Patterns

Let us consider an attractor pattern emerging in an awake human. The pattern may describe a reaction pattern to a force of the visual world. Figure 6.1 illustrates an example in this regard. It is assumed that a certain BA pattern emerges in a person who is standing in front of a house and subjected to the external forces exerted by the house (via electromagnetic waves of the visual spectrum hitting the person's eyes). Let us denote the emerging pattern as visual world house attractor reaction pattern (see Sect. 5.4.1) and its corresponding attractor as visual world house attractor. In the absence of external forces, the attractor is assumed to exist under appropriate conditions as discussed in Sect. 5.5. In general, the attractor will differ to some extent from the original attractor. Nevertheless, characteristic features should remain intact. In fact, experimental research supports this point of view to some extent [193–195], see also Chap. 8 on "Perception without sensation" in Ref. [192]. Consequently, in the absence of external forces a pattern similar to the visual world house attractor pattern can emerge when the brain state is in the basin of attraction of the visual world house attractor pattern. For sake of simplicity, the emerging attractor pattern will be named in the same way irrespective whether or

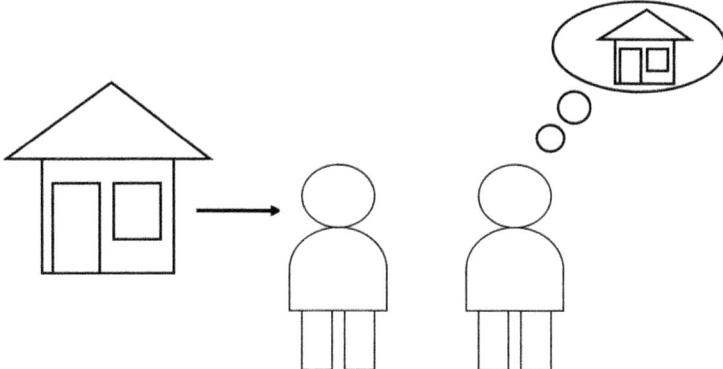

Fig. 6.1 Two step approach for understanding the emergence of a visual world house BA attractor pattern in the human isolated system. Step 1 (left): The visual world house BBA attractor pattern emerges under the impact of external forces. Step 2 (right): In the isolated human brain the corresponding visual world house BA attractor pattern emerges. See also Fig. 5.16

Fig. 6.2 Emergence of a acoustic world house attractor pattern in the isolated human brain (right) in analogy to the formation of house BBA attractor patterns as reaction to the spoken word house (left) and associated with the verbal production of the word house (middle)

not there is an external force. Visual world attractor patterns of the human isolated system have been considered as "graphical thoughts". In fact, in the isolated human brain typically a plenitude of visual world attractor reaction patterns can emerge. For example, visual world attractor reaction patterns of a house, car, airplane, the moon, a friend, a cat or a dog can emerge in the isolated brain of healthy humans. In analogy to such "graphical thoughts", "verbal thoughts" can be addressed.

Figure 6.2 exemplifies an individual in which an acoustic world house attractor reaction pattern has emerged. When exposed to the sound of somebody speaking the word house, it is assumed that a particular attractor pattern emerges in certain brain areas of the individual. It is further assumed that when the individual produces forces appropriate for speaking out the word house that a similar attractor reaction pattern is formed. The two attractor patterns will differ due to the fact that in the former case forces are involved that are related to sensory body activity, whereas in the latter case forces are involved that are related to muscle body activity. While acknowledging possible differences between those patterns, for our purposes the focus is on what the patterns and their corresponding attractors have in common. In what follows the attractors will be regarded as two variations of the acoustic world house attractor. The attractor is assumed to exist in some modified form in the absence of external forces and when the individual is not involved in force production. That is, the attractor is assumed to exist in the isolated human brain as discussed in Sect. 5.5. If the ongoing brain state happens to be in the basin of attraction of the acoustic world house attractor, then the corresponding attractor reaction pattern emerges. The emergence of that pattern may be interpreted as an individual having the "verbal thought" of the word house. The visual and acoustic world house attractors may be considered as parts of a single attractor: the house attractor.

Table 6.2 summarizes the reaction patterns emerging in the isolated human brain of an awake person discussed so far. As argued in Sect. 5.4.5, in general, it is plausible to assume that there are various not further specified BA attractor patterns. If the ongoing brain activity happens to fall in the basin of attraction of an attractor of such a BA pattern, then the pattern will emerge. The individual experiences an unspecified noise or has some kind of unspecified graphical impression. Patterns of this kind emerging in the isolated human brain may be interpreted as "unsymbolized thoughts" [171].

Table 6.2 Examples of BA patterns emerging in the human isolated system of awake individuals

Condition	Subcondition/related to . . .	Pattern formation terminology	Other terminology
Awake	Visual world forces	Visual world isolated BA patterns of the awake person	"Graphical thoughts"
Awake	Forces of own speech	Own speech isolated BA patterns of the awake person	"Inner speech", "verbal thoughts"
Awake	Forces of speech of somebody else	Others speaking isolated BA patterns of the awake person	"Voice imaginations"
Awake	Forces exerted by other sounds	Acoustic world isolated BA patterns of the awake person	"Sound imaginations"
Awake	Force production	Force production isolated BA patterns of the awake person	"Movement imagination", "Imagined muscle contractions"
Awake	Not related to external forces or muscle force production	Not further specified isolated BA patterns of the awake person	"Unsymbolized, non-verbal thoughts"

Humans exhibit different sleep stages. A particular sleep stage is the rapid-eye-moment (REM) stage [192]. In this stage, humans move their eyes although they are asleep and have their eyes closed. Humans that wake up during REM sleep typically report visual or acoustic experiences. This can be explained by assuming that during REM sleep visual and acoustic world attractor patterns emerge just as in the awake human. In general, the attractors of an individual during REM sleep will not be identical to the attractors of that individual, when he or she is awake. The reason for this is that the neurophysiological parameters of an awake person and a person in REM sleep differ from each other. However, despite the differences in conditions, it is plausible to assume that under appropriate conditions an attractor that exists in an awake person can still exist in some modified form in the person under REM sleep. For example, the attractors of the aforementioned visual world and acoustic world house attractor patterns should exist under appropriate sleep conditions. If so, the corresponding patterns can emerge. We may say that a person dreams about a house or hears the word house in a dream. Note that the physics perspective of humans as pattern formation systems points out that there is no principle difference between patterns emerging in the awake human or in humans under REM sleep. We will return to this issue below in the context of sequences and cycles of pattern formation.

Brain activity reaction patterns may emerge in humans under special conditions under which they usually do not emerge. Diseases may lead to the emergence of patterns that would not emerge in healthy adults given all circumstances held constant except for the disease. Likewise, certain attractor reaction patterns may emerge in individuals under the impact of drugs, while the would not emerge in "normal" individuals [38, 39]. In this context, visual and acoustic worlds reaction patterns may be interpreted as disease-related or drug-induced "visual and acoustic hallucination". Notably, so far, we focused on reaction patterns related to forces acting on humans via the visual and acoustic sensory system. Other sensory modalities, for example, the sense of touch, could be discussed by analogy. If somebody is kissed or hit, it is assumed that certain attractor patterns emerge in particular regions of the brain. If the corresponding attractors exist at least in somewhat modified versions in the absence of the external forces, then the being kissed or being hit attractor patterns can emerge in the isolated human brain. Awake humans may "imagine" or "recall" being kissed or hit. Humans in REM sleep stage may dream about being kissed or hit. Patients suffering from certain psychological disorders may "think" being hit although they are actually not hit and in this sense experience "tactile hallucinations". Table 6.3 summarizes the BA patterns of the isolated human brain discussed in the context of REM sleep and drug and disease conditions. Note that the examples listed in Table 6.3 are just a few examples of such patterns. The reader might be in a condition such that as a reaction to reading this text the brain activity of the reader will be put in a state that leads in the reader's brain to the emergence of isolated BA patterns related to further examples.

Table 6.3 Further examples of BA patterns emerging in the human isolated system

Condition	Subcondition/Related to . . .	Pattern formation terminology	Other terminology
REM sleep	Visual and acoustic worlds forces	Visual and acoustic worlds isolated BA patterns under REM sleep	"Graphical and acoustic dream experiences"
Drug-induced	Visual and acoustic worlds forces	Drug-induced visual and acoustic worlds isolated BA patterns	"Drug-induced hallucinations"
Disease-related	Forces acting on the visual, acoustic, and tactile sensory system	Disease-induced visual, acoustic, and tactile worlds isolated BA patterns	"Disease-induced hallucinations"

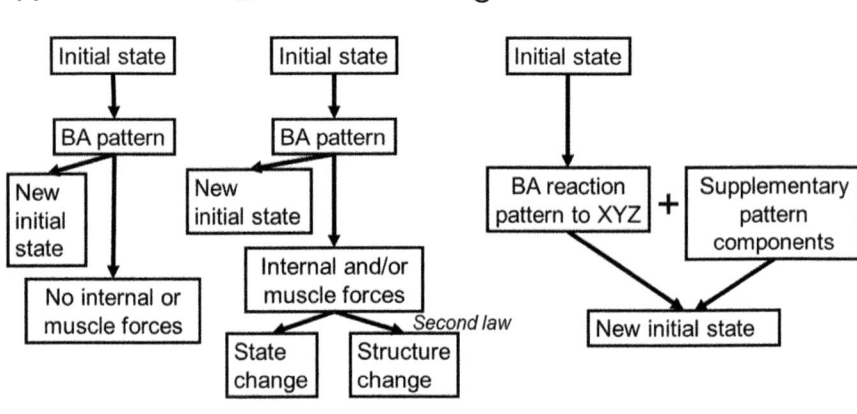

Fig. 6.3 Possible implications of an emerging BA pattern

6.1.2   External and Internal Forces Produced by Attractor Patterns

The emergence of a BA attractor pattern implies that the brain state changes from a particular initial state to a new state defined by the BA attractor pattern. That pattern may serve as an initial state for the formation of another attractor pattern. This situation is illustrated in panel (a) of Fig. 6.3.

In addition, a BA pattern in a certain brain area may exert forces on the brain activity in the same or different brain areas. Such forces are referred to as internal forces. Internal force can lead to further state changes as indicated in panel (b) of Fig. 6.3. That is, while the emerging BA pattern describes a transition between different brain states, the BA pattern that has emerged can induce further changes of brain activity. Internal forces can also lead to changes of structure (see panel b again). Therefore, internal forces produced by a certain BA attractor pattern may be interpreted as external force once they have been generated. In line with this interpretation, the state changes and structural changes caused by internal forces may be considered as application of the second law.

If a BA pattern comes with muscular force production, then the pattern is part of a muscular activity and BA pattern (as pointed out in panel (b)). The muscular activity may alter external circumstances and, in doing so, external forces. External forces induced in this way may lead again to state and structural changes. While in this section on the isolated human brain, muscular force production is not considered and likewise structural changes are not considered, this mechanism as illustrated in panel (b), in general, plays an important role as will be shown in subsequent sections of this chapter.

6.1.3 Supplementary Pattern Components and Stable Basis Pattern

As far as pattern formation in the ordinary state of the isolated human brain is concerned, panel (c) of Fig. 6.3 shows a special case of panel (b). When an attractor pattern as reaction to a force XYZ emerges, then the emerging pattern may involve supplementary components that are related to forces other than the force XYZ [127, 157]. Two examples are shown in Fig. 6.4 in this regard.

Panel (a) of Fig. 6.4 illustrates that when a visual world house attractor pattern emerges, then in addition an acoustic world house attractor pattern in English and in other languages such as Italian and French may emerge. Moreover, the emerging attractor pattern may have components that are related to the English spelling of the

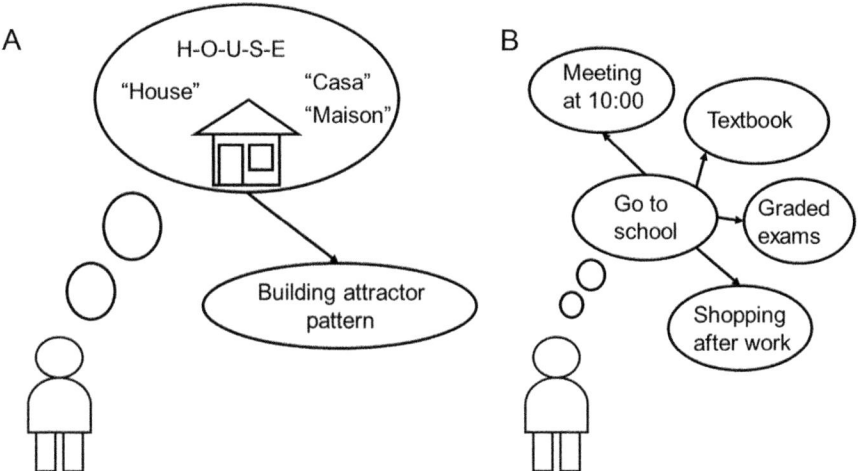

Fig. 6.4 Illustrations of supplementary pattern components

word house as indicated in panel (a) of Fig. 6.4 by the broken word H-O-U-S-E. In this context, the emergence of supplementary components has been referred to as "associative memory" [157]. Panel (b) of Fig. 6.4 illustrates the case of a teacher in which either the acoustic, own speech go-to-school isolated BA pattern or some kind of visual world go-to-school BA pattern emerges. Such a visual world pattern could be a pattern similar to a pattern that would emerge as reaction to the visual forces to which the teacher typically is exposed on his or her way to school (e.g. forces from certain street scenes, buildings, traffic lights, etc). In the example, the emergence of the go-to-school pattern involves the emergence of other patterns or components of other patterns. For example, the own speech meeting-at-10:00 pattern emerges. Furthermore, visual world patterns of a certain textbook, graded exams, and other things that the teacher needs to take to school emerge. The own speech shopping-after-work pattern is assumed to emerge in this example as well. The visual world textbook attractor pattern may makes the teacher to form a body pattern (i.e., a BBA pattern) such that he/she grasps the textbook. If so, we are dealing with a sequence of emerging patterns.

Interestingly, supplementary pattern components may correspond to stable basis patterns as discussed in Chap. 4. When stable basis patterns are pumped by unstable basis patterns, then the amplitudes of the stable basis pattern can increase and the patterns can become part of the emerging attractor pattern, see Sect. 4.3.8 and Fig. 4.11. For example, a house attractor pattern as shown in Fig. 6.4a may be composed of a visual world house *unstable* basis pattern, and three acoustic world *stable* basis patterns house, casa, maison about how houses are called in English, Italian, and French.

6.1.4   Type 1 Pattern Sequences and Cycles

Sequences of pattern formation are sequences in which patterns are formed one after another. Cycles are a special case of sequences. Cycles are sequences that are closed such that after a certain number of patterns a sequence repeats itself. Type 1 pattern sequences do not involve structural changes. Type 2 pattern sequences involve structural changes. In what follows, type 1 pattern sequences will be considered.

According to the human pattern formation reaction model, type 1 sequences of pattern formation occur in the multistable human brain. There are various possible mechanisms leading to type 1 sequences. Let us consider a mechanism according to which an attractor pattern produces a force such that its amplitude decays to zero or a relatively low magnitude. Figure 6.5 illustrates such a mechanism. Let us consider an awake person. We still consider the isolated human brain. That is, the person is not subjected to external forces. A healthy adult person is considered that should feature under appropriate circumstances a visual world box attractor. (That is, if the person is subjected to the forces of the visual world exerted by a box then in the symmetric case the brain activity would converges to a particular attractor, which is the visual world box attractor.) We consider the situation in

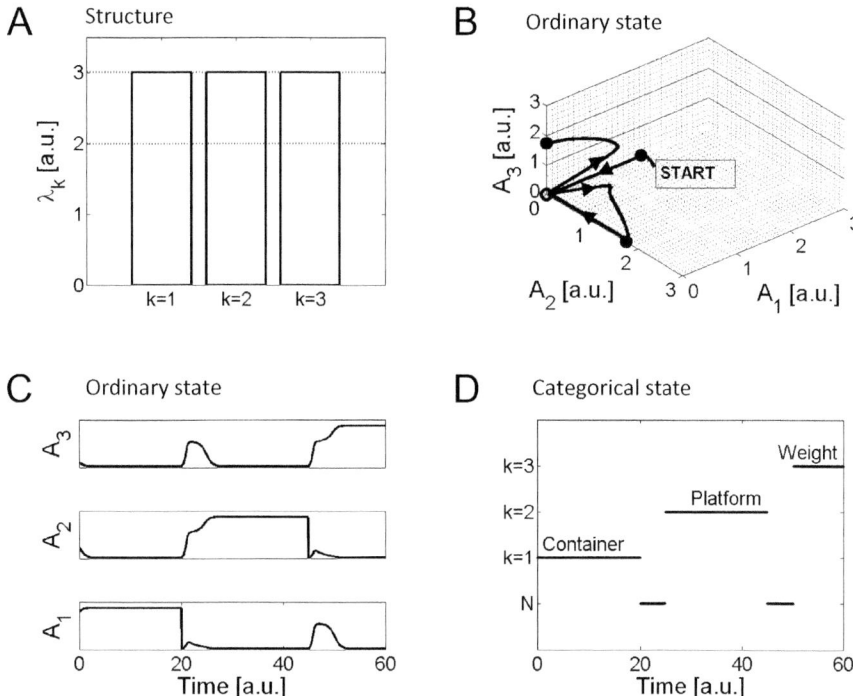

Fig. 6.5 Type 1 pattern sequences as computed from the LVH model for an individual in which BA attractor patterns related to the three object functions (container, platform, weight) of a box emerge in succession. Panel (**a**): Structure in terms of eigenvalues. Panel (**b**): Trajectory of the ordinary state (here: brain state) described in terms of unstable pattern amplitudes (i.e., the reduced amplitude space is shown). Panel (**c**): As in panel (**b**) but showing the time course of the unstable pattern amplitudes. Panel (**d**): Categorical brain (or ordinary) state over the course of time

which there is no box. However, it is assumed that at that time point of interest for the person under consideration the eigenvalue of the visual world box unstable basis pattern is sufficiently large such that the corresponding visual world box attractor exists. Moreover, it is assumed that the ongoing brain activity of the person under consideration is in the basin of attraction of the visual world box attractor. Consequently, the box attractor pattern emerges. According to Fig. 6.5 the box attractor pattern triggers the formation of subsequent patterns. Three patterns related to functions of a box are considered. A box may be used as container, platform, or weight. It is assumed that there is no bias towards any of those functions. The pattern formation system (modeled by the LVH model) is assumed to be symmetric. That is, the eigenvalues of all three basis patterns assume the same value, see panel (a) of Fig. 6.5.

The reduced amplitude space spanned by the basis pattern amplitudes exhibits three stable fixed points located on the amplitude axes. The origin is an unstable fixed point. In the simulation presented in Fig. 6.5 it is assumed that while there is no bias with respect to the structure of the neuronal network under consideration, the box attractor pattern leads to a biased shift of the brain activity such that the amplitude A_1 of the container basis pattern is large relative to the amplitudes A_2 and A_3 of the two other box function basis patterns. Consequently, the amplitude A_1 converges to its finite stationary value, whereas the amplitudes A_2 and A_3 decay to zero, see panel (c). The container basis pattern and the corresponding container attractor pattern (that may involve only the basis pattern or several stable patterns in addition to the basis pattern, see Chap. 4) emerges. After a certain duration, as mentioned above, the attractor pattern produces a force that results in the decay of the amplitude of the attractor pattern that is present. In the simulation, the amplitude A_1 is put to a negligibly small value, see panel (c) again. Since the amplitudes A_2 and A_3 of the remaining box function basis patterns are typically larger than that small value, they dominate the amplitude A_1 such that A_1 cannot emerge again. The amplitudes A_2 and A_3 complete with each other, see panel (c). In the simulation shown in Fig. 6.5 the amplitude A_2 wins the competition process. The platform basis pattern and the corresponding platform attractor pattern emerge. After a short duration, the aforementioned damping process takes place again: A_2 is put to a negligibly small value. A_1 and A_3 compete with each other. A_3 wins the competition. The basis pattern and the attractor pattern related to the weight object function of a box emerge. Panel (b) not only shows the fixed point attractors but also the trajectory of the pattern formation sequence. Panel (b) also illustrates that the structure, here given in terms of the fixed point attractors, does not vary during the whole sequence. In contrast, the state does vary. In doing so, panel (b) points out that type 1 sequences do not involve structural changes as part of the emergence of attractor patterns. Finally, panel (d) shows the brain state in terms of the attractor patterns that are present at any given time point. The label N stands for nothing again, meaning, that during those episodes indicated by N no clear brain pattern was formed (see also Sect. 1.3 and Fig. 1.2b). From the pattern formation perspective of humans, Fig. 6.5 illustrates the case of an isolated human brain that forms a sequence of brain activity patterns. The brain under consideration forms a visual world box attractor pattern (not shown) and subsequently forms due to self-damping a sequence of three visual BA attractor patterns, which are the BA patterns of the container, platform, and weight box functions. We may say Fig. 6.5 illustrates a person that "thinks" about a box and subsequently "thinks" about three box functions, which are the container, platform, and weight box functions.

According to our consideration made above, type 1 sequences as illustrated in Fig. 6.5 (and likewise cycles) may occur in the awake human person. Such type 1 sequences in the awake human may be used to explain what has been called in the literature "thinking" and "imagination". As mentioned above, we have initially considered an awake human person. However, it is plausible to assume that type 1 sequences as discussed above can occur in the REM sleep stage as well. In addition, they might occur due to the intake of drugs or under certain disease conditions. Table 6.4 summarizes those examples. Table 6.4 may be regarded as generalization of Tables 6.2 and 6.3 to the case of sequences and cycles.

Finally note that an engineered example of a pattern formation system that exhibits pattern formation cycles is the Belousov-Zhabotinsky hand [357]. In this

Table 6.4 Type 1 sequences of pattern formation in the human isolated brain

Condition	Subcondition/related to …	Pattern formation terminology	Other terminology
Awake	Visual and acoustic worlds forces	Type 1 sequences of visual and acoustic worlds isolated BA patterns in the awake person	"Thinking", "imagination"
REM sleep	Visual and acoustic worlds forces	Type 1 sequences of visual and acoustic worlds isolated BA patterns under REM sleep	Dreaming
Drug-induced	Visual and acoustic worlds forces	Type 1 sequences of drug-induced visual and acoustic worlds isolated BA patterns	"Drug-induced hallucinations"
Disease-related	Forces acting on the visual, acoustic, and tactile sensory system	Type 1 sequences of disease-induced visual, acoustic, and tactile worlds isolated BA patterns	"Disease-induced hallucinations"

system a robot hand is coupled to a chemical reaction. The chemical reaction, the so-called Belousov-Zhabotinsky reaction, leads to the emergence of spirals in a two dimensional plane. The emerging spirals determine the finger movements of the robot hand. The finger movement change the chemical concentrations in the two dimensional plane. In this sense, the finger movements shift the state of the chemical system. That is, we deal with a pattern formation system that shifts its own state by means of a feedback loop. Oscillatory finger movements are observed under appropriate conditions. However, it is open for debate whether or not the system exhibits transitions between attractor states as considered in the example above.

6.2   Pattern Formation Involving Muscle Force Production: B1 Systems

Human brain activity can exert forces on the muscular system, which, in turn, leads to muscle force production. Human systems of the class B1 describe minimal systems that feature muscle force production. By inspection of the component-component interactions in B1 systems, see Fig. 5.5, the pattern formation is assumed to take place in the brain state only. The muscular system is a driven system. In B1 systems there are no direct or indirect feedback loops from the muscular system to the brain state. Class B1 systems capture an idealized situation. In general, muscle activity is connected to the proprioceptive system that in turn exerts forces on the brain state (direct feedback) [179, 326]. Likewise, when a human forms a moving BBA attractor pattern (i.e., when a person is moving from a location to another location) the body movement typically changes external circumstances, which, in turn, changes the external forces acting on the human (i.e., the human pattern formation system). Human B1 pattern formation systems are idealized systems in which this kind of feedback can be neglected (just as the proprioceptive feedback). Consequently, examples of attractor patterns of B1 systems are all kind of physical activity patterns such as walking, running, or reaching provided that feedback does not play a role for the emergence of those activity patterns. For example, it is assumed that certain rhythmic body movements are generated by means of so-called central pattern generators [144, 145, 147, 172, 298]. Central pattern generators are units (or subsystems) in the human body that produce rhythmic output signals in the absence of rhythmic input signals. The output signals act as driving forces of muscle activity. Central pattern generators that involve only so-called feed forward connections [200] are an example of B1 systems.

Type 1 sequences and cycles of pattern formation involving muscle force production can be discussed by analogy to the preceding section. For examples, an attractor pattern involving muscle forces may produce internal forces that lead to the decay of its attractor pattern amplitude. If the system under consideration is multistable, the amplitude decay may result that the brain state (or ordinary state) is shifted to the basin of attraction of another attractor. If so, a transition between two attractor patterns occurs. For example, when a person forms a running pattern, after a while the running pattern may break down due to self-damping of the pattern amplitude. A walking pattern or even a standing still pattern may emerge.

6.3   Patterns As Reactions to External Forces Inducing Initial Conditions: B2 Systems

In this section we consider pattern formation in B2 systems. Accordingly, external forces shift the brain state and, in doing so, put the human brain into a particular initial state. Depending on that initial state and on the structure of the brain, a particular pattern emerges as reaction to the forces. Neither the external forces nor the emerging pattern changes the structure. However, the emerging pattern may affect the brain state as discussed in the pervious section. In doing so, an external force can trigger a type 1 sequence of emerging patterns.

6.3.1   Reactions to Objects

A fundamental phenomenon is how the human brain reacts to forces exerted by objects for which under idealized symmetric conditions the corresponding object attractors exist. This phenomenon has also been called "object recognition". In what follows, a simplified description of the pattern formation approach [61, 95, 97, 125,

127, 157, 299] to understand the reaction of humans to objects will be presented. More details can be found in the original studies.

Let us illustrate human reactions to objects by means of the reactions to the written capital letters E and H. As shown in Fig. 6.6 (and already anticipated in Fig. 5.10a), it is assumed that under symmetric conditions the adult human brain exhibits letter attractors such as the letter E and letter H attractors. Brain activity at the location of those fixed point attractors forms letter E and H attractor patterns, respectively. As such, these patterns can emerge without the presence of external forces, for example, as letter E or letter H isolated BA patterns under REM sleep or in the awake person (see Sects. 5.5 and 6.1.1).

In contrast to the mechanisms discussed in the previous sections that hold for A1 and B1 pattern formation systems, in the context of B2 pattern formation systems it is assumed that the visual world external forces of the letters E and H affect the ongoing brain activity and shift it to certain initial states [61, 95, 97, 125, 127, 157, 299]. Figure 6.7 illustrates this mechanism in more detail.

First of all, the symmetric case is considered in which the letter E and H basis patterns exhibit the same eigenvalues, see panel (a). According to the LVH

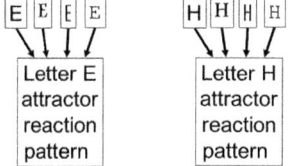

Fig. 6.6 Attractor patterns emerging as reactions to the forces exerted by various types of capital letters "E" and "H"

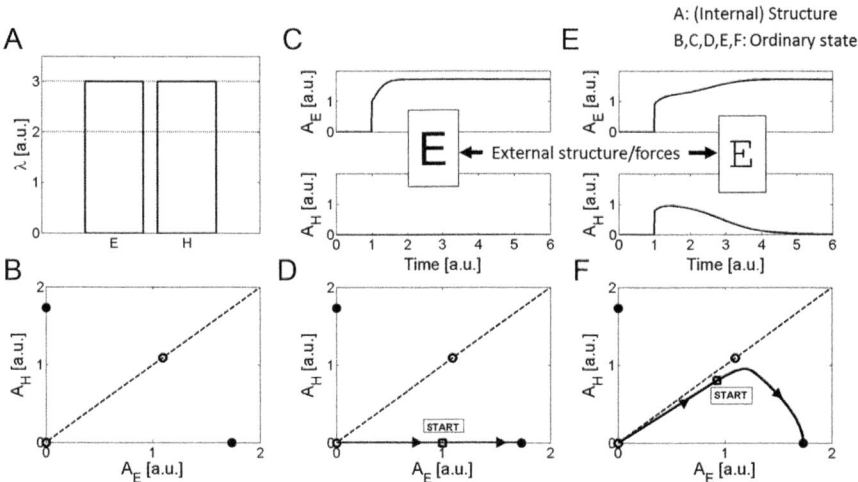

Fig. 6.7 Formation of BBA attractor patterns as reaction to letters E and H. See text for details

amplitude equations, in the reduced amplitude space spanned by the amplitudes A_E and A_H of the basis patterns E and H shown in panel (b), there are two stable fixed points on the axes of the reduced amplitude space and there are two unstable fixed point (the co-existence point at which both amplitudes are finite and the origin at which both amplitudes are zero). The fixed points on the axes correspond to the letter E and letter H attractors, respectively. The diagonal separates the basins of attraction of the two attractors. That is, for any initial brain state in the lower (upper) half, the brain activity converges towards the fixed point of the letter E attractor (letter H attractor). Panel (c)/(d) and (e)/(f) illustrate two different cases of reactions to letter E forces. Panels (c)/(d) refer to the case in which the letter E exerts forces that only increase deviations from the reference brain state that correspond to the letter E basis pattern. Consequently, in the two-dimensional reduced amplitude space the brain state is shifted only along the amplitude axis of the letter E basis pattern. In other words, the letter E basis pattern increases in amplitude due to the impact of the external force, while the amplitude of the letter H remains at the reference level. This is shown in panel (c) (both subpanels) at the 1 unit time point. In panel (d), the shift is indicted by a square. That square describes the initial brain activity as relevant for the formation of a pattern and, consequently, has been labeled by START. Not surprisingly, according to the LVH amplitude dynamics, the amplitude A_E of the letter E basis pattern increases towards its stationary finite value defined by the letter E attractor, while the amplitude A_H of the letter H attractor remains on the reference level (that corresponds to a letter H basis pattern amplitude of zero), see panels (c) and (d). Panels (e)/(f) illustrate the reaction to forces that set the initial brain state to a location that does not fall on one of the two axes of the reduced amplitude space. For the letter E as written in the Courier style shown in panel e it is assumed that the initial state is close to the diagonal (due to the fact that the Courier style letter E exhibits more vertical segments that are similar to the letter H). Panel (e) shows the shift at time point of 1 unit of the amplitudes A_E and A_H of the letter E and H basis patterns. In panel (f) this shift is indicated by the square labeled START. As can be seen in panel (f), the initial state is located in the lower half of the reduced amplitude state. Accordingly, the Courier style written letter E puts the brain activity in the basin of attraction of the letter E attractor. As a result, the brain activity converges over time to the letter E attractor. This implies that the letter H basis pattern amplitude A_H decays to zero, while the letter E basis pattern amplitude A_E increases towards its stationary, finite value. In both cases (c)/(d) and (e)/(f) the letter E attractor pattern emerges as a result of the forces exerted by the differently written letters E.

Let us turn to the letter E attractor pattern. The letter E attractor pattern is assumed to have as dominant component the letter E basis pattern. As discussed in Chap. 4, it may involve other stable basis patterns (i.e., basis patterns with negative eigenvalues). As discussed above in Sect. 6.1.3, the letter E attractor pattern may exhibit supplementary components (that may or may not correspond to stable basis patterns). For example, it might be possible that a combined acoustic and visual worlds attractor E pattern emerges. That is, a pattern that would emerge when the person would be exposed to the visual forces of a letter E and the acoustic forces of somebody speaking the letter E (i.e., when the person would see the letter E and hear the sound of a spoken letter E). In this case, that is, if only the letter E is shown to a person but the combined acoustic and visual worlds pattern emerges, then the person would "perceive" the letter E and simultaneously would have a "verbal thought" of the letter E.

Supplementary pattern components could also correspond to patterns that would emerge when the person is exposed to objects that (assuming a native English speaker) would begin with the letter E like eye, ear, eggs, etc. or names of persons starting with the letter E like Elizabeth or Elvis. If the person happens to be a fan of the late rock singer Elvis Presley, then the emerging letter E attractor pattern, may result in the shift of brain activity such that the brain activity is put into the location of a basin of attraction of an attractor related to famous Elvis Presley songs. In doing so, we could get for example the following pattern formation sequence. First, the letter E attractor pattern emerges. Subsequently (or almost simultaneously [157]), an acoustic attractor pattern of the word Elvis emerges, and finally, an acoustic world attractor pattern related to the song "Your are always on my mind" emerges.

6.3.2 Reaction Patterns

The phrase to form a brain and body activity pattern (i.e., a BBA pattern) as reaction to something may be abbreviated by the phrase to form a reaction pattern. For example, when a person forms a brain and body activity pattern as reaction to a car, then we may say the person under consideration forms a car reaction pattern.

6.3.3 Multistable, Winner-Takes-All Human Pattern Formation Systems Subjected to Several Forces

As we have seen in Chap. 4, in general, self-organizing systems are multistable. For the human brain under the impact of several external forces multistability in the context of a winner-takes-all dynamics has the following implications. If the human brain exhibits under certain conditions several attractors, say, the attractors X and Y, and if the brain activity happens to form the attractor pattern X, then (by definition) the brain does not form the attractor pattern Y as long as the pattern X is present. That is, the property of winner-takes-all systems that patterns are mutually exclusive implies an "either-or" situation, in which either one pattern or another pattern can emerge.

Let us consider an individual exposed to the forces exerted by the scene shown in panel (a) of Fig. 6.8. Panel (a) of Fig. 6.8 shows a house with a roof, a door, a window and the house number 1. Let us assume that the individual under consider has a visual world house attractor as well as several visual world attractors for numbers.

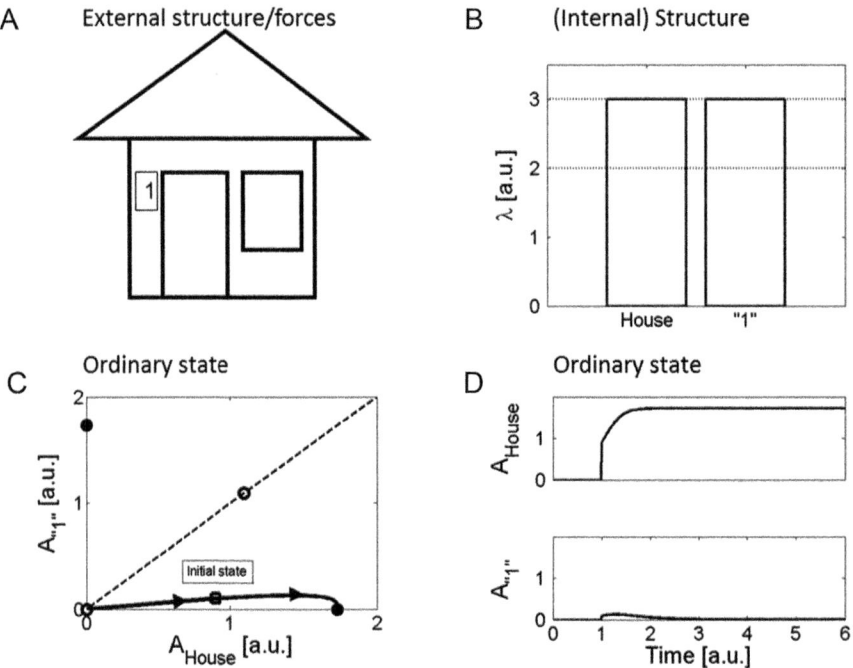

Fig. 6.8 Formation of a visual world house BBA attractor pattern under the impact of external forces exerted by several objects. Panel (**a**): Illustration of relevant external forces and objects: house and house number 1. Panel (**b**): Eigenvalue spectrum (structure). Panel (**c**): Evolution of the pattern formation system in the reduced amplitude space. Panel (**d**): As in panel (**c**) but the time course of the amplitudes is shown

In what follows, only the number-1 attractor will be considered. The brain of the individual under consideration is assumed to be bistable in the sense that the house attractor and the number-1 attractor both exists.

Typically, brain activity will converge to the house attractor rather than the number-1 attractor. Let us discuss three scenarios describing this situation. The first scenario is a B2 system scenario that assumes that the external forces do not affect the brain structure of the individual. They only set the initial brain state for the formation of a BA pattern as in B2 systems. Panels (b)–(d) of Fig. 6.8 illustrate this case. It is assumed that the structure is symmetric. That is, the eigenvalues λ_{House} and $\lambda_{"1"}$ of the house and number-1 basis patterns assume the same value, see panel (b). In this symmetric case, from the LVH model it follows that in the reduced amplitude space spanned by the amplitudes A_{House} and $A_{"1"}$ of the two basis patterns, there are two stable fixed points on the axes corresponding to the house and number-1 attractors and there are two unstable fixed points (the origin and a co-existence point on the diagonal). The diagonal separates the basins of attraction of the two attractors. As such the situation is the same as in our previous example on the formation of visual world letter BBA attractor patterns (compare Figs. 6.7 and 6.8). In the first (B2 system) scenario it is assumed that the scene shown in panel (a) exerts forces that shift the brain state to a location in the basin of attraction of the house attractor. In the example shown in panels (c) and (d), the individual is exposed to the scene at 1 time unit and at that moment the forces exerted by the scene shift the amplitude A_{House} of the house basis pattern by 0.9 units, while they shift the amplitude $A_{"1"}$ of the number-1 basis pattern only by 0.1 units. Consequently, as shown in panels (c) and (d) the brain activity as described in terms of the basis pattern amplitudes $A_{House}(t)$ and $A_{"1"}(t)$ converges to the house attractor and the house attractor pattern emerges. Importantly, as long as the pattern formation system is located in the attractor, that is, the brain activity forms the house attractor pattern, the system cannot form the number-1 attractor pattern. In order to form the number-1 attractor pattern, the house attractor pattern must induce a shift of brain activity (as discussed in Sect. 6.1.4 in the context of type 1 pattern sequences) or result in a structural change that decreases its own eigenvalue and/or increases the number-1 eigenvalue (see examples in Chap. 7).

Using terminology that is actually not needed from a scientific point of view (see Sect. 6.3.4 below), we would say that as long as the house attractor pattern is formed the person under consideration "focuses" or "pays attention" to the house and therefore is "blind" to or cannot "perceive" the house number 1.

Figure 6.9 illustrate the second fundamental scenario that leads to the emergence of the house attractor pattern. In this scenario the forces exerted by the scene affect the amplitudes of the house and number-1 basis patterns in the same way. However, the forces also affect the structure of the brain and are assumed to do so in an asymmetric way. Consequently, we are dealing with pattern formation in a D2 system (see Fig. 5.5). Panel (a) shows the reduced amplitude space before the forces act on the individual. It is assumed that the eigenvalues of both basis patterns are the same. Consequently, in the reduced amplitude space the attractors and unstable fixed point are located just as in the previous example (compare panel (a) of Fig. 6.9 with panel (c) of Fig. 6.8). At 1 time unit the individual is confronted with the house/number-1 scene. Both amplitudes A_{House} and $A_{"1"}$ are shifted by 0.5 units. At the same time the eigenvalue λ_{House} of the house basis pattern is increased by 1 unit. Panel (c) shows the eigenvalues λ_{House} and $\lambda_{"1"}$ as functions of time and illustrates the shift by 1 unit of the eigenvalue λ_{House} of the house basis pattern. Due to the structural change (i.e., the shift of the eigenvalue) the topology of the reduced amplitude space changes from panel (a) to panel (b). In panel (b) it is shown that the attractor location of the house attractor is shifted to a somewhat higher value. This is a consequence of the increased eigenvalue (see Sects. 4.6.5 and 4.7.10). Importantly, the co-existence point and the separation line shift in the direction of the number-1 attractor. That is, the basin of attraction of the number-1 attractor becomes smaller, whereas the basin of attraction of the house attractor becomes larger. Although both amplitudes are shifted by the same amount the initial state of brain activity thus obtain is not on the separation line between the basins of attraction of the two attractors. Rather, the initial state falls in the basin of attraction of the house attractor. Consequently, as shown in panels (b) and (d)

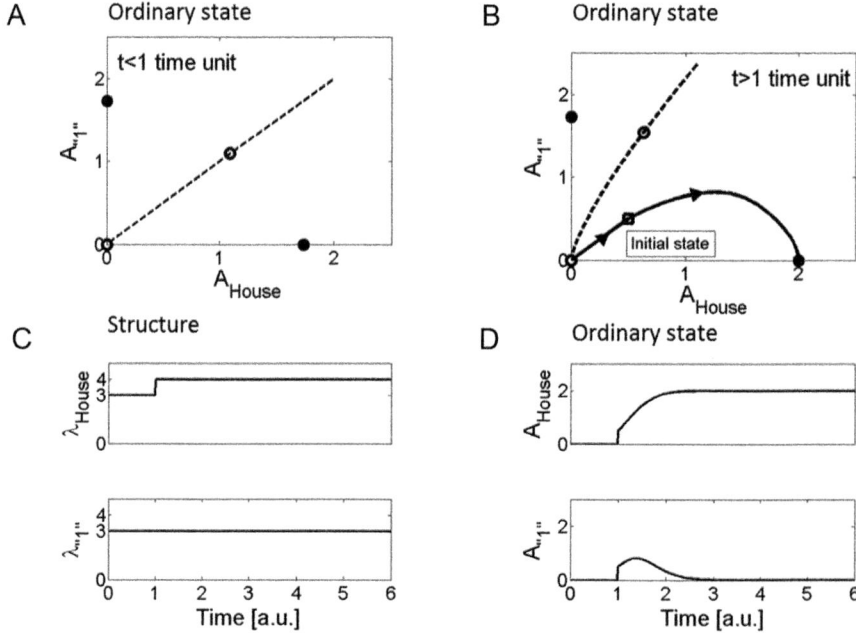

Fig. 6.9 Scenario of formation of a BBA attractor pattern under the impact of several forces that involve force-induced structural changes. Panels (**a**) and (**b**): Phase portraits before and after a human is exposed to the house/number-1 scene shown in panel (**a**) of Fig. 6.8. Panel (**c**): Structural change induced by the external forces exerted by the house. Panel (**d**): Evolution of the brain state as described by the relevant unstable pattern amplitudes in the reduced amplitude state: the convergence of $(A_{house}, A_{"1"})$ to $(A_{house}, A_{"1"}) = (2, 0)$ indicates the emergence of the visual world house BBA attractor pattern

the brain activity (when depicted in terms of the amplitudes $A_{House}(t)$ and $A_{"1"}(t)$ of the basis patterns) converges to the house attractor. The house attractor pattern emerges. Just as in the first scenario, as long as the brain activity forms this pattern, it does not form the number-1 pattern (and, using again unnecessary terminology, in this sense the individual exhibits "selective attention" to the house feature of the scene and "ignores" the number-1 feature).

The third scenario is a mixture of the first (B2 system) and second (D2 system with unbiased brain shifts) scenario and takes place in D2 systems again. Accordingly, the forces exerted by the house/house-number-1 scene shift the brain activity to a biased initial state (e.g., biased towards the house attractor pattern) and at the same time cause biased changes of the brain structure (e.g., such that the basin of attraction of the house attractor increases).

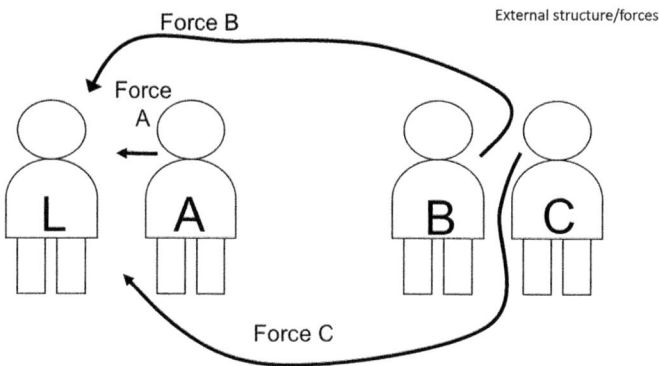

Fig. 6.10 Set-up at a cocktail party. Two groups of two people in each group stand together. Each pair is involved in a conversation. Person L is listening. Persons A, B, and C are talking

The Cocktail Party Example

Let us illustrate the consequence of the multistability of the human brain by an example that has been frequently discussed in the literature: the cocktail party phenomenon. At cocktail parties, people may stand in groups together. People standing together in a given group are able to talk about their topic although they are surrounded by other people talking about other things. Figure 6.10 illustrates the basic scenario. In this scenario, there are four people arranged in two pairs of people. Without loss of generality, we describe the effect from the perspective of the first person in the first group, who is assumed to listen only. That is, the person does not do any talking. Accordingly, the individual will be referred to as listener (and has been labeled L). In the group of the listener there is another person, say person A. The members of the second group are denoted accordingly as persons B and C. Persons A, B, C are assumed to do the talking. In doing so, they exert forces on the listener L.

The reference state of the brain of listener L is assumed to become unstable when joining the cocktail party. Let us consider the symmetric case first. At the instability point, under symmetric conditions the brain structure is such that the brain as a pattern formation system exhibits several unstable basis patterns of brain activity with the same eigenvalues. Among those basis patterns there are three basis patterns that are of particular interest in what follows. Under symmetric conditions, according to the LVH model, the basis patterns give rise to three attractors that correspond to winner-takes-all fixed points. If the ongoing brain activity is in the basin of attraction of one of the attractors then the brain activity converges to the attractor and the corresponding attractor pattern emerges. In the idealized case, if only person A speaks and persons B and C are quiet, the external forces of the sound waves produced by A shift the brain activity to a location in the basin of attraction of one of the attractors. The brain activity converges to the fixed point of that attractor and the corresponding attractor pattern emerges (which may involve stable basis patterns, see Chap. 4, and comes with supplementary pattern components, see Sect. 6.1.3 above). Using the terminology introduced in Sect. 5.4.1, the particular attractor pattern will be referred to a listen-to-A BBA attractor pattern. The corresponding basis pattern and the corresponding attractor are the listen-to-A basis pattern and attractor, respectively. The amplitude of the listen-to-A basis pattern is the listen-to-A amplitude denoted by A_A. Likewise, the two remaining basis patterns can be identified as listen-to-B and listen-to-C basis patterns giving rise to listen-to-B and listen-to-C attractors that come with listen-to-B and listen-to-C BA attractor patterns.

Figure 6.11 illustrates the cocktail party phenomenon for the aforementioned symmetric case. Panel (a) shows the eigenvalues of the basis patterns. Panel (b) shows the three-dimensional reduced amplitude space spanned by the three aforementioned basis pattern amplitudes A_A, A_B, and A_C. The listen-to-A, B, C attractors correspond to locations on the axes, as indicated in panel (b). The reference state given as the origin corresponds to an unstable fixed point. When listener L and person A come together (i.e., form pair standing together), the forces exerted by person A are assumed to dominate the forces exerted by persons B and C such that the brain state of listener L is shifted to a location in the basin of attraction of the listen-to-A attractor. Panels (b) and (c) illustrate that initial state. The brain activity then converges to the listen-to-A attractor, see panels (b) and (c) again.

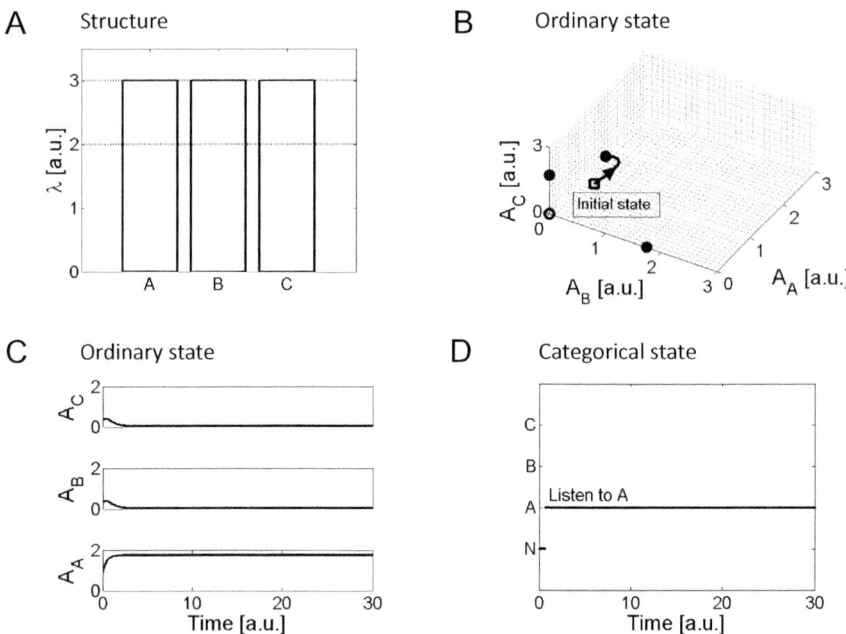

Fig. 6.11 Baseline situation in which listener L forms the listen-to-A BBA attractor pattern. See text for details

The listen-to-A attractor pattern emerges. When the pattern has emerges, we may say that listener L is in the categorical state of listening to A. Panel (d) illustrates the evolution of the human pattern formation system of our hypothetical listener L in terms of the categorical states of listening to A, B, C and the state of doing nothing (see Fig. 6.5 above). As long as the brain activity forms the listen-to-A attractor reaction pattern, the brain activity does not form the listen-to-B or listen-to-C attractor patterns.

Using terminology that is frequently used in the literature but leads to a mystification of humans and animals (see Sect. 6.3.4 below), one could (but maybe better should not do so) say that listener L "focuses" or "pays attention" to person A.

Figure 6.12 demonstrates a switch from one attractor pattern to another assuming that the listener L corresponds to a B2 system. It is assumed that at a particular point in time person C starts laughing loud. The forces produced by C (in terms of the mechanical forces of sound waves hitting the listener L ear receptors that are transduced into chemical driving forces and electric forces at synapses, axons, dendrites, and neuron bodies) are assumed to shift the ongoing brain activity (measured e.g. in terms of axonal pulse rates and dendritic currents in appropriately

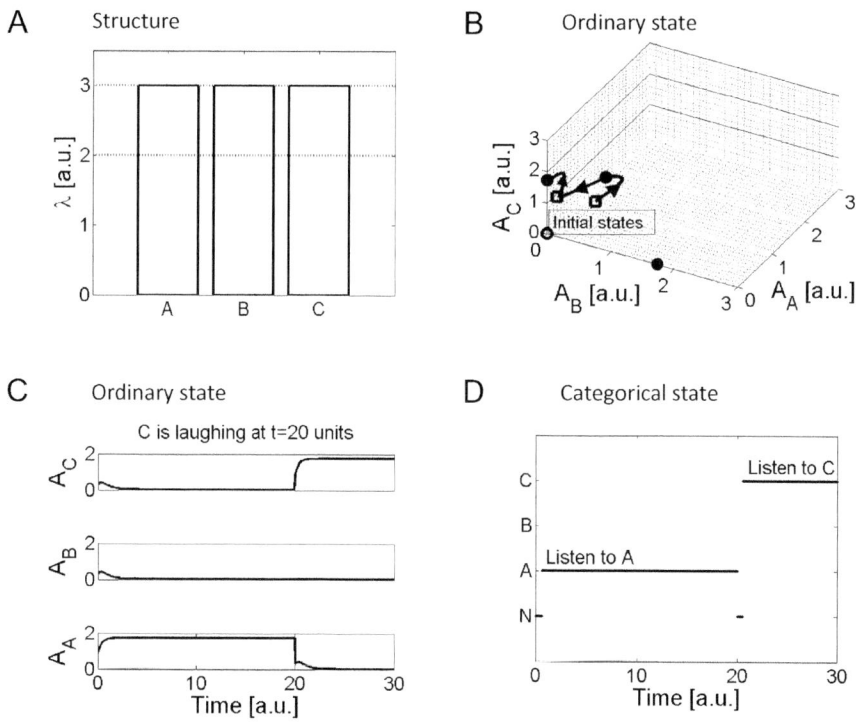

Fig. 6.12 Switch from a listen-to-A BBA attractor pattern to a listen-to-C attractor pattern due a shift of the ordinary state caused by the external force exerted by speaker C (B2 system scenario). See text for details

defined brain areas) out of the listen-to-A attractor to a location in the reduced amplitude space that is in the basin of attraction of the listen-to-C attractor. Panel (a) of Fig. 6.12 is a replication of panel (a) of Fig. 6.11 and just reminds the reader that the symmetric case is considered. Panels (b) and (c) demonstrate the assumed shift of the brain activity at the time point at which person C starts laughing (which in the simulation happens at 20 time units). The shifted brain activity describes the initial state for the formation of the listen-to-C BBA attractor pattern that takes place subsequent to the brain activity shift. Panel (d) illustrates that as far as the categorical states of listener L are concerned, he/she is first listening to A and subsequently listens to C. Again, one could (but maybe better should not do so) say that listener L shifts "attention" or "attentional resources" from person A to person C.

In the two examples about the cocktail party phenomenon depicted in Figs. 6.11 and 6.12 it is assumed that the structure of the brain of listener L is fixed throughout the period under consideration and that the structure assumes symmetry leading to multistability. However, the brain structure may exhibit asymmetry and that asymmetry may change over the course of time, in particular, due to the impact of external forces (e.g. the laughing of person C). Figure 6.13 illustrates the cocktail party phenomenon for a listener L exhibiting a time-varying asymmetric brain structure. In doing so, a D0 or D2 system is considered.

When listener L joins the party together with persons A, B, C it is assumed that due to the fact that listener L and person A stand together as a pair, the brain structure of listener L in fact is or becomes asymmetric. In this scenario, the eigenvalue λ_A of the listen-to-A basis pattern is assumed to be large relative to the eigenvalues λ_B and λ_C of the two other basis patterns, see panel (a). In particular, the eigenvalue is

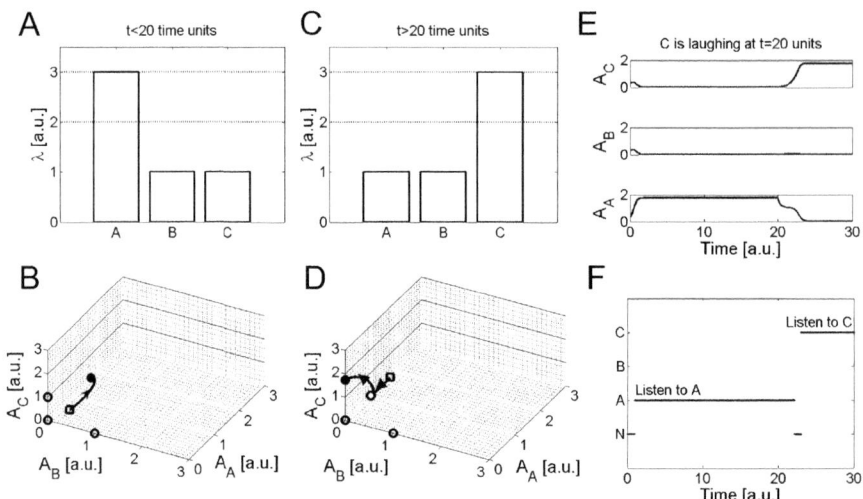

Fig. 6.13 Bifurcation from a listen-to-A BBA attractor pattern to a listen-to-C attractor pattern due to a structural change caused by the external force exerted by speaker C (D0/D2 system scenario). See text for details

sufficiently high such that the eigenvalues of the two other basis patterns drop out of the stability band (see Sect. 4.6.3). Consequently, the listen-to-B and listen-to-C basis patterns do not give rise to attractors but to unstable fixed points. The brain activity does not converge to those fixed points. Listen-to-B and listen-to-C attractor patterns cannot emerge.

Panel (b) of Fig. 6.13 shows the reduced amplitude space and the stable and unstable fixed points. The brain of listener L as a pattern formation system is monostable and exhibits only the listen-to-A fixed point attractor. For an arbitrary initial state the brain activity converges to the listen-to-A attractor and the corresponding pattern emerges. Panel (e) shows the emergence of the listen-to-A attractor patterns in terms of the dynamics of the basis pattern amplitude A_A, A_B, and A_C. Panel (f) describes the categorical state of the listener A over time. Panels (e) and (f) show that the brain activity of listener L settles down in the listen-to-A attractor pattern (panel e) and accordingly listener L can be said to listen to person A (panel f). At 20 time units it is assumed again that person C starts laughing. The forces produced by C are assumed to affect the brain structure, see panel (c). More precisely, they are assumed to switch the asymmetry (compare panels a and c). The eigenvalue λ_C of the listen-to-C basis pattern is assumed to become large, while the eigenvalue λ_A of the listen-to-A basis pattern drops back to the baseline level. As a result, the listen-to-A attractor disappears. It turns into an unstable fixed point and the location of the fixed point shifts along the amplitude axis A_A to a smaller value. Conversely, the unstable fixed point related to the listen-to-C basis pattern becomes an attractor and the fixed point location shifts to a larger value. Panel (d) shows the stable and unstable fixed points after the switch in asymmetry in the human brain of listener L caused by the forces exerted by person C on the listener L. Simulation of the LVH amplitude equations reveals that the brain activity exhibits a dramatic drop in the amplitude A_A of the listen-to-A BBA basis pattern (see the bottom subpanel of panel (e) at 20 time units), which corresponds to the shift of fixed point A along the A_A axis in the reduced amplitude space, see panel (d). In more detail, the location of the listen-to-A attractor that was present before person C started to laugh serves as initial brain state (indicated by a square in panel d). From this initial state the brain activity shifts to the unstable fixed point of the listen-to-A basis pattern, which corresponds to the first phase of rapid A_A amplitude decay shown in bottom subpanel of panel (e). Since this fixed point is unstable, the brain activity subsequently converges to the stable listen-to-C fixed point, that is, to the listen-to-C attractor. This contributes to the second phase of the A_A amplitude decay shown in panel (e). At the same time, the A_C amplitude increases over time. In summary, panels (d)–(f) demonstrate the switch. Panel (f) shows the switch from listening to A to listening to C in categorical terms. Note that there is a delay in the emergence of the listen-to-C BBA attractor pattern. That delay depends on the structural details of the brain as a dynamical system and pattern formation system.

6.3.4   Multistability, Winner-Takes-All Dynamics, and "Selective Attention"

Self-organizing systems of the animate and inanimate worlds can exhibit different stable and unstable states and patterns at the same time. Let us first consider multistable systems, see Table 6.5. Ferromagnetic materials are self-organizing equilibrium systems (see Sect. 3.2.3). If they assume rod shaped forms, then at sufficiently low temperatures they exhibit two stable state. They show either a magnetic north pole on one face-side of the rod and the south pole on the other side or show the south pole on the one side and the north pole on the other side. Convection roll patterns in gases and fluids heated from below show bistability. The rolls can rotate clockwise or counter-clockwise. Convection roll patterns are non-equilibrium patterns. Certain lasers under appropriate conditions can exhibit for the same pumping current two types of laser light. Roughly speaking, laser light in general comes with a directional property called polarization. Let us consider a three dimensional Euclidean coordinate system in which the laser light travels in the z-direction. Then bistable lasers can exhibit for the same pumping current light that is polarized in x-direction or in y-direction [16, 48, 244]. Importantly, if bistable lasers produced x-polarized laser light, then they do not produced y-polarized light at the same time. They can produce only either of the two types of light at a particular time point. Lasers are self-organizing non-equilibrium systems. Gas discharge systems fall in the same category as laser. They are self-organizing non-equilibrium systems. For a given voltage gap across the plates of a gas discharge system, the system can exhibit different discharge patterns [263, Sec. 2.3]. Cells can exhibit bistability in the sense that they can exhibit bistable signaling pathways. Those pathways among other things feature two biochemical substances, say X and Y. In the bistable domain, for fixed reaction rates the signaling pathways can be either in a stable state for which X assumes a high and Y a low concentration or in a stable state for which X assumes a low and Y a high concentration [111, 233, 257]. Cells are living self-organizing non-equilibrium systems.

Humans are assumed to exhibit bistability as well (and in general show not only bistability but multistability). As far as our example about the house/house-number-1 scene in Fig. 6.8 is concerned, the forces exerted by a house alone (i.e., without the house number) are assumed to lead to emergence of a house attractor

Table 6.5 Examples of multistable systems in the animate and inanimate worlds

System	Equilibrium or non-equilibrium	Animate or inanimate
Ferromagnetic materials	Equilibrium	Inanimate
Convection rolls	Non-equilibrium	Inanimate
Bistable lasers	Non-equilibrium	Inanimate
Gas discharge systems	Non-equilibrium	Inanimate
Cells	Non-equilibrium	Animate
Humans and animals	Non-equilibrium	Animate

reaction pattern. If a number-1 alone (i.e., without context) is presented to the same hypothetical individual, then the forces exerted by that number-1 picture are assumed to lead to the emergence of a number-1 BBA attractor pattern. That is, depending on the initial brain state and the forces that are assumed to affect the initial brain state, the human pattern formation system can exhibit after a transient period several, different attractor patterns. As far as the cocktail party phenomenon under assumed fixed and symmetric brain structure as demonstrated in Fig. 6.11 is concerned, the brain is assumed to exhibit three stable states (given by the listen-to-A, B, C BBA attractor patterns) and the brain activity can correspond to any of those states.

From the physics perspective of the nature of humans, the fact that a human individual as a multistable pattern formation system is in a particular stable state does not mean anything else than that the individual is in that particular state and not in any other state. One could (but maybe better should not do so) use phrases like "focusing on", "ignoring", "attention", or "selective attention". These phrases are not needed when taking the physics perspective of the world that treats animate and inanimate systems on an equal footing. When a rod-shaped ferromagnetic shows on one face-side a south pole and not a north pole then the ferromagnetic does not "focus" to produce the south pole on that face-side. The ferromagnet does not "ignore" the state in which on that face-side a north pole could be. The ferromagnet does not "pay attention" to the production of the south pole on that face-side. Likewise, a roll pattern that rotates clockwise does not "focus" on that activity and does not "ignore" the counterclockwise rotation pattern. The roll pattern or the fluid or gas system as a whole does not "pay attention" to what it is doing. Bistable lasers do not pay "selective attention" to their polarized laser states. Bistable signaling pathways that are in a stable state characterized by a high concentration of X and a low concentration of Y do not "focus" on that stable state and do not "ignore" the other possible stable state. Likewise, humans from the physics perspective of self-organization, pattern formation system do not "focus" on anything, do not "ignore" anything, do not "pay attention" to anything, and do not exhibit some kind of superpower of "selective attention" that would make humans different from other self-organizing, pattern formation systems.

The concepts of "attention" and "selective attention" belong to a package of religions and ethical concepts that involves other concepts like "free will", "choice", "control", and "decision making". The author as a private person (not as a scientist) believes that those concepts are important concepts that have great value in religion and ethics. For example, the three largest religions Christianity, Islam, and Hinduism assume that a human being has a soul that makes the human different from other animate and inanimate self-organizing systems such as ferromagnets, lasers, and cells. Given this religious motivated qualitative difference between humans and other systems it makes sense to discuss within those religious frameworks concepts such as "free will", "choice", "control", "decision making", etc. Likewise, Buddhism states that while life, death, and re-birth is determined to some extent by cause-and-effect-chains, humans can overcome those cause-and-effect-chains. Again, in doing so, according to Buddhism, humans have properties (powers) different from animate and inanimate self-organizing systems like ferromagnets, lasers, and cells, which motivates to talk in the context of Buddhism about "free will", "choice", "control", "decision making", etc. However, in the physics perspective to understand the nature of humans there is no place for such concepts. That is, they are not needed. Note that as mentioned in Chap. 2, this does not mean that the physics perspective negates the value of those concepts for non-scientific perspectives. To re-iterate, the physics perspective does not make any statement about the use of the concept "attention" outside the physics perspective of the human nature (e.g., in Christianity, Islam, Hinduism, and Buddhism).

6.3.5   Moving Dots Brain Activity Patterns in Monkeys

Let us return to the example introduced in Sect. 1.8.2 (see Fig. 1.10) about brain activity measured on single neurons of monkeys under the impact of external forces from the visual world. The experiment was conducted by Shadlen and Newsome [296, 297] and will be briefly reviewed from a simplified view. For details the reader is referred to the original papers [296, 297]. The experiment involved two parts: a brain restructuration part and an experimental part. During the restructuration part the brain of the monkeys was restructured such that when the monkeys were presented with a field of dots that moved leftwards then the moneys reacted by moving their eyes leftwards (i.e., the gaze) from a certain reference point to a target point on the left of the reference point. In contrast, when the dots moved rightwards then the monkeys moved their eyes rightwards from the reference point to a target point on the right of the reference point. Note that the brain of the monkeys was structured such that the monkeys performed the eye moments only after a certain trigger signal was given. In the experimental part, the same kind of setup composed of the moving dots, the reference point and the two target points was used.

Figure 6.14 illustrates the experiment and the modeling of the observed brain activity in terms of an emerging BA pattern. Panel (a) shows schematically the movie presented to the monkeys. There was a dot field composed of several dots. Some of them moving either leftwards or rightwards. The remaining dots appeared for short periods at random locations and, in doing, so created a noisy background. Shadlen and Newsome introduced a measure for the degree to which the dots as a whole (including the randomly appearing dots) moved in one direction and called that measure coherence. Zero percent coherence means that all dots moved in a random fashion. 100% means that there was no randomness at all and all dots moved in the same direction. In what follows the data from the experimental condition with 6.4% coherence taken from Ref. [296] will be discussed. Panel (b) summarizes (1) the external forces acting on the monkeys during the experiment, (2) the recorded brain activity, and (3) the body movement reaction (i.e., the eye movement). During a period of 2 s the dot field was presented. The dot field was switched off and after a delay of about 1 s a trigger signal was presented. The monkey moved the eyes leftwards or rightwards depending on the movement direction of the dots. During

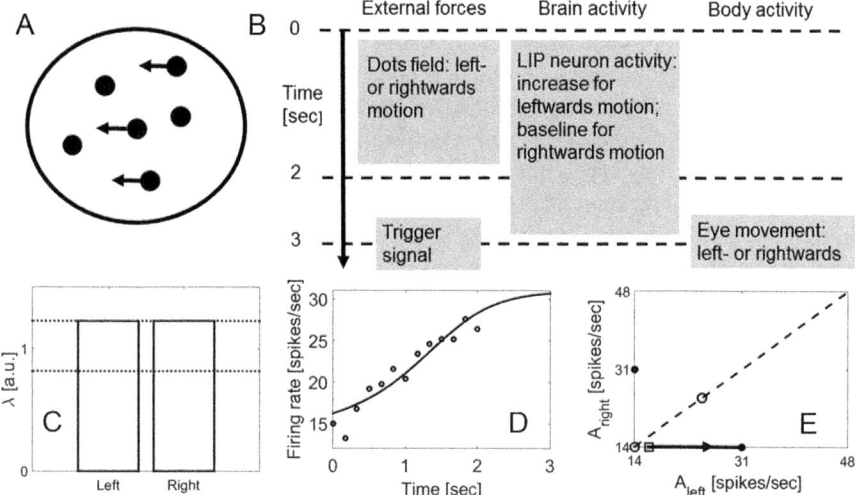

Fig. 6.14 Brain activity recordings from monkeys reacting to moving dot fields. Panels (**a**) and (**b**): The experimental design involved a dot field (panel **a**) and a multistep schedule (panel **b**). Panels (**c**)–(**e**): Modeling of experimental data (panel **d**) using the human pattern formation reaction model. See text for details

the total period of 3 s brain activity from certain neurons in the lateral intraparietal (LIP) regions of the monkeys was recorded by means of implanted electrodes. The LIP region is part of the parietal lobe. The neurons will be referred to as LIP neurons. The activity of the LIP neurons was measured in spikes per second. Shadlen and Newsome used LIP neurons that showed increased spiking activity in trials when the monkeys performed leftwards eye movements but showed only baseline spiking activity when the monkeys performed rightwards eye movements. Among other things, it was found that when the monkeys were exposed to the forces of the leftwards moving dots then the brain activity increased monotonically as shown in panel (d) (which has already been shown in panel (c) in Fig. 1.10 of Chap. 1) for trials in which the monkeys also performed leftwards eye movements (i.e., for "error"-free trials). According to the pattern formation approach to understand human and animal reactions to forces of the visual world, the increasing brain activity reflects the emergence or build-up of an attractor pattern. Accordingly, the observed brain activity can approximately be considered as a measure for the amplitude of a leftwards-moving dots basis pattern in the parietal lobe. It is assumed that forces exerted by the dots field make a reference state given by baseline spiking activity unstable. At the instability point, the neuronal pattern formation of the parietal lobe exhibits (among other unstable and stable basis patterns) two unstable basis patterns of interest for the Shadlen and Newsome experiment: a leftwards-moving dots basis pattern and a rightwards-moving dots basis pattern. Using the LVH model, it is assumed that the basis patterns create two winner-takes-all attractors in the reduced amplitude space of the two basis patterns. For sake of simplicity, we assume symmetry. That is, as shown in panel (c) the eigenvalues are assumed to have the same value. If so, the two attractors are located at the same distance away from the unstable reference point as shown in panel (e). The diagonal separates the basins of attraction of the two attractors. When the dots of the dot field move to the left then the electromagnetic forces of the light coming (or reflected) from the moving dots and acting on the monkeys eye receptors are converted into electrical forces and chemical driving forces that eventually shift the brain activity of the parietal lobe into the basin of attraction of the leftwards-moving dots attractor. That is, the dot field is assumed to set the initial state such that the initial state is located in the lower half of the reduced amplitude space (see Sect. 6.3.1).

Panel (d) shows the data (see circles) from the study by Shadlen and Newsome [296]. For sake of simplicity, in order to fit the LVH model to the data it is assumed that for leftwards-moving dots the amplitude of the rightwards-moving dots basis pattern can be neglected. Under this assumption, the LVH amplitude equations can be solved analytically. That is, an analytical expression for the increase of the amplitude A_{left} of the leftwards-moving dots basis pattern can be obtained. The solid line in panel (d) shows the best fit curve thus obtained. The increase of the amplitude A_{left} is also shown in the reduced amplitude space as straight line along the horizontal axis, see panel (e). Figure 6.14 does not address the production of the body movements (i.e., the eye movements). They can be modeled by assuming that the leftwards-moving dots BA attractor pattern in combination with the trigger signal leads to a structural change of the monkey brain that in turn induces another bifurcation and the emergence of a BBA pattern (i.e., a brain activity and body activity pattern) describing the eye movement and the associated brain activity. We will return to this issue in Sect. 7.5.5 when discussing pattern formation of grand states.

6.4   Reaction Patterns Involving Force Production to External Forces: B and D Systems

6.4.1   Shoe Example Revisited Again: B1, B3, D1 Systems Approach

Let us return to the shoe example [112] discussed in Sects. 2.5.2 and 4.6.1. In Sect. 4.6.1 the BBA attractor pattern of putting on a particular pair of shoes (either pair A or B) has been explained in the context of two different scenarios (1) and (2). The first scenario (1) assumes either asymmetric, monostable B1 systems or systems of type D1 in which the external forces exerted by the shoes break the symmetry of the human pattern formation system reacting to the shoes (i.e., the forces exerted by the shoes). In contrast, the second scenario (2) assumes symmetric, bistable B1 or B3 systems. For the systems classification B1, B3, D1 see Fig. 5.5. For sake of simplicity, let us begin with the second scenario.

According to the second scenario, when an individual is exposed to the visual world forces of the two pairs of shoes the current reference state becomes unstable and there are two basis patterns (putting pair A on and putting pair B on) with equal eigenvalues. Consequently, as shown in the panels of cases A(ii) and B(ii) of Figs. 4.18 and 4.19, respectively, the pattern formation system exhibits a two-dimensional reduced amplitude space with two attractors located at equal distances from the unstable reference point on the two axes of the reduced amplitude space. The diagonal separates the basins of attraction of the two attractors. Let us assume that the external forces exerted by the shoes have no effect at all on the brain state. If so, we are dealing with a B1 system. The initial brain state is given by the ongoing brain activity. If the initial brain state happens to be located in the lower part of the reduced amplitude state (i.e., in the basin of attraction of the putting-shoes-A-on attractor) then the brain activity converges to the putting-shoes-A-on attractor, see case A(ii) and the person under consideration puts on pair A. In contrast, if the ongoing brain activity happens to be located in the upper half of the reduced amplitude space, then the brain activity converges to the shoes B attractor, the putting-on-shoes-B BBA attractor pattern emerges, see case B(ii). The person puts on pair B.

However, the forces exerted by the shoes may shift the ongoing brain activity to a particular initial state (see Sect. 6.3.1). If so, we are dealing with a B3 system. Accordingly, the forces shift the initial state into the basin of attraction of one of the two attractors and, as a result, the brain activity converges to the corresponding attractor.

Let us turn next to the first scenario (1) that involves monostable pattern formation systems. According to the monostability scenarios described by the cases A(i) and B(i) of Figs. 4.18 and 4.19, it is assumed that the human pattern formation system exhibits either initially strong asymmetry that leads to monostability or the external forces exerted by the shoes lead to structural changes that result in asymmetry and monostability of the pattern formation system. In the former case, we are dealing with a B1 system, in the latter case with a D1 system. In the case of a B1 pattern formation system, when the individual is confronted with putting on the shoes, the reference state becomes unstable—just as in the scenario (2) cases addressed above. Unlike the scenario (2) cases addressed above, the pattern formation system is assumed to exhibit two basis patterns with different eigenvalues λ_A and λ_B. The difference in the eigenvalues is assumed to be sufficiently large such that only one of the two eigenvalues is in the stability band (see Sect. 4.6.3). The other eigenvalue is below the stability band. Consequently, the system is monostable. The ongoing brain activity is assumed to serve as initial state. Due to the monostability of the system, for any arbitrary initial brain state the brain activity converges to the fixed point of the unique attractor. Figures 4.18 and 4.19 provide examples in this regard for monostable systems featuring a putting-on-shoes-A attractor (case A(i)) or putting-on-shoes-B attractor (case B(i)).

In the case of a D1 pattern formation system, the impacts of the forces exerted by the shoes are twofold. First, they lead to the instability of the reference state. Second, they lead to a structural change in terms of a change of the eigenvalue spectrum.

In doing so, the forces bias the structure towards one of the two basis patterns. It is assumed that the forces exerted by the shoes make that the difference between the eigenvalues becomes sufficiently large such that the pattern formation system under consideration becomes monostable. Consequently, just as for the asymmetric, monostable B1 systems, the brain activity converges to the unique, monostable attractor irrespective of the initial brain state.

In general, the shoe example may involve various generalization of the two scenarios (1) and (2) discussed so far. For example, B1 and B3 systems may exhibit asymmetry but still feature bistability. In this case, the two eigenvalues λ_A and λ_B differ by a relatively small amount such that both eigenvalues are in the stability band. Likewise, D1 systems are not necessarily monostable. External forces may affect the brain structure only weakly such that the eigenvalues λ_A and λ_B differ by a certain amount but nevertheless belong both to the stability band. In general, a mixture of scenarios (1) and (2) should be considered. Accordingly, in line with the second law (see Sect. 5.1.2), the forces exerted by the shoes lead to changes of the initial state and structural changes. If so, we are dealing with D3 systems rather than D1 systems. Such D3 systems could exhibit either a weak or strong asymmetry leading to bistability or monostability, respectively. If they are monostable, the shift of the brain state would have no effect at all. That is, while the external force would shift the brain activity to some initial state, that shift would not matter at all. The brain activity would converge to the unique, monostable attractor irrespective of the brain activity shift.

Finally, note that in the introduction to this chapter, we stated that in this chapter D1 and D3 systems will not be discussed because D1 and D3 systems typically are involved in pattern formation of grand states. In fact, if we consider the episode of putting shoes on as part of a pattern formation sequence that involves structural changes (e.g., the morning ritual to get to work), then the sequence when considered as an overall pattern describes a pattern that is formed on the grand state of an individual. We will return to this issue in Chap. 11.

6.4.2 Why Physical Systems Do Not Make "Decisions"

As mentioned already in Sect. 4.6.1, in the literature frequently the concept of "decision making" has been used. For example, it is frequently assumed that humans "decide" which particular pair of shoes they want to put on, when there are several pairs of shoes available. On the one hand, as mentioned throughout previous chapters and, in particular, in Sect. 6.3.4 above, there are various systems that show categorical changes due to bifurcations and phase transitions (see Table 3.1) and systems that show bifurcations in the context of multistability (see Table 6.5). On the other hand, humans and animals continuously exhibiting switches between categorical different states (such as standing in front of several pairs of shoes and then putting one pair on or sitting on a chair and then standing up). From a physics perspective, humans and animals when exhibiting switches between categorical different states satisfy the same kind of laws as all the other physical systems (water, ferromagnets, heated gas layers, atmospheric systems producing hurricanes, tornadoes and lightning, gas discharge systems producing discharge lightening, lasers, chemical systems producing spirals and stripe patterns, etc). If humans make "decisions" when they put on shoes, then water molecules make a "decision" when they turn from water into ice and atmospheric layers make "decisions" when they produce hurricanes. However, physics does not need the concept of "decision making" in order to explain that water turns into ice or that hurricanes emerge. Likewise, physics does not need the concept of "decision making" to explain why humans and animals do what they do.

To be clear, the concept of "decision making" as such is not rejected. "Decision making" as such is not condemned as an illusion. The author as a private person (not as a scientist) believes that the concept of "decision" as such is a useful one outside of the scientific perspective. If we assume that humans have a soul, then this makes them different from other self-organizing systems such as lasers and cloud patterns. For example, Christianity assumes that humans have a soul. In particular, God created Adam and gave him a soul. The bible tells us the story about Adam, who decided to take a fruit of the tree of knowledge although God had forbidden him to do so. In that context, the author of the book as a private person (not as a scientist) believes that the concept of decision making is an important concept (just as the concept of knowledge and the notion of a tree of knowledge). Importantly, the physics perspective of the nature of humans as discussed throughout this book (and in this section) does not make any statement about biblical stories like the one about the tree of knowledge and Adam's decision. In general, the physics perspective does not make any statement about the concept of decision making as used in religions such as Christianity, Islam, Hinduism, or Buddhism.

6.4.3 Reactions of Tolman's Rats to Mazes

As reviewed in Sect. 2.2 (see Fig. 2.2) in a study by Tolman [319] rats were restructured such that a particular color of a door caused the rats to go through that door. For example, when confronted with two doors, one painted with a white frame and the other painted with a black frame, after restructuring the rats, the rats passed through the door with the white frame but not through the door with the black frame. In this section, the experiment by Tolman is modeled from the perspective of B3 pattern formation systems. Figure 6.15 illustrate this B3 system approach.

In line with the B3 system modeling approach, it is assumed that the structure changes in the animal brains only concern the transduction of external forces into the forces that set the initial brain activity for the formation of BBA patterns. Accordingly, it is assumed that putting an animal into the experimental setup involving the two differently colored doors makes an appropriately defined reference state unstable. For our purposes, it is sufficient to assume that the reference state is the standing still or keeping the same location state. Consistent

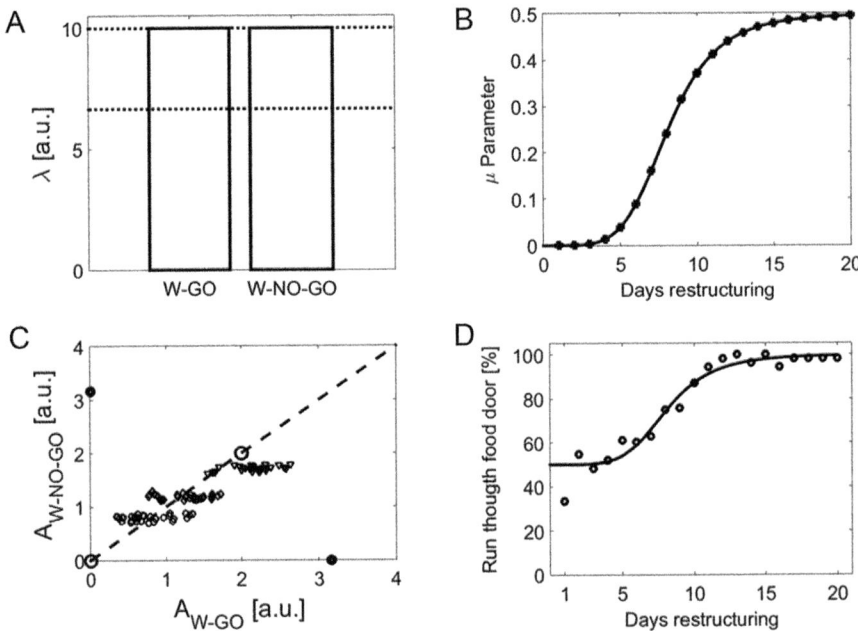

Fig. 6.15 Change of rat reactions to mazes due to restructuring explained in terms of a B3 pattern formation system. A brain state shift model is proposed. While the structure is assumed to be fixed and symmetric with respect to the eigenvalues (panel **a**), the mechanism that leads to brain state shifts induced by the forces exerted by the maze changes during the restructuration period (panels **b**–**d**). Panel (**d**): Experimental data [319] and fit of the model. See text for details

with the observation of the animal dynamics, it is plausible to assume that at the bifurcation point two mutually exclusive unstable basis patterns exist. One pattern is a white-door-and-walking-through-that-door basis pattern (W-GO). That is, the basis pattern has elements from a visual world pattern and elements from patterns involving muscle force production. It is a pattern that emerges as reaction to visual world external forces from doors with white doorframes and involves as supplementary components a movement pattern throughout the door and the corresponding movement-related brain activity. The other pattern is a white-door-and-walking-away-from-that-door basis pattern (W-NO-GO). Just as in the previous case, the pattern emerges as reaction to visual world external forces exerted by white doorframes but it involves as supplementary pattern components movements patterns directed away from the door and the corresponding movement-related brain activity patterns. For sake of simplicity, it is assumed that if the rat goes away from the white door, then the animal goes through the other door, that is, the black door. More sophisticated models could be constructed in this regard (in fact, a more advanced model will be presented in Chap. 9)

The structure in terms of the eigenvalues of the basis patterns is assumed to be symmetric, as shown in Fig. 6.15 (panel a). Therefore, in the reduced amplitude space shown in panel (c) there are two attractors at equal distances from the unstable reference points: the W-GO and W-NO-GO attractors. As far as the mechanism that determines the initial state of pattern formation is concerned, it is assumed that the mechanism features initially the same kind of symmetry as the pattern formation system. Forces exerted by the white door set the brain activity in a state such that there is a 50% chance that the W-GO basis pattern amplitude is larger than the W-NO-GO basis pattern amplitude and vice versa. Note that the chance element is assumed to reflect the impact of ongoing brain activity. Consequently, if we would model the ongoing brain activity explicitly, then the whole model would be deterministic (see deterministic perturbations in Sect. 2.4.4 and Fig. 2.3). In order to model the chance element, the center parameter μ is introduced. The center parameter describes the center point or mean value of the distribution of the differences between the W-GO and W-NO-GO initial amplitudes. The differences are assumed to be uniformly distributed in an interval of one unit. At the beginning of the experiment, that is, before the animals were restructured, the parameter is at zero. Accordingly, differences are scattered around zero in the interval from -0.5 units to 0.5 units.

Panel (b) shows the center parameter μ as function of days of restructuring. Panel (c) shows for day 1 of the experiment the initial states for 30 computer simulated animals (circles) in the reduced amplitude space spanned by the W-GO and W-NO-GO amplitudes. Half of the initial values fall in the basin of attraction of the W-GO attractor. The remaining half of the initial values fall in the basin of attraction of the W-NO-GO attractor. Accordingly, there is a 50% chance that the animals walk through the door that leads to the food. This predicted probability is shown in panel (d) as solid line. Panel (d) also show the data from the original study by Tolman [319] as reviewed earlier in Sect. 2.2. Every time an animal walks through the door that leads to the food (here the white door), the forces from the food reward are assumed to change the structure of the animal with respect to the center parameter μ. The parameter μ is assumed to increase monotonically. The increase of the parameter is shown in panel (b) (and was obtained by a best fit of a sigmoid function to the data shown in (d)). For example at days 5 and 10 we have $\mu = 0.04$ and $\mu = 0.41$, respectively. As a result, at days 5 and 10 there is a 54% and 91% chance that the W-GO amplitude initially is larger than the W-NO-GO amplitude. In panel (c) the diamonds and triangles correspond to initial values obtained from simulations for days 5 and 10, respectively, for 30 animals again. While at day 5 there are only a few more initial conditions that fall in the basin of attraction of the W-GO attractor as compared to the basin of attraction of the W-NO-GO attractor, for day 10 the situation is quantitatively different. Almost all initial conditions fall in the basin of attraction of the W-GO attractor. In summary, as a consequence of the sigmoidal increasing function of the center parameter μ shown in panel (b), the probability that animals walk through the food door is predicted to increases in the sigmoid fashion as shown in panel (d).

The modeling approach sketched in this section is about the movement pattern as reaction to the external forces exerted by the doorframe color. With respect to this animal reaction, the pattern formation model belongs to the class of B3 systems.

An external force sets the initial state of a multistable animal pattern formation system and depending on the initial state a particular BBA attractor pattern emerges. However, when the food-rewarding W-GO attractor pattern emerges, then the pattern changes the circumstances of the animal such that it gets a reward. That is, the animal is exposed to chemical driving forces related to the food intake. Those external (chemical driving) forces are assumed to change the structure of the pattern formation system (here: shift of the structure parameter μ). Therefore, when looking over the course of the restructuring period from day 1 to day 20, the pattern formation systems of the animals correspond to D3 systems (see Fig. 5.5 for the system classes). When considering the 20 days episode as a whole, we are dealing with animals exhibiting pattern formation sequences that come with structural changes. That is, the (life-) sequences taken as a whole correspond to temporal patterns defined on the grand states of the animals (see Chaps. 7 and 11).

In this context, note that the labeling on the horizontal axes ("days restructuring") in panels (b) and (d) of Fig. 6.15 is misleading. From a theoretical point of view, it is not so much about time (i.e., about "days") but about the emergence of the W-GO attractor patterns and the fact that those patterns come with forces related to a food reward that in turn leads to structural changes. Note that one could talk about "reinforcement" in this context. We return to the concept of "reinforcement" in Chap. 9.

6.5   Reaction Patterns to Forces That Change Structure: D Systems

Pattern formation systems of the classes D and E (see Fig. 5.5) account for external forces that affect the structure of pattern formation systems. BBA patterns emerging in class D systems do not directly affect the structure of the systems (they may do so indirectly by changing external forces). In contrast, for systems of the class E the structure of the systems is affected both by external forces and emerging BBA attractor patterns. In what follows, D1 and D3 systems will be considered for which pattern formation takes place primarily in the corresponding D0 and D2 system components. That is, we will study humans and animals under the impact of the forces of their environments and involved in various body activities. However, the human and animal body will be considered as a driven system. That is, we consider pattern formation phenomena for which the body activity and associated changes in the external forces do not play a crucial role. In this sense, the pattern formation phenomena considered in this section take place in the ordinary states of humans and animals.

6.5.1   Categorical Reactions to Objects About Force Production

So-called psychophysical experiments have produced a large body of experimental data about how humans react to objects with respect to the possible body patterns that they could form that would involve those objects. Frequently, the reactions are of categorical nature and, consequently, are reported by means of verbal categorical statements like "Yes" or "No".

In a study by Warren [343] participants were presented with several stairways with steps that had the same step-height for each stairway but different step heights across the set of stairways. Participants were asked to react to the stairways by producing a verbal statement whether or not they would be able to produce a body movement pattern of climbing up the stairways "in the normal way without using hands or knees". The step-height was varied over a range of values. Typically, adult humans can climb up conveniently stairways with sufficiently small step heights in the "normal way". They are not able to do so if the step height is relatively large. For example, hands and feet must be used to climb up relative large obstacles. Accordingly, for sufficiently small step-heights participants produced positive statements like "I can climb up", whereas for step-heights exceeding critical, subject-dependent values participants produced negative statements like "I cannot". Warren [343] showed that while the critical values differed from individual to individual, when rescaling the scores appropriately, all participants exhibited approximately the same critical value. The rescaling was done by dividing the critical step height with the leg length of the participants. The ratio thus obtained will be considered as bifurcation parameter α. At the critical value of about 0.89 of the bifurcation parameter the reaction switched from a "Yes, I can climb up" to "No, I cannot climb up in the normal way".

Note that in the experiment participant did not climb up the stairs. They "just" produced verbal statements. In fact, the experimental design resembles an everyday situation in which an individual reacts to an external force with respect to a body activity pattern XYZ associated to that force but without producing the muscle activity component of that pattern XYZ. In such situations individuals correspond to D0 and D2 systems. External forces affect structure (D0) or structure and brain state (D2) such that a BA pattern emerges. Although the BA pattern is about a pattern that involves force production, the BA pattern does not lead to or induce the corresponding muscular activity. Schematically speaking, there is no connection from the BS box to the F box in D0 and D2 systems, see Fig. 5.5.

Figure 6.16 describes the experiment by Warren [343] from the perspective of D0 and D2 pattern formation systems. Panel (a) of Fig. 6.16 shows the experimental results in terms of the percentage of participants who reacted with positive statements (i.e., "Yes, I can" statements) to forces exerted by stairways with certain bifurcation parameters. The circles connected by solid lines correspond to the experimental data (of the so-called tall group, for details see the original study [343]). The dashed line shown in panel (a) shows a fit to the data based on the theory of pattern formation as will be discussed below.

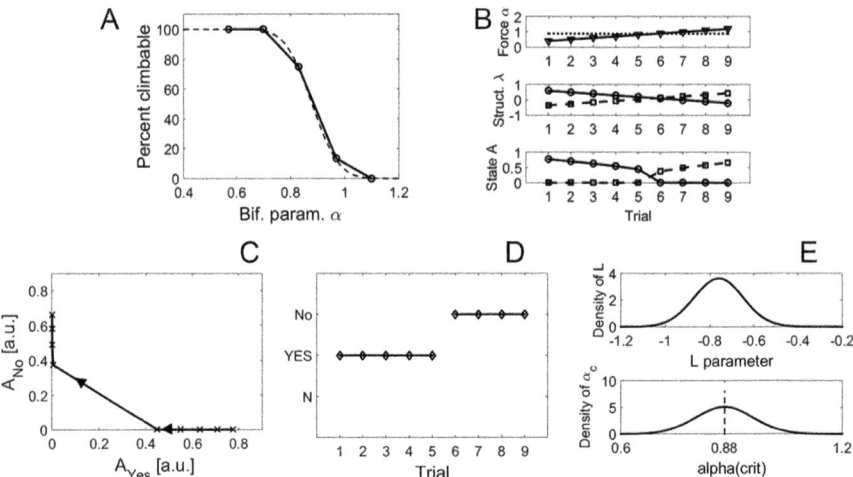

Fig. 6.16 Human categorical reactions to objects about body movement patterns related to the objects that are not performed, here exemplified for the study by Warren [343]. In the study by Warren [343] the emergence of D0/D2-system stair-climbing BBA patterns were examined. Panel (**a**): The percentage of observed positive D0/D2-system stair-climbing BBA patterns associated with verbal "Yes, I can climb up" supplementary components for various step heights (solid line) and the best-fit model predictions (dashed line) are shown. Panels (**b**)–(**d**): Formation of positive and negative D0/D2-system stair-climbing BBA patterns in an individual participant. Panel (**e**): Model predictions for a population of participants leading to the predicted curve shown in panel (**a**). See text for details

According to the D0/D2 human pattern formation reaction model, in general, and the LVH model, in particular, it is assumed that the external forces of the stairways in combination with the forces of the experimental setup (e.g., instructions by the experimenter to the participants) make an appropriately defined reference state unstable. As reference state we may use an isolated human system BA pattern of doing nothing. Moreover, it is assumed that the forces exerted by the steps of the stairways on a participant with a particular leg length can be quantified to some degree by the aforementioned bifurcation parameter α, the ratio of step height and leg length. Panel (b) of Fig. 6.16 shows the external forces (in terms of α), the brain structure (in terms of the relevant eigenvalues), and the brain activity state (as described by the relevant pattern amplitudes in the reduced amplitude space) for a sequence of nine trials with different step heights. Note that in the experimental study by Warren [343] stairways with different bifurcation parameters were presented in a random fashion, while in the modeling approach we will for sake of simplicity consider the ascending sequence of nine trials shown in panel (b). As will be argued below, in view of the particular LVH model discussed here, the presentation order has no impact on the model predictions.

As mentioned above, the experimental setup results in a primary bifurcation that makes the reference state unstable. At that bifurcation point, the human pattern formation system is assumed to exhibit two basis patterns related to positive (climbing up in the normal way) or negative (climbing up in a not normal way) reactions, respectively. Those basis patterns will for sake of simplicity referred to as YES and NO basis patterns, respectively. For the LVH model under consideration in this application, the pattern formation system exhibits symmetry (i.e., features the same eigenvalues for both basis patterns) for a bifurcation parameter α equal to the critical value $\alpha(crit)$ of 0.88. The stairways forces (as modeled by α) are assumed to affect the symmetry such that for smaller than critical step height values ($\alpha < \alpha(crit)$) the eigenvalue λ_{YES} of the YES basis pattern is relatively large, while for larger than critical step height values ($\alpha > \alpha(crit)$) the eigenvalue λ_{NO} of the NO basis pattern is relatively large (see Sect. 4.4.2 and Fig. 4.14b). Using linear relationships between the eigenvalues λ_{YES} and λ_{NO} and the bifurcation parameter α and fitting the linear relationships to the critical value of 0.88, we obtain the graphs shown in panel (b). The LVH model presented here involves a coupling parameter g slightly above unity. Consequently, the model does not exhibit a bistability domain except for values very close to the critical value of 0.88. In other words, for a coupling parameter $g \approx 1$ the bistability domain can be neglected. The model is effectively monostable for all bifurcation parameters. For bifurcation parameters smaller than the critical value only the YES BBA reaction attractor exists. Likewise, for bifurcation parameters larger than the critical value only the NO BBA reaction attractor exists. Therefore, the initial value is irrelevant for the formation of patterns. This also implies that the model predictions do not depend on the presentation sequence of the trials and without loss of generality, for illustration purposes, the ascending sequence shown in panel (b) (top subpanel) can be used. Let us consider panel (b) from top to bottom. The step height is assumed to increase from trial to trial, which implies that the bifurcation parameter (reflecting the forces of the stairway steps) increases gradually. The eigenvalues (reflecting brain structure) decrease (YES basis pattern, see solid line and circles) and increase (NO basis pattern, see dashed line and squares) accordingly. For bifurcation parameters smaller than the critical value, the YES eigenvalue λ_{YES} exceeds the NO eigenvalue λ_{NO}. For bifurcation parameters larger than the critical value, the opposite is true. As a consequence of the coupling parameter g being only slightly above unity, if the YES eigenvalue λ_{YES} exceeds the NO eigenvalue λ_{NO}, only the YES attractor exists (as mentioned above). Consequently, the amplitude A_{YES} of the YES basis pattern converges to a stationary finite value, whereas the amplitude A_{NO} of the NO basis pattern converges to a zero value. This relationship holds for the first five trials, as shown in the bottom subpanel of panel (b). When the bifurcation parameter exceeds the critical value, the NO eigenvalue exceeds the YES eigenvalue and only the NO attractor exists. The BBA attractor pattern as described in terms of the two basis pattern amplitudes, A_{YES} and A_{NO}, switches from $A_{YES} > 0, A_{NO} = 0$ to $A_{YES} = 0, A_{NO} > 0$. That is, a bifurcation takes place at trial 6 and the NO BBA attractor pattern emerges. The NO BBA attractor pattern remains for the remaining trials 7, 8, and 9.

In the two-dimensional reduced amplitude space of the pattern formation system shown in panel (c), the brain and body activity occupies the states indicated by the crosses. Since the brain structure changes (in terms of the eigenvalues λ_{YES} and λ_{NO}) from trials 1 to 5, the fixed points change their locations in the reduced amplitude space (and, in general, the ordinary state of the pattern formation system) and move along the A_{YES} axis towards the origin. At the bifurcation point (trial 6), the brain and body activity switches from a location on the A_{YES} axis to a location on the A_{NO} axis. Subsequently, that is, for trials 7, 8, and 9, the brain and body activity moves along the A_{NO} axis away from the origin. Panel (d) shows the corresponding categorical brain and body activity states. The hypothetical participant described by the LVH model produces verbal YES patterns for trials 1, 2, 3, 4, 5 and verbal NO patterns for trials 6, 7, 8, 9.

Panels (b)–(d) show the application of the LVH model for an individual participant whose structure parameters are such that the bifurcation takes place at the value of 0.88. A detailed analysis shows that the LVH model only features one free parameter, the offset parameter L. For $\alpha(crit) = 0.88$ the offset parameter is given by $L = -0.76$. However, the critical bifurcation parameter of 0.88 in fact corresponds to a mean value. That is, on average, participants exhibit a bifurcation around 0.88. Assuming for the moment that structure parameters in general are normally distributed, it has been suggested that the critical bifurcation parameters exhibit a normal distribution as shown in panel (e) [120]. The normal distribution in turn is assumed to be a consequence of a normally distributed offset parameter L as shown in panel (e) as well. More precisely, with respect to the experiment by Warren, it is assumed that humans exhibit an offset parameter L normally distributed around a mean of -0.76. This implies that the critical bifurcation parameters are normally distributed around a mean of 0.88. The normal distribution of the critical bifurcation parameters leads to the sigmoid function of percentage of Yes reactions shown in panel (a) as dashed line [120]. Note that this model for the population of humans involves a single free parameter, the standard deviation of the normal distribution (of L). This parameter has been fitted to best fit the experimental data shown in panel (a). Note also that the assumption that the structure parameter L is normally distributed is not essential for the arguments made above. Any unimodal but not normal distribution (e.g., a symmetric distribution with heavy tails or cut-off tails, or a non-symmetric, skewed distribution) could be used and would qualitatively yield a sigmoid function similar to the one shown in panel (a).

Various psychophysical studies have been conducted in the past in which participants were examined under D0/D2 system conditions. That is, participants produced BA patterns about body activities related to objects (or forces exerted by objects) without performing the activities. In some sense, those studies examined the object function BBA patterns as addressed in Sect. 6.3.1. For example, participants have been exposed to sitting panels with certain heights. The emerging BBA reaction patterns about sitting on those panels and involving verbal supplementary components ("Yes, I can sit" or "No, I cannot sit") were recorded [164, 232, 312]. In a study by Putfall et al. [264] participants were exposed to relatively small obstacles on the floor and the emerging BBA patterns about stepping over those obstacles (again without performing the stepping) involving verbal supplementary components ("Yes, I can step over" or "No, I cannot") were examined. Likewise,

BBA reaction patterns of standing upright on tilted surfaces [83, 122, 336, 337] and BBA reaching patterns [43, 338] were studied.

6.5.2   X0 and X2 System BBA Patterns and "Judgments"

In view of the fact that awake persons and persons under REM sleep can produce isolated BA patterns that are about body movements related to objects (e.g., sitting on a chair or stool), the question arises to what extent BA patterns that are about whole body and limb movements but do not involve the movements are similar to the corresponding BBA patterns that involve the movements. Human pattern formation under those circumstances corresponds to pattern formation in systems of the classes B2, C2, D2, E2 and D0 and E0, see Fig. 5.5. They can be summarized as X0 and X2 systems (excluding the A2 system class that does not involve external forces). X0 and X2 systems involve external forces that either act on the brain state or the structure or both. Importantly, the emerging BA patterns do not involve the production of forces related to the movement patterns.

From a pattern formation perspective, so-called "judgments" may be considered as BA attractor patterns about force production emerging in such X0 and X2 systems, that is, systems that do not exhibit a force production component. If a customer in a bar "wonders" if another customer would engage with him or her in a conversation, then the customer produces an isolated BA attractor pattern that is incomplete in the sense that the pattern does not involve the motor component of saying "Hello" to the other customer. When in my brain a graphical isolated BA pattern emerges about throwing a small stone across a small river, then depending on the circumstances this means I "think" (awake state) or "dream" (REM sleep) about throwing a stone across a river. When there is actually a small stone and a small river and in my brain a BA pattern emerges about throwing the stone across the river but the pattern does not involve the body activity of throwing the stone, then somebody could say that I "imagine" throwing the stone over the river or "judge" whether I would be able to throw the stone across the river. Again, the isolated BA pattern about force production occurs in a system that does not feature a force production component.

BA patterns about movements do not necessarily have a categorical character. They may refer to graded, scaled, or continuous properties (see Chap. 8). For example, when I am exposed to the forces of a city map that shows a railway station and a hotel and the forces of the map lead to the emergence of a BA pattern about me walking from the station to the hotel, then I am actually not at the station and I do not walk to the hotel. On the one hand, the BA pattern could have a categorical character and express something categorical like "Yes, I can walk" versus "No, I cannot walk that distance. I should take a taxi". On the other hand, the BA pattern could have a continuous, graded aspect for example about the time it takes to walk from the station to the hotel. Such a X0/X2 system BA pattern would describe that

I make a "judgment" about or "estimate" the time it takes to go from the station to the hotel.

In experimental studies as those reviewed in Sect. 6.5.2 participants typically produced verbal categorical reactions. Consequently, in such psychophysical experiments, the human systems under consideration typically involve a muscle force component related to the reporting activity. Instead of dealing with X0 and X2 systems, we are actually dealing with X1 and X3 systems, where X in the X0→X1 re-classification stands for D and E and X in the X2→X3 re-classification stands for B, C, D, E. However, the forces produced by the participants in order to report to the experimenters are not related to the produced force involving the objects under consideration (stairways, seat panels, etc). Therefore, the produced force related to the reporting activity may be ignored or considered as supplementary pattern components (as we did above). Interestingly, neuroscience research allows us to measure the reactions of participants on the brain activity level in a way that they do not need to make a verbal report. For example, in a study by Kleinschmidt et al. [188] (see Sect. 6.5.4 below) participants viewed a screen in which a letter appeared from the background and, subsequently, disappeared again. Participants performed two experimental tasks. In the first task, participants had to press a button to indicate when they saw the letter popping out and when it disappeared. This task involved a force production component as a means to communicate with the experimenters. However, in a second task (the so-called passive viewing task) participant just watched the screen with the appearing and disappearing letter and did not produce a manual reaction. During the performance of both tasks brain activity was recorded. From the pattern formation perspective taken in this book, by means of the second task Kleinschmidt et al. [188] could experimentally study the emergence of BBA patterns in human X0 and X2 systems.

6.5.3   *Hysteresis, Bistability, and the Human Pattern Formation Reaction Model for Hysteresis*

Bifurcations describe qualitative changes that occur at particular critical bifurcation parameters. Hysteresis in our context describes bifurcations between two states that occur at different bifurcation parameters when the parameters are gradually increased, on the one hand, and gradually decreased, on the other hand. Let us denote the two states by A and B, respectively. Let us denote the bifurcation parameter by α. Accordingly, when the bifurcation parameter is increased gradually the bifurcation from A to B takes place at a certain critical value of α. When the bifurcation parameter is gradually decreased the bifurcation from B to A does not take place at the first critical value of α. Rather, it frequently takes place at a smaller critical value.

There are two issues that should be pointed out. First, frequently it has been observed that the critical value of the decreasing condition is smaller than the critical value of the increasing condition. For this reason, the critical value of the increasing condition is denoted by $\alpha(crit, 2)$, while the critical value of the decreasing condition is denoted by $\alpha(crit, 1)$. That is, the "first" critical value of the increasing condition is denoted by $\alpha(crit, 2)$. The "second" critical value of the decreasing condition is denoted by $\alpha(crit, 1)$. Actually, in experiments the increasing condition is not necessarily the first condition. Typically, experiments are counterbalanced such that some participants perform the increasing condition first and the decreasing condition second, while other participants perform the decreasing condition first and the increasing condition second. The second issue to note is that in this scenario the difference between the two critical values when measured like $\alpha(crit, 2) - \alpha(crit, 1)$ is positive. The difference is referred to as hysteresis size. This kind of hysteresis that comes with a larger critical value for the increasing condition than for the decreasing condition exhibits a positive hysteresis size. Accordingly, it has also been called positive hysteresis in order to distinguish the phenomenon from other hysteresis-related phenomena (see Chap. 7).

Positive hysteresis has been found in several experiments on human reactions to external forces that will be reviewed in Sect. 6.5.4 below. While in the experiment by Warren [343] reviewed above on human reactions to stairways trials with different bifurcation parameters were presented in a random order, in experiments revealing positive hysteresis the methodology was such that bifurcation parameters were gradually increased and decreased. Let us address next the phenomenon of positive hysteresis within the framework of the pattern formation reaction model of humans. Figure 6.17 summarizes the basic elements [120].

First of all, the pattern formation systems under consideration must be affected by external forces such that external forces change the structure of the systems, panel (a). This features only applies to systems of the classes D and E (see Fig. 5.5). As mentioned at the beginning of Sect. 6.5, we will ignore systems of the class E for the moment and as far as systems of the class D are concerned, only D1 and D3 systems will be considered. Panel (b) points out the second ingredient of the human bistable pattern formation reaction model proposed here. In the case of bistable systems, the initial state (in combination with the basins of attraction of the attractors at hand) determines which BBA attractor pattern emerges. The emerging BBA attractor pattern determines the ongoing brain state after a particular trial with a particular value of the bifurcation parameter is completed. That ongoing brain state serves as initial state for the subsequent formation of patterns. That is, the (top-down feedback) loop from the emerging pattern to the initial state (introduced in Chap. 1, see Fig. 1.5) becomes relevant. In doing so, the modeling approach is qualitatively different from the modeling approach of the study by Warren [343]. While for modeling the study by Warren [343] the initial state was ignored, for the bistable model it is taken into account: the initial state for a given trial is determined by the pattern that emerged in the previous trial.

For sake of simplicity, only a bistable model is considered that involves two attractor patterns A and B (multistable models that involve more than two attractor

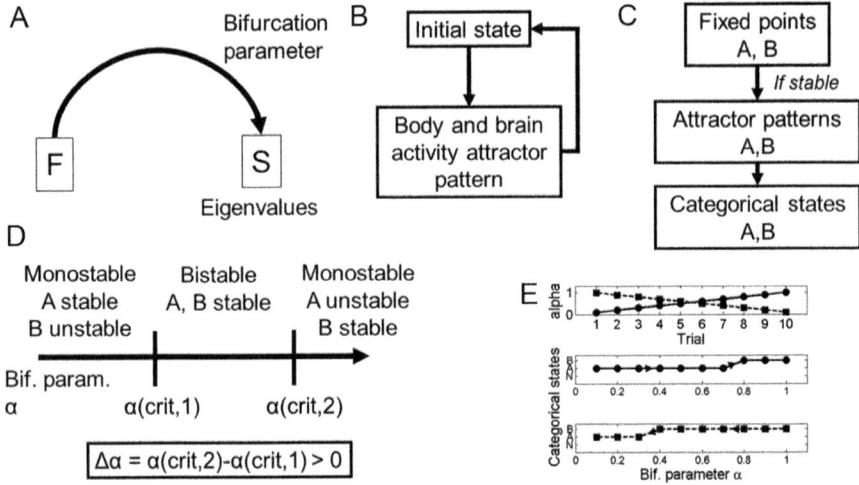

Fig. 6.17 Human pattern formation reaction model for categorical reactions exhibiting positive hysteresis. Panels (**a**) and (**b**): Key elements on the component (panel **a**) and systems dynamics (panel **b**) level that lead to hysteretic reactions. Panels (**c**) and (**d**): Hysteresis follows from bistability of the underlying dynamical (pattern formation) system. Panel (**e**): Categorical states as functions of the bifurcation parameter α as obtained from the stability chart shown in panel (**d**) for the increasing (solid lines, circles) and decreasing (dashed lines, squares) conditions

patterns can be considered in a similar way). In this case, the reduced amplitude space of the pattern formation model is two dimensional. In general, there are two fixed points A and B in the reduced amplitude space (in some exceptional cases only one fixed point might exist), see panel (c). If the fixed points are stable they correspond to attractors. When the brain and body activity converges in the reduced amplitude space to one of the attractors then the corresponding attractor pattern emerges. The attractor patterns A and B may be identified as categorical states A and B. Panel (d) shows the fundamental relationship between structure and forces leading to hysteresis. The bifurcation parameter α is shown on the horizontal axis. In the space of the bifurcation parameter the pattern formation system exhibits three qualitatively different domains: a monostability domain with only fixed point A is stable, a bistability domain with both fixed points are stable, and a monostability domain with only fixed point B stable. More precisely, if the bifurcation parameter is smaller than the smaller critical value, $\alpha(crit, 1)$, then only fixed point A is stable. Fixed point B is unstable or in exceptional cases does not even exist. Conversely, if the bifurcation parameter is larger than the larger critical value, $\alpha(crit, 2)$, then only fixed point B is stable. Fixed point A is unstable or does not exist. For bifurcation parameters in the interval between the two critical values, both fixed points are stable. The fundamental relationship shown in panel (d) implies the dynamics of the system shown in panel (e). For demonstration purposes, it is assumed that the decreasing and increasing conditions exhibit ten trials each, see top subpanel in panel (e). The bifurcation parameter is changed gradually in ten steps of 0.1 from

0.1 to 1.0 (increasing condition) or from 1.0 to 0.1 (decreasing condition). In panel (e) it is assumed that the smaller and larger critical values are $\alpha(crit, 1) = 0.35$ and $\alpha(crit, 2) = 0.75$. The middle subpanel of panel (e) shows the categorical state obtained from the stability chart displayed in panel (d) for the increasing condition. For $\alpha < 0.35$, which is the case in the first trials with $\alpha = 0.1$, the system is monostable with only fixed point A stable. Consequently, irrespective of the initial condition, the BBA attractor pattern A emerges. According to panel (b), the BBA pattern sets the initial state for the subsequent trials. For trials 2 and 3 with $\alpha = 0.2$ and $\alpha = 0.3$ the initial condition is irrelevant because the system is still monostable. Consequently, as shown in the middle subpanel of panel (e) the system is in the state A exhibiting the BBA attractor pattern A. For trials 4, 5, 6, 7 the bifurcation parameters are 0.4, 0.5, 0.6, and 0.7 and the system is bistable. Since the initial state is close to the fixed point of attractor A it is assumed that it is in the basin of attraction of attractor A and not in the basin of attraction of attractor B. Consequently, for trials 4, 5, 6, 7 the attractor pattern A remains. However, in trial 8 the bifurcation parameter changes from 0.7 to 0.8 and exceeds the critical value $\alpha(crit, 2) = 0.75$. Therefore, the system becomes monostable with fixed point B the only stable fixed point. The brain and body activity converges towards the fixed point B and the attractor pattern B emerges. For trials 9 and 10 this situation does not change. In short, the bifurcation from A to B takes place around $\alpha(crit, 2) = 0.75$ between the trials with $\alpha = 0.7$ and $\alpha = 0.8$. The decreasing condition is described in the bottom subpanel of panel (e). The bottom subpanel should be read from right to left as indicated by the arrows. That is, trial 1 comes with $\alpha = 1.0$, trial 2 comes with $\alpha = 0.9$, and so on. In the decreasing condition, irrespective of the initial state, for trials 1, 2, 3 (i.e., $\alpha = 1.0, 0.9, 0.8$) the attractor pattern B emerges because B is the only stable fixed point. Decreasing the bifurcation parameter further, we enter the bistability domain for trials 4, 5, 6, 7 (i.e., $\alpha = 0.7, 0.6, 0.5, 0.4$). Since the initial state is assumed to be close to the fixed point B it is assumed that it falls in the basin of attraction of attractor B and consequently the attractor pattern B remains present for trials 4, 5, 6, 7. At trial 8 the bifurcation parameter drops from 0.4 to 0.3 and assumes a value smaller than the critical value $\alpha(crit, 1) = 0.35$. The fixed point B becomes unstable (or disappears entirely). Fixed point A is the only stable fixed point. Consequently, pattern A emerges. As a result, the switch from B to A takes place in the decreasing condition around the critical value of $\alpha(crit, 1) = 0.35$ between the trials with $\alpha = 0.4$ and $\alpha = 0.3$

Let us illustrate the human pattern formation reaction model for hysteretic reactions by a simulation of the LVH model. As in Fig. 6.17 the critical parameters are assumed to be at $\alpha(crit, 1) = 0.35$ and $\alpha(crit, 2) = 0.75$. Moreover, the bifurcation parameter is assumed again to change in steps of 0.1 in the range from 0.1 to 1.0. The LVH model comes with two unknown parameters. The coupling strength g and the offset parameter L mentioned above in the context of the study by Warren [343]. The unknown parameters g and L can be determined from the given critical values (see mathematical notes). Figure 6.18 shows a simulation of the LVH model for the model parameter thus obtained. Panel (a) shows the bifurcation parameter α as a function of trials in the increasing (solid line) and decreasing

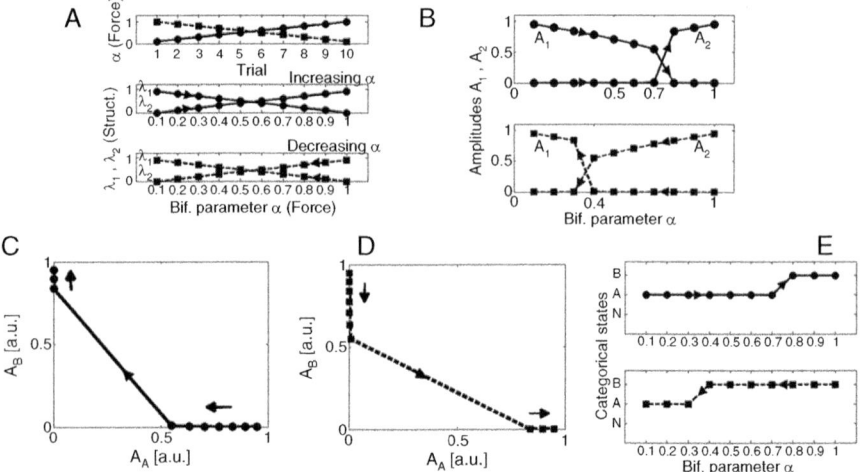

Fig. 6.18 Simulation of the human pattern formation reaction model by means of the corresponding LVH amplitude equations. See text for details

(dashed line) condition. Panel (a) also shows the brain structure as described in terms of the eigenvalues λ_1 and λ_2 of the basis patterns A and B, respectively, for the increasing condition (middle subpanel) and the decreasing condition (bottom subpanel). The eigenvalues are shown as functions of the bifurcation parameter α that in turn is assumed to be related to the external forces acting on the individual under consideration.

It is assumed that the eigenvalue λ_1 of the basis pattern A decays as a function of α. In contrast, the eigenvalue λ_2 of the basis pattern B is assumed to increase as a function of α. These dependencies taken together make that the fixed point A is unstable at values of α larger than $\alpha(crit, 2)$, while the fixed point B is unstable for sufficiently small values of the bifurcation parameter, namely, values of α smaller than $\alpha(crit, 1)$. As mentioned above, the middle and bottom subpanels in panel (a) describe the brain structure in terms of λ_1 and λ_2 for the increasing and decreasing conditions, respectively. Since for any value of the bifurcation parameter α, the eigenvalues λ_1 and λ_2 assume particular values that are independent of whether the bifurcation parameter is increased or decreased, the graphs in the two subpanels are identical. The only difference is that in the increasing condition the structure is changed according to the graphs when we follow them from the left to right as indicated by the arrows. In contrast, the structural changes during the decreasing condition are described by the graphs when we follow them from the right to the left (see arrows again).

Depending on the initial condition and the structure, the amplitudes A_A and A_B of the basis patterns A and B of the LVH model converge to the fixed point of an attractor. If the system is monostable the initial conditions do not play a role. In contrast, in the bistable domain, the initial conditions determine which attractor pattern emerges. Panel (b) shows the simulation results for the increasing (top subpanel) and decreasing (bottom subpanel) conditions. For the increasing condition, we need to read the top subpanel from left to right. For trials 1 to 7 the amplitude A_A of the basis pattern A is finite, while the amplitude A_B of the basis pattern B is zero, indicating that the attractor pattern A related to the basis pattern A emerges. At the critical value $\alpha(crit, 2)$ the eigenvalue λ_1 of the basis pattern A becomes critically smaller than the eigenvalue λ_2 of the basis pattern B. This implies that the eigenvalue λ_1 falls out of the stability band (see Sect. 4.6.3). The attractor A becomes a repellor. That is, the fixed point becomes unstable. Consequently, the amplitude dynamics converges to the fixed point attractor B. The amplitude A_B becomes finite, while the amplitude A_A becomes zero.

In the decreasing condition we need to follow the graphs shown in the bottom subpanel of panel (a) (eigenvalues) and in the bottom subpanel of panel (b) (amplitudes) from the right to the left. For trial 1 the bifurcation parameter equals 1. The eigenvalue λ_2 of the basis pattern B is much larger than the eigenvalue λ_1 of the basis pattern A (which actually equals zero for $\alpha = 1$ in the special case considered here), see panel (a). The system is monostable. The amplitude A_B of the basis pattern B is finite, see panel (b). The amplitude A_A of the basis pattern A equals zero. Consequently, the attractor pattern B emerges. The situation remains the same for trials 1, 2, 3, that is, for bifurcation parameters $\alpha = 1.0, 0.9, 0.8$. For trials 4, 5, 6, 7 both eigenvalues λ_1 and λ_2 assume comparable values such that they both are in the stability band. The system is bistable. However, for trial 4 ($\alpha = 0.7$) the initial condition is given by the amplitude relationship shown for trial 3 ($\alpha = 0.6$). The initial condition is located in the basin of attraction of the attractor B. Therefore, pattern B remains present. This situation does not change for trials 5, 6, 7. At trial 8 ($\alpha = 0.3$) the eigenvalue of the basis pattern B becomes critically small and drops out of the stability band. The fixed point B becomes unstable. The system converges to the attractor A and the corresponding attractor pattern emerges. Panels (c) and (d) show the evolution of the ordinary state of the pattern formation system in the reduced amplitude state. Note that panels (a) and (b) or panels (a), (c), and (d) depict the evolution of the grand state of the bistable human pattern formation system: forces, structure, and ordinary state. Panel (e) shows the categorical states derived from panel (b) (or from panels (c) and (d)).

Panel (e) of Fig. 6.18 is identical to panel (e) of Fig. 6.17. The difference between the categorical states shown in Figs. 6.17 and 6.18 is that in Fig. 6.17 they are derived from theoretical considerations on bistable models in general, while in Fig. 6.18 they are obtained from simulations of an explicit bistable model: the LVH amplitude equation model.

Finally, note that in Sect. 4.4.1 we discussed pattern formation under symmetry leading to bistability and presented several examples of systems that are assumed to exhibit under appropriate conditions symmetry in terms of a homogeneous spectrum of their eigenvalues, see in particular Fig. 4.12. From Fig. 6.18, panel (a), it follows that for $\alpha = 0.55$ both eigenvalues assume the same value. That is, the bistable system exhibits structural symmetry for $\alpha = 0.55$. In this context, panel (a) of Fig. 6.18 describes the mechanism leading to an inhomogeneous eigenvalue spectrum via symmetry-breaking forces as discussed in Sect. 4.4.2, see Fig. 4.14b. That is, the bifurcation parameter α is related to the asymmetry parameter discussed in the context of Fig. 4.14b.

6.5.4   Examples of Human Hysteretic Systems

Let us review a few examples of studies in which positive hysteresis in reactions to visual world forces exerted by scenes and objects has been found. These hysteretic reactions are considered as bistable reactions because humans can exhibit in a certain domain (with respect to the scenes and objects and their forces) for the same scenes or objects two different reactions that both correspond to stable patterns according to the LVH model reviewed in the preceding section.

Hysteretic Reactions to Objects and Scenes of the Visual World

Human reactions to pictures and scenes have been examined that exhibit two features that will be denoted by A and B. The pictures and scenes do not only exhibit two feature but also have a special property. They can be morphed gradually between two extreme conditions related to the two features. The degree of morphing towards either of the two extremes can be quantified by means of a parameter, which in our context, correspond to the bifurcation parameter of the human system and its environment. In both extreme conditions the human pattern formation system is monostable such that in the feature A extreme condition the feature A attractor pattern and in the feature B extreme condition the feature B attractor pattern emerges. For a set of values of the bifurcation parameter between the two extreme values the human pattern formation system is bistable. For example, in the study by Gori et al. [139] mentioned in the introduction, see Sect. 1.6.3, the two features were four dark-painted squares located in corners of a larger light-colored square, on the one hand, and a light-colored cross on a dark background, on the other hand, see Fig. 1.7. Panels (a) and (c) of Fig. 1.7 show the two extreme monostable conditions. Panel (b) shows one of the medium bistable conditions.

In general, in experimental paradigms that reveal bistable reactions to aspects of the visual world, participants are exposed to two types of sequences of pictures and scenes: sequences that start with A extremes and end with B extremes and sequences that start with B extremes and end with A extremes. Participants produce some kind of forces (typically make certain verbal statements) to indicate when a switch between the attractor patterns A and B takes place or to indicate which attractor pattern is present at every picture or scene presented in the given sequence of pictures and scenes. In terms of the bistable human pattern formation reaction model presented above, the presentation of the pictures and scenes in combination with the experimental instructions are assumed to make an appropriately defined reference state unstable. At that primary bifurcation point two basis patterns are assumed to exist that are related to the features A and B (see Sect. 6.5.3). Depending on the eigenvalues of the basis patterns, the basis patterns give rise to attractors and repellors in the reduced amplitude space of the pattern formation systems. As discussed in the preceding section, the eigenvalues are assumed to depend differently on the bifurcation parameter. Without loss of generality the eigenvalue of the feature A basis pattern is assumed to decrease monotonically as a function of the bifurcation parameter, while the eigenvalue of the feature B basis pattern is assumed to increase monotonically as a function of the bifurcation parameter. Consequently, when participants are put under the forces of the sequences characterized by increasing and decreasing bifurcation parameters the feature A and B attractor patterns switch at different values of the bifurcation parameter. The A-to-B and B-to-A transitions exhibit hysteresis.

Gori et al., in their study on reactions to figure-ground features (i.e., study on figure-ground "perception"), used pictures as shown in Fig. 1.7 that involved as two features four black squares, on the one hand, and a pale cross, on the other hand. The four black squares were located in the four corners of a squared area. Pictures, for which the four squares were relatively small, as shown in panel (a), typically led to the emergence of a four squares attractor pattern and the corresponding verbal force production reaction pattern. In contrast, pictures for which the four squares were relatively large, as shown in panel (c), a visual world cross attractor pattern emerged with an appropriate supplementary verbal reaction pattern component. However, participants were not put in a situation that made them to react to every presented picture. In fact, Gori et al. [139] did not present a discrete sequence of pictures. Rather, they presented a movie that showed an image morphing gradually from one extreme to the other extreme. Participants were set up such that when a switch between the two attractor patterns (either a four-squares to cross or a cross to four-squares transition) occurred then they produced the word STOP. As bifurcation parameter the ratio of the area of the four squares and the area of the cross was used. In doing so, each frame of the movie could be assigned to a bifurcation parameter. Small bifurcation parameters indicated that frames showed small squares and accordingly the extreme four-square picture had a small bifurcation parameter of 0.01. In contrast, frames with relatively large bifurcation parameters showed a cross on a background given by the four squares. Accordingly, the extreme cross picture had a bifurcation parameter of $\alpha = 1.65$. Hysteretic reactions were found in all individual participants. That is, the critical bifurcation parameters of four-squares-to-cross transitions in increasing sequences were larger than the critical bifurcation parameters of cross-to-four-squares transitions in the decreasing sequences. For the sample of participants from the data presented by Gori et al. the critical bifurcation parameters of $\alpha(crit, 1) \approx 0.2$ (decreasing condition) and $\alpha(crit, 2) \approx 0.7$ (increasing condition) can be obtained. Accordingly, hysteresis size was positive and about 0.5. It was statistically significant larger than zero (for details see the original study [139]).

In the same study, Gori et al. also discuss another experiment based on the so-called Petter's effect. The experiment involved switches between two categorical reactions and a gradually changing bifurcation parameter that was either increased

or decreased. Again, individual subjects showed hysteretic reactions when comparing their switching points of the increasing and decreasing conditions. Again,
hysteresis size as computed from the sample of participants was positive and
statistically significant larger than zero.

A study by Kim and Frank [187] replicated to some extent the first experiment
of the study by Gori et al. [139] about human reactions to four-squares and cross
figure-ground images. In contrast to the study by Gori et al. [139], in the study
by Kim and Frank [187] a discrete set of pictures was presented. That is, one
picture was presented after another. Participants reacted to each picture depending
on their ongoing brain states and the forces exerted by the picture by pressing one
of two keys. That is, from the perspective of pattern formation, participants were set
up such that the emergence of the visual world four-square BBA attractor pattern
induced the emergence of the motor activity pattern of pressing the key "1" on a
keyboard placed in front of them (–just like a hurricane can induce the emergence
of tornadoes). In contrast, the emergence of the visual world cross BBA attractor
pattern made participants pressing the key "0" on that keyboard. Irrespective of that
difference, in the study by Kim and Frank [187] again hysteresis with a positive
hysteresis size was found when averaged across all participants. In particular, the
hysteresis size was statistically significant different from zero (for details see the
original study [187]).

In a study by Poltoratski and Tong [262], the authors report among other things
from an experiment that involved images that, on the one hand, featured a particular
object (like a car or coffee mug) but, on the other hand, featured a whole scene
as well (like a street scene in a city or a living room). Importantly, the objects of
interest were part of the scenes (e.g., the car was part of the street scene and the
coffee mug was part of the scene that showed the living room). Again sequences of
images were used with two extreme images. The extreme images showed either the
object in magnified view and with little surrounding or the whole scene including
the object on a smaller scale. The images in-between the extreme images were
zooming-in images into the scene extreme images. As in the study by Gori et al.
[139], a movie was used rather than a presentation of individual pictures. Let us
refer to the zooming-out condition, in which the images were gradually morphed
from the object view to the scene view, as the increasing condition. Conversely, the
zooming-in condition, in which the images were morphed from the scene view to
the object view, will be referred to as decreasing condition. Due to the experimental
instructions, participants were structured such that when there was a transition
between the object and scene attractor patterns induced by the movie images then
that transition made them to pressed a key on a keyboard. Different movies with
objects and scenes were used. All movies lasted 16 s. When averaging across
participants, the object attractor pattern was present for a longer time period in the
increasing condition than in the decreasing condition. That is, when considering
frame number as bifurcation parameter, the critical frame number $\alpha(crit, 2)$ of the
increasing condition was larger than the critical frame number $\alpha(crit, 1)$ of the
decreasing condition. Accordingly, hysteresis was found and hysteresis size was
positive. Hypothesis testing showed that it was statistically significant different

from zero (for details see Ref. [262]). In the same study, Poltoratski and Tong [262] presented several variants of the object-scene experiment. In those additional experiments again hysteretic reactions to the movies were found.

Hysteretic reactions were also observed in a study by Kleinschmidt et al. [188] in which participants were exposed to a movie in which a letter appeared and disappeared from the background. The letter appeared and disappeared by changing the contrast between letter and background. The experiment involved two tasks: an active and a passive one. In the active task, participants reported by pressing a button when the letter appeared and disappeared. Let us denote the conditions in which the contrast was increased and decreased as increasing and decreasing conditions. Furthermore, let us regard the contrast of the letter as bifurcation parameter α. It was found that the critical bifurcation parameter at which letters appeared, $\alpha(crit, 2)$, was larger than the critical bifurcation parameter at which letters disappeared, $\alpha(crit, 1)$. That is, $\alpha(crit, 2) > \alpha(crit, 1)$ or positive hysteresis was observed. In the passive task, participants just watched the movie with the increasing and decreasing conditions. In both the active and the passive tasks brain activity was recorded using so-called fMRI measurements. In both tasks, the hysteresis phenomenon was also observed on the brain activity level. Interestingly, the brain activity increased in a similar way as it was found in the animal experiment by Shadlen and Newsome displayed in Fig. 6.14d, see Sect. 6.3.5. Therefore, the increase of brain activity observed in the study by Kleinschmidt et al. [188] is consistent with the pattern formation perspective according to which emerging visual world letter BBA patterns come with increasing brain activity during the formation of the patterns (see Sect. 6.3.1 and panels (c) and (e) in Fig. 6.7). Finally note that hysteresis on the level of brain activity has also been found in other fMRI studies (see e.g. Ref. [291]).

Hysteretic Reactions Involving Small Body Movements

In the remainder of this section, let us consider bistable human reactions that involve limb movement patterns involving objects or whole body movements. More precisely, studies on grasping, on the one hand, and gait transitions, on the other hand, will be reviewed.

Humans grasp small objects, like a pencil, typically with one hand. Bigger objects, like a small table, are typically grasped with two hands. Medium sized objects, like a book, are grasped with either one hand only or with two hands. As discussed in detail in Sects. 2.5.1 and 4.4.1, such objects of medium size challenge to some extent the concept of physics-based determinism. If medium sized objects are sometimes grasped with one hand but on other occasions the very same objects are grasped with two hands, then how can there be a deterministic law that determines human grasping patterns? The answer to the question is that for medium sized objects the human pattern formation system is assumed to be bistable (and close to be symmetric, see Sect. 4.4.1). This answer is supported by experimental studies revealing hysteretic grasping reactions (when assuming that hysteresis is evidence for bistability—as it has been done above). That is, several studies have shown that grasping of objects is subjected to hysteresis related to the size of the objects. Van der Kamp et al. [329] asked children at the age of 5–9 years to grasp cubes of different sizes and to move them away from one location to another location over a distance of about 30 cm. While small cubes were grasped and moved away with one hand, larger cubes were typically grasped with two hands. The experiment involved an increasing condition in which cube size was gradually increased in steps of 1 cm from 2.2 cm to 16.2 cm. Moreover, there was a decreasing condition in which cube size was decreased from 16.2 cm to 2.2 cm. For each child, the transition point at which the child switched from one hand to two handed grasping or vice versa was noted. While cube size could have been used as bifurcation parameter to describe the transition points, the experimenters showed that a more appropriate bifurcation parameter was a body-scaled measure: the ratio of the size of a cube and the finger span of a participant. Finger span was defined as the distance from the end of the thumb to the end of the index finger. Looking at the performance of individuals, most of the participants exhibited hysteretic reactions. That is, the critical bifurcation parameter $\alpha(crit, 2)$ at which the transition from one-hand grasping to two-handed grasping took place in the increasing condition was larger than the critical bifurcation parameter $\alpha(crit, 1)$ at which the transition from two-handed grasping to one-hand grasping occurred. When taking the data from all participants together, the sigmoid percentage functions shown in panel (a) of Fig. 6.19 were obtained. The graphs show the percentage of one hand grasps performed by all participants given a particular bifurcation parameter (i.e., cube size to finger span ratio). Two graphs are shown: one describing the increasing cube size condition, the other describing the decreasing condition. The graphs demonstrate the hysteresis phenomenon on the level of the participant sample. When increasing cube size gradually children not only performed one hand grasps for relatively small cubes but also for medium sized cubes. They switched to grasp cubes with two hands only when cube size was relatively large. In contrast, when decreasing cube size gradually children not only performed two-handed grasping movements for the larger cubes but also for medium sized cubes. The switches back to the one-hand grasping pattern occurred for cubes that were relatively small. According to the (bistable) human pattern formation reaction model for hysteretic reactions discussed in the previous section, in the increasing condition children formed one-hand grasping BBA attractor patterns in the bistable parameter domain, while in the decreasing condition children formed two-handed grasping BBA attractor patterns in the bistable domain.

Richardson et al. [269] and Lopresti-Goodman et al. [214] reported about hysteretic grasping reactions to planks. Several planks that differed in length were used. By means of experimental instructions, participants were structured such that they grasped the planks on the end sides only either with two hands or with one hand. When grasping planks with one hand, participants had to put the thumb on one end side and the 4th and 5th fingers of the same hand on the other end side. Consequently, short planks could be picked up by one hand. In contrast, planks that exceeded in length the hand span of participants could not be picked up with

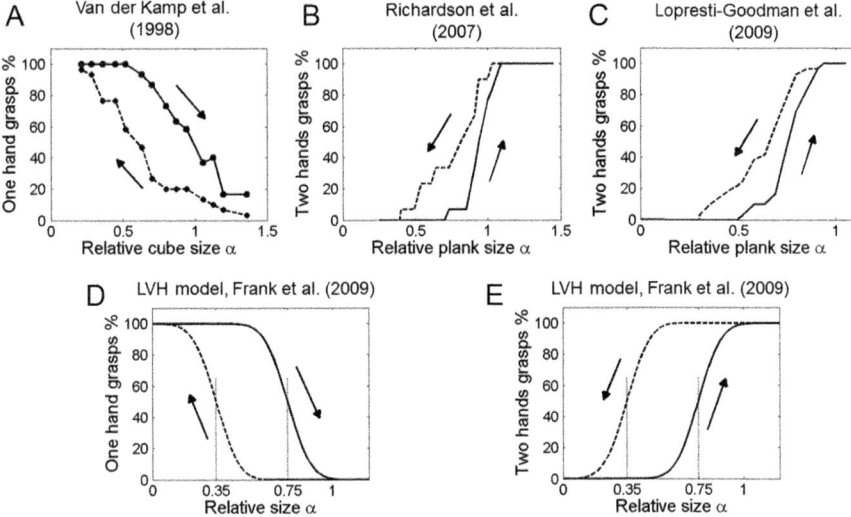

Fig. 6.19 Change in the grasping type as a function of object size and presentation order as observed in experiments and predicted by the LVH model. Panel (**a**): Percentage of performed one hand grasps when relative cube size was increased and decreased, respectively, as reported by Van der Kamp et al. [329]. Panels (**b**) and (**c**): Percentages of two-handed grasps of planks with different relative plank sizes when size was increased and decreased, respectively, as reported by Richardson et al. [269] and Lopresti-Goodman et al. [214], respectively. Panels (**d**) and (**e**): Percentage of one hand grasps (panel **d**) and two-handed grasps (panel **e**) as computed from the LVH model (see Frank et al. [120]) when object size is increased and decreased, respectively. The critical parameters $\alpha(crit, 1) = 0.35$ and $\alpha(crit, 2) = 0.75$ taken from Fig. 6.18 have been used as population mean values. They indicate the 50% points on the vertical axes of the graphs in panels (**d**) and (**e**)

one hand in the required grasping style. Hand span was defined in the same way as finger span in the study by van der Kamp et al. [329], see above. In the studies by Richardson et al. [269] and Lopresti-Goodman et al. [214] several experiments were described. In what follows, only Experiment 2 of Richardson et al. [269] and Experiment 1 of Lopresti-Goodman et al. [214] will be considered. Moreover in Experiment 2 of Richardson et al. [269] a short and a long set of planks was used. Only the short set of planks containing planks ranging from 2 to 35 cm will be considered. In Experiment 1 of Lopresti-Goodman et al. [214] the impact of the speed with which planks were presented was tested. For sake of brevity, the details of the impact of this speed factors will be ignored. Only the effect of the presentation condition (increasing versus decreasing) will be reviewed.

Planks were presented in two different ways, in an increasing order from short to long planks and in a decreasing order from long to short planks. Planks were presented one at a time. In the increasing condition, the shortest plank of the set of planks was presented first and subsequently longer planks were presented. In the decreasing condition, the longest plank was presented first and subsequently shorter planks were presented. In line with our comment made above, participants typically grasped the relatively short planks with one hand and the relatively long planks with two hands. In this context, it should also be noted that the mean hand span of the participants in the study by Lopresti-Goodman was 21.32 cm. Therefore, a great portion of the participants was physically not able to grasp the longest plank (i.e., the plank with 24.5 cm in size) at the two end sides with a single hand because the length of the longest plank exceeded their hand span. They had to grasp the longest plank with two hands. Consequently, it does not come as a surprise that in both experimental conditions (increasing and decreasing) participants typically switched from one grasping type to another while planks were presented. That is, from a pattern formation perspective, participants typically showed transitions between different grasping BBA attractor patterns. In both studies participants grasped planks and moved them away. For each plank the grasping movement pattern was recorded. The bifurcation parameter α was defined as relative plank size given in terms of the plank length divided by the hand span of a participant. In both studies, it was found that on average participants performed one-hand grasps in the increasing condition for medium sized planks for which they used two-handed grasps in the decreasing condition. That is, in terms of the bifurcation parameter both studies found that the critical bifurcation parameter $\alpha(crit, 2)$ at which in the increasing condition transitions from one-hand grasping BBA attractor patterns to two-handed BBA attractor patterns occurred was larger than the critical bifurcation parameter $\alpha(crit, 1)$ at which in the decreasing condition transitions from two-handed grasping BBA attractor patterns to one-hand grasping BBA attractor patterns took place. Positive hysteresis was found. The hysteresis size was found to be statistically significant different from zero in both studies. Panels (b) and (c) in Fig. 6.19 show the percentage values of observed two-handed grasping BBA attractor patterns as functions of the bifurcation parameter separately for the increasing and decreasing conditions. The functions shown in panels (b) and (c) demonstrate graphically the hysteresis phenomenon on the level of the participant samples. In the increasing conditions of both studies the graphs start to increase only at relative high values of the bifurcation parameter α indicating that participants continued to grasp even relatively large planks with one hand. Likewise, in the decreasing conditions of both studies the graphs start to decrease only for relatively small values of the bifurcation parameter α indicating that participants grasped even relatively small planks with two hands. Overall, with respect to the bistable human pattern formation reaction model for hysteretic reactions the graphs show that in the increasing conditions participants formed one-hand grasping BBA attractor patterns in the bistability domain, whereas in decreasing conditions participants formed two-handed grasping BBA attractor patterns in the bistability domain.

Finally, panels (d) and (e) of Fig. 6.19 presents percentage graphs as predicted by the bistable human pattern formation reaction model. The graphs have been computed using the LVH model essentially in the same way as described above in the context of the study by Warren [343], see also panel (a) of Fig. 6.16. However, while in the study by Warren [343] we were only concerned with the distribution of a single critical bifurcation parameter, in the context of the reviewed studies on hysteresis phenomena there are two critical bifurcation parameters $\alpha(crit, 1)$ and

$\alpha(crit, 2)$ and, consequently, two distributions. From those two distributions of the critical bifurcation parameters (that are in fact again related to the distribution of the offset parameter L of the LVH model and may even related to the distribution of the coupling parameter g) the sigmoid percentage graphs shown in panels (d) and (e) of Fig. 6.19 can be constructed [120].

Hysteretic Reactions Involving Whole Body Movements

While grasping objects and moving them away may be considered as a relatively small-scaled body activity, walking and running are whole body activities. Just as transitions between different ways of grasping (e.g. one-hand and two-handed grasping) have been studied, transitions between walking and running have been examined in experimental studies (as mentioned briefly in Sect. 4.4.1). Here walking is characterized as a gait for which at any time there is at least one foot in contact with the ground. In contrast, running involves a flight (or aerial) period. During the flight period, the body loses contact with the ground. Typically, humans tend to walk at low locomotion speeds, whereas at high locomotion speeds running is the preferred gait. In this regard, walk-run gait transitions in humans belong to the general class of gait transitions in animals. For example, walk-trot gait transitions in horses have been addressed briefly in Sects. 3.2.3 and 4.4.1.

Walk-run gait transitions have frequently examined in laboratory studies by putting participants on moving treadmills. In such studies the treadmill speed is manipulated (increased or decreased) while participants move on a treadmill. For example, in a study by Diedrich and Warren [64] treadmill speed was gradually changed from about 1 m/s to about 4 m/s and back to 1 m/s. When averaging the walk-to-run and run-to-walk transition speeds it was found that the gait transitions took place at a speed of 2.07 m/s. In another study by Diedrich and Warren [65] somewhat higher values of transitions speeds were found at 2.19 m/s (see the "control" conditions in Experiments 1 and 2 reported there). In general, it seems that values somewhat above 2 m/s are typically critical transition speeds for walk-to-run and run-to-walk transitions. For example, Hreljac [167] reports a transition speed of 2.05 m/s for a sample of 28 participants. Li [211] found transition speeds in the range of 2.2–2.4 m/s (see Figure 3 of Li [211]).

Based on theoretical reasoning, it has been suggested that transition speed should increase with leg length. Therefore, a rescaled parameter, the so-called Froude number or, more precisely, the walking Froude number, has been introduced to eliminate the impact of leg length and allow for better comparisons between participants that differ in leg length. The walking Froude number is a dimensionless variable and defined by the speed squared divided by leg length and the earth gravitational constant[1] [3]. Transition values measured in Froude numbers rather

---

[1] The walking Froude number is proportional to the square of the speed. In contrast, there is another Froude number that is defined by the speed divided by the square root of leg length and the earth

than locomotion speeds should be independent of the leg length of participants and, for this reason, are considered more appropriate bifurcation parameters. In the context of the human pattern formation reaction model, the Froude number as bifurcation parameter has the advantage that it comes with a constraint. At Froude numbers of 1 and larger, it is physically impossible to walk. The centrifugal force will lift the whole body from the ground such that there is a flight period such that permanent contact with the ground, which is the characteristic feature of walking, is not possible. In the aforementioned study by Diedrich and Warren [64], the authors also report an overall critical Froude number of 0.49 for walk-to-run and run-to-walk transitions. Similarly, in Experiments 1 and 2 of the study by Diedrich and Warren [65] critical Froude numbers of 0.54 and 0.55 are reported. Frank [109] reports from a treadmill experiment with a baseline condition for which a critical Froude number of 0.43 was found obtained by averaging critical Froude numbers of walk-to-run and run-to-walk transitions.

In order to find support for the phenomenon of bistable human reactions and support the model-based hypothesis that external forces of everyday objects (object being viewed or to be grasped) and seemingly simple circumstances (walking or running on a treadmill in a gym) can change the structure of human pattern formation systems, increasing and decreasing speed conditions of treadmill experiments need to be considered separately. In fact, Diedrich and Warren [64] found a walk-to-run critical Froude number of 0.50, while the run-to-walk critical Froude number was slightly lower at 0.48. That is, positive hysteresis was observed. The relative small difference of 0.02 between the two Froude number was statistically significant. Hreljac et al. [168] studied gait transitions on inclined treadmills. Three different slope conditions (zero slope, medium slope, and large slope) were considered. For all slope conditions the walk-to-run transition speeds were about 0.07 m/s higher than the run-to-walk transition speeds. That is, again, participants as a group showed a performance consistent with the bistable pattern formation system described in the previous section. The differences between the critical transitions speeds were statistically significant. The values of 0.02 in terms of Froude numbers and 0.07 in terms of locomotion speeds are relatively small. The size and even its sign seems to depend on the precise circumstances of the experimental setup (see below).

According to the pattern formation reaction model described in Sect. 6.5.3, in the increasing condition treadmill speed is increased such that at certain critical locomotion speeds or Froude numbers the BBA walking pattern becomes unstable and a transition to the BBA running pattern occurs. Likewise, in the decreasing condition treadmill speed is decreased while participants initially form BBA running attractor patterns until the patterns become unstable and the human body pattern and the corresponding brain activity switches to a BBA walking attractor pattern.

---

gravitational constant. This classical Froude number is proportional to speed. The walking Froude number is just the square of the classical Froude number.

The speed with which treadmill speeds are changed, that is, the belt acceleration and de-acceleration can be manipulated in some range. Li [211] found for relatively high acceleration and de-acceleration values of treadmill speed that the walk-to-run transition speeds were considerably larger than the run-to-walk transition speeds (see Figure 3 of Li [211]). That is, for high acceleration and de-acceleration values positive hysteresis was found. However, at relatively low acceleration and de-acceleration values the transition scores were the same or the effect was even reversed. This discussion illustrates that there is some experimental evidence for the bistability of human pattern formation system of walking and running but it is a challenge to identify the precise conditions for bistability.

6.5.5 Relational Character of Bifurcation Parameters

Throughout this book it has been pointed out that bifurcations and phase transitions take place at certain critical parameter values (i.e., critical values of bifurcation parameters). For example, in Chap. 3 it has been mentioned that iron exhibits a phase transition to the magnetic state at a particular critical temperature. It has been stated that spatially homogeneous fluid and gas layers make bifurcations to convection roll patterns when the temperature difference between the bottom and top layers reaches a critical value. It has been stated that horses bifurcate from walk to trot at critical locomotion speeds. However, in general, phase transition points and bifurcation points are not properly characterized by a single physical quantity like a temperature or velocity value. The critical conditions at which phase transitions and bifurcations take place are described by several physical quantities or by parameters (i.e., bifurcation parameters) that are functions of several physical quantities.

What matters for a bifurcation to occur is how large a certain physical quantity is with respect to other physical quantities. Relational properties determine bifurcation points. Therefore, bifurcation parameters are more adequately be described in terms of variables describing the relationship between different physical properties. This holds for self-organizing systems as well as for systems that do not belong to the class of self-organizing systems. Table 6.6 illustrates several examples and also demonstrates that this feature of bifurcation parameters is a universal one that holds in all kind of systems exhibiting bifurcations.

Let us begin our discussion with a mechanical system that does not belong to the class of self-organizing system: the inverted pendulum with an oscillating base. The inverted pendulum as such is a stick with a mass at the one end, which is put upside down such that the mass is on the top. If the inverted pendulum is placed on a table, it will skip to either side as soon as the stick has a little bit of a tilt or as soon as the table is shaken a little bit. That is, the upside down state of the inverted pendulum is an unstable fixed point. Let us refer to the side of an inverted pendulum that points down (i.e., does not carry the mass) as the base of the inverted pendulum. The inverted pendulum with an oscillatory base is an inverted pendulum that is positioned upside down (i.e., with the mass pointing up) and has a base that

Table 6.6 Examples of relational bifurcation parameters, that is, bifurcation parameter expressing relations between several physical quantities

System	Type of system	Bifurcation parameter
Inverted pendulum	Classical mechanics system (not self-org. system)	Base oscillation frequency divided by square root of pendulum length
Pure materials showing aggregate phase transitions	Equilibrium systems	Reduced state variables
Cavity lasers	Non-equilibrium self-org. systems (inanimate world)	Pumping rate over decay rate of pumped energy levels
Convection cells/cloud patterns	Non-equilibrium pattern formation systems (inanimate world)	Temperature difference divided by diffusivity and viscosity (Rayleigh number)
Bistable signaling pathways	Non-equilibrium self-org. systems (animate world)	Ratio of forward and backward reaction speeds
Humans stair climbing	Non-equilibrium pattern formation systems (animate world)	Ratio of stair step height over leg length
Humans grasping	As above	Relative cube size/relative plank size
Humans making gait transitions	As above	Speed squared rescaled by leg length (walking Froude number)

oscillates up and down with a certain frequency and amplitude. Importantly, the mechanisms that moves the base up and down does not restrict the pendulum from tilting to either side. For very low frequencies, the oscillating base does not have any effect on the stability of the upside down state. That is, if the stick is perturbed out of the perfect vertical orientation, then the pendulum skips to a side and falls down. However, if the oscillation frequency is sufficiently high, then the unstable fixed point is turned into a stable fixed point. That is, if the pendulum is tilted out of the perfect vertical by a not too large angle, then it does not fall down but after a transient period the pendulum returns to the perfect vertical alignment. That is, for sufficiently high oscillation frequencies the inverted pendulum with oscillating base has a fixed point attractor with a certain basin of attraction. Therefore, the base oscillation frequency may be considered as bifurcation parameter. However, in fact, base oscillation frequency as such is not a good bifurcation parameter. If pendulums with different length are used than the critical oscillation frequencies at which the upside down states become stable differ from each other. Long pendulums have a relatively high critical frequencies. Short pendulums have a relatively low critical frequencies. A detailed analysis shows that the bifurcation parameter is the oscillation frequency divided by the square root of the pendulum length. Using this bifurcation parameter, pendulums of different length exhibit the transition to stability at the same critical value. In other words, it is not the oscillation frequency as such that matters but how large the oscillation frequency is relative to the square root of the pendulum length. The bifurcation parameter given in terms of the ratio of base oscillation frequency over the square root of pendulum length is a relational parameter that relates different physical quantities to each other.

As mentioned in Sect. 1.5, materials exist in different aggregate states as solids, liquids, and gases. They change from one state to another at equilibrium aggregate phase transitions. Such phase transitions feature the general feature of all phase transitions and bifurcations, namely, that there is state variable that exhibits a kink or discontinuity at the bifurcation point or phase transition point (see Sects. 3.2.1 and 3.2.3 and Figs. 3.4 and 3.5). For example the noble gases Argon, Krypton, and Xenon turn from liquids into gases at boiling point temperatures of -186, -153, and $-108\,^\circ$C, respectively (under standard pressure). While these critical absolute temperature values differ from each other, the phase transitions lines of the noble gases at which the liquid-to-gas and solid-to-liquid phase transitions take place fall all onto each other when relative state variables are used. They are called reduced temperature, reduced pressure, and reduced density [334]. For example, the reduced temperature is the absolute temperature measured in Kelvin divided by a particular substance property: the temperature in Kelvin at the so-called critical point. That is, when the states of the noble gases are described in absolute state variables such as temperature, pressure, and density, then phase transitions for different noble gases take place at different temperature, pressure, and density values. However, when the states of noble gases are characterized in terms of relative variables (reduced temperature, reduced pressure, and reduced density) then the phase transitions (bifurcations) of the different gases take place under the same conditions, see panel (a) of Fig. 6.20. Again, the relevant state variables (bifurcation parameters) to understand phase transitions (bifurcations) are not particular physical quantities but variables expressing relations between physical quantities.

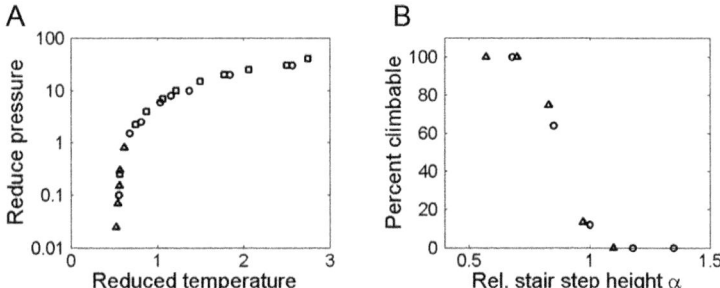

Fig. 6.20 Panel (**a**): Melting curve of noble gases shown in the two dimensional phase diagram spanned by temperature and pressure [334]. Reduced thermodynamic variables are used. Symbols refer to neon (circles), argon (squares), and xenon (triangles). The melting curves of the noble gases fall on one single universal curve when using reduced variables. Panel (**b**): Percentage of observed D0/D2 climbing-up BBA patterns emerging in participants for different stair step heights for two groups (triangles: long leg group, circles: short leg group) of participants [343]. Relative, body-scaled step height is used. When using the relative, body-scaled measure, then the experimental results obtained from both participant groups fall on a single universal curve

Let us briefly address two self-organizing systems of the inanimate world showing non-equilibrium phase transitions and bifurcations: cavity lasers and convection cell systems. The latter have been introduced earlier in Sect. 3.2.3. Cavity lasers show a bifurcation from a non-lasing state to a state, in which they produce laser light, when the pumping current or the rate of pumping exceed particular critical values. The critical values depend among other things on the laser active materials used in the laser devices. That is, the critical values differ across lasers exhibiting different laser active materials. However, if we take into account that the atomic levels of different laser active materials also come with different decay rates, then a unique bifurcation parameter can be constructed: the pumping rate divided by the decay rate of the laser active material levels [156]. With respect to this relational bifurcation parameter, different lasers exhibit the same critical bifurcation parameter. Note that in this consideration cavity losses are assumed to be negligible. If this is not the case, then the bifurcation parameter may be modified again to take into account that different lasers may exhibit different cavity losses. The bifurcation thus obtained is a function of the pumping rate, decay rate, and cavity losses.

As discussed in Sect. 3.2.3, in fluid and gas layers heated from below roll patterns emerge when temperature differences ΔT across layers become sufficiently large. That is, roll patterns emerge at critical values $\Delta T(crit)$. The critical values $\Delta T(crit)$ depend on various factors. For example, fluids and gases that come with relatively large thermal diffusivity parameters tend to exhibit relatively large critical values $\Delta T(crit)$. This observation may be explained heuristically by arguing that fluids and gases do not tend to produce convection rolls "in order to" transport heat when heat can be transported quickly by diffusion. Likewise, substances that exhibit relatively large viscosity coefficients typically come with relatively large critical values $\Delta T(crit)$. Again, heuristically this may be explained by arguing that the large viscosity (i.e., some sort of friction) works against the emergence of convection rolls (i.e., cells that move on circles upwards and downwards in space). In contrast, fluids and gases with small viscosity and diffusivity parameters, typically exhibit low critical temperature differences $\Delta T(crit)$ at which roll patterns emerge. Consequently, the temperature difference ΔT as such is not the key variable that determines whether or not roll patterns emerge. At issue is how large the temperature difference ΔT is relative to other physical properties such as thermal diffusivity and viscosity. A detailed examination of those pattern formation systems shows that the adequate bifurcation parameter is a rescaled temperature, called the Rayleigh number [252, Chap. 6]. The Rayleigh number is a variable that involves several physical quantities. For our purposes it is sufficient to note that the Rayleigh number can be computed as the temperature difference divided by the thermal diffusivity coefficient and the viscosity coefficient multiplied by a factor that depends on other physical quantities.

The systems considered so far and listed in Table 6.6 were systems of the inanimate world. Let us turn to self-organizing systems of the animate world exhibiting bifurcations. Various cells exhibit bistable signaling pathways that switch from monostability to bistability when bifurcations parameters exceed certain critical values. The bifurcations parameters are typically variables composed of various parameters describing the biochemical reactions of the signaling pathways. For example, in a relative simple model for a bistable signaling pathways proposed in a series of studies [111, 233, 257] the bifurcation parameter is the ratio of certain chemical forward and backwards reaction rates. That is, again, it is not of primary issue how large the reaction rate of a single biochemical reaction is. What matters is how large a reaction rate is relative to the other reaction rates characterizing the self-organizing system.

Finally, in Sect. 6.5.4 several bifurcation parameters have been introduced that describe relations between physical quantities. For stairways climbing, it is not the size of the stair steps as such that makes adults and children to go up stairways or makes them to refuse to go up stairways. What matters is the step height relative to the leg length of the adults or children under consideration. This issue was demonstrated by Warren [343] with the graph shown in panel (b) of Fig. 6.20. Just as in Fig. 6.16a, the graph in Fig. 6.20b show percentage functions of YES reactions of participants to stairways with different stair step heights. While in Fig. 6.16a only one percentage function of one particular participant group is presented, in Fig. 6.20b results obtained from two participant groups are shown. The participants in the two groups differed with respect their body anatomy. One group consisted of participants with relatively short legs. In the other group of participants had relatively long legs. The triangles in Fig. 6.20b are obtained for the group with long legs (i.e., the tall group). This graph has been shown earlier in Fig. 6.16a. The circles in Fig. 6.20b describes data obtained for the group of participants with short legs. When using the relative stair height (as defined above as stair step height divided by leg length) as bifurcation parameter, then the results from both groups fall on top of each other—just as the phase transition lines of noble gases fall onto each other when the noble gases are described by relative variables (reduced state variables), compare panels (a) and (b) of Fig. 6.20. Therefore, it is plausible to assume that human reactions to stairs are determined by a relational bifurcation parameter as introduced by Warren [343].

Likewise, experimental research has shown that grasping transitions are not as such determined by the sizes of the to-be-grasped objects (e.g., cube or planks). Rather, grasping transitions depend on how large objects are relative to the physical properties of the humans that are about to grasp the objects. Accordingly, the adequate bifurcation parameters are cube size divided by hand span and plank length divided by hand span [248, 249, 329]. Finally, for walk-run gait transitions the locomotion speed as such seems to be a less useful bifurcation parameter. Rather, the Froude number introduced in Sect. 6.5.4 that takes the leg length of individuals into account is a more adequate bifurcation parameter to describe gait transitions between walking and running [112].

6.5.6 Bifurcation Parameters and "Information Variables"

Within the framework of ecological psychology [133], studies on human reactions to objects have used the concept of "information". In particular, it has been argued that bifurcation parameters such as stair step height relative to leg length or plank size relative to hand span carry or reflect some sort of "information" and should be considered as "information variables". From a physics perspective these notions and concepts of "information" and "information variables" are not needed and only add to the mystification of humans and animals. For example, an inverted pendulum that oscillates around its base with a certain frequency does not pick up any kind of "information" that "guides" the dynamics of the pendulum and makes that the pendulum falls down or not. A cavity laser does not pick up "information" about the rate with which the laser device is pumped relative to the decay rate of the laser active material levels. A cavity laser does not "decide" based on that "information" to produce or not to produce laser light. In the environment of a fluid or gas layer heated from below there is not such a thing as "information" sitting around about the temperature difference between the bottom and top sides of the layer and how large that temperature difference is relative to the thermal diffusivity and viscosity of the fluid or gas molecules. In particular, the layers are not "connected" to such "information" and the formation of patterns in the layers is not "guided" by such "information". In short, the Rayleigh number is a bifurcation parameter but not an "information variable". Bistable signaling pathways in cells do not pick up "information" about the chemical driving forces of the chemical reactions determining the pathways. Signaling pathways do not use "information" about the corresponding backwards and forwards reaction rates. A child standing in front of a stairway with stairs of a particular step height is not connected to some kind of "information". The child does not refuse to go up the stairs or is happy to go up the stairs because of some sort of "information" available to the child or because of a "information variable" "guiding" the "behavior" of the child. According to the physics perspective of humans, humans do not "decide" to grasp objects with one hand or two hands because of some kind of spiritualistic energy called "information" sitting around the to-be-grasped objects. The emergence of human grasping patterns does neither involve "choice", nor "decision making", nor "information". BBA attractor movement patterns like grasping with one hand or two hands, walking or running are not "guided" by "information"—just as the emergence of stripe patterns on the skin of Zebras is not influenced by "information". Noble gases do not turn from solids into liquids because there is "information" about the reduced temperature. "Information" does not make noble gases to switch their matter of state from a solid to a liquid.

From a scientific perspective, systems showing bifurcations and phase transitions satisfy all the same principle. They are subjected to forces from their environments. The forces as such do not matter. The forces in relation to the properties of the systems matter. Relevant questions are for example: How large—with respect to the pendulum length of an inverted pendulum—is the oscillation frequency of the force acting on that inverted pendulum? How large is the temperature of the heat bath surrounding a noble gas with respect to the atomic (structural) properties of the gas as characterized by the gas temperature at the so-called critical point? In other words, we may ask the question: What is the impact of the forces of the heat bath on the noble gas (i.e., the impact of the elastic collisions between heat bath molecules and noble gas molecules) as captured by the heat bath temperature relative to the structural properties of the noble gas? How large is the rate of pumping (induced by external forces on a laser) relative to the decay rate (structural laser property) of the atomic or molecular laser active levels of the laser? How large is the temperature difference across a gas layer with respect to the gas viscosity and thermal diffusivity? That is, how large is the impact of the collisions between the heating material molecules and gas molecules at the bottom side of the layer relative to the conditions on the top side taking the structural properties of the gas into account? What is the impact of the electromagnetic forces in the light reflected by the stairs of a stairway on the pattern formation system of a child given the structural properties of the child such as leg length? What is the impact of the forces exerted by objects on human pattern formation systems relative to the structural properties of the human pattern formation systems?

Having said that, the physics perspective to understand human and animal brain and body activity is just one perspective out of a plenitude of perspectives about the human nature (see Sects. 2.4.1 and 2.7.2). There are various non-scientific perspectives (in the sense of perspectives that are not based on physics and violate the laws of physics to some extent) about the human nature. In particular, there are religious perspectives and spiritualistic perspectives. For example, in the two largest religions, Christianity and Islam, God and Allah are frequently considered as the sender of information or as the source of information. Humans are considered to be the recipients of that kind of information. The Bible and the Quran (or Koran) are assumed to contain information. Humans have a free will and can decide to use the information that is contained in religious texts like the Bible and the Quran to guide their actions.

As pointed out several times throughout this book, concepts like "free will" and "decision making" and likewise the notion of "information" and "information variables" may or may not be useful concepts in perspectives different from the scientific perspective based on physics. In fact, the author of this book as a private person (not as a scientist) believes that it is important for humans to explore religious and spiritualistic perspectives and their concepts (see also Chap. 12). Importantly, the physics perspective of the animate and inanimate worlds does not make any statement about other perspectives and about the value of concepts used in other perspectives. In short, while within the physics perspective of humans and animals concepts like "free will", "choice", "decision making", "control", "information" and "information variables" are not needed, these concepts might be very powerful and fundamental concepts in other (non-scientific) perspectives of the human nature.

6.6   Mathematical Notes

6.6.1   LVH Model of Type 1 Sequences

The LVH amplitude equations for the example discussed in Sect. 6.1.4 reads

$$\frac{d}{dt} A_k = \lambda A_k - g A_k \sum_{j \neq k, j=1}^{3} A_j^2 - A_k^3 \tag{6.1}$$

for the three amplitudes $k = 1, 2, 3$ of the object function BBA attractor patterns related to the container, platform, and weight properties of a box. Equation (6.1) holds as long as there is no force that shifts the brain state. A BBA pattern has emerged if brain activity is close to the respective attractor, which means if the amplitude satisfies $A_k \geq \theta A_{k,st}$, where $A_{k,st} = \sqrt{\lambda}$ is the fixed point value and θ is a threshold value close to 1. When a BBA pattern is present for a period $\tau > 0$ then it is assumed that it causes a brain state shift such that its own amplitude decays. Following Eq. (5.9) this force-induced change is modeled like

$$\exists \, [t - \tau, t] \; : \; A_k(t') \geq \theta A_{k,st}(t') \text{ for all } t' \in [t - \tau, t] \implies A_k(t + \epsilon) = \Phi^{A_k} = \delta , \tag{6.2}$$

where the function Φ^{A_k} corresponds to a constant δ (here: $\delta > 0$) that is small and $\epsilon > 0$ is a small time shift. The graphs shown in Fig. 6.5 have been computed by solving Eqs. (6.1) and (6.2) numerically with an Euler forward scheme using a single time step $\Delta t = 0.001$ and the simulation parameters $\lambda = 3$, $g = 1.5$, $\epsilon = \Delta t$, $\theta = 0.9$, $\delta = 0.001$, $\tau = 20$, and a noise term (reflecting deterministic perturbations, see Sect. 2.4) added to the deterministic dynamics given by the time-discrete version of Eq. (6.1) at every time step with uniform noise in $[0, Q]$ with noise amplitude $Q = 0.003\sqrt{\Delta t}$.

6.6.2   Formation of Winner-Takes-All BBA Patterns in Reactions to Several Objects

As discussed in Sect. 6.3.3 humans exposed to scenes like the house with the house number 1 in Fig. 6.8a are subjected to several forces related to different features of the scenes. In Sect. 6.3.3 two scenarios have been discussed about how a multistable, winner-takes-all human pattern formation system may react to such scenes.

The Brain State Shift Model: House Example

Let A_1 and A_2 denote the amplitudes of the house and house-number-1 BBA attractor patterns. According to the LVH model with cubic nonlinearity, the pattern amplitudes satisfy

$$\frac{d}{dt} A_k = \lambda A_k - g A_k A_j^2 - A_k^3 \tag{6.3}$$

for $k = 1, 2$. Here, $k = 1 \Rightarrow j = 2$ and $k = 2 \Rightarrow j = 1$. In the brain state shift model, the pattern formation system exhibits a homogeneous eigenvalue spectrum as indicated by the fact that there is only one eigenvalue λ in Eq. (6.3). The pattern formation system exhibits symmetry. The diagonal in the reduced amplitude space separates the basins of attraction, see Sect. 6.3.3. Before the individual is exposed to the scene the amplitudes are assumed to be at the origin like $A_1 = A_2 = 0$. At $t = t_{start}$ the individual under consideration is confronted with the scene. The forces from the scene are assumed to shift the brain state into the basin of attraction of the visual world house attractor like

$$A_1(t_{start}) = \Phi^{A_1}, \quad A_2(t_{start}) = \Phi^{A_2}, \Phi^{A_1} > \Phi^{A_2}, \tag{6.4}$$

where Φ^{A_1}, Φ^{A_2} are constants. Consequently, the brain state as described in terms of the amplitudes A_1 and A_2 converges to the visual world house attractor and the visual world house BBA attractor pattern emerges. The simulation results in Fig. 6.8 were computed from Eqs. (6.3) and (6.4) with $t_{start} = 1$ time unit, $\lambda = 3$, $g = 1.5$, $\Phi^{A_1} = 0.9$, and $\Phi^{A_2} = 0.1$.

The Structure Shift Model: House Example

Figure 6.9 illustrates the structure shift model for the example of human reactions to the house scene. The spectrum of eigenvalues is inhomogeneous and it is assumed that the forces from the scene affect the spectrum. The pattern amplitudes are assumed to satisfy the cubic LVH model

$$\frac{d}{dt} A_k = \lambda_k A_k - g A_k A_j^2 - A_k^3 \tag{6.5}$$

with $k = 1 \Rightarrow j = 2$ and $k = 2 \Rightarrow j = 1$, again. At $t = 0$ we have $A_1 = A_2 = 0$ and $\lambda_1 = \lambda_2$. The human under consideration is exposed to the scene at $t = t_{start}$. The forces of the scene are assumed to shift the brain state in an unbiased way like

$$A_k(t_{start}) = \Phi^{A_k} \text{ for } k = 1, 2 \text{ with } \Phi^{A_1} = \Phi^{A_2}. \tag{6.6}$$

However, the forces are assumed to cause a change of structure as described by Eq. (5.13). Importantly, the structure is change in a asymmetric way, see Fig. 6.9b. In our example, for $k = 2$ we have $d\lambda_2/dt = 0$ for all t. However, for $k = 1$ integration of Eq. (5.13) over a short interval $\epsilon > 0$ leads to

$$\lambda_1(t_{start} + \epsilon) = \lambda_1(t_{start}) + \int_{t(start)}^{t(start)+\epsilon} f_1(\alpha = \text{House scene}) \, dt \, . \tag{6.7}$$

Assuming that $f_1 > 0$ is constant, we obtain

$$\lambda_1(t_{start} + \epsilon) = \lambda_1(t_{start}) + C_1 \, , \tag{6.8}$$

where $C_1 > 0$ is a constant again. The simulation results presented in Fig. 6.9 were computed from Eqs. (6.5), (6.6), and (6.8) with $t_{start} = 1$ time unit, $\lambda_1(t = 0) = 3$, $\lambda_2 = 3$, $g = 1.5$, $\Phi^{A_1} = \Phi^{A_2} = 0.5$, $C_1 = 1$, $\epsilon = \Delta t$, where $\Delta t = 0.01$ was the single time step of the Euler forward method used to solve Eq. (6.5).

Cocktail Party Phenomenon

Let us turn next to the cocktail party phenomenon discussed in Sect. 6.3.3. The pattern amplitudes of the three relevant BBA attractor patterns are assumed to satisfy the cubic LVH amplitude equations

$$\frac{d}{dt} A_k = \lambda_k A_k - g A_k \sum_{j \neq k, j=1}^{3} A_j^2 - A_k^3 \tag{6.9}$$

with eigenvalues λ_k that in general can differ from each other. Here $k = 1, 2, 3$ refer to A, B, C. For the baseline scenario described in Fig. 6.11 it is assumed that the human pattern formation system of the listener L is symmetric such that $\lambda_k = \lambda$ for all $k = 1, 2, 3$. However, when listener L meets with speaker A at time t_0, the brain state is shifted into the basin of attraction of the listen-to-A attractor. That is,

$$A_k(t_0) = \Phi^{A_k} \text{ for } k = 1, 2, 3 \wedge \Phi^{A_1} > \Phi^{A_2}, \ \Phi^{A_1} > \Phi^{A_3} \, . \tag{6.10}$$

The results shown in Fig. 6.11 were computed from Eqs. (6.9) and (6.10) for $t_0 = 0$, $\lambda = 3$, $g = 1.5$, $\Phi^{A_1} = 0.87$, and $\Phi^{A_2} = \Phi^{A_3} = 0.3$.

The transition from listen-to-A BBA pattern to the listen-to-C BBA pattern shown in Fig. 6.12 was modeled in line with the brain state shift model described in this section above. Accordingly, the forces exerted by speaker C (when laughing) on listener L at $t_1 = 20$ time units were modeled like

$$A_k(t_1) = \Phi^{A_k} \text{ for } k = 1, 2, 3 \wedge \Phi^{A_3} > \Phi^{A_1}, \ \Phi^{A_3} > \Phi^{A_2} \tag{6.11}$$

with $\Phi^{A_3} = 0.87$, and $\Phi^{A_1} = \Phi^{A_2} = 0.3$.

The transition from listen-to-A BBA pattern to the listen-to-C BBA pattern shown in Fig. 6.13 was modeled in line with the structure shift model described in this section above. Accordingly, we assumed that speaker A exerts at $t = t_0$ forces on listener L that change the eigenvalue spectrum such that it becomes inhomogeneous as shown in Fig. 6.13a and characterizes a monostable pattern formation system. That is, we assumed that

$$t < t_0 \; : \; \lambda_k = \lambda_{baseline} \text{ for } k = 1, 2, 3$$

$$t = t_0 \; : \; \lambda_k(t_0 + \epsilon) = \lambda_k(t_0) - C \text{ for } k = 2, 3 \qquad (6.12)$$

with $\lambda_{baseline} - C < \lambda_{baseline}/g \Rightarrow C > \lambda_{baseline}(1 - 1/g)$ (which implies monostability, see our discussion of the stability band in Chap. 4). Similarly, at $t = t_1$ the forces exerted by speaker C result in (see also Eq. 5.14)

$$\lambda_3(t_1) = h_3(\alpha = \text{speaker C laughing}) = \lambda_{baseline} \, ,$$

$$\lambda_k(t_1) = h_k(\alpha = \text{speaker C laughing}) = \lambda_{baseline} - C \text{ for } k = 1, 2 \, .$$

$$(6.13)$$

The simulation results presented in Fig. 6.13 were computed from Eqs. (6.9), (6.12), and (6.13) for $t_0 = 0$, $t_1 = 20$, $\lambda_{baseline} = 3$, $g = 1.5$, $C = 2$, and $A_k(t_0) = 0.3$ for $k = 1, 2, 3$. At every simulation step, noise was added to the deterministic dynamics described by Eq. (6.9) using a noise term uniformly distributed in the interval $[0, Q]$ with $Q = 0.003\sqrt{\Delta t}$, where Δt was the Euler forward single time step with $\Delta t = 0.001$.

6.6.3 Event-Related/Force-Induced Brain Activity: Pattern Formation Perspective

The force-induced or event-related brain activity related to the emergence of a BBA attractor pattern can be describe by a monotonous increase of an amplitude $A(t)$ from $A(t_0)$ at $t = t_0$ towards the fixed point value A_{st} for $t \to \infty$. Neglecting the impacts of all other amplitudes, the dynamics of A satisfies the LVH model

$$\frac{d}{dt} A(t) = \lambda A(t) - A^q(t) \, , \qquad (6.14)$$

with nonlinearity exponent q. As pointed out in Sect. 4.6.6 the nonlinearity exponent q determines the fixed point value A_{st}. However, the parameter q also determines the temporal evolution of $A(t)$. If we fix q, then Eq. (6.14) exhibits only a single parameter λ. This implies that the temporal and stationary features of the amplitude dynamics both depend on λ and, consequently, are closely related to each other.

In general, this is not the case. As discussed in Sect. 4.7.7, the term $-A^q$ actually comes with a coefficient that has been eliminated in Eq. (6.14). Therefore, the more general amplitude equation model reads

$$\frac{d}{dt}A(t) = \lambda A(t) - CA^q(t) \tag{6.15}$$

and involves two parameters λ and $C > 0$ in addition to q. If we fix q, then those two parameters determine the stationary and temporal properties of the formation of a brain and body activity pattern under consideration. Equation (6.15) can be solved analytically [90]. Using the analytical solution, the parameters λ and C can be fitted to data from experimental studies, in general, and data describing the emergence of brain activity patterns, in particular. The result of such a fitting procedure is shown in panel (d) of Fig. 6.14.

6.6.4 Human Pattern Formation Reaction Model for Reactions with Positive Hysteresis

The human pattern formation reaction model for reactions with positive hysteresis was introduced in the context of grasping transitions [120, 215] and gait transitions [105, 109, 110]. The two pattern amplitudes of interest are denoted by A_k with $k = 1, 2$ or $k = A, B$. The amplitude equations for the cubic case read

$$\frac{d}{dt}A_k = \lambda_k A_k - g A_k A_j^2 - A_k^3 \tag{6.16}$$

with $k = 1 \Rightarrow j = 2$ and $k = 2 \Rightarrow j = 1$. The eigenvalues λ_k are related to the bifurcation parameter α in terms of the linear relationships

$$\lambda_1 = 1 - \alpha \; , \; \lambda_2 = L + \alpha \; , \tag{6.17}$$

where L denotes an offset parameter. The model involves the unknown parameters L and g. Given the critical values $\alpha(crit, 2) \in [0, 1)$ and $\alpha(crit, 1) \in [0, 1)$ with $\alpha(crit, 2) > \alpha(crit, 1)$ (positive hysteresis), the parameters L and g can be determined. A detailed calculation shows that [120]

$$g = \frac{1 - \alpha(crit, 1)}{1 - \alpha(crit, 2)} \; \Rightarrow \; g > 1 \; , \; L = 1 - \alpha(crit, 1) - \alpha(crit, 2) \; . \tag{6.18}$$

Vice versa, when the parameters g and L are given the critical bifurcation parameters $\alpha(crit, 1)$ and $\alpha(crit, 2)$ are given by [120]

$$\alpha(crit, 1) = \frac{1 - gL}{1 + g} \; , \quad \alpha(crit, 2) = \frac{g - L}{1 + g} \; . \tag{6.19}$$

The simulation results in Fig. 6.18 were obtained from Eqs. (6.16), (6.17), and (6.18) for $\alpha(crit, 1) = 0.35$ and $\alpha(crit, 2) = 0.75$ ($\Rightarrow g = 2.6$, $L = -0.1$) using an Euler forward scheme with single time step $\Delta t = 0.001$. A noise term was added to the Euler forward discrete version of Eq. (6.16) at every simulation step using uniformly distributed noise in the interval $[0, Q]$ with $Q = 0.0001\sqrt{\Delta t}$.

Chapter 7
Pattern Formation of Grand States

This chapter is concerned with pattern formation of grand states. That is, pattern formation is considered that involves structural changes. The structural changes are not due to external forces that act as driving forces as discussed in Chap. 6. Rather, the structural changes are part of the pattern formation. In this chapter, various possible explicit scenarios are discussed related to the classification scheme of human and animal A,B,C,D,E systems proposed in Chap. 5, see Fig. 5.5.

First of all, brain activity attractor patterns may affect structural components supporting the emergence of the brain activity patterns such that temporal sequences emerge of structural changes and emerging attractor patterns (as in certain A and C systems). More explicitly, a given brain activity pattern may induce certain structural changes which in turn leads to the emergence of another brain activity pattern. From the system classification scheme shown in Fig. 5.5 it follows that this kind of grand state pattern formation can occur in human isolated systems (A2 systems) or human systems producing muscle forces without being affected by external forces (C1 systems). Furthermore, brain activity patterns may induce changes in brain structure in systems in which the brain activity (but not the structure) is also affected by external forces but does not lead to force production (C2 systems). If there exists a feedback loop between brain state and structure (as in A2 systems) and, in addition, (1) external forces affect the brain state of the human or animal system under consideration and (2) the system produces forces, then we deal with pattern formation of grand states in C3 systems.

Grand state pattern formation can also be due to external forces affecting structural components such that brain activity attractor patterns emerge that lead to force production that, in turn, changes the external forces that affect the structural components (as in certain D and E systems). In systems of this type, again, temporal sequences can emerge that involve emerging and disappearing brain activity attractor patterns, changes of external forces, and structural changes. In particular, in human and animal D1 and D3 systems external forces result in structural changes just as discussed in Chap. 6. However, unlike the cases discussed

© Springer Nature Switzerland AG 2019
T. Frank, *Determinism and Self-Organization of Human Perception
and Performance*, Springer Series in Synergetics,
https://doi.org/10.1007/978-3-030-28821-1_7

in Chap. 6, the external forces are not considered to be external driving forces. Rather, in the context of grand state pattern formation, situations are consider in which human D1 and D3 systems produce forces that change the environmental circumstances of the systems which affects the external forces acting on the D1 and D3 systems. Therefore, the external forces are part of the pattern formation phenomena under consideration.

Finally, in this chapter self-organizing human and animal systems will be considered whose structure is changed both by the emerging brain activity patterns and the external forces, where the external forces are changed or affected by the produced forces related to the emerging brain activity patterns (E3 systems). This kind of systems involve the feedback loop between structure and brain activity of A2 systems and the feedback loop between external forces, produced forces, and brain activity of D1 and D3 systems.

7.1 Preliminary Remarks

7.1.1 Type 2 Pattern Sequences and Cycles

Type 2 pattern sequences and cycles are given by a series of emerging and disappearing patterns just as type I pattern sequences and cycles. In contrast to type 1 pattern sequences and cycles, type 2 sequences and cycles involve structural changes. The basic scheme of type 2 pattern sequences and cycles is illustrated in Fig. 7.1.

According to Fig. 7.1, it is assumed that a particular attractor pattern A (i.e., a BA or BBA attractor pattern) has emerged. The pattern results in structural changes. For example, the emerging pattern may affect the eigenvalue spectrum that determined the emergence of the pattern A in the first place. As mentioned

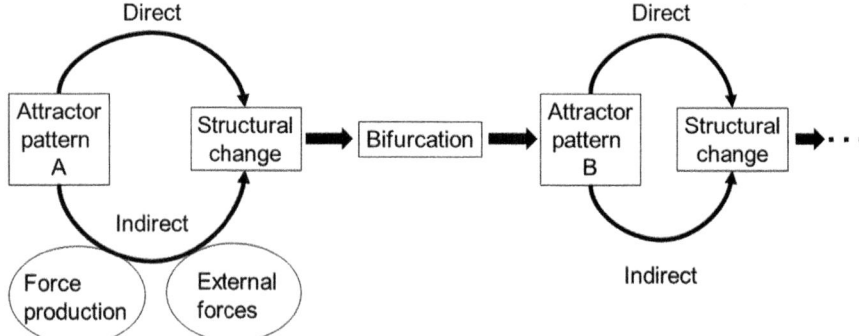

Fig. 7.1 Type 2 sequences of patterns involve direct paths or indirect paths of restructuration or both

above, the structure may be affected in two different ways. First, the structure may be affected by the brain state (direct path of restructuration). For example, in human and animal isolated A2 systems, BA patterns may affect the structure that determines the emergence of BA patterns. Likewise, according to Fig. 5.5, in C systems the structure is directly impacted by the emerging BA or BBA patterns. Second, the structure may be affected by external forces as in D and E systems, see Fig. 5.5 again. In this case, type 2 pattern sequences and cycles exhibit an indirect path of restructuration, in which BBA patterns exhibit a force production component that results in changes of external forces (see Fig. 7.1). These changes of the external circumstances are assumed to result in changes of the structure.

In general, pattern cycles are special cases of pattern sequences in which after a certain number of BA or BBA patterns sequences repeat themselves.

7.2 Dynamics of Isolated Human A2 Systems

7.2.1 Type 2 Pattern Sequences in A2 Systems

Human A2 systems are characterized by an interaction between brain state and structure (see Fig. 5.5). On the one hand, the structure determines the evolution of the brain state. On the other hand, the brain state affects, that is, changes the structure. That is, the structure is a dynamical system just as the brain state. In isolated human systems external forces do not affect brain dynamics and structure dynamics. Likewise, attractor patterns emerging in the grand state spanned by brain state variables and structure variables are not involved in the production of muscle forces. Just as A1 systems, A2 systems describes pattern formation of brain activity of awake humans and humans in the REM stage related to what has been called "thinking" and "dreaming". Unlike A1 systems, A2 systems address the situation in which the brain exhibits some kind of plasticity such that its structure changes over time. However, the brain structure does not depend explicitly on time. Rather, the brain structure depends on the brain state that in turn depends on time.

Let us address the example of a sequence of box functions attractor patterns introduced in Sect. 6.1.4. A box exhibits several object functions. For example, it can serve as a container, platform, weight, as a tool to hide something behind it or to cover something, and so on. Just as in Sect. 6.1.4, we consider the isolated human brain. That is, we consider an awake person or a person in the REM stage. As starting point, it is assumed that the visual world box attractor pattern has emerged in a particular brain area. The awake person "thinks" about a box. The person in REM stage "dreams" about a box. The box attractor pattern is assumed to make a certain reference point unstable. The reference point is related to brain activity attractor patterns about object functions. Furthermore, without loss of generality, let us assume that the reference pattern does not describe any specific object function but emerges as attractor pattern as a consequence of the box attractor pattern and

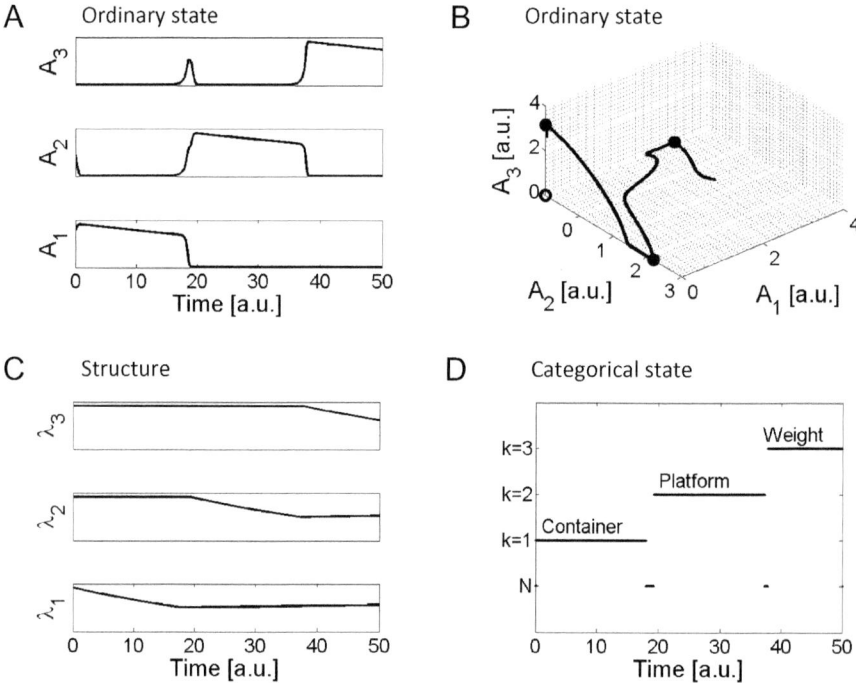

Fig. 7.2 Pattern sequence emerging in the A2 system brain state of a person (i.e., a "thinking" or "dreaming" person) about possible functions of a box. Panels (**a**) and (**b**): brain state described by the reduced amplitude state of the amplitudes of the unstable container, platform, and weight basis patterns. Panel (**c**): Evolution of the brain structure in terms of evolving eigenvalues. Panel (**d**): Categorical states

subsequently becomes unstable to give rise to a set of unstable basis patterns that are associated with object function attractor patterns. For example, the reference point fixed point may correspond to objects that are completely useless in the sense that they have no object functions at all (if such objects exist at all). As in Sect. 6.1.4 we consider three unstable basis patterns at the reference point related to the three object function attractor patterns container, platform, and weight.

Figure 7.2 illustrates the formation of a pattern sequence in the grand state spanned by brain state variables and structure variables. The simulation shown in Fig. 7.2 is based on the LVH model. Accordingly, the brain state variables used are the amplitude variables of the three unstable basis pattern constituting the reduced amplitude state. The structure variables are the eigenvalues of the unstable basis patterns. Figure 7.2 follows from Fig. 7.1 for A2 systems. Accordingly, initially, the container attractor pattern emerges. This is due to the fact that in the simulation it is assumed that initially the amplitude of the container attractor basis pattern is the largest. Moreover, all eigenvalues assume the same value. Therefore, the system initially exhibits symmetry and the basis pattern with the

largest initial amplitude wins the competition process. Panel (a) shows the basis
pattern amplitudes A_1 (container), A_2 (platform), and A_3 (weight) as functions of
time. Panel (b) shows the trajectories in the reduced amplitude space. When the
amplitude $A_1(t)$ of the unstable container basis pattern reaches its fixed point value,
the corresponding container attractor pattern (that may or may not involve further
stable basis pattern components and other supplementary components, see Chap. 4)
has emerged. Following Fig. 7.1 for A2 systems it is assumed that the container
attractor pattern changes the structure. In our example, the attractor pattern reduces
the eigenvalue of its unstable basis pattern. The eigenvalues λ_1 (container), λ_2
(platform), and λ_3 (weight) of the basis patterns are shown as functions of time
in panel (c). As shown in panel (c) (bottom subpanel), the eigenvalue λ_1 of the
container basis pattern decays. The decay of the eigenvalue λ_1 has quantitative and
qualitative implications. First, the decay of the eigenvalue implies quantitatively that
the fixed point amplitude $A_{1,st}$ decays as well (compare the bottom subpanels of
panels (a) and (c) of Fig. 7.2 and see Sect. 4.6.5). Second, at a certain critical value
the eigenvalue of the container basis pattern drops out of the stability band, see
Sect. 4.6.3. The container attractor pattern turns into a repellor pattern. However,
the eigenvalues of the two other unstable basis patterns (platform and weight)
are relatively large such that they are in the stability band. Therefore, the system
exhibits a bifurcation, as has been stated in general in Fig. 7.1. In the simulation, the
platform and weight basis patterns compete with each other which can be seen by the
temporary increase of the amplitudes of both patterns (around $t = 18$ time units, see
panel (a) of Fig. 7.2). In the simulation, the initial conditions at the time point when
the container attractor pattern becomes unstable are such that the platform basis
pattern wins the competition process. The amplitude A_2 of the platform basis pattern
increases towards its finite fixed point value $A_{2,st} > 0$, whereas the amplitude A_3 of
the weight basis pattern decays to zero. The platform attractor pattern emerges. With
regard to Fig. 7.1 the platform attractor pattern corresponds to pattern B, whereas the
container attractor pattern corresponds to pattern A. Figure 7.2 illustrates explicitly
by means of the LVH model the general scheme introduced in Fig. 7.1 for A2
systems. As indicated in Fig. 7.1, the process that leads from the emergence of
one attractor pattern to the emergence of another attractor pattern is assumed to
take place again and again. With respect to the example presented in Fig. 7.2, when
the platform attractor pattern has emerged, the pattern is assumed to damp its own
eigenvalue (i.e., the eigenvalue λ_2). In contrast, the eigenvalue of the disappearing
attractor pattern (i.e., λ_1) increases back to its baseline level. At a particular critical
value, the eigenvalue λ_2 drops out of the stability band. The platform attractor
pattern becomes unstable. The attractor becomes a repellor. A bifurcation takes
place (at about $t = 38$ time units, see panels a and d). In our simulation the (grand
state) conditions at the bifurcation point in terms of the eigenvalues (structure) and
the amplitudes (brain state) are such that the system bifurcates to the weight attractor
pattern. That is, the amplitude A_3 of the unstable weight basis pattern increases and
converges towards it finite fixed point value $A_{3,st} > 0$. Panels (a) and (c) illustrate
the pattern formation sequence in the grand state spanned by brain state variables
(panel a) and structure variables (panel c). Finally, panel (d) presents the categorical

brain states related to the container, platform, and weight BA attractor patterns. The reader "may take some time" to compare Figs. 7.2 and 6.5 of the previous chapter. The two figures illustrate how the attractor pattern sequence container-platform-weight can occur in the human brain as type 1 (Fig. 6.5) or type 2 (Fig. 7.2) sequence. Note that "may take some time" actually means that the reader will do so when his/her grand state conditions are appropriate. Otherwise, the reader will skip this step of comparing Figs. 6.5 and 7.2. That is, the reader has no "choice" in this regard.

7.2.2 Hierarchical Structures in Self-organizing Systems and Freud's Concepts of "Consciousness", "Unconsciousness", and "Drive"

Pattern formation systems with mutually exclusive unstable basis pattern exhibit at reference points a hierarchical structure as shown in panel (a) of Fig. 7.3. In general, as discussed in Chap. 4, at bifurcation points systems exhibit a plenitude of basis patterns with negative eigenvalues, which are called the stable basis patterns. In addition, there is a set of basis patterns with positive eigenvalues, the unstable basis patterns. In the amplitude space of all basis patterns the stable basis patterns do not come with fixed points and, consequently, cannot give rise to attractor patterns (one caveat: they may contribute to attractor patterns, see below and Chap. 4). In contrast, unstable basis patterns come with fixed points and, in doing so, can lead to fixed point patterns in terms of attractor and repellor patterns. Unstable basis patterns and their fixed point patterns are listed within the triangle shown in Fig. 7.3a. We can distinguish between three types of unstable basis patterns. First, some of the unstable basis patterns have positive small eigenvalues that fall out of the stability band. Those basis patterns can give rise only to repellors in the reduced amplitude space. In particular, those patterns cannot emerge as attractor patterns. The remaining basis

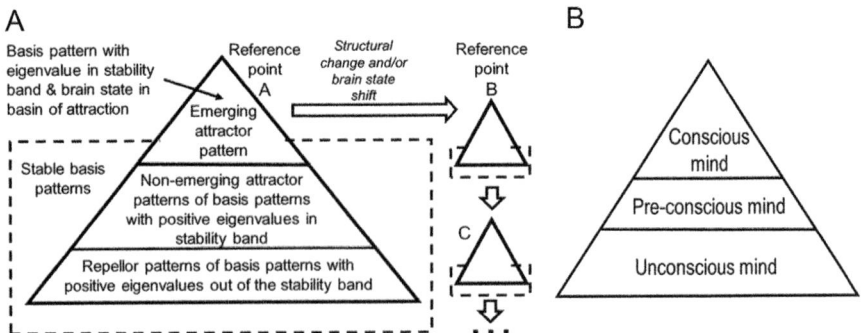

Fig. 7.3 Pattern hierarchy from a physics perspective (panel (**a**)) and Freud's conceptualization of the human mind as a hierarchically structured entity (panel (**b**))

patterns exhibit eigenvalues in the stability band. For each of basis pattern there exists an BBA or BA attractor pattern. Consequently, the unstable basis patterns with eigenvalues in the stability band give rise to a multistable dynamical system. Each attractor exhibits a basin of attraction. Therefore, we obtain as a second set of attractor patterns those attractor patterns that have basins of attractions that do not contain the current brain state. As a result, the brain state does not converge to any of those attractors. The patterns do not emerge. In Fig. 7.3 they have been called non-emerging attractor patterns. Since the basis patterns are assumed to be mutually exclusive there is exactly one basis pattern that gives rise to a particular attractor whose basin of attraction contains the current brain state. The brain state converges to that attractor and the corresponding BBA or BA attractor pattern emerges.

There are various interactions between the groups of patterns of the pattern hierarchy shown in Fig. 7.3a. The basis patterns with positive eigenvalues and the corresponding attractor patterns compete with each other. Moreover, stable basis patterns can damp the emerging attractor pattern and can be pumped by the emerging attractor pattern such that they emerge as part of the attractor pattern, see Chap. 4.

As mentioned in the introduction and discussed in Sect. 2.7, the physics perspective of the world and of human and animals in particular addresses any aspect of this world and any aspect of human and animal life. This also implies that all observation made in the context of Freud's psychoanalytical approach to humans can be explained within the physics perspective. For example, Freud assumed that there is a hierarchical structure of the "mind" as illustrated in panel (b) of Fig. 7.3. There is the "conscious mind", the "pre-conscious mind" and the "unconscious mind". According to Freud, the "unconsciousness" can affect the "behavior" of humans. For example, a person who has witness as a child the death of a school friend may be affected by that experience as an adult "without knowing it".

From a physics perspective, it is plausible to assume that dramatic events like witnessing the death of a friend come with forces that change the structure of a person's brain. Such structural changes can persist over the whole lifespan of a person. As a result, at certain unstable reference points, brain activity patterns similar to those induced by the forces at the time of a dramatic event can exist in various forms as stable basis patterns or unstable basis patterns in any of the three groups listed within the triangle hierarchy shown in panel (a). Let us assume that at a particular reference point (e.g., a person gets out of bed in the morning and the state of standing there and doing nothing becomes unstable) the pattern of the dramatic event exists as a stable basis pattern or an unstable basis pattern with a small eigenvalue (i.e., an eigenvalue out of the stability band). Then, the pattern cannot lead to an attractor pattern of its own. However, it can affect the emergence of other attractor patterns. For example, it can slow down the formation of emerging pattern. The person becomes slow in "getting things done" (i.e., to take a shower, getting dressed, etc.). Moreover, as a stable basis pattern the pattern of the dramatic event may emerge as part of an attractor pattern. For example, if a person under consideration is verbally introduced to somebody and that person happens to have the same first name as the childhood friend who died in the aforementioned hypothetical dramatic event, then according to the physics perspective the acoustic force exerted by the spoken name makes a particular reference state unstable such that a particular acoustic world attractor pattern related to the name emerges. The emerging acoustic world name BBA pattern may involve as supplementary component the stable basis pattern related to the death of the friend. If so, that component in turn can come with negative feelings. As a result, when the person X under consideration is introduced to a person Y, who happens to have the same name Y as the childhood friend, then our person X may experience negative feelings.

The fact that the emergence of an attractor pattern can involve not only the basis pattern of the attractor pattern but also other basis patterns is a general feature of all kind of pattern formation systems. Air, gas, and fluid layers forming cloud patterns, roll patterns, and other patterns exhibit the hierarchical structure shown in Fig. 7.3 and feature interacting basis and attractor patterns. Laser devises operating at the laser threshold at which laser light emerges exhibit the hierarchy shown in Fig. 7.3. According to the physics perspective, the human brain and human body as a pattern formation system exhibits the hierarchy shown in Fig. 7.3. The elements of the hierarchy shown in panel (a) should not be considered as counterparts of the elements of Freud's hierarchy shown in panel (b). That is, emerging attractor patterns are not physical counterparts of "consciousness". They are not physical correlates of "consciousness". Likewise, stable basis patterns are not physical counterparts or physical correlates of the "unconscious mind". If we would state that certain basis patterns of brain activity of the hierarchy shown in Fig. 7.3 emerging in the human brain reflect "consciousness" or "unconsciousness" and at the same time would deny that other self-organizing systems (fluid layers, lasers, etc.) that exhibit (just like humans) the pattern hierarchy shown in Fig. 7.3a have "consciousness" or "unconsciousness", then we would make a difference, where there is no difference. We would mystify humans over the non-living thingbeings.[1] Therefore, in physics there is no place for the concepts of "consciousness" and "unconsciousness". This argument supports the notion that "consciousness" and "unconsciousness" are non-scientific, spiritualistic, and religious motivated concepts.

Having said that the author of this book as a private person not as a scientist believes that non-scientific, spiritualistic approaches such as Freud's psychoanalytical approach involving the concepts of "consciousness" and "unconsciousness" are useful approaches in their own merit. Within the psychoanalytical (non-scientific) approach, rituals have been design that have help in the past generations of people to improve their lives. However, the fact that an approach has improved the lives of people and has help patients to overcome their problems does not mean that the approach is a scientific one. In summary, any laser device and any air layer producing a cloud pattern has in common with any human being and animal the

---

[1]The author of this book is tempted to coin the slogan "Consciousness for everyone or for no-one", which would be polemics. The matter-of-fact statement should read "Consciousness for any pattern formation system or for no pattern formation system" and would hold only within the physics perspective.

hierarchy shown in Fig. 7.3a. If humans would have "consciousness" and "unconsciousness" (which they do not have from a scientific perspective) because of the fact that human systems exhibit the hierarchy property illustrated in Fig. 7.3a, then lasers and air layers would have "consciousness" and "unconsciousness" as well. "Consciousness" and "unconsciousness" are non-scientific concepts. They may or may not be useful in the context of non-scientific approaches for understanding human beings. Physics does not make any statement about the usefulness of those concepts just as physics does not make any statement about the usefulness of non-scientific concepts in general.

In the context of Freud's psychoanalytical approach, "drive theory" has been developed. Accordingly, human beings have "drives" that make them to "behave" in certain ways. For example, according to drive theory, a heterosexual female single adult who happens to get into some contact with a male individual may initiate some kind of flirting (or human mating) "behavior" because of "sexual drive". As shown in Figs. 7.1 and 7.3a (see the sequence of emerging hierarchy-triangles) and discussed in the context of type 1 and 2 sequences and cycles, according to the physics perspective, emerging attractor patterns in general trigger bifurcations that result in the emergence of other attractor patterns. In the previous example, the female is exposed to external forces related to the male individual which results in an instability of some reference point and in the emergence of a particular attractor pattern. For example, if the male stands in front of the female and the female happens to see the male then in certain brain areas a visual world male individual BBA attractor pattern is assumed to emerge. That pattern can make a reference point (of doing nothing or standing still and saying nothing) unstable such that a flirting BBA attractor pattern emerges in which the female moves her body in certain ways and starts a conversation (with respect to flirting see also Sect. 13.3 at the end of this book).

As mentioned in the introduction, a hurricane under appropriate conditions can lead to the emergence of billow clouds on top of its hurricane spiral, see Sect. 5.2.3 and Fig. 5.7c. Likewise, hurricanes frequently induce the formation of tornadoes. Just as air layers can produce hurricanes that in turn produce billow clouds and tornadoes, a female-male system can produce a certain BA and sensory body activity pattern in the female that produces in a second step certain BA and body dynamics pattern. If the female would possess some "drive" property (which she does not possess from a scientific perspective), then the air layer or hurricane would possess the same kind of "drive" property. The hurricane would be "driven" to produce tornadoes (should meteorologists talk about a "tornado-drive" just like psychologists talk about a "sex-drive"?). In this context, note that self-organizing systems on a material level typically differ from each other. For example, humans have genes. Hurricanes do not have genes. However, the principles of pattern formation hold in all kind of self-organizing systems and do not depend on the material constitution of the systems under consideration [160]. This line of thoughts suggests that "drive theory" is not part of a physics-based scientific approach to understand human beings. From physics it follows that humans are pattern formation systems and that all aspects of human beings and human life can be understood in terms of emerging attractor patterns that change circumstances and, in doing so, lead to the emergence of other attractor patterns. There is no room for concepts like "drive" or "consciousness" or "unconsciousness". Again, the reader should note that classifying "drive theory" as a non-scientific approach does not mean that any value is attached to "drive theory". In other words, the fact that from physics it follows that self-organizing pattern formation systems including humans do not possess a "drive" property does not say anything about the value of "drive theory" in general.

7.2.3   Physics Perspective Versus Freud's Psychoanalytical Goal to Become "Master of Your Life"

Given the "pre-conscious mind" and the "unconscious mind" that affect human "behavior" according to Freud's psychoanalytical approach to humans (see panel b of Fig. 7.3), according to Freud, human individuals are frequently not "under control" of their lives. Human beings are not the "masters of their lives". According to Freud, a goal of therapy should be to gain "control" over one's life. As argued in Sects. 2.5 and 3.2.2 in the context our considerations "free will", "choice", and "control parameters", there is no physical system in the universe that exhibits "control" over anything. "Control" is not a concept of physics. It is not part of a scientific perspective because it is at odds with the fundamental laws of physics. In particular, the considerations on physics-based determinism in Chap. 2 and more explicitly, Figs. 7.1 and 7.3b illustrate that human beings as pattern formation systems do not possess any "control" property at any time of their lives. Grand state pattern formation comes with sequences of attractor patterns. These pattern sequences follow laws and are affected by forces just like a pen falling down to the floor satisfies Newton's law and is subjected to forces, in particular, the gravitational force. In the aforementioned example of the female that forms a flirting BBA attractor pattern in the presence of a male, at no point in time the female is under "control" of her life. The female does not make a "decision" to start a conversation with the male. The bifurcation leading to the emergence of the flirting BBA attractor pattern happens just as phase transitions from water to ice happen when the conditions are appropriate. The H_2O molecules that constitute ice and water do not possess any "decision" or "control" property. Likewise, humans when considered from a scientific perspective as bio-machines (see Skinner [300]) do not possess any "decision" or "control" power. Therefore, from a scientific perspective it is impossible for a person to gain "control" over his or her life. Within the scientific perspective of human beings, Freud's goal either involves a misconception of the human beings and therefore cannot be achieved or it simply cannot be formulated because it uses terminology inappropriate for a scientific perspective.

Within the physics perspective the goal to gain "control" over one's life cannot be addressed because there does not exist such as thing as "control". This does not mean that as such Freud's goal would be useless or doomed to fail. In particular, the author of this book does not deny the usefulness of the concept of "control" in general. The author as a private person (not as a scientist) believes that non-scientific concepts such as Freud's psychoanalytical concepts capture important aspects of human life. The author as a private person (not as a scientist) also believes that humans have a soul (as it is assumed by the three main religions Christianity, Islam, and Hinduism). In summary, when approaching humans and animals from religious and non-scientific perspectives, the concept of "control" and the goal to "master one's life" may play important roles. However, such a debate is not part of this book.

7.3   Pattern Formation Involving Force Production Neglecting External Forces: C1 Systems

Human pattern formation systems that are characterized by a circular causality loop that involves brain state and structure and as a by-product (but not as part of the circularity loop) produce forces correspond in the classification scheme proposed in Chap. 5 to C1 systems, see Fig. 5.5. Just like A2 systems, they describe type 2 sequences and cycles in which pattern sequences emerge due to structural changes induced by the emerging patterns. Unlike A2 systems, C1 systems are not isolated. They involve a force component related to muscular body activity. For example, the person under consideration is talking or moving. As far as external forces are concerned, C1 systems correspond to situations in which external forces acting on human beings can be neglected. Let us address a few examples.

7.3.1   The Lonely Speaker

The lonely speaker is a person who talks to somebody or an audience but is not affected by forces from the environment (given in terms of the addressee of his/her talk or the audience). That is, there are no external forces acting on the speaker or there are external forces but they do not affect the dynamics of the speaker as a pattern formation system. The speaker may cover a single topic or several topics. While talking it is assumed that the lecturing (i.e., talking) pattern changes the brain structure that was supporting the emergence of the lecturing (i.e., talking) pattern in the first place. More precisely, the lecturing pattern is assumed to decrease its own eigenvalue. Consequently, after a while, the topic becomes unstable and the speaker switches to another topic. If there is no topic left, then the speaker stops talking. Figure 7.4 illustrates a lonely speaker for two topics. Panel (a) shows the amplitudes of two basis patterns giving rise to two attractor patterns describing the two topics. Panel (b) shows the eigenvalues of the basis patterns. Figure 7.4 presents

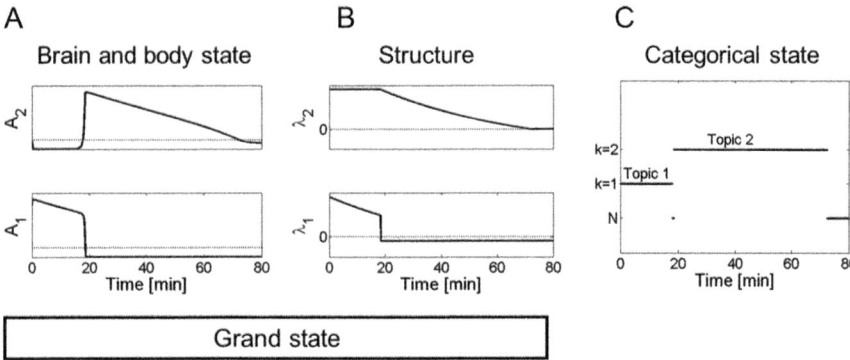

Fig. 7.4 Brain and body activity dynamics (i.e., grand state dynamics) of the lonely speaker who talks about two topics. Panels (**a**) and (**b**) show the evolution of the grand state (brain state/ordinary state in panel (**a**) and structure in panel (**b**)). Panel (**c**) shows the categorical state. Simulation results have been obtained from an appropriately defined LVH model

simulation results from the LVH model. Taken together, panels (a) and (b) show the evolution of the grand state of the speaker. Panel (c) describes the categorical state.

At time $t = 0$ a reference state becomes unstable. For example, the speaker meets with a person and the state of being silent and not talking to the person becomes unstable. For sake of simplicity, it is assumed that at the reference point there are only two unstable basis patterns related to two topics labeled 1 and 2. It is assumed that the system is symmetric at $t = 0$ (but becomes asymmetric later on) such that the eigenvalues λ_1 and λ_2 assume the same value at $t = 0$ (see panel b). It is assumed that the ongoing brain activity at $t = 0$ is such that the brain state falls into the basin of attraction of the topic 1 attractor. Accordingly (see panel a of Fig. 7.4), at $t = 0$ we have $A_1(t = 0) > A_2(t = 0)$, where A_1 and A_2 are the amplitudes of the topic 1 and 2 basis patterns, respectively. The brain activity converges to the topic-1 attractor with $A_1 = A_{1,st} > 0$ and $A_2 = 0$, see panel (a). The talking-topic-1 BBA attractor pattern is assumed to decrease its own eigenvalue λ_1, see panel (b). Consequently, at a certain point in time (here at about $t = 18$ time units), the eigenvalue λ_1 drops out of the stability band and the talking-topic-1 fixed point pattern becomes unstable. The pattern formation systems undergoes a bifurcation. The system switches form the talking-topic-1 BBA pattern to the talking-topic-2 BBA pattern as indicated by the amplitude dynamics shown in panel (a). That is, the amplitude A_2 increases to its finite stationary value, while A_1 decays to zero. It is assumed that the bifurcation also puts the eigenvalue λ_1 to a negative value. In doing so, the talking-topic-1 pattern will not emerge any more until due to other circumstances (e.g., the speaker meets with another person, talks to another audience, the brain goes into REM stage, etc.) the eigenvalue λ_1 is increased to a positive value. While the speaker is talking about topic 2, the eigenvalue λ_2 decays (see panel b). This implies quantitatively that the fixed point (or stationary) value $A_{2,st}$ of the talking-topic-1 pattern decays (see panel a). It is assumed that when the amplitude A_2 falls beyond a certain level (e.g. a "noise" level) the eigenvalue λ_2 is put to zero. The amplitude A_2 drops to zero. The speaker stops talking.

7.3.2   Freud's Slips of the Tongue

Which attractor patterns occur in a sequence of attractor patterns depends on the detailed conditions at the bifurcation points that connect attractor patterns with each other. In what follows this will be illustrated for the production of verbal pattern sequences for two particular sequences. The second of the two sequences may be interpreted as an example of a "slip of a tongue".

Let us first consider panels (a) and (b) of Fig. 7.5. In what follows, for the sake of simplicity but without loss of generality, it is assumed that each word comes with its own attractor pattern. We consider a speaker that produces the phrase "English course" composed of two words. Accordingly to panels (a) and (b), the grand state of the system is described by means of a reduced amplitude space with amplitudes of three basis patterns (panel a) and a three dimensional space describing

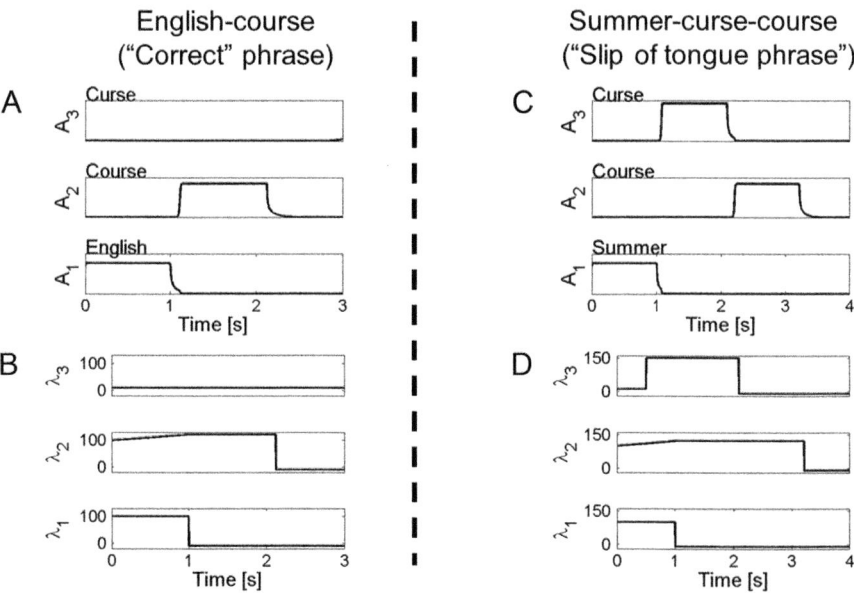

Fig. 7.5 Production of a two word phrase and slip of the tongue phenomenon modeled by the LVH amplitude equations. Panels (**a**) (ordinary state/brain state) and (**b**) (structure) describe the grand state evolution of the production of the two-words phrase "English course" in the presence of the unstable basis pattern of the word "curse". Panels (**c**) and (**d**) describe the grand state evolution of the production of the three-words phrase "summer curse course" that might be regarded as a slip of the tongue example. See text for details

the brain structure in terms of the three eigenvalues. The amplitudes A_1, A_2, A_3 and eigenvalues $\lambda_1, \lambda_2, \lambda_3$ correspond to the basis patterns related to the words "English", "course", and "curse". Panel (a) shows the amplitudes $A_k(t)$ as functions of time. Panel (b) shows the functions $\lambda_k(t)$.

The "curse" basis pattern ($k = 3$) is assumed to exhibit a fixed (i.e., time-independent) positive eigenvalue λ_3. However, the eigenvalue is assumed to be relatively small such that it falls out of the stability band (see panel b; note that all three eigenvalues are drawn in the same scale). In contrast, the eigenvalues of the word basis patterns for "English" and "course" are relatively high and assume the same value. The speaker regarded as a pattern formation system is assumed to exhibit initially a brain state in the basin of attraction of the attractor of the "English" BBA attractor pattern. As demonstrated in the simulation (using an appropriate LVH model) the amplitude A_1 of the "English" word production basis pattern increases over time to its finite fixed point value. The BBA pattern emerges and the speaker speaks out the word "English". For sake of simplicity, it is assumed that speaking out (i.e. verbally producing) words takes about 1 s. In general, after a word has been produced, it is assumed that the corresponding word attractor pattern induces a structural change and decreases rapidly its own eigenvalue. In the simulation, the eigenvalue is put to a negative value. Moreover, during word production it is assumed that the word "English" increases the eigenvalues of words that typically occur in sentences subsequent to the word "English". For example, words like "people", "teacher", "tea", and "course" would qualify as such related words. In the simulation, only the word "course" is considered. The eigenvalue λ_2 of the "course" basis pattern increases gradually during the presence of the "English" word production BBA attractor pattern (see middle subpanel of panel b). When the production of the word "English" has been completed and the eigenvalue λ_1 drops to a negative value, the system undergoes a bifurcation. The basis patterns of the words "curse" and "course" exhibit positive eigenvalues λ_2 and λ_3, respectively, but as mentioned above the eigenvalue λ_3 of the "curse" basis pattern is too small to belong to the stability band. Therefore, at the bifurcation point the system is monostable and the brain activity and muscular activity converges to the BBA pattern of producing the word "course". In total, the speaker produces the two words phrase "English course".

The situation is different in panels (c) and (d). In panels (c) and (d) we consider the verbal production of a phrase that starts with the word "summer". While initially we are assuming exactly the same situation as in panels (a) and (b) except for the fact that the word "English" is replaced by the word "summer", the situation is assumed to differ when about half of the word "summer" has been produced. It is assumed that the structure of the speaker is such that the production of the word "summer" (and the related brain activity) increases dramatically the eigenvalue λ_3 of the word "curse". The structure might exhibit that property due to forces that have acted on the speaker previously. For example, forces related to negative experiences during the summer time may have changed the structure of the speaker such that the summer BBA pattern increases eigenvalues of BA patterns and BBA patterns related to expressions describing unpleasant situations. When the word "summer" has been produced completely then the corresponding eigenvalue λ_1 drops to a negative value, just as in the case illustrated in panels (a) and (b). The system undergoes a bifurcation, just as in panels (a) and (b). Unlike the situation shown in panels (a) and (b), the conditions are such that there are two unstable basis patterns $k = 2$ and $k = 3$ with eigenvalues λ_2 and λ_3 in the stability band. That is, the system is bistable. In the simulation shown in panels (c) and (d), the brain and body state at the bifurcation point is in the basin of attraction of the attractor of the "curse" word production attractor pattern. Therefore, the system converges to the "curse" word production BBA attractor pattern. The speaker speaks out the word "curse". Once the word "curse" has been produced completely, its eigenvalue λ_3 drops to a negative value. The system undergoes another bifurcation and the "course" word production BBA attractor pattern emerges. The speaker speaks out the word "course". In total, the speaker produces the phrase "summer curse course". The sequence may be interpreted by a listener as a slip of tongue example. A speaker "intends" to say "summer course" but produces the phrase "summer curse course".

7.3.3   Child Play

In this section, a child performing a sequence of leisure activities is described in terms of a human pattern formation system producing a type 2 sequence of attractor patterns. Without loss of generality, three activities are considered: TV watching ($k = 1$), reading ($k = 2$), and puzzle solving ($k = 3$). The model can be generalized to account for an arbitrary number of activities. Each activity is associated with a BBA attractor pattern. In line with the amplitude equation description of pattern formation systems, the emergence of each leisure activity pattern is determined by the corresponding pattern amplitude A_k. If one of the three pattern amplitudes is at its finite fixed point value $A_{k,st} > 0$, while the other amplitudes are close to zero, then the BBA pattern k exhibiting the finite amplitude has emerged. The child performs the corresponding leisure activity. Importantly, it is assumed that performing a certain activity affects the brain structure and, in doing so, affects the process that supported the emergence of the activity in the first place. For example, when a child is solving a puzzle composed of a relative small set of pieces, after a certain amount of time the puzzle will be solved completely. Having solved the puzzle is assumed to decrease the eigenvalue λ_3 of the puzzle solving basis pattern. This self-damping mechanism may be modeled as an abrupt change (like in the word production example). In what follows, a slightly different approach will be used. It is assumed that a BBA attractor pattern once it has emerged decreases its own eigenvalue in a continuous, gradual manner (as in the lonely speaker example). For example, as far as the puzzling activity is concerned, the child may have several small puzzles available and solves one after another while the puzzling basis pattern eigenvalue decreases. In a different scenario, the child is solving a large puzzle without finishing it and while doing so the eigenvalue of the puzzle solving basis pattern decreases.

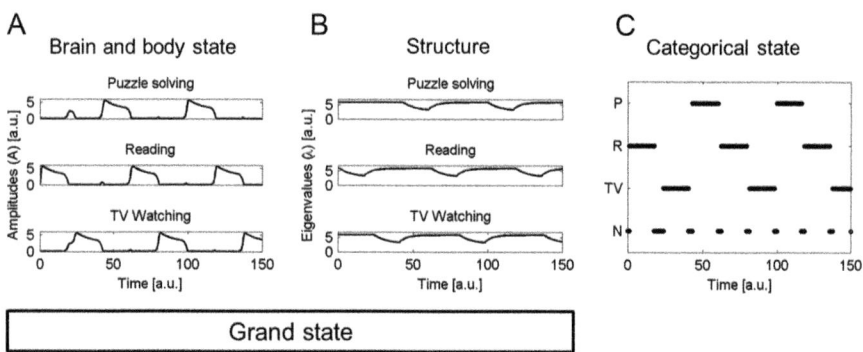

Fig. 7.6 Simulation of the LVH model for a type 2 pattern formation sequence that describes a playing child. Panel (**a**): Amplitudes of three leisure activity BBA patterns as functions of time. Panel (**b**): Eigenvalues as functions of time. Panel (**c**): Categorical state of the child as function of time as derived from the amplitudes shown in panel (**a**)

In general, when a BBA attractor pattern k decreases the eigenvalue λ_k of its corresponding basis pattern k, then at a certain point in time the eigenvalue λ_k drops out of the stability band. The attractor of the pattern k turns into a repellor. The system undergoes a bifurcation. With respect to the child this implies that the child switches from one activity to another activity.

Let us demonstrate these issues by means of a computer simulation based on the LVH model as discussed earlier in Frank [103, 104, 112]. The simulation results are presented in Fig. 7.6. Panel (a) shows the evolution of the amplitudes A_k of the basis patterns related to puzzle solving, reading, and TV watching and, in doing so, describes the brain and body state. Panel (b) shows the dynamics of the eigenvalues λ_k and, in doing so, describes the evolution of the structure over time. Panels (a) and (b) taken together describe the evolution of the grand state over time. Panel (c) describes the categorical states related to the brain and body states presented in panel (a).

First of all, in the computer simulation it was assumed that initially the system exhibited structural symmetry. Accordingly, all eigenvalues assumed the same value. However, the initial brain and body state was assumed to exhibit a bias: the amplitude of the reading basis pattern was larger than the amplitudes of the other patterns. Consequently, the brain and body state was initially located in the basin of attraction of the reading attractor. Therefore, the corresponding BBA attractor pattern emerged (see middle subpanel of panel (a)). Subsequently, the eigenvalue λ_2 of the reading activity decayed as a result of self-damping (see middle subpanel of panel (b)). In contrast, the eigenvalues λ_1 and λ_3 of the two other activities stayed at constant levels. At about 20 time units the eigenvalue λ_2 of the reading activity dropped out of the stability band. The reading activity as a spatiotemporal BBA pattern became unstable. At that time point, there was a competition process between the puzzle solving and the TV watching activity. The competition between the two patterns can be seen in panel (a) as the temporary increase of both amplitudes A_1 and A_3 (see top and bottom subpanels) at about 20 time units. The initial conditions at that time point were such that the TV watching activity was the winner of the competition. The amplitude of the TV watching basis pattern approached its finite fixed point value, whereas the amplitude of the puzzle solving basis pattern decayed to zero. The self-damping process of the eigenvalue (now with respect to λ_1) started again (see bottom subpanel of panel (b)). In contrast, the eigenvalue λ_2 of the reading activity relaxed back to its baseline level (see middle subpanel of panel (b)). At about 40 time units the eigenvalue λ_1 of the TV watching activity dropped out of the stability band and the TV watching activity became unstable. At that second bifurcation point there was a competition between reading and puzzle solving. As can be seen from panel (b), the eigenvalue of the reading activity did not yet reach the same baseline value as the eigenvalue of the puzzle solving basis pattern. This feature in combination with the initial values of the amplitudes at the time of the second bifurcation made that the puzzle solving activity turned out as winner of the amplitude-amplitude competition process. The brain and body state (as described in terms of basis pattern amplitudes) converged to the attractor of the BBA puzzle solving attractor pattern. Once the puzzle solving BBA pattern emerged, the self-damping process of its eigenvalue λ_3 took place and the eigenvalue λ_3 decayed as a function of time—more precisely as a function of the duration of the performed puzzle solving activity (see top subpanel of panel (b)). Furthermore, the eigenvalue λ_1 of the TV watching basis pattern relaxed back to the baseline value. At about 60 time units the eigenvalue λ_3 of the puzzle solving activity dropped out of the stability band and the fixed point of the activity became unstable. At that third bifurcation point, the pattern formation system switched to the BBA pattern associated with reading and the three-activities cycle that we have just discussed started again.

Panel (c) is derived from panel (a). An activity was performed when the corresponding amplitude was at its finite fixed point value, while the amplitudes of the other activity patterns were close to zero. Note that as can be seen in panel (a) the finite fixed point values of the amplitudes exhibited slowly decaying drifts. Those drifts are due to the fact that the fixed point values are determined by the eigenvalues, see Sect. 4.6.5. When the eigenvalues decayed slowly due to self-damping, then the corresponding amplitudes decayed on the same slow time scale as well.

As pointed out by Frank [103] the LVH amplitude equations model for child play eventually produces a periodic temporal pattern characterized by a three-activities cycle. In the simulation presented in Fig. 7.6 the three-activities cycle is given by the sequence: reading, TV watching, puzzle solving. The observation of the three-activities cycle helps our understanding of the pattern formation in grand states: grand state pattern formation may come with attractors such as limit cycle attractors that exist on higher hierarchical levels. However, in the real world a child is likely to perform only a single three-activities cycle. After having performed all three activities, the child will likely not repeat them again. Rather, the child will engage in other activities (e.g., play with a friend or sibling, play with the family cat—if there is one, eat something, etc.). Consequently, when modeling real data only the first cycle might be of interest. In this context, we would like to re-iterate that the model can be generalized for an arbitrary number of activities. Accordingly, the first cycle can consist of as many activities as we would like to consider. In particular, the cycle may describe the whole day of a school child from getting up in the morning to getting to bed and falling asleep at the end of the day (see Chap. 11).

Finally note that the impact of external forces was neglected in this example. In order to take external forces into consideration, the C1 model may be replaced by a C3 model (see below) or E1 model. For example, when a child watches TV and there is a commercial, the commercial may exert certain forces on the child that shift the brain state to another basin of attraction (C3 model) or change the structure such that TV watching becomes unstable (E1 model). In both cases the child stops watching TV. The child may go to the kitchen looking for something to eat.

7.4 Patterns as Reactions to External Forces in C2 and C3 Systems

In this section, experiments on humans and animals will be considered in which frequently brain activity has not been measured. Rather, verbal reactions or other kind of body activities were measured (e.g. a key press). Likewise, animal movements were recorded. Therefore, as such C3 systems will be discussed, see Fig. 5.5. However, for the experiments discussed in Sects. 7.4.1, 7.4.2, and 7.4.3 the body is considered to be a driven system that in a first approximation is assumed to play a negligible role for the formation of the relevant brain activity patterns (in contrast to what will be discussed in Sect. 7.5.3 below). In this sense, C2 pattern formation systems will be considered that produce patterns with supplementary force production components that can be neglected for the build-up of the patterns under consideration. In principle, the experiments reported in Sects. 7.4.1, 7.4.2, and 7.4.3 could be conducted by only measuring brain activity. The participants would physically "just" sit around, while in their brains under the impact of certain external forces BA patterns and sequences of BA patterns would emerge (e.g., as in the study by Kleinschmidt et al. [188], see Sect. 6.5.2). An exception in this regard is the maze-runner experiment by Tolman that will be re-discussed in Sect. 7.4.4. In this case, the rats need to go physically through the doors of the maze.[2]

---

[2]In fact, future technology probably will allow us to conduct experiments in which rats would be rewarded not on the basis of the physical movements that they produce (i.e., running through the food door) but on the basis of the brain activity patterns that they produce. In such experiments, force production would be negligible. For studies on monkeys such technology is already in use [208].

7.4.1 Emerging Patterns Increasing Eigenvalues: State-Induced Restructuring and "Retrieval-Induced Forgetting"

In the literature, it has frequently been stated that "experiences" can affect "perception" and "behavior". The phenomenon has been given several names like "priming", "conditioning", "learning", etc. From a physics perspective, "priming" experiments are experiments in which materials that happen to be humans are restructured. From a physics perspective, all systems of the animate and inanimate worlds exposed to forces that change the structure of the systems typically show a change in the way they evolve or react to forces (see second law). Therefore, physics makes the somewhat trivial statement that if the structure of thingbeings is changed (i.e., if thingbeings are restructured), then thingbeings typically react or evolve differently as they did prior to the structural change. Let us show that this statement is consistent with the observations made in so-called "priming" experiments. To this end, the human pattern formation reaction model and more explicitly the LVH model will be applied to model the brain activity dynamics (and the related body dynamics) of individuals participating in a typical "priming" experiment.

In a typical priming experiment [192, Chap. 6], participants are exposed to two word lists: a study-list and a test-list. A study-list of ten words may read: Science, Home, Car, Child, Flower, Money, Water, Sun, Meal, Nose. Each word (one after the other) of the study-list is shown to the participants on a computer screen. Due to the instructions given by the experimenter (i.e., forces exerted by the experimenter on the participants), the structure and brain state of the participants is such that the participants read the words as quickly as possible. The observable in the experiment is the reaction time defined as the time between presentation of the word on the screen and the utterance. After a few days or a week, the participants repeat the experiment with another set of ten words, which is the test-list. An example of a test list could be: Table, Clock, Light, Science, Sun, Tiger, Car, Home, Chair, Flower. The test list is constructed such that it contains some words of the study list. In our example, the study and test list have the following words in common: Science, Car, Home, Flower, Sun. In the literature, these words are typically referred to as "primed" words. In contrast, the new words in the test-list are referred to as "non-primed" words. In what follows, the words in the test list that have been practiced by participants during the study part will be called the previously read words. The new words in the test-list will be called the new words. Typically, the reaction times are shorter for the previously read words as compared to the new words [192, Chap. 6]. That is, from a physics perspective this means that participants produce verbal BBA attractor patterns faster when they have produced those verbal BBA attractor patterns previously during the study part of the experiment.

Following earlier work [90, 98], in order to discuss the aforementioned state-induced restructuring experiment in the context of the LVH model, $m = 15$ words will be considered. Those words are the 15 words of the study and test lists combined and read: Car, Chair, Child, Clock, Flower, Home, Light, Meal,

Money, Nose, Science, Sun, Table, Tiger, Water. In line with our considerations on B2 systems (see Sect. 6.3) it is assumed that when a word is presented to a participant, then a reference state becomes unstable such that the system becomes multistable and exhibits unstable basis patterns corresponding to all kind of words. For modeling purposes, only the 15 unstable basis patterns will be considered that are related to the 15 aforementioned words of the study and test lists. The forces exerted by a word do not only make a reference state unstable, it is also assumed that the particular word given to a participant shifts the brain state in the basin of attraction of the attractor related to that word (i.e., related to the unstable basis pattern of that word). Consequently, the brain state converges to the corresponding word BA attractor pattern. As far as the amplitude dynamics is concerned, the amplitude k of the relevant word basis pattern increases to its finite fixed point value, whereas the amplitudes $j \neq k$ of all other word basis patterns decay to zero. The time it takes for the amplitude k to reach approximately its finite fixed point value can be used as a measure for the time it takes for the participant to react to the word. It can be shown that the reaction time thus defined depends on the eigenvalue of a word basis pattern. The larger the eigenvalue, the shorter the reaction time [90]. Consequently, consistent with the aforementioned experimental findings, it is assumed that the emerging BA patterns during the study part of the experiment increase their own eigenvalues.

In what follows results from computer simulations of participants performing study- and test-trials will be presented. In the computer simulations the ten words of the study list were assigned to the basis patterns (and attractors) $k = 1, \ldots, 10$. The ten words of the test list were assigned to the basis patterns (and corresponding attractors) $k = 6, \ldots, 15$. The order in which the words were presented during study and test was randomized. First a single participant was simulated for an idealized situation. For the participant we assumed that all eigenvalues $\lambda_1, \ldots, \lambda_{15}$ had the same value before study. Each word of the study list was presented once. The amplitude of the word basis pattern increased to its finite fixed point value indicating that the corresponding word attractor pattern emerged. The attractor pattern was interpreted as brain activity and body activity attractor pattern (i.e., as BBA pattern) that described that the participant produced verbally the word. It was assumed that the emerging BBA attractor pattern increased its corresponding eigenvalue. Consequently, at the end of the study part, the eigenvalues $\lambda_1, \ldots, \lambda_{10}$ of the first ten basis patterns were larger than the eigenvalues $\lambda_{11}, \ldots, \lambda_{11}$ of the basis patterns k with index $k = 11, \ldots, 15$. Note that this feature is the characteristic difference between the C2 (or C3) system considered here and the B2 system considered in the previous chapter. In C2 and C3 systems emerging BA or BBA patterns affect the structure that determines the evolution of the brain and body state and the emergence of patterns. In contrast, in B2 systems that structure is not affected by the brain state and is fixed.

Let us return to the simulations. After having completed the study part, we simulated the test part. We conducted ten test trials using the ten words of the test list. In those trials we determined the reaction times to (latencies of) the words. The ten reaction times (the first five for previously read words and the second

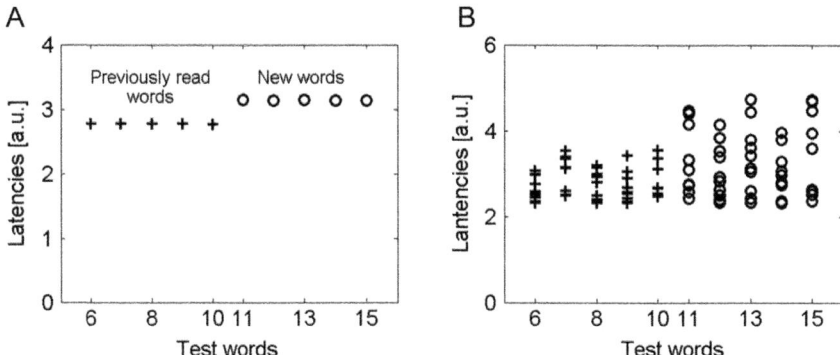

Fig. 7.7 Simulation results of the LVH model applied to the state-induced restructuring phenomenon (also called "priming"). Panel (**a**): Latencies of ten test words produced by a simulated participant under idealized conditions are shown. Plus signs and circles refer to previously read words and new words, respectively. Panel (**b**): Latencies produced by ten simulated participants assuming more realistic conditions as in the simulation shown in panel (**a**). Accordingly, the eigenvalues of all word basis patterns initially varied in a certain range. Plus signs and circles refer to previously read words and new words, as in panel (**a**)

five for new words) are shown in panel (a) of Fig. 7.7. Recall that eigenvalues determine the temporal dynamics of amplitudes, see Sect. 4.5.1 and Fig. 4.15. Large eigenvalues imply that patterns form quickly. Small eigenvalues imply that pattern formation takes more time. Given this feature of pattern formation systems and the assumed state-induced increase of eigenvalues when the system settles down in a certain pattern (i.e., state), we expected that the reaction times were shorter for the previously read words. In fact, as it is clear from Fig. 7.7a words that have been read during the study part exhibited shorter reaction times than new words.

Humans are individuals (although the term "individuality" should be avoided when taking a science perspective, see Sect. 13.2). Structure components across individuals typically exhibit differences. Therefore, in a second step we simulated a group of ten participants. For each participant the eigenvalues $\lambda_1, \ldots, \lambda_m$ of the $m = 15$ word basis patterns were uniformly distributed in the interval $[0.5, 1]$. For each participant we conducted ten study trials and subsequently ten test trials. In the test trials the reaction times to the test words were determined. The reaction times thus obtained are shown in panel (b) of Fig. 7.7. The plus signs and circles indicate reaction times of previously read and new words, respectively. It was found that the reaction times to the five previously read words on average were shorter than the reaction times of the five new words ($M = 2.8$ time units versus $M = 3.2$ time units). In experimental studies, those averages are typically subjected to hypothesis testing in order to show that differences are statistically significant. In fact, the simulations presented here on the basis of the LVH model can yield sample mean values that differ (consistent with experimental findings) in a statistically significant sense [90, 98].

Why Humans and Animals Are Not "Primed" Just Like Pencils Are Not "Primed"

When we use a soft pencil to draw a line on a piece of paper two things happen. First, we get a line on the paper. Second, the pencil becomes shorter because it is used. From a physics perspective a force is applied to the pencil such that the pencil is moving over the paper. As a result, a line appears on the paper. At the same time, the pencil becomes shorter. From a pattern formation perspective, human and animal activities can be discussed in analogy. In the experiment discussed above, when forces from certain words act on the human brain, then these forces are assumed to make that certain word attractor patterns emerge—just as the force in the pencil example makes that the pencil moves on the paper. The emerging attractor patterns are assumed to change the brain structure—just as the movement of the pencil on the paper changes the structure of the pencil and makes the pencil shorter. In physics, drawing a line with a pencil on a piece of paper is not mystified by saying that we "prime" the pencil to become shorter. Likewise, in physics, there is no need to mystify humans and animals performing certain activities under the impact of certain forces. From a physics perspective, under certain circumstances certain forces acting on humans and animals can make that humans and animals form certain BBA patterns that in turn change the structure of those humans and animals under consideration—just like the movement of a pencil over a paper makes a pencil shorter.

In more general terms, under certain conditions when systems are in certain states (e.g., pattern formation systems exhibit certain attractor patterns), then those states lead to changes of the structure of the systems. The pencil is an example in this regard. Another example is a passenger plane. When a passenger plane is flying it loses weight due to the fact that it uses up fuel. Of course, when the airplane is just standing on the ground with all engines switched off, then nothing happens. That is, the structural change of losing weight is induced by the state of flying. Clearly, neither pilots nor airline companies "prime" their planes to lose weight during flight phases.

"Retrieval Induced Forgetting"

Anderson and Spellman among other researcher discovered that state-induced restructuring (i.e., "priming"), can result in "forgetting" of "prime-related" material [7, 8]. Anderson and Spellman showed that "recalling" items from "memory" may induce "forgetting" of items that are different from but related to the "recalled" items. This phenomenon is referred to as "retrieval-induced forgetting". "Retrieval-induced forgetting" has been studied in the context of "memory tasks" involving words and word pairs [7, 8]. The phenomenon has also been observed when participants had to "memorize" visual material [53]. Furthermore, "retrieval-induced forgetting" has clinical applications [245].

From a physics perspective, the phenomenon is consistent with the human pattern formation reaction model assuming (just as in the previous example) that emerging attractor patterns can change structure. Since from a physics perspective, on the one hand, "memory" does not exist, and, on the other hand, "forgetting" is an unnecessary concept, in what follows, the "retrieval-induced forgetting" phenomenon will be considered as a special case of human grand state dynamics that leads to changes in the relationship between external forces and brain state dynamics. That is, we are concerned with brain and body activity induced changes that remove certain brain activity reactions to external forces from the repertoire of reactions of a human being. This terminology is in line with a car mechanic in an auto-repair shop who changes a shuttering car engine by fastening some screws. After the mechanic made the changes, the shuttering of the car engine is removed. Of course, it is inappropriate to say that the car in this example did "forget" the shuttering dynamics. Likewise, from a scientific perspective there is no reason to say that humans and animals do "forget" anything.

Let us describe a typical experiment on brain and body activity induced changes that lead to the removal of certain brain activity reactions to external forces. The experiment typically consists of three phases: pre-structuring, re-structuring, and testing. We will describe here only the key elements of these phases that are relevant to the human pattern formation reaction model (for further details the reader is referred to the original studies [7, 8]). During the pre-structuring phase participants are studying a relatively small set of acoustic word pairs. Examples of word pairs are Fruit-Banana, Fruit-Grape, Fruit-Lemon, Fruit-Orange, Fruit-Pineapple, Fruit-Strawberry. In these examples, we are dealing with pairs consisting of a category word (Fruit) and a category example (e.g., Banana). From the pattern formation perspective, the pre-structuring phase is supposed to change the structure of the brain of the participant with respect to the word pairs such that if a category word is given (e.g., Fruit) then an instability occurs at which word pairs (e.g., Fruit-Banana) emerge as BA attractor patterns. In the re-structuring phase, participants are asked to study only a subset of the acoustic word pairs. This is done by presenting incomplete word pairs and asking participants to complete the pairs. For example, a participant may be exposed to the three phrases: Fruit-Ba, Fruit-Gr, and Fruit-Le. The participant is supposed to complete those phrases to Fruit-Banana, Fruit-Grape, and Fruit-Lemon. From a pattern formation perspective, the re-structuring phase introduces a bias because only a subset of the items of the pre-structuring phase are addressed. Some items are neglected. Subsequent to the re-structuring phase, the participant is tested. To this end, the participant is put in a condition in which he or she produces a sequence of word pairs consistent with his or her brain structure as obtained during the pre-structuring and re-structuring phases. That is, from a pattern formation perspective, the instructions of the experiment produce again an instability that comes again with basis patterns corresponding to word pairs. The brain state evolves towards the attractors of the word pair basis patterns (with eigenvalues in the stability band). The brain state visits the attractors one at a time (like in the previous examples of pattern sequences) and, in doing so, produces a sequences of BA attractor patterns.

For methodological reasons participants are typically asked to produces forces (e.g., they are asked to give verbal responses). That is, the produced patterns are brain activity and muscular body activity patterns. Taking those forces into account, we would deal with C3 pattern formation systems rather C2 systems. For sake of simplicity, the produced forces will be neglected in what follows as explained in the beginning of Sect. 7.4.

The fundamental observation is that the word pairs practiced in the re-structuring phase have a higher probability to occur than word pairs that have been neglected in the re-structuring phase [7, 8]. In our example, we would find that the items Fruit-Banana, Fruit-Grape, Fruit-Lemon occur when averaged across participants with a higher percentage than the items Fruit-Orange, Fruit-Pineapple, Fruit-Strawberry. In addition to the aforementioned three phases, a "control" experiment is conducted that involves only the pre-structuring and testing phases. Not surprisingly, it is found that the items used in the re-structuring phase have higher probability to be named by the experimental group of participants that performed the re-structuring phase as compared to the "control" group of participants that only participated in two phases. In our example, this means that the Fruit-Banana, Fruit-Grape, and Fruit-Lemon word pairs are more frequently named by the experimental group as compared to the "control" group. The key issue is that the word pairs neglected in the re-structuring phase are named with a lower probability by the experimental group as compared to the "control" group. That is, with respect to our example the items Fruit-Orange, Fruit-Pineapple, and Fruit-Strawberry would be produced less frequently by the experimental group as compared to the "control" group. This observation may come as a surprise because when looking over the three experimental phases of the experimental group and the two experimental phases of the "control" group then both groups are dealing with those items (i.e., the neglect items) in the same way. With respect to the human pattern formation model, the findings can be explained (consistent with the explanation of "priming" phenomena) by assuming that the BBA patterns emerging in the re-structuring phase increase their eigenvalues. If the increase is sufficiently large, then the eigenvalues of the basis patterns of the remaining (i.e., neglected) items drop out of the stability band. The corresponding attractors turn into repellors. As a result, those BA attractor patterns are removed from the repertoire of patterns that can emerge for appropriate brain activity initial conditions. In fact, detailed mathematical considerations and numerical simulations support that line of argument [98]. The removal of certain BA or BBA attractor patterns X,Y,Z from the repertoire of patterns that can be observed in humans can be due to the fact that eigenvalues of other patterns A,B,C become critically large such that the attractor of the X,Y,Z patterns turn into repellors.

Let us briefly return to the aforementioned example of a car mechanic repairing a car engine. Assuming that the pattern formation approach captures the basic mechanics that determines human reactions in "retrieval-induced forgetting" experiments, then in a nutshell the situation is as follows. The experimental procedure of such experiments acts via forces on the participants such that certain BBA patterns emerge that in turn change the brain structures of the participants. As a result of those structural changes certain attractors become temporarily unstable, that is, they disappear. This implies that effectively the forces exerted by the experimental procedure result in structural changes in the human brains of the participants and remove certain attractors. When the aforementioned hypothetical car mechanic repairs an engine by fastening some screws, then the mechanic exerts forces on the car engine that in turn change the structure of the engine. As a result of the structural changes, the shuttering dynamics is removed. In both situations, forces are exerted on thingbeings that lead to structural changes such that the thingbeings do no longer exhibit a certain dynamics. To call these situations a "car repair" in one context and "forgetting" in the other context creates irrelevant differences and makes it difficult to understand the relevant cause-effect relationships.

7.4.2 Emerging Patterns Decreasing Eigenvalues: From Scene Decomposition to "Functional Fixedness"

Scene Decomposition

Scenes by definition are composed of several features. Humans can react to scenes in various ways. Some issues in this regard have been discussed already in Sect. 6.3.3. While in Sect. 6.3.3 we discussed implications of the winner-takes-all dynamics of multistable pattern formation systems under the forces exerted by scenes, in what follows, we consider the case in which scenes give rise to the emergence of sequences of BBA attractor patterns. The individual BBA patterns in a sequence are assumed to correspond to the features of a scene under consideration. In doing so, the sequence decomposes the scene into chunks given in terms of BBA attractor patterns.

The human pattern formation reaction model for scene decomposition on the neuronal network level has been discussion in Fuchs and Haken [127] and Haken [157]. In what follows, only the pattern amplitude level will be considered. Let us consider the example introduced in Sect. 6.3.3. Panel (a) of Fig. 7.8 displays a scene given in terms of a house featuring a front door, a window, a roof, and the house number 1. According to the human pattern formation reaction model, the scene makes a reference state unstable and the unstable reference state is assumed to exhibit several unstable basis patterns related to the features (house as a whole, front door, window, roof, house number 1) of the scene. In the illustration presented in Fig. 7.8 only two basis patterns are considered. The basis patterns for the whole house and the house number 1. Panel (b) presents the dynamics of the structure given in terms of the evolution of the eigenvalues of the two basis patterns. Panels (c) and (d) present the evolution of the basis pattern amplitudes A_1 (whole house) and A_2 (house number 1) as two functions of time (panel c) and as a single trajectory in the reduced two-dimensional amplitude space (panel d). Panels (b) and (c) taken together describe the grand state evolution.

It is assumed that before the individual is exposed to the scene (i.e., at time point $t = 0$) the structure exhibits no bias (i.e., shows symmetry) such that both

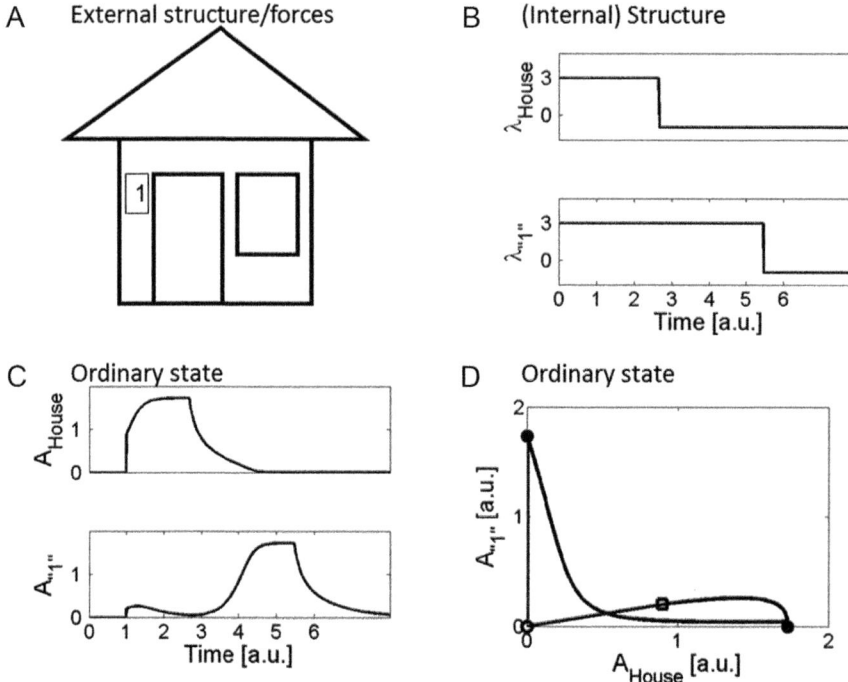

Fig. 7.8 Demonstration of scene decomposition. Panel (**a**): Scene given by the external circumstances of an individual. Panel (**b**): Evolution of the brain structure of the individual in terms of eigenvalues λ_{House} and $\lambda_{"1"}$ of the house BBA pattern and the number-1 BBA pattern, respectively. Panel (**c**): Evolution of the brain state of the individual in terms of the amplitudes A_{House} and $A_{"1"}$ of the aforementioned BBA patterns. Panel (**d**): Evolution of the brain state in the reduced amplitude space. The square indicates the assumed brain state shift due to the forces exerted by the scene. Note that the two winner-takes-all fixed points indicated by the full circles only exist initially and disappear due to self-damping caused by the emerging BBA patterns

eigenvalues assume the same value. The individual is confronted with the scene at $t = 1$ time units. It is further assumed that at that time point (of $t = 1$ time units) the external forces exerted by the scene put the brain state in the basin of attraction of the whole-house attractor (see the square in panel d). Consequently, the amplitude A_{House} of the whole-house basis pattern increases towards its finite fixed point value, whereas the amplitude $A_{"1"}$ of the number-1 basis pattern decays to zero. The emerging BBA pattern is assumed to decrease its own eigenvalue. For sake of simplicity, the eigenvalue is dropped abruptly to a negative value. Without loss of generality, it is assumed that there is a certain period between emergence of a BBA pattern and the action of the BBA pattern on the brain structure involved in the formation of patterns. In the simulation, a period of 1 time unit was used. As can be seen in panel (c) (top subpanel), the amplitude A_{House} of the whole-house basis pattern remains at its fixed point value for about 1 time unit. Subsequently, the eigenvalue λ_{House} is decreased abruptly to a negative number (see top subpanel of panel b). The whole house attractor stops to exist. Since the eigenvalue is negative, it does not even exist as a repellor. Consequently, the dynamical system exhibits a bifurcation. From panels (c) and (d) it follows that the brain state (given in terms of the amplitude variables) converges to the number-1 fixed point attractor. The amplitude $A_{\text{"1"}}$ reaches its finite fixed point value, whereas A_{House} converges to zero. The visual world whole-house attractor pattern disappears. The visual world number-1 attractor pattern emerges. After a period of about 1 time unit, the number-1 attractor pattern causes an abrupt self-damping of its own eigenvalue such that its eigenvalue drops towards a negative value. The number-1 attractor pattern disappears. The pattern formation system returns to its baseline state of doing nothing or doing something unrelated to the forces exerted by the scene.

"Functional Fixedness"

Problem solving is generally understood as the process of finding a solution to a given problem [75, 246, 256]. It has been suggested that problem solving depends on how humans (and animals) "perceive" the world. This was demonstrated for example by Maier's classical two-string problem [225, 226] and Duncker's box-candle problem [75] that has been replicated by several authors [2, 136]. In Duncker's experiment participants were instructed to place a small, lightweight candle on a door. The task was to put the candle in a vertical, upright orientation such that the candle's wick was on the top. To solve the problem, the participants were allowed to use a number of objects that were placed on a table, see Fig. 7.9.

There were matches, a small box, several thumbtacks, and the candle. The solution was to use the box as a platform for the candle. That is, in order to solve the assembly tasks, participants had to fix the candle to the box by melting some wax. Subsequently, participants were supposed to tack the box to the door. In doing so, participants used the box as a candle holder or platform. When an empty box

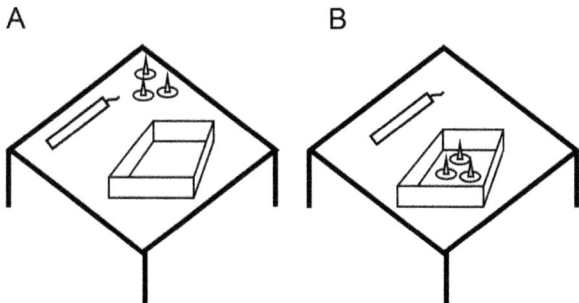

Fig. 7.9 Setup in Duncker's box-candle experiment. A candle, a box, tacks, and matches (not shown) were presented on a table. The task was to mount the candle against a door (not shown). Panels (**a**) and (**b**) show two conditions tested in the experiment. In condition (**a**) the box was empty. In condition (**b**) the box was used as a container

was provided and the tacks were placed at some distance to the box (see panel a of Fig. 7.9) participants arrived at the aforementioned solution relatively quickly. In contrast, when the tacks were put inside the box (see panel b), that is, when the box was used as a container for the tacks, then it typically took more time for participants to solve the assembly task or they were unable to find the solution. In general, participants had difficulties to "perceive" the platform function of the box that was needed in the assembly task, when the box was used as a container. It has been suggested that they "perceived" the container function of the box and that the "perception" of this function had a negative impact on the perception of other box functions. In general, in the literature, it has been suggested that "perceived" functions of objects have a negative impact on the "perception" of additional, possible object functions. In other words, the "perception" of an object function blinds the observer such that he or she cannot "perceive" other object functions [219]. The term "functional fixedness" has been coined in this context [6, 75]. Accordingly, people are fixated on a particular object function and as a result of this kind of fixation "fail to perceive" other functions of the same object.

Let us discuss Duncker's box-candle experiment from the physics perspective of humans as pattern formation systems following an earlier work by Frank [106]. To this end, let us simulate a participant in the experiment with the help of the human pattern formation reaction model. Without loss of generality, $m = 6$ different box functions are considered. A box may be used as a container, platform, weight, base (to put something upon), tool for closing an opening, or tool for hiding something. For sake of simplicity, it is assumed that under appropriate conditions the forces exerted by a box make a particular reference point unstable such that at the instability point basis patterns emerge that are related to the aforementioned $m = 6$ box functions. It is assumed that there is an unstable basis pattern k for each box function. According to the LVH model, unstable basis patterns with eigenvalues in the stability band give rise to attractors. Consequently, in the symmetric case (when all eigenvalues are the same) there are $m = 6$ attractor patterns $k = 1, \ldots, 6$ related to the $m = 6$ box functions. The attractor patterns can be characterized in terms of pattern amplitudes $A_1, \ldots, A_6$, see Table 7.1. If the brain state happens to fall into the basin of attraction of a particular attractor then the brain state converges to

Table 7.1 Description of some BBA attractor patterns of object functions relevant for Duncker's problem solving experiment in terms of pattern amplitudes

k	Object function BBA attractor pattern	Amplitudes
1	Container	$A_1 > 0$ and $A_j = 0$ for all $j \neq k$
2	Platform	$A_2 > 0$ and $A_j = 0$ for all $j \neq k$
3	Weight	$A_3 > 0$ and $A_j = 0$ for all $j \neq k$
4	Base	$A_4 > 0$ and $A_j = 0$ for all $j \neq k$
5	Tool for closing an opening	$A_5 > 0$ and $A_j = 0$ for all $j \neq k$
6	Tool for hiding	$A_6 > 0$ and $A_j = 0$ for all $j \neq k$

the corresponding attractor and the corresponding BBA pattern emerges. While the symmetric case is a useful hypothetical case to label attractors and basis patterns (see Chap. 4) in general the eigenvalue spectrum is assumed to be inhomogeneous.

Let us consider first the baseline condition in which an empty box is presented to the participants (see panel (a) of Fig. 7.9). Given the fact that the reference point is about putting a candle upright against a door, it is plausible to assume that the basis pattern $k = 2$ of the platform function exhibits the largest eigenvalue. For sake of simplicity, in what follows, it is assumed that this eigenvalue λ_2 dominates the whole eigenvalue spectrum such that the human pattern formation system becomes monostable. An explicit example of a spectrum with dominant eigenvalue of the platform function basis pattern is shown in panel (a) of Fig. 7.10. Only the eigenvalue λ_2 of the platform function basis pattern belongs to the stability band. Consequently, the brain and sensory body state of the participant evolves as follows. First, the participants looks around (where the looking around dynamics is assumed to be determined by a dynamical system whose explicit details are not of interest at the moment). Second, the gaze of the participant falls on the box. The forces exerted by the box induce the emergence of a box BBA attractor pattern in certain areas of the brain and sensory body system of the participant. The box BBA attractor pattern makes a reference state (e.g., of doing nothing with the box) unstable—just like a hurricane induces an atmospheric instability that leads to the emergence of a tornado. At that instability point the aforementioned $m = 6$ basis pattern exist and exhibit the eigenvalue spectrum shown in panel (a) of Fig. 7.10. Accordingly, the system is monostable and the brain state converges to the platform-function attractor $k = 2$ related to the platform-function basis pattern. This, in turn, initiates a sequence of further BBA patterns that describe how the participant takes and handles

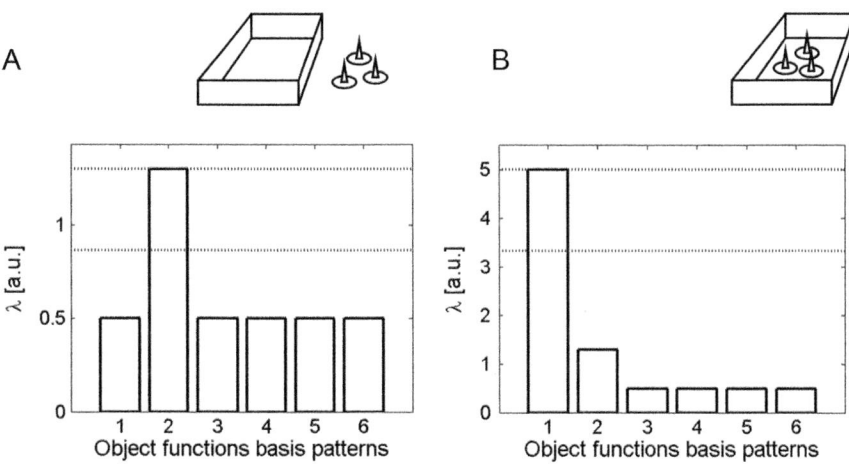

Fig. 7.10 Panels (**a**) and (**b**): Eigenvalue spectrums of six basis patterns for the two conditions (**a**) and (**b**) (see Fig. 7.9) tested in Duncker's experiment. Upper and lower dotted lines indicate the boundaries of the stability band

objects in order to assemble the candle against the door. However, the details of this final BBA pattern sequence is not of interest here.

The situation is different for a participant who is exposed to a filled box (see panel b of Fig. 7.9). In this case, it is assumed that the pre-utilization of the box as a container leads to a dominance of the eigenvalue λ_1 of the container-function basis pattern. Accordingly, eigenvalue λ_1 is large and pushes the stability band to high levels such that λ_2 and all other eigenvalues drop out of the stability band. An example of such an eigenvalue spectrum is shown in panel (b) of Fig. 7.10. The system is monostable again. In particular, the platform-function basis pattern does not give rise to an attractor. The corresponding platform-function BA attractor pattern cannot emerge. The participant becomes "blind" to all object functions except for the container function.

In Duncker's box-candle experiment [75] and related studies [2, 136] some participants were able to solve the task even when they were exposed to the box filled with tacks. As mentioned above, those participants typically took longer to complete the task as compared to participants to which the empty box was shown. This observation is consistent with the assumption that during the experiment participants of the experimental group were subjected to gradual structural changes. Following the aforementioned study by Frank [106], with respect to the LVH model it is assumed that the gradual changes were related to the spectrum of eigenvalues. That is, the emerging BBA attractor patterns during the course of the experiment turned a spectrum as shown in panel (b) of Fig. 7.10 in several steps into a spectrum as shown in panel (a). Let us dwell on this point.

Figure 7.11 describes a hypothetical participant in terms of a simulated pattern formation sequence. The participant is assumed to exhibit some kind of looking dynamics. The details of that dynamics is not of interest at this stage. For sake of simplicity, we consider a discrete time grid. As shown in panel (b) of Fig. 7.11 we consider seven looking events in which the participant looks at objects A,B,C,D and the box. In the second, fourth, and seventh event the participant looks at the box. In all other cases, the participants looks at other objects (labeled A,B,C,D). Note again that the details of the looking dynamics is not of primary interest here. Any other sequence could be used to demonstrate the main point. When ignoring the forces produced by the participant for looking and searching, then the participant corresponds to a C2 pattern formation system. According to panel (a) of Fig. 7.11, (1) the external forces exerted by the box results in shifts of the brain state, (2) the brain state as defined in terms of pattern amplitudes A_1 and A_2 evolves under the impact of the external (box) forces and the structure given in terms of the eigenvalues λ_1 and λ_2, and (3) the structure in terms of λ_1 is affected by the brain state in terms of A_1. Let us discuss these aspects of the C2 system model in more detail.

Every time the participant looks at the box, it is assumed that the forces exerted by the box lead to the instability described above for the baseline experiment. An instability occurs that involves the $m = 6$ unstable basis pattern listed in Table 7.1. The initial eigenvalue spectrum is shown in panel (b) of Fig. 7.10. Consistent with the aforementioned experimental results, it is assumed that when the brain state converges to a brain activity pattern that does not lead to the final assembly

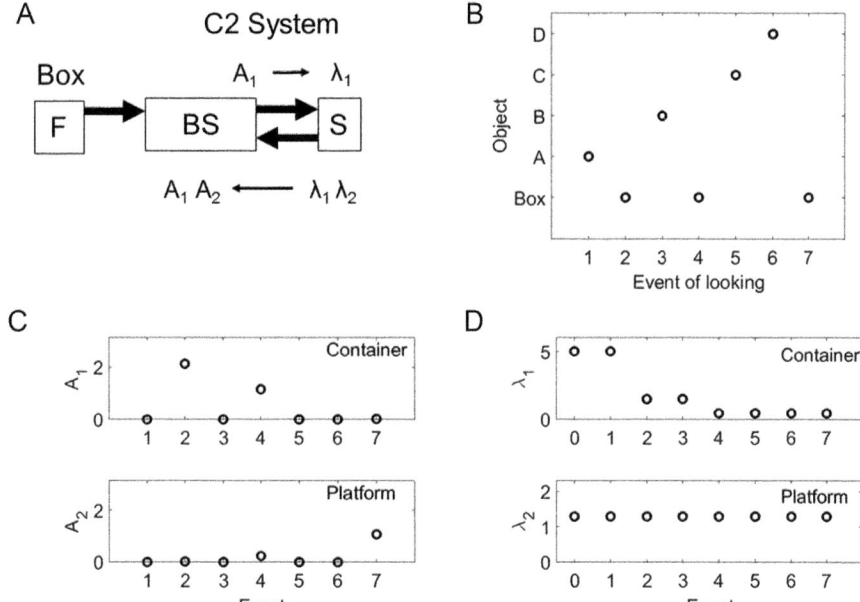

Fig. 7.11 Type 2 pattern formation sequence in the grand state of a hypothetical participant. Panel (**a**): The participant is considered as C2 pattern formation system (when ignoring forces produced for looking and searching). Panel (**b**): Looking sequence assuming that there are only five objects: the box and four other objects not further specified and labeled as A, B, C, D. Panel (**c**): Evolution of brain state in terms of stationary values of amplitudes A_1 (of the container function BBA pattern) and A_2 (of the platform function BBA pattern) as computed from the LVH amplitude equations given the eigenvalue spectrum λ_1 and λ_2 shown in panel (**d**). Panel (**d**): Evolution of brain structure in terms of eigenvalues λ_1 and λ_2. Eigenvalue λ_1 changes due to self-damping mechanism $A_1 \rightarrow \lambda_1$ as indicated in panel (**a**)

sequence (i.e., to the solution), then the corresponding eigenvalue is decreased. That is, we assume a self-damping mechanism as in previous examples discussed in this chapter. In particular, it is assumed that the eigenvalue of the container-function basis pattern decays every time the container-function BBA attractor pattern emerges when the participant is looking at the box. Panel (d) of Fig. 7.11 shows the eigenvalues λ_1 and λ_2 for each looking event after a BBA pattern has emerged and after the BBA pattern in the case of the container-function pattern reduced the eigenvalue λ_1. In doing so, panel (d) shows the decay of the eigenvalue λ_1 at the relevant looking events.

In total, using the LVH amplitude equations for the the human pattern formation reaction model the following grand state dynamics can be obtained. When the participant looks at the box (events 2, 4, and 7) the amplitudes A_1 (container function) and A_2 (platform function) evolve according to the LVH model given the current eigenvalues λ_1 (container function) and λ_2 (platform function). The stationary values of A_1 and A_2 obtained from the simulation of the LVH model are shown in panel (c). They indicate whether the platform-function BBA attractor pattern or the container-function BBA attractor pattern emerges. When the container-function pattern emerges, the corresponding eigenvalue is reduced (as discussed above). Panel (c) shows that (in the simulation) for looking events 2 and 4 the container-function pattern emerges. Since this pattern does not initiate the final assembly sequence, the pattern produces internal forces that reduce its own eigenvalue as shown in panel (d). The eigenvalue λ_1 drops at events 2 and 4 to lower values. Consequently, at the looking event 7, when the hypothetical participant happens to look at the box again, the eigenvalue λ_1 is much smaller than the eigenvalue λ_2 such that the eigenvalue λ_1 is no longer in the stability band. In contrast, the eigenvalue λ_2 of the platform function belongs to the stability band. In fact, for the eigenvalues of the basis patterns $k = 3, 4, 5, 6$ shown in panel (b) of Fig. 7.10 it follows that at the event 7 the system is monostable. In summary, due to the repetitive reduction of its own eigenvalue the container-function attractor turns into a repellor. At the same time, the platform-function repellor turns into an attractor and becomes the only attractor of the system. Consequently, when at the event 7 the participant is exposed to the forces of the box again, then the participant forms a platform-function BBA attractor pattern. Just as under the baseline condition discussed above, it is then assumed that the platform-function pattern triggers the final assembly sequence.

We have shown that the observations made in Duncker's experiment can be explained by means of the human pattern formation reaction model. By analogy, findings of similar experiments on "functional fixedness" may be explained. In particular, the human pattern formation reaction model and the underlying physics perspective of thingbeings can do without concepts like "fixedness" and "blindness". In fact, pattern formation in general has the in-build property that there is competition between different unstable basis pattern such that some of the unstable basis patterns give rise to attractors whereas other give rise to repellors. This has been explicitly demonstrated for the LVH model in Chap. 4 with the help of the stability band, see Sect. 4.6.3. Accordingly, only basis patterns with positive eigenvalues that are sufficiently large and fall in the stability band can lead to attractors such that under appropriate initial conditions the human pattern formation system can produce the BBA attractor patterns of those attractors. Laser devices, the skin of animals producing stripe or dot patterns, air layers producing clouds and hurricanes share this property with human beings and animals. According to the physics perspective of the animate and inanimate worlds, pattern formation systems are not "fixated". They are not "blind" against anything. In particular, the fact that the skin of a zebra produce a stripe patterns with a particular spatial period does not mean that the zebra skin is "fixated" to this spatial pattern. It also does not mean that the zebra skin is "blind" to all kind of periodic patterns with larger or smaller periods. In this context, see also the discussion in Sect. 6.3.3.

7.4.3   Oscillatory Human Reactions to the Visual World: Emerging Patterns Decreasing Eigenvalues in Bistable Systems

As discussed in Sects. 5.4.6 and 6.5.3, theoretical considerations and experimental data suggest that in general there is not a one-to-one mapping between the external circumstances of individuals (as given in terms of the external forces acting on the individuals) and the BA and BBA attractor patterns emerging in individuals or formed by individuals. For example, the same external circumstances can give rise to different (brain activity and body) patterns. Pattern formation in humans can be multistable. An important special case are circumstances that in a first, immediate step make the human pattern formation system of individuals bistable such that there are two stable states A and B and, subsequently, give rise to an oscillatory instability of the grand state of the pattern formation system such that the system oscillates between A and B.

In this context, a number of benchmark systems have been studied extensively in the literature such as the Necker cube, the spinning dancer, the 3D rotating cloud movie, the motion-induced blindness paradigm, and the two-dots in a square apparent motion paradigm. The Necker cube is a stationary figure that induces two different reactions [9, 33] that (from the pattern formation perspective pursued in this book) are assumed to be related to two different attractor patterns. On the one hand, the cube can induce a reaction similar to the reaction of a human to a real three-dimensional cube that points down to the left. On the other hand, the cube can induce a reaction similar to a human reaction to a real cube that points upwards to the right. Individuals react to the Necker cube in one or the other way for a particular period. Subsequently, the reaction switches from the cube-pointing-down reaction to the cube-pointing-up reaction or vice versa from the cube-pointing-up reaction to the cube-pointing-down reaction. The reaction is maintained for a certain period. After that the reaction switches again and so on. As a result, the Necker cube in a first step makes the human pattern formation system bistable. In a second step, the system settles down into an oscillatory dynamics.

Frequently, the human brain is considered as a computer (which is metaphor not needed when considering humans from a physics perspective because humans are self-organizing systems and computers are not) and it is assumed that the human brain would "compute" or "process" something [138, 229, 230]. In particular, it is assumed the brain would process "information" (which is an unnecessary concept, again, see Sect. 6.5.6). Accordingly, it is assumed that there would be "ambiguous" circumstances of the visual world that do not contain sufficient "information" such that "computational processes" cannot yield unique solutions [229, 230]. In physics, there are no "ambiguous" circumstances (which follows from the discussion on determinism in Chap. 2, in particular, the discussion about "choices" in Sect. 2.5.2). As will be shown below, the experimentally observed oscillatory dynamics is consistent with the assumption that the brain exhibits plasticity and that BA and

BBA patterns can change brain structure such that pattern formation takes place in an oscillatory fashion on the level of grand states of human beings.

Before we discuss modeling issues, let us mentioned a few more experimental paradigms that cause humans to react in an oscillatory fashion. While the Necker cube and similar figures [157, 208, 209] that lead to oscillatory reactions are stationary figures, there are dynamic (i.e., non-stationary) circumstances of the visual world that make human reactions bistable and eventually oscillatory. Some examples of such non-stationary circumstances are the spinning dancer movie, the motion-induced blindness paradigm, and the two-dots in a square apparent motion paradigm. The spinning dancer movie displays a three-dimensional silhouetted of a female dancer [114, 213]. The dancer is shown in black and rotates in front of a gray background. In this context, one may ask the question in which direction does the dancer rotate: clockwise or counterclockwise? Asking several individuals we obtain different answers. Some individuals report that the figure rotates clockwise, other individuals report that the figure rotates counter-clockwise. Importantly, individuals typically report that the rotation direction switches after a certain period and then continuous to switch from time to time. That is, the rotation switches from clockwise to counter-clockwise and back again to clockwise. To be clear, the movie itself plays the same frames again and again. That is, while the dynamic visual world circumstances do not change, individuals react to those circumstances by switching between two categorical states in an oscillatory fashion. Liu et al. [213] tested participants under the spinning dancer paradigm. Liu et al. found that when participants had the gaze on the feet of the dancer the rotation direction switched more frequently as compared to an experimental condition in which participants had the gaze on the body of the dancer. Moreover, consistent with earlier work on the Necker cube perception [191], Liu et al. found that the way in which participants were instructed had an affect on the frequency of the switches.

The 3D rotating cloud is a movie of a cloud composed of dots that forms more or less the surface of a ball [209, 330]. The cloud or ball is rotating. Again, the question arises in which direction does it rotate? The cloud movie is constructed in such a way that it makes the human visual system bistable. Humans either produce BBA patterns similar to those when they watch clockwise rotating objects or BBA patterns similar to those when they watch counter-clockwise rotating objects. Moreover, the human system switches back and forth between those two types of patterns in an oscillatory fashion.

Above and in what follows, the phrase oscillatory is used to characterize the reactions of humans. In fact, in the Necker cube, the spinning dancer, and the 3D rotating cloud experiments an alternation between two states can be observed just like if someone is switching a light on and off repeatedly. Nevertheless, in what follows the phrase oscillatory will be used because it is assumed that the underlying brain activity changes in an oscillatory fashion like a pendulum. These discrete and continuous pictures of the same issue are not inconsistent with each other. The reader may "think" (i.e., produces an appropriate BA pattern of his/her isolated system) of a pendulum that swings from left to right. Then, let us assume that there is an observer who switches a light on and off every time when the pendulum reaches its most left or most right position, respectively. If so, the continuous oscillations of the pendulum produce an alternating temporal pattern between two discrete states (the on and off states).

Motion-induced blindness (MIB) is a figure-ground perception phenomenon in which the background rotates, while the foreground is stationary [31]. In a typical MIB experiment, a field of dots is rotating in the background, while a small number of dots with different color are stationary in the foreground. Participants viewing the foreground pattern experience that parts of the foreground pattern disappear for certain periods of time. In this sense, those participants become temporarily "blind"[3] to details of the foreground figure presented to them. The MIB phenomena has been examined under various conditions [31]. In particular, it has been shown that the number of disappearances per time increases when the rotation speed of the background pattern increases. Moreover, it has been shown that the rate of MIB disappearances (the number of disappearances per time) is positively correlated with the rate of switches in a perceptual rivalry experiment [44]. Perceptual rivalry, in turn, has been suggested to arise due to competitive interaction between activity of the left and right hemispheres of the human brain [261]. The MIB phenomenon— just as the spinning dancer phenomenon—is considered an oscillatory phenomenon. Participants react in two qualitatively different ways to the movie and switch from one reaction to the other in an alternating sequences.

Another experimental paradigm leading to alternation between two reactions to the same dynamical visual circumstances is the two-dots-in-a-square apparent motion paradigm [199, 209]. In this paradigm two dots appear and disappear on the corners of a square. The dots appear and disappear simultaneously on the corners that connect the diagonals of the square. That is, they appear on the top left and bottom right corners, disappear, subsequently appear on the top right and bottom left corners, disappear again, and finally appear on the top left and bottom right corners again, which starts a new cycle. That is, each cycle consists of four steps. If the intervals between the four steps are sufficiently small, then human observers report to see the dots moving in a continuous fashion between the corners. That is, the dots are not reported as blinking dots or as flashing dots. Rather, humans report to see dots that move in some directions. From a pattern formation perspective, the appearing and disappearing dots are assumed to induce brain activity patterns similar to the ones that are triggered by actually moving dots. Since the experimental design exhibits symmetry in the vertical and horizontal direction, some individuals report that the dots move horizontally between the corners, while other individuals report that the dots move vertically between the corners. That is, the dynamical visual environment of the apparent motion paradigm induces two qualitatively different reactions (or BBA attractor patterns) that are related to left and right moving dots, on the one hand, and up and down moving dots, on the other hand. For sufficiently "fast moving" dots participant exhibit one of the two reactions to the movie only for a certain amount of time. Then the type of reaction switches from

---

[3] See however the previous Sect. 7.4.2.

seeing horizontal to seeing vertical movements or vice versa from seeing vertical to seeing horizontal movements. The new reaction is maintained for a while before the reaction switches again. In the end, participants report that the dots move up and down for a while, then left and right for while, then up and down again, and so on. The overall reaction to the apparent motion movie is oscillatory in the sense described above.

Not only forces from the visual world can induce oscillatory reactions. The so-called verbal transformation effect is an example in which forces from the acoustic world induce an oscillatory reaction [69, 342].

Various studies have determined characteristic features of brain activity during visually-induced oscillatory reactions. For the Necker cube paradigm, brain activity before and after switching events was compared to brain activity during periods of constant reactions [40, 190]. Likewise, brain activity has been examined for another stationary experimental design based on "perceptual rivalry" that induces oscillatory reactions [208, 209, 309]. For the apparent motion paradigm, features of brain activity have been identified that are characteristic for switches to occur [240, 241]. Certain bio-chemical substances (so-called neurotransmitter) have been identified in the visual cortex that seem to play a role in the emergence of the aforementioned oscillatory reactions in humans [330]. However, animals studies on monkeys suggest that areas other than the visual cortex also play a crucial role for understanding this kind of phenomena [208]. According to the human pattern formation reaction model, the brain activity observed in those studies reflects brain activity patterns that emerge due to instabilities.

In general, it has been acknowledged [261] that oscillatory reactions as reported above are likely to be caused by bistable brain oscillators (in the sense of the aforementioned continuously oscillating pendulum that produces discrete on and off states). In particular, Ditzinger and Haken [67] and others [68, 102, 114, 157] have introduced such bistable brain oscillators on the basis of the theory of pattern formation as discussed in this book.

Let us discuss the application of the human pattern formation reaction model to oscillatory instabilities as reviewed above. To this end, let us consider human reactions to the spinning dancer movie. However, analogous considerations can be made for oscillatory reactions as observed in any of the aforementioned reviewed experiments revealing human oscillatory reactions.

According to the modeling approach within the framework of synergetics and the human pattern formation reaction model [67, 68, 102, 114, 157], the spinning dancer movie creates external circumstances such that the human individuals exposed to the movie correspond to bistable pattern formation systems. The human system can exhibit two kinds of brain activity patterns: one pattern related to a clockwise spinning dancer, the other pattern related to a counter-clockwise spinning dancer. In what follows, we ignore the neuronal activity in the sensory body system and only consider the brain activity. Consequently, we refer to the emerging patterns as BA patterns (rather than BBA patterns).

Brain activity patterns will be qualitatively described in terms of firing rates of spatially distributed neuron populations. In Fig. 7.12b a group of four neuron pop-

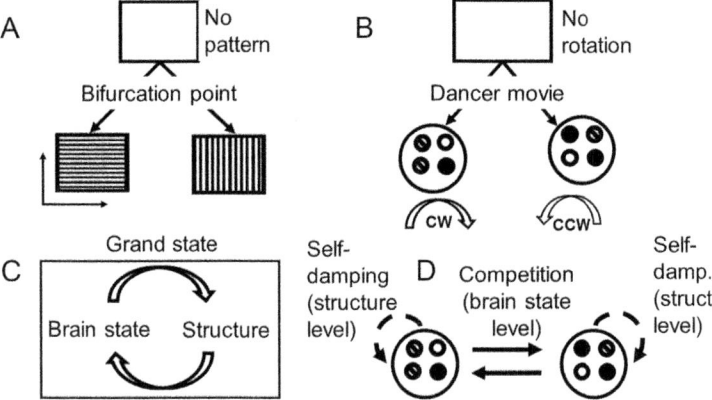

Fig. 7.12 Panels (**a**) and (**b**): Bistability hypothesis. Panel (**a**): The symmetric design of a chemical pattern formation system leads to a bistability for chemical substance concentrations such that if the system operates above the bifurcation point then it exhibits two stable chemical concentration patterns. Panel (**b**): The neuronal network of humans reacting to rotating objects is assumed to be approximately in a symmetric condition when the dancer movie is presented. Accordingly, brain activity patterns associated with clockwise (CW) and counterclockwise (CCW) rotating objects can emerge. The brain activity patterns are illustrated by firing rates of four neuron populations. The patterns are shown for illustration purposes only. The patterns shown in panel (**b**) are not related to any experimental data. Panels (**c**) and (**d**): Overview of the components and the component-component interactions of the oscillatory neuronal network under consideration. Panel (**c**): Brain state and structure depend each other. Panel (**d**): Key interactions between the components of the neuronal pattern formation system

ulations is considered. Two possible firing rates patterns are illustrated qualitatively (where full circles, circles with a bar, and empty circles stand for high, medium, and low firing rates). These patterns describe BA patterns. Note that the patterns depicted in Fig. 7.12b are given for illustration purposes only. The firing rates do not reflect any experimental data. Due to the assumed symmetry of the human pattern formation system exposed to the spinning dancer movie two BA patterns can emerge. The two patterns correspond to reaction patterns that have similarities to BA patterns that would emerge if the individual under consideration is exposed to a clockwise (CW) or a counter-clockwise (CCW) rotating object in the real world. The BA patterns are mutually exclusive such that only one pattern can emerge at a time. The bistable human pattern formation system is seen in analogy to the example of the bistable chemical pattern formation system in a squared dish discussed in Sect. 4.4.1 and shown in panel (a) of Fig. 7.12. In the chemical systems stripe patterns can emerge when appropriate concentrations of the chemical substances are used (i.e., when the system operates above the bifurcation point). Due to the symmetry of the system it follows that when a stripe pattern in the vertical direction can emerge, then a stripe pattern in the horizontal direction can emerge as well and vice versa.

Which pattern emerges in the bistable chemical systems depends on the initial conditions. Likewise, which brain activity pattern emerges first when an individual is exposed to the spinning dancer movie at the beginning of the experiment depends on the initial conditions of the individual given in terms of the ongoing brain activity of the individual (and maybe even in terms of the initial brain structure). In order to address oscillatory human reactions, we need to add a second component to the theoretical framework: a self-damping mechanism [157] as used in various applications discussed previously. In order to do so, it is useful to distinguish between processes that take place on two different time scales [102]. In line with the definition of structure and state proposed in Sect. 1.6.1 it is assumed that the state evolves on a relative fast time scale on which structure changes only slowly, see Table 1.2. Accordingly, the patterns form in the brain state on a relatively fast time scale. In contrast, the self-damping mechanism changes structure on a relatively slow time scale. Structural components correspond to time-dependent variables that vary on a time scale that is slow relative to the time scale on which state variables evolve. Importantly, the structural changes on the slow time scale are assumed to break the symmetry property of the system and create a bias. More precisely, consistent with the observed oscillations, it is assumed that an emerging BA pattern results in self-damping of its eigenvalue. Therefore, after a certain duration, the fixed point BA pattern becomes unstable. For example, if the dancer induces the emergence of a visual world BA attractor pattern of clockwise rotations, then the eigenvalue of the corresponding clockwise rotations basis pattern is damped and decays such that the BA attractor pattern of clockwise rotations becomes unstable. As a result, a bifurcation takes place from the BA attractor pattern of clockwise rotations to the BA attractor pattern of counterclockwise rotations. While the BA attractor pattern of counterclockwise rotations is present the eigenvalue of the clockwise rotation BA basis pattern increases again and relaxes back to its baseline value, whereas the eigenvalue of the counterclockwise BA basis pattern is subjected to self-damping and decreases. At a certain point in time this mechanism leads to instability of the counterclockwise rotations BA attractor pattern such that another bifurcation takes place in which the clockwise rotations BA attractor pattern emerges again. The cycle composed of two BA attractor patterns (and their corresponding basis patterns and) is completed and a new cycle begins.

Panels (c) and (d) of Fig. 7.12 illustrate graphically how brain state and structure are assumed to depend on each other and give rise to an oscillatory reaction. Accordingly, a given BA pattern affects the structure such that it becomes unstable due to the competition between the two BA patterns under consideration. Consequently, the other pattern emerges. Then the process repeats itself.

From a mechanistic perspective, the brain state may correspond to firing rates of neuron populations, while the structure may correspond to synaptic weights connecting different neuron populations. For considerations on the neuronal network level of the pattern formation reaction model the reader is referred to Haken [157] and others [38, 39, 61, 95, 97, 125, 127, 157, 299].

Figure 7.13 present simulation results of the LVH model (for mathematical details see Sect. 7.8.3). Panel (a) describes the oscillator system as C2 pattern formation system. Panel (c) presents the evolution of the brain state in terms of the pattern amplitudes of the CW and CCW basis patterns. In contrast, panel (b)

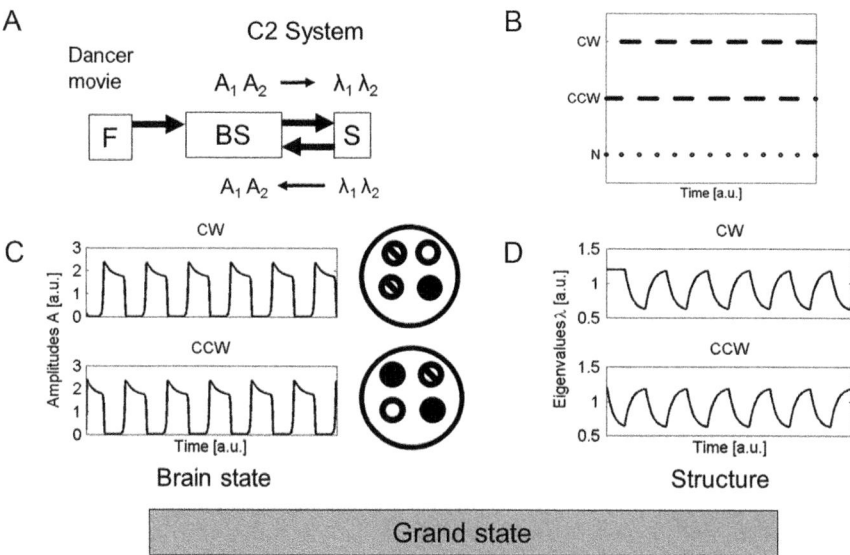

Fig. 7.13 Simulation results obtained from the oscillatory human pattern formation reaction model defined by the LVH model with structure dynamics as described in this figure and in the main text (see also Sect. 7.8.3 for mathematical details). Panel (**a**): Model overview in terms of a C2 system. Panels (**c**) and (**d**): Pattern amplitudes and eigenvalues as functions of time. Panels (**c**) and (**d**) taken together describe the grand state dynamics as indicated. Panel (**b**): Categorical brain state derived from the amplitudes shown in panel (**c**)

describes the brain activity in terms of three categorical states. Either the CW or CCW pattern is present or both patterns are absent. The latter case is denoted by N for "nothing". The evolution of the structural parameters, that is, the eigenvalues, for the CW and CCW patterns is shown in panel (d). The eigenvalue of a given pattern decays as long as the pattern is present. Otherwise, the eigenvalue relaxes back to its baseline level.

Note that the consequences of a bifurcation from one attractor pattern to the other pattern are twofold. First, the eigenvalue of the pattern that just disappeared increases again towards its baseline level. Second, the eigenvalue of the pattern that has just emerged begins to decay due to the hypothesized self-damping mechanism. If the eigenvalue of the absent basis pattern has increased to a sufficiently large value relative to the parameter of the present basis pattern, then the present basis pattern becomes unstable.

7.4.4   Pattern Formation in C3 Systems: A Second Look at Tolman's Rats

While in Sect. 6.4.3 the dynamics of rats in Tolman's maze has been described in terms of a B3 pattern formation system, in what follows, a somewhat more

comprehensive point of view will be presented. According to the B3 model proposed
in Sect. 6.4.3, the experimental setup exerts forces on a rat under consideration. As
reaction to those forces a BA attractor pattern emerges in certain brain areas of the
rat or the rat as whole forms a certain BBA pattern. In particular, as a result of the
emerging attractor pattern the rat produces a particular muscular activity pattern that
in turn determines the movement dynamics of the animal. The B3 model comes with
a pattern that accounts both for the forces of the visual world and the movement
dynamics. Let us split this pattern into two patterns that might emerge one after
another and might also emerge in different areas of the animal brain and body. The
two patterns are related to the forces of the visual world, on the one hand, and
the body pattern produced, on the other hand. A pattern formation C3 system is
considered in which the brain activity of the first attractor pattern related to the
external forces changes the structure of the brain and determines the emergence of
a second attractor pattern related to the produced forces.

Figure 7.14 illustrates the key elements of the C3 system model for a restructured
animal (i.e., an animal at the end of the 20 days restructuring period). First of
all, the visual world forces exerted by the experimental setup (in particular the
forces exerted by the black and white doors) are assumed to make a reference state
unstable such that two unstable basis pattern emerge related to the two differently
colored doors. As illustrated in panel (a), it is assumed that there is a symmetric
brain structure such that the eigenvalues of both basis patterns assume the same

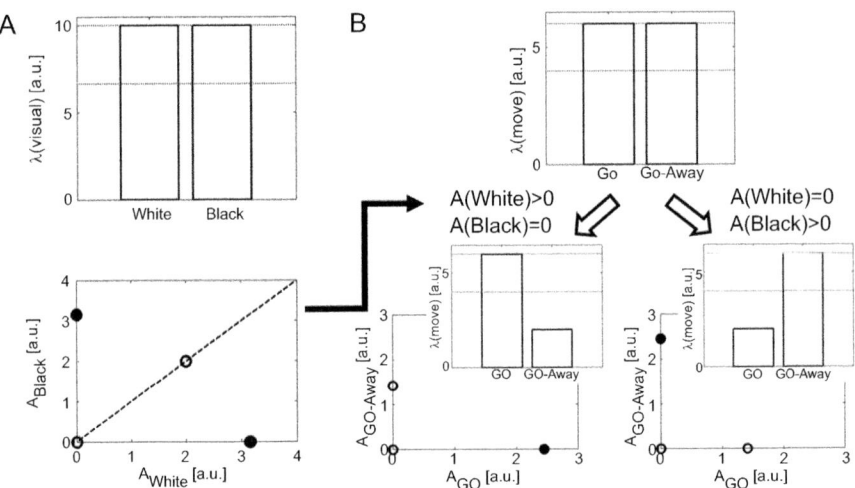

Fig. 7.14 Two coupled pattern formation systems after restructuring rats such the restructured rats
walk in any case through the food door, which is the white door in the example. Panel (**a**): Top and
bottom subpanels describe the structure and the reduced amplitude space of the pattern formation
system reacting to the visual world forces of the doors. Panel (**b**): Scheme of the movement pattern
formation systems. The movement system is affected by the visual system such that emerging
white and black door BBA patterns affect the structure of the movement system in terms of the
relevant eigenvalue spectrum for walking through or walking away from a door

value. Consequently, in the reduced amplitude space, there are two attractors that correspond to the white and black door, respectively, see panel (a), see bottom part. The attractors are defined (see Sect. 5.4) by the property that the forces exerted by the white door shift the brain state into the basin of attraction of the attractor of the visual world white door BA attractor pattern, while the forces exerted by the black door shift the brain state into the basin of attraction of the attractor of the black door BA attractor pattern. In addition to the pattern formation system reacting the external forces of the visual world, according to the C3 model there is another pattern formation system producing patterns related to body movements. Panel (b) illustrates this second system. For sake of simplicity, only two movement patterns are considered. The movement through a door (called GO) and the movement away from a door (called GO-Away). Accordingly, two unstable basis patterns are considered. As such the system exhibits structural symmetry such that both patterns exhibit the same eigenvalues, see top part of panel (b). However, the structural symmetry is assumed to be broken by the emerging visual world patterns. In panel (b) it is assumed that the white door is the food door. If the white door attractor pattern emerges then the brain activity changes the structure such that the movement pattern formation system becomes monostable and only the GO (through) attractor exists. In panel (b) (lower part) it is assumed that the brain activity decreases the eigenvalue of the GO-Away basis pattern such that it drops out of the stability band. Note that increasing the eigenvalue of the GO (-through) basis pattern by a sufficiently large amount would have the same effect (see our discussion about "functional fixedness" and "retrieval induced forgetting"). In summary, if the animal stands in front of the white door and gazes on the door, in a first step the visual world white door BA attractor pattern emerges. The BA pattern makes the movement pattern formation system monostable such that the brain and body state of that system converges to the GO (through) BBA attractor pattern. The animals goes through the white door. In a similar vein, the black door BA attractor pattern is assumed to change the brain structure just in the opposite way as the white door attractor pattern. According to panel (b), the black door BA attractor pattern decreases the eigenvalue of the GO (through) basis pattern such that the eigenvalue drops out of the stability band and the corresponding attractor becomes a repellor. Consequently, the brain and body activity of the animal converges to the GO-Away attractor. The animal abandons the black door.

The two-level model discussed here is a C3 system in the sense that emerging brain activity patterns induce structural changes. Let us extend this line of modeling to a mace that involves not doors but several T links. That is, let us consider a maze that involves not one but several junctions at which the animal can turn left or right. In this case, at every T junction, the junction leads to the emergence of one of two visual world attractor patterns (either a left branch-off or right branch-off BA pattern). The emerging visual world attractor pattern is then assumed to affect the movement pattern formation system by changing the eigenvalues appropriately. As a result, the animal turns left or right. However, the movement will bring the animal to another T link. In doing so, the movement will affect the external forces acting on the animal. In summary, a type 2 pattern sequence is obtained: external forces of a T link result in the emergence of a visual world T link BA pattern that results (via structural changes) in the emergence of left-turn or right-turn BBA movement pattern that brings the animal to the next T link that exerts again forces on the animal.

In closing this section, let us point out that the model described here describes animals at the end of the restructuring period. Those animals walk through the food door once they are subjected to the forces of the food door. In contrast, at the beginning of the restructuring period (i.e., on day 1, see Fig. 6.15) the visual world BA patterns are assumed to have no effect on the eigenvalues of the movement pattern formation system. Consequently, given the assumed symmetry of the system, animals facing a particular door either walk through the door or go away from the door depending on the initial brain and body activity. This applies to both doors. As a result, when averaging over several animals, there is a 50% chance that the animals walk through the food door. In the context of the proposed two-level system, it is assumed that the restructuring period makes that the visual world attractor patterns affect the structure (i.e., the eigenvalues) of the movement pattern formation system.

7.5 Pattern Formation of Grand States in D1 and D3 Systems

7.5.1 Pattern Sequences of D and E Systems That Cannot Be Produced by A,B,C Systems

Sequences of patterns formed in D and E systems involve structural changes due to external forces. In contrast, if pattern sequences of A,B,C systems involve structural changes, then those changes are due to the emerging BA and BBA patterns. As such the human brain of an individual can act as any kind of system: A,B,C,D,E. In particular, a pattern that emerges when the brain operates as D or E system may emerge in a similar form when the brain operates as A,B, or C system.

For example, a person may stand in front of a house and be exposed to the visual world forces of the house. Let us assume the brain of the person operates as B2 system, see Sect. 6.3. Among other things, the external forces shift the brain state into the basin of attraction of the visual world house attractor and the corresponding BA attractor pattern emerges (for details see Sect. 6.3.3). Let us turn from the awake person to the same person during REM sleep. It is assumed that there are unstable reference points that come with basis patterns similar to those of the awake person. Let us assume that the structure of the brain (in a particular brain area) is such that there is an unstable reference point with an unstable basis pattern similar to the visual world house unstable basis pattern. Furthermore, the eigenvalue of that basis pattern is assumed to be in the stability band. If the ongoing brain state happens to fall in the basin of attraction of the attractor of the basis pattern then the brain state converges to the attractor and the pattern emerges. The person "dreams" about the house.

However, consistent with experimental data, it is plausible to assume that there are situation in which certain BA and BBA patterns can only emerge if there are appropriate external forces that lead to appropriate changes of the brain structure. If so, those patterns only emerge as part of pattern sequences in D and E systems. They cannot emerge when the brain operates as a A,B, or C system. In particular, in A systems, that is, in the isolated human system of an awake person or a person in REM sleep, the patterns cannot emerge. When studying humans and animals from the classical two-step perspective of physics outlines in Chap. 5 (step 1: force-free, isolated system; step 2: system under force), then it becomes clear that there might be circumstances in which the external forces acting on the humans and animals are essential for patterns to form.

In pattern formation sequences involving external forces, external forces are frequently the result of produced forces (e.g., certain body movements). If so, there are patterns and pattern sequences that only occur if the human or animal bodies are formed into appropriate brain activity and body patterns, where the body patterns induce appropriate external forces acting back on the humans and animals under considerations.

Figure 7.15 exemplifies this point. On the left hand side the scheme of a D1 system is shown. The D1 system is a minimal D or E systems in which patterns or pattern sequences can emerge that involve structural changes induced by external forces. The latter are type 2 sequences. Other D and E systems that have this property are D3, E1 and E3 systems (see Fig. 5.5 for the system classification scheme). As far as grand state patter formation is concerned, only D1, D3, E1, E3 systems can lead to type 2 pattern sequences with structural changes in which the external forces are part of the pattern formation. That is, D1, D3, E1, and E3 system can produce forces that lead to the change of external forces (e.g., by moving somewhere or directing the gaze to some point) that in turn affect the brain structure and induce bifurcations inducing switches between BA and BBA attractor patterns.

Let us return to Fig. 7.15. On the right hand side the scheme of A2 systems and the scheme of the class of Cx systems (i.e., C1,C2, and C3 systems) is shown. The lowercase letter x in Cx stands for 1,2,3. In contrast, the symbol "x" in the scheme stands for the three possible interaction types between brain state and structure characterizing the C1, C2, and C3 systems. As discussed in the previous Sects. 7.2, 7.3, and 7.4, in A2 and Cx systems type 2 sequences that involve structural changes induced by the brain state can emerge. Comparing the D1 systems on the left hand side with the A2 and Cx systems on the right hand side, it is plausible to assume that there are conditions in which the structural changes induced by the

Fig. 7.15 Illustration of components and interactions in D1 systems (left) and A2 and Cx systems (right). Certain patterns involving restructuration may exclusively be formed in D1 and similar systems (i.e., D3, E1, and E3 systems) because of the lack of impacting external forces in A2 and Cx systems

external forces in the D1 systems cannot be induced by the brain state as in A2 and Cx systems. Roughly speaking, A2 and Cx system cannot "mimic" D1 systems (or D1,D3,E1,E3 systems). If so, some patterns and pattern sequences can only occur when individuals produce forces and are exposed to the resulting external forces.

Let us have a look at the frequently discussed example of bike riding and two more examples: guiding a person to a certain place and typing-in a phone number on a cell phone. Children and adults typically have little problems to ride a bike. However, they often have problems to explain how to ride a bicycle.[4] Likewise, a person A might be able to show a person B the way to a certain place, where B wants to go, by guiding person B to that place. However, the guiding person A may not be able to describe verbally the path. That is, our guide A has to walk the way to the target place in order to get there. Likewise, the author of this book can easily type in certain phone numbers on a key pad of a cell phone but the author of this book cannot produce the numbers without performing the finger movements. From the physics perspective described above, the bike riding example is not a good example. People certainly "dream" about bicycle riding. Therefore, the riding-a-bicycle BBA pattern can emerge at least in some approximation as a BA pattern in the isolated human system. In contrast, the leading a person to a place and typing-in phone number examples are more useful examples provided that in the brain of the guiding person or the person who is typing-in numbers the related BA patterns cannot emerge without the produced forces and external forces associated to the produced forces. In particular, we need to assume that the guide in the guiding example does not "dream" about walking a person to the target place. Likewise, the phone number typing person does not "dream" about the phone numbers or about typing-in the phone numbers.

7.5.2 "Travel Light" Hypothesis or the Hypothesis of Life Determined by Environment-Body Interaction Forces

The "travel light" hypothesis is that human and animal brain and body activity to a large extent depends on the restructuring of the human and animal brains by means of external forces. Accordingly, the outside world by means of external forces that lead to restructuration crucially determines human and animal pattern formation in grand states. If so, humans and animals operate during considerable parts of their lives as D and E pattern formation systems. The "travel light" hypothesis implies that when humans and animals go through their lives they actually do not "carry"

---

[4]In this context, frequently the catch-phrase "know how" is used that involves the unnecessary concept of "knowledge", see Sect. 7.5.3. That is, it frequently has been stated that people "know" how to ride a bike but they often cannot verbalize that "knowledge". However, the concept of "knowledge" is not needed in physics and only adds to the mystification of humans, see Sects. 7.5.3 and 9.4.

a lot in their brains and bodies. They are like passengers with light suitcases. Note that the phrase "travel light" mystifies human life as a "journey". From a science perspective, the hypothesis may be formulated that what matters are the interactions between the external forces (i.e., the environment) and the structure of an individual. Accordingly, the hypothesis may be called the hypothesis of life determined by environment-body interaction forces.

7.5.3 Physics Versus "Tacit Knowledge" and "Embodied Cognition"

The theoretical considerations made above in Sect. 7.5.1 are consistent what has been called in the literature "tacit knowledge" and "embodied cognition". Note that the phrase "knowledge" is not needed in physics. A child, who is writing the letter A and, in doing so, is forming a particular body pattern is like a ferromagnet that forms a magnetic field or like water that turns into ice. Ferromagnets and water and humans and animals alike are thingbeings who/that produce self-organized states and patterns and make phase transitions and bifurcations between those states and patterns given the circumstances at hand and given their structure. As illustrated in this book, phenomena that can be observed in the animate and inanimate worlds can be explained within the physics perspective without the need to introduce the concept "knowledge". This has the advantage that all kind of physical thingbeings (humans, ferromagnets, water, etc.) are treated on an equal footing. When taking this unified, general perspective, a child does not "know" how to write the letter A, just like a ferromagnet does not "know" how to form a magnetic field and water does not "know" how to turn into ice. Likewise, as explained in Sect. 5.2.2 the concept of "cognition" is an unnecessary concept that only adds to the mystification of humans and animals. In short, experimental findings reported in the literature on "tacit knowledge" and "embodied cognition" can be explained from a scientific point of view by acknowledging—as in Sect. 7.5.1—that there is a fundamental difference between D and E pattern formation systems, on the one hand, and A,B,C pattern formation systems, on the other hand.

7.5.4 Skinner's Three Components Model as a D1, D3, E1, or E3 System

Skinner suggested a three components model to describe cause-and-effect relationships in animal reactions and dynamics [300]. Accordingly, as a first component the model states that there are operations performed on animals (or organisms in general) from outside. For example, an animal is not given water for a substantial period of time. The second component is the inner condition of the animal under consideration. In the example, not providing water leads to an inner condition of water deprivation. The operation on the animal is the cause. The inner condition the effect. Not giving water leads to water deprivation. The third component is the dynamics or reaction of the animal (using Skinner's terminology: the animal "behavior"). In Skinner's three components model, the dynamics of an animal is caused by the inner condition of the animal. Among other things, the model is used to argue that the inner condition can be neglected when discussing cause-and-effect relationships. Since it is assumed that the operation on the animal from outside causes the inner condition that in turn causes the animal dynamics, effectively, the outside operation is the cause of the animal dynamics. Skinner's argument and the relevance of inner conditions (and internal states) has been discussed in detail in Sect. 2.6.

The three components model is consistent with pattern formation in D1, D3, E1, and E3 systems as will be shown next. Figure 7.16 illustrates this for D1 systems. Since D3, E1, and E3 systems include all component interactions of D1 systems, the following considerations apply to those systems as well. As shown in Fig. 7.16, external forces changing the structure of human and animal pattern formation systems describe among other things operations performed on organisms from the outside that affect the structure. The structure is considered as counterpart to the inner condition. Finally, as suggested in Fig. 7.16 animal "behavior" is the result of the production of forces.

In D1 systems, external forces affect structure but not the brain state. When the grand state (i.e., brain state and structure) is regarded as counterpart to Skinner's inner condition, then we may take into consideration that the dynamics of the brain state is affected by the structure. Therefore, D1 systems describe how operations performed on an animal given in terms of external forces affect the inner condition of the animal given in terms of the brain state and structure. Taking a different point of view, D3 and E3 system describe that operations performed on an animal exhibit a direct impact on the inner condition of the animal, when the inner condition is given in terms of the properties of the brain state and the structure of the animal. In summary, depending on the interpretation of the inner condition of an animal, D1 and E1 (structure as inner condition) or D3 and E3 (grand state as inner condition) are consistent with Skinner's three components model.

Fig. 7.16 Three components model proposed by Skinner interpreted as a D1 pattern formation system

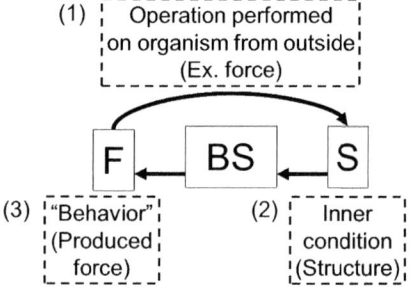

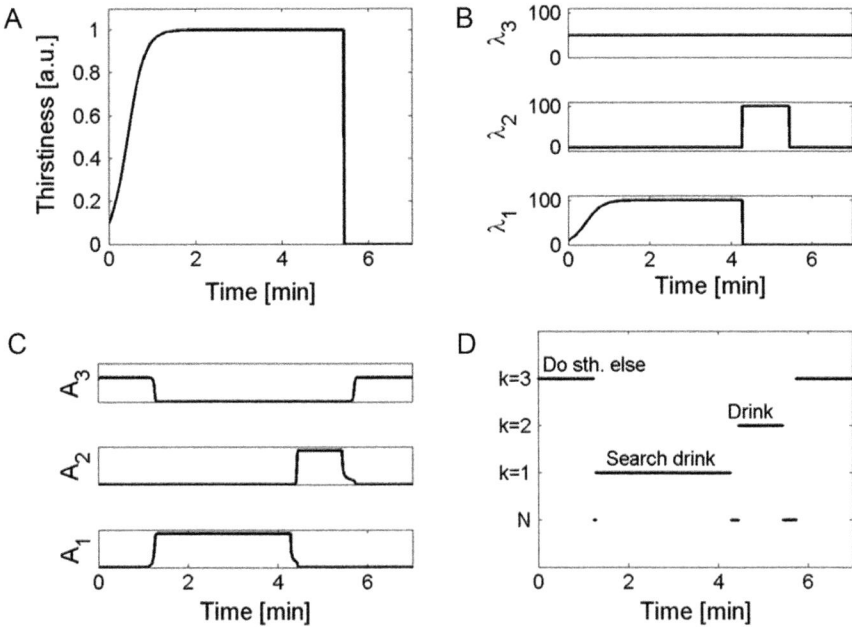

Fig. 7.17 Simulation of a thirsty person looking for something to drink. Simulation based on the LVH amplitude equations. Panel (**a**): Degree of thirstiness over time. Panels (**b**) and (**c**): Evolution of the grand state of the person described in terms of structure (eigenvalues) and ordinary state (amplitudes) with $k = 1$ searching, $k = 2$ drinking, and $k = 3$ doing something else. Panel (**d**): Categorical state as function of time

Let us illustrate pattern formation in a D1 system by means of an example. In the example, a thirsty adult drinks something and, in doing so, changes his or her structure. In this context, the structural change is indicated by the fact that the adult is no longer thirsty. Figure 7.17 provides details of the example. The degree of thirstiness is described by a relative variable that ranges from 0 to 1. A zero score means the person under consideration is not thirsty at all. In contrast, a score of 1 means that the person is extremely thirsty. It is assumed that the degree of thirstiness increase over a period of time. According to the graph in panel (a), the degree of thirstiness increases within a minute to its maximum level. This is a somewhat unrealistic case but simplifies the presentation. It is assumed that the person under consideration can be regarded as a pattern formation system at an unstable reference point that comes with several unstable basis patterns. In what follows only three basis patterns will be considered. The basis pattern $k = 1$ gives rise to an attractor of the BBA pattern for searching for something to drink provided the eigenvalue λ_1 of the $k = 1$ basis pattern is in the stability band. The basis pattern $k = 2$ is associated to (i.e., is the dominant part of) the drinking BBA attractor pattern that describes that the person is actually drinking. The basis pattern $k = 3$ captures a whole class of basis patterns that describe all kinds of reactions and body patterns different from searching for something to drink and drinking. Panel (b) depicts the eigenvalues λ_1, λ_2, and λ_3 of the three basis patterns as functions of time. Panel (c) shows the basis pattern amplitudes A_1, A_2 and A_3 as functions of time, while panel (d) shows the categorical state of the person under consideration.

At the beginning of the simulation, the person is assumed to do something different form drinking and searching for something to drink. A_3 is at its finite fixed point value, while A_1 and A_2 are at zero, see panel (c). The eigenvalue λ_3 of the basis pattern for doing something else is assumed to be constant. The eigenvalue λ_1 of the basis pattern related to the searching attractor is assumed to increase with the degree of thirstiness. In our simulation, λ_1 is identical to the degree of thirstiness except for a constant proportionality factor (compare panel (a) and the bottom subpanel of panel (b)). Consequently, when the degree of thirstiness increases, then the eigenvalue λ_1 increases as well. As a result, λ_1 at a certain point in time exceeds λ_3 by a critical amount such that λ_3 drops out of the stability band. A bifurcation takes place. The eigenvalue λ_2 of the drinking basis pattern is at a (small) negative value as long as nothing to drink is available (see middle subpanel of panel b). Therefore, at the bifurcation point when λ_3 drops out of the stability, the system is monostable and bifurcates to the searching pattern. The person stops doing whatever he or she is doing at that moment and starts searching for something to drink, see panels (c) and (d). The switch towards to searching-for-something-to-drink BBA attractor pattern takes place in the simulation at about 1.2 min.

It is assumed that the person is searching for something to drink for about 3 min. After 3 min he or she finds something to drink. At that moment, the eigenvalue λ_1 of the searching basis pattern drops to zero and the eigenvalue λ_2 of the drinking basis pattern jumps to a sufficiently large value such that all other eigenvalues drop out of the stability band (see panel b). That is, the external visual world forces exerted by the beverage (water, juice, etc.) are assumed to change the brain structure of the person under consideration. In a D1 system this is done directly. In a C3 system the change in structure may occur indirectly via an emerging BBA pattern (the forces exerted by the beverage shift the brain state into the basin of attraction of a beverage BBA attractor; the emerging pattern subsequently changes the brain structure).

When λ_1 and λ_2 switch their values, that is, λ_1 drops to zero, while λ_2 jumps to a relatively large value, a bifurcation occurs. The person stops searching and drinks the beverage that he or she has just discovered. In the simulation, it is assumed that drinking takes about 1 min. While the degree of thirstiness actually decays during drinking, in the simulation, for sake of simplicity, at the end of the 1 min period the thirstiness variable is put to zero indicating that the person is no longer thirsty. Drinking involves produced forces but also external forces in terms of biochemical substances entering the human body and initiating chemical reactions. Therefore, the intake of the beverage comes with chemical driving force that act on the person under consideration and change the structure of that person. The chemical driving forces taken together as a single force acting on the drinking person are similar to Skinner's proposed first component that describes an operation performed on an organism. When the thirstiness score drops to zero, the eigenvalue λ_2 of the drinking basis pattern drops to zero. A bifurcation occurs in which the attractor pattern of doing something else emerges, see panels (c) and (d).

7.5.5 Brain Activity Patterns as Reactions to Moving Dots: A Second Look

In Sect. 6.3.5 the increase of brain activity measured in monkeys when they look at moving dots was described in terms of an emerging BA attractor pattern and the increase of the corresponding attractor pattern amplitude. Consistent with experimental data, the model presented in Sect. 6.3.5 assumes that an animal under consideration has been structured such that forces exerted by moving dots shift the brain state of the animal into the basin of attraction of one of two attractors. The two attractors describe BA and sensory body activity patterns emerging when the monkey is exposed to movies of left or right moving dots. Panel (d) of Fig. 6.14 shows both the experimental data of measured brain activity from LIP neurons in units of spikes/seconds and the brain activity modeled via the LVH amplitude equations. The emergence of visual world left-moving/right-moving dots BBA patterns has been discussed in Sect. 6.3.5 in the context of a B2 systems.

As mentioned in Sect. 6.3.5, the animals were not only subjected to moving dots but also to a trigger signal. In the absence of the trigger signal the animals starred on a fixation location. However, when the trigger signal was given, then the signal in combination with the moving dots made the animals to move their eyes (i.e., perform a saccade) to the left or right in correspondence to the movement direction of the dots. The activity of the LIP neurons was recorded during the whole experiment involving the presentation of the moving dots, the presentation of the trigger signal, and the eye movements. Schematically, the brain activity showed a graph as depicted in panel (a) of Fig. 7.18. During the first phase (before the trigger was given), the activity increased as discussed already in Sect. 6.3.5. When the trigger signal was given, the activity increased even further (second phase). It reached a maximum slightly before the eye movement was performed. Subsequently, the activity decayed rapidly (third phase).

Let us explain the LIP neuron activity during the whole observation time by means of the pattern formation reaction model. Consistent with the experimental data, the animal pattern formation system is considered to be a D3 or even E3 system. Let us split the whole observation period in three phases. The first phase (before the trigger was given) was addressed already in Sect. 6.3.5 and above. The increase of LIP activity for that first phase is also shown in the top subpanel of panel (b) in Fig. 7.18. The second phase describes LIP activity from the moment when the trigger signal is given to the moment when the LIP activity starts to decay. In the context of the pattern formation model it is assumed that the trigger signal as an external force makes the attractor unstable that determines the dynamics during the first phase. Returning to panel (e) of Fig. 6.14 in Chap. 6, it is assumed that during

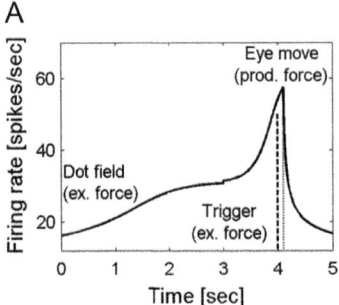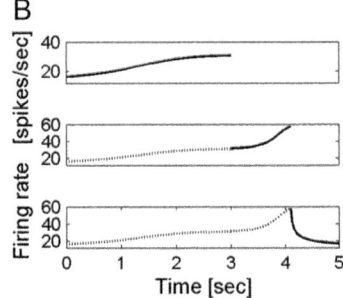

Fig. 7.18 LIP neuron activity of monkeys watching a moving dot field and reacting to the field by performing an appropriate eye movement. Panel (**a**): Firing rate computed from the LVH model. Panel (**b**): Firing rates in three distinct phases of the observation period

the first phase of an experimental trial an attractor emerges (due to an instability of an appropriately defined reference state), which in our example is located at 31 spikes/seconds along the horizontal axis in Fig. 6.14e (which is the axis of the basis pattern for left moving dots). The trigger signal is assumed to affect the brain structure such that this attractor becomes unstable. At that instability point several basis patterns are assumed to exist. Two of them are the basis pattern of moving the eyes left and right (i.e., making a saccade to the left or right). Consistent with the experimental data and the experimental procedure (treatment of the animals), it is plausible to assume that the animals have been structured such that the system is monostable. There is only the attractor for left eye movements. Accordingly, the brain activity converges to the attractor and the BBA attractor pattern of moving the eyes to the left emerges. The emergence of that pattern corresponds to the increase of brain activity as observed in the second phase. Simulation of the LVH model yields the solid line as shown in the middle subpanel of panel (b) of Fig. 7.18. Note that it has also been assumed that the trigger signal shifts the brain activity slightly which is the reason for the small step or jump in the graph shown in panel (a) of Fig. 7.18. When the eye-moving-left BBA attractor has emerged, it is assumed that the animal body produces the appropriate forces to perform the eye movements. The dashed vertical line is a hypothetical line at which the amplitude of the eye-moving-left BBA pattern reaches a critical threshold that indicates that the eye-moving-left BBA pattern has emerged. In contrast, the dotted vertical line describes the time point when the eye movement is performed. That is, it is assumed that there is a short delay between the BBA pattern reaching an appropriately defined threshold and the movement. Note that from a modeling point of view, this delay can be made arbitrarily short or long. The third phase of the graph shown in panel (a) describes the decay of the LIP neuron spiking activity. It is assumed that the eye-moving-left attractor becomes unstable or disappears entirely after a short period. A possible mechanism has been discussed above in the context of scene decomposition. Accordingly, the BBA pattern puts its eigenvalue to a negative value. Another possible mechanism is that the eye movement changes the external forces acting on the system. That is, when the eye movement to the left has been completed, the force related to the new gaze position or the forces related to the completion of the eye movement affect the structure (in terms of the aforementioned eigenvalue) and make the eye-moving-left BBA pattern unstable. In both cases, when putting the eigenvalue to a negative value, simulation of the LVH model yields the graph shown as a solid line in the bottom subpanel of panel (b). Taking the three phases simulated by the LVH model and shown as solid lines in the subpanels of panel (b) together, we obtain the graph shown in panel (a).

The pattern formation system proposed here to explain the LIP neuron activity during an entire experimental trial as shown in panel (a) belongs either to the class of D3 or E3 systems (see Fig. 5.5 for the relevant system classes). The trigger signal is assumed to change the structure of the system by making the visual world dots moving left attractor unstable. Therefore, we are dealing with a D or E system. The brain state is affected by external forces and involved in the production of external forces. Therefore, we are dealing with a D3 or E3 system. Whether or not we deal with a D3 or E3 system depends on the mechanism that leads to the decay of the LIP neuron activity in the third phase of an experimental trial. If the scene decomposition mechanism applies, that is, the BA pattern decreases its own eigenvalue and, in doing so, affects structure, then we are dealing with an E3 system (because there is a BS-affects-S interaction). If the completion of the eye movement results in external forces that affect the brain structure and make that the eigenvalue of the eye-moving-left basis pattern decays, then we are dealing with a D3 system (because there is only the F-affects-S interaction and not the BS-affects-S interaction).

7.6 E1/E3 Systems and Negative Hysteresis

7.6.1 Positive and Negative Hysteresis

In Sect. 6.5.3 hysteresis was introduced as a phenomenon about bifurcations between two attractor patterns A and B. When a bifurcation parameter is gradually increased such that at a particular critical value a bifurcation from A to B occurs and, subsequently, the parameter is gradually decrease such that again at a particular critical value a bifurcation from B to A occurs, then there is hysteresis if the two critical values differ from each other. In contrast, if the A to B and B to A bifurcations take place at the same bifurcation parameter value, then the pattern formation system does not exhibit hysteresis. In Sect. 6.5.3 a particular type of hysteresis was considered: positive hysteresis. Positive hysteresis is characterized by a critical value of the bifurcation parameter in the increasing condition that is larger than the critical value of the bifurcation parameter in the decreasing condition.

Panels (a) and (b) of Fig. 7.19 schematically demonstrate the phenomenon of positive hysteresis for systems that switch from states A to B when a bifurcation parameter α is increased and vice versa switch from B to A when α is decreased. In panel (a) the states A and B are drawn on the bottom and top, respectively. In

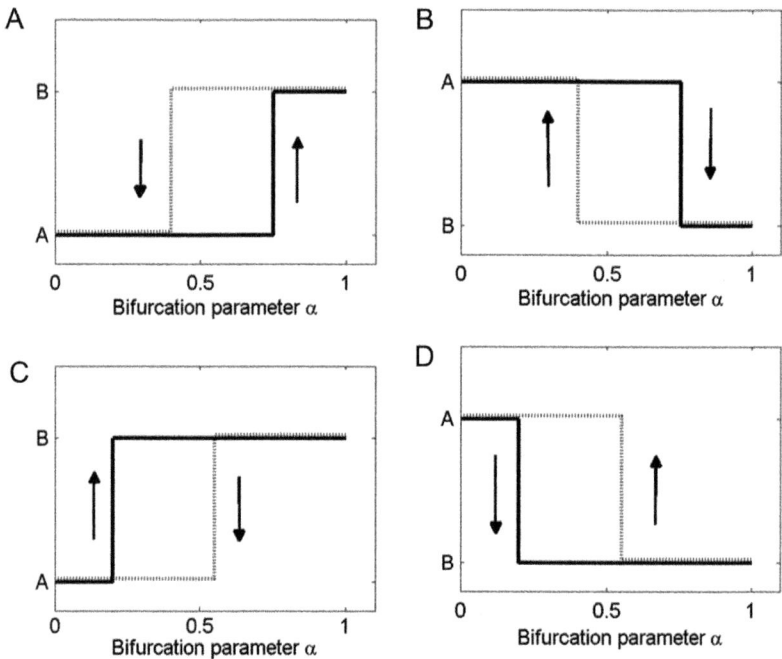

Fig. 7.19 Demonstrations of positive and negative hysteresis in systems exhibiting two states A and B. In all panels solid and dotted lines refer to the increasing and decreasing conditions, respectively. See also the arrows. Panels (**a**) and (**b**): Positive hysteresis. Panels (**c**) and (**d**): Negative hysteresis

panel (b) the presentation order is reversed. In panel (a) the switch from A to B occurs at a relatively large critical value $\alpha(crit, 2)$. In contrast, when the bifurcation parameter is decreased, the switch back from B to A occurs at a relatively small critical value $\alpha(crit, 1)$ such that $\alpha(crit, 2) > \alpha(crit, 1)$. The same phenomenon holds in panel (b). The bifurcation from A to B in the increasing condition takes place at $\alpha(crit, 2)$, while the bifurcation from B to A in the decreasing condition takes place at $\alpha(crit, 1)$ with $\alpha(crit, 2) > \alpha(crit, 1)$. That is, the order in which the states are presented on the vertical axis does not play a role.

As discussed in Sect. 6.5.3, positive hysteresis is consistent with bistability. In the parameter domain between $\alpha(crit, 1)$ and $\alpha(crit, 2)$ the system is assumed to be bistable such that both patterns A and B can emerge. In the increasing condition, the pattern that emerges for small parameter values also occurs in the bistability domain. In the decreasing condition, the pattern that emerges for large parameter values also occurs in the bistability domain. Consequently, the pattern that is initially present in a given presentation condition can be observed for a larger range of bifurcation parameters in a given presentation condition as compared to the pattern that was initially absent. For example, let us assume that the bifurcation parameter ranges from 0 to 10 and we have $\alpha(crit, 1) = 2$ and $\alpha(crit, 2) = 7$. Then, in the increasing sequences the pattern that is present at 0 is also present in the whole range from 0 to 7. In contrast, that pattern can only observed in the interval from 0 to 2 in the decreasing sequence. Furthermore, for positive hysteresis, hysteresis size $\Delta\alpha$ as defined by $\Delta\alpha = \alpha(crit, 2) - \alpha(crit, 1)$ is positive.

Negative hysteresis is illustrated schematically in panels (c) and (d). In the increasing condition, the bifurcation from A to B occurs at a relatively small critical bifurcation parameter, while for the decreasing bifurcation parameter the bifurcation from B to A occurs at a relatively large critical value. Therefore, a pattern that is initially present in the increasing condition, disappears "relatively quickly" as compared to the decreasing condition. Likewise, a pattern that is initially present in the decreasing condition, disappears "relatively quickly" as compared to the increasing condition. Using the same notation as for the positive hysteresis, then $\alpha(crit, 2)$ denotes the critical value in the increasing condition, whereas $\alpha(crit, 1)$ denotes the critical value in the decreasing condition. For negative hysteresis we have $\alpha(crit, 2) < \alpha(crit, 1)$. Consequently, hysteresis size $\Delta\alpha = \alpha(crit, 2) - \alpha(crit, 1)$ is negative. In panel (c) the states A and B are drawn on the bottom and top, respectively. In panel (d) the presentation order is reversed. Irrespective of the way the states are presented on the vertical axis, the hysteresis size $\Delta\alpha$ is negative for negative hysteresis phenomena.

With respect to hysteresis size, we may distinguish between three cases. If hysteresis size equals zero, then the two critical bifurcation parameters assume the same value. There is no hysteresis. If hysteresis size is positive then there is positive hysteresis and patterns "stay longer" in the sense discussed above. If hysteresis size is negative then there is negative hysteresis and patterns "disappear earlier" as described above.

In the literature, the positive hysteresis phenomenon (i.e., the observation that a pattern "stays longer") has also been referred to as a tendency of the brain to "maintain" its state [188], as illustrations how the human brain "maintains percepts" [10] and has been referred to as "persistency" tendency of the brain or as indication for "perceptual inflexibility" [234], see also Sect. 10.5.2.

Research on humans frequently reports the critical bifurcation parameters when averaged across samples of participants. These mean values may be compared to determine whether the data suggest that the human system under consideration exhibits no hysteresis, positive hysteresis, or negative hysteresis. As discussed in Sect. 6.5.4 in the context of positive hysteresis, several studies also report the percentage of observed categorical states (or reactions) for each tested bifurcation parameter. In doing so, percentage functions are obtained as shown in Fig. 6.19 of Sect. 6.5.4.

Figure 7.20 exemplifies percentage functions of states or reactions for positive hysteresis (panels a and b) and negative hysteresis (panels c and d). The graphs refer to idealized situations in which for all participants individually we have $\alpha(crit, 2) > \alpha(crit, 1)$ in the positive hysteresis case and $\alpha(crit, 2) < \alpha(crit, 1)$ for the negative hysteresis case. Just as in Fig. 7.19 it is assumed that the human systems under consideration exhibit two states or reactions A and B. Moreover, the systems bifurcate from A to B for increasing bifurcation parameters and from B to

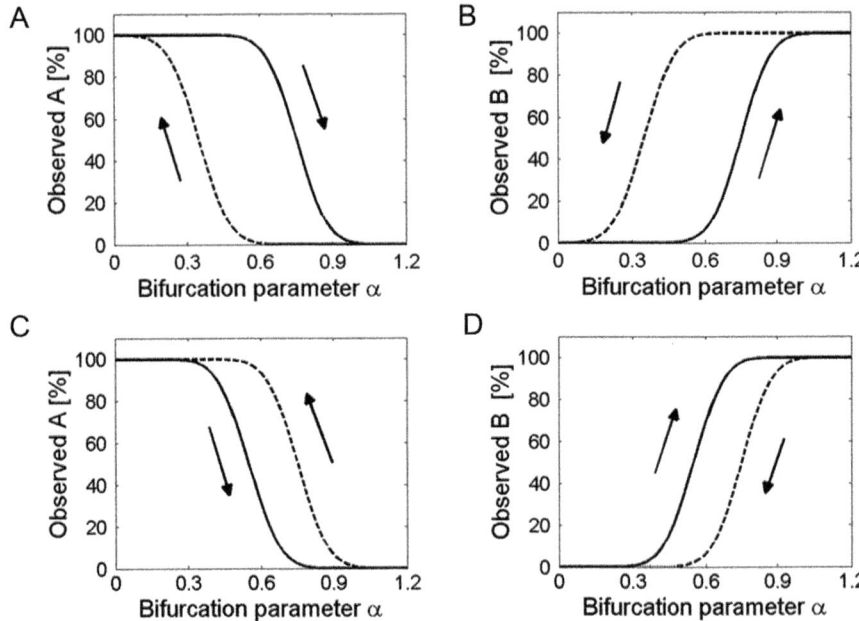

Fig. 7.20 Examples of idealized percentage functions to observe certain BBA patterns, states or reactions for positive hysteresis systems (panels (**a**) and (**b**)) and negative hysteresis systems (panels (**c**) and (**d**)). The systems only exhibit two patterns, states, or reactions A and B and make A-to-B and B-to-A transitions when the bifurcation parameter α is increased or decreased, respectively. Solid and dotted lines refer to the conditions in which α is increased or decreased, respectively. Panels (**a**) and (**c**): Percentage curves for observing the BBA pattern, state, or reaction A. Panels (**b**) and (**d**). Percentage curves for observing the BBA pattern, state, or reaction B

A for decreasing bifurcation parameters. Solid lines refer to increasing conditions, whereas dotted lines refer to decreasing conditions. Panel (a) of Fig. 7.20 shows the percentage of observed states or reactions A (i.e., the percentage of observed BBA attractor patterns associated with the reaction A) for a hypothetical experiment. For both graphs we see that for small bifurcation parameters the percentage of observed states (reactions) A is high, while for large bifurcation parameters the percentage of observed states (reactions) A is relatively low. The graphs of the increasing and decreasing conditions differ with respect to the fact that when following the percentage function for the increasing condition (solid line) from the left to the right then the function decreases at relatively large values, whereas when following the function for the decreasing condition (dotted line) from the right to the left then it starts to increase at relatively small values. This feature is due to the fact that we consider the idealized case of positive hysteresis (i.e., we assume that for all participants the individual critical values $\alpha(crit, 2)$ and $\alpha(crit, 1)$ satisfy $\alpha(crit, 2) > \alpha(crit, 1)$). In other words, the two graphs are characterized by the fact that for any given bifurcation parameter α the percentage of participants showing state or reaction A is higher in the increasing condition as compared to the decreasing condition. The graph of the increasing condition "lies above" the graph of the decreasing condition. Panel (b) shows the percentage of observed states or reactions B. Since the percentage values of states (reactions) A and states (reactions) B add up to 100%, the graphs $p_B(\alpha)$ in panel (b) can be computed from the graphs $p_A(\alpha)$ in panel (a) like $p_B(\alpha) = 100 - p_A(\alpha)$. This implies if we draw a horizontal line in panel (a) at a level 50% and then mirror the graphs shown in panel (a) at that 50% line, then we obtain the graphs in panel (b). In this sense, the graphs in panel (b) are mirror-images of the graphs in panel (a).

As such, the percentage of observed patterns B (or states, reactions B) is relatively small for small bifurcation parameters and the percentage of observed patterns B is relatively high for large bifurcation parameters for both the increasing and decreasing conditions. The graph of the increasing condition "lies below" the graph of the decreasing condition for the idealized case that for all participants individually the critical values $\alpha(crit, 2)$ are larger than the critical values $\alpha(crit, 1)$. For any given bifurcation parameter there are fewer participants exhibiting pattern, state, or reaction B in the increasing condition as compared to the decreasing condition.

A comparison of panels (a) and (b) reveals that some caution is needed when interpreting percentage functions of observed states and reactions. In panel (a) the percentage graph of the increasing condition is above the graph of the decreasing condition. In panel (b) the opposite is true. The reason for this is that panel (a) refers to the percentage of observed states A, while panel (b) refers to the percentage of observed states B. Both panels present pairs of percentage functions that describe positive hysteresis. The characteristic feature that the pair of graphs have in common across panels (a) and (b) is that the graph for the increasing condition makes its change (increase or decay) relatively "late".

Pairs of percentage functions that describe negative hysteresis are shown in panels (c) and (d) of Fig. 7.20. In panel (c) of Fig. 7.20 the graphs overall show the same sigmoidal decaying shape as the graphs presented in panel (a). The difference between negative and positive hysteresis is reflected in the relationship between the graphs. For positive hysteresis and A-to-B transitions induced by an increasing bifurcation parameter it follows that the percentage of observed patterns (states, reactions) A is larger in the increasing condition as compared to the decreasing condition. Therefore, (1) the decay of the graph of the increasing condition starts "later" and (2) the graph of the increasing condition is "located above" the graph of the decreasing condition. In contrast, as shown in panel (c), in the case of negative hysteresis and A-to-B transitions induced by an increasing bifurcation parameter it follows that (1) the decay of the graph of the increasing condition (solid line) starts "earlier" and (2) the percentage of observed patterns (states, reactions) A is smaller in the increasing condition as compared to the decreasing condition and, consequently, the graph of the increasing condition is "located below" the graph of the decreasing condition. Property (1) reflects that in negative hysteresis systems pattern (or states, reactions) "disappear early", as mentioned above. In the increasing condition, pattern A "disappears early", which is the reason why the percentage of observed patterns B is relatively high in the increasing condition as compared to the decreasing condition.

Finally, panel (d) shows schematically percentage functions for observing states B (reactions B or B-patterns) in the increasing and decreasing conditions for systems exhibiting negative hysteresis. The graphs are mirror-images (in the sense described above) of the graphs displayed in panel (c). The graphs in panels (c) and (d) have in common that the graphs of the increasing condition show an "early" change (increase or decay) when reading the horizontal axis from left to right. Likewise, the graphs of the decreasing condition show an "early" change (increase or decay) when reading the horizontal axis from right to left. However, while in panel (c) the graph for the increasing condition is located "below" the graph for the decreasing condition, the opposite is true for the the the graphs shown in panel (d). Irrespective of this difference, the pairs of graphs shown in panels (c) and (d) describe negative hysteresis.

7.6.2   Experimental Studies on Negative Hysteresis

Several studies with humans on negative hysteresis bifurcations have been conducted. Negative hysteresis has been found in reactions of humans to forces of the acoustic [325] and visual [83, 109, 164, 186, 216, 269, 331] world. Negative hysteresis has also been found when participants performed body movements and switched gaits between walk and run [211]. Let us briefly review those studies from a pattern formation perspective.

Note that in this section we will describe experimental findings from the perspective of patterns emerging in X0/X2 systems as discussed in Sect. 6.5.2. The X0/X2 systems description of negative hysteresis addresses the emergence of individual BBA patterns (and, in doing so, tries to shift the brain state of the reader to an attractor about single events). However, as such negative hysteresis only occurs in the context of sequences of observations (e.g., sequences of states, reactions, patterns). When taking the detailed mechanisms leading to the negative hysteresis phenomenon in pattern sequences into account, then from a modeling perspective the patterns are considered as patterns emerging in E1 and E3 pattern formation systems—as will be shown in the subsequent section (Sect. 7.6.3). Accordingly, the E1/E3 systems description addresses the whole context in which the negative hysteresis phenomenon occurs (and, in doing so, tries to shift the brain state of the reader to an attractor about chains of events). Note that the E1/E3 systems description includes the X0/X2 systems description (e.g., E3 systems exhibit all component interactions of X0/X2 systems, see Fig. 5.5).

Negative Hysteresis Bifurcations Induced by Forces of the Acoustic World

In an experiment by Tuller et al. [325], participants were exposed to the spoken word "say" that was manipulated such that there was a short gap between the letter "s" and the letter group "ay". As illustrated in panel (b) of Fig. 5.10 in Chap. 5, in general, reactions to the word "s..ay", where ".." indicates the gap, fall into the two categories: "say" and "stay". Accordingly, in the experiment by Tuller et al. [325], it was observed that when the gap was sufficiently small, participants reacted to the word as if they were exposed to (the forces of) the ordinarily spoken word "say". In contrast, when the gap was relatively large, participants reacted to the words as if they were exposed to the word "stay". The gap length was used as bifurcation parameter. Participants were tested under conditions in which the bifurcation parameter was increased and decreased. Say-to-stay and stay-to-say transitions were observed when the bifurcation parameter (gap length) was increased and decreased, respectively. Nine participants were tested. One of the nine participants showed negative hysteresis in all experimental repetitions. Three participants showed negative hysteresis in most of their performed experimental repetitions (i.e., trials). From the remaining five participants one participant showed no hysteresis and four participants showed positive hysteresis.

Negative Hysteresis Bifurcations Induced by Forces of the Visual World

In a study by Hirose and Nishio [164] a seat pan was presented to participants and participants were asked to "estimate" their maximal height of the seat pan that would allow them to sit on the pan in the normal way (i.e., with both feet on the floor). In terms of the pattern formation approach, Hirose and Nishio verbally instructed (brain restructuring phase or brain state shifting phase by means of acoustic external forces) the participants to produce X0/X2 systems BBA patterns (see Sect. 6.5.2) about sitting on the seat pan in the normal way, while not to perform the sitting movement patterns. The experiment involved an increasing and a decreasing condition. In the increasing condition, the seat pan was slowly moved up, while in the decreasing condition the seat pan was slowly moved down. The reaction to indicate the maximal height was to produce verbally the word STOP. In addition to the change of the seat pan, the experimenter manipulated the overall height of the participants. To this end, in one condition participants were standing (and walking around) on some kind of blocks, while in the baseline conditions participants were standing (and walking around) with socks on their feet. Both in the normal (without blocks) and height-increased (with blocks) conditions, negative hysteresis was found. The sample mean values for the normal condition (without blocks) are shown in panel (a) of Fig. 7.21. The bifurcation parameter $\alpha(crit, 2)$ of the increasing condition was on average smaller than the bifurcation parameter $\alpha(crit, 1)$ of the decreasing condition.

Panel (d) reports results from an experiment by Fitzpatrick et al. [83] in which participants were tested with respect to their reactions to a slope of a ramp. The ramp was placed on the floor and participants were positioned in front of the ramp. The inclination angle of the ramp was varied from small angles (flat ramp) to relatively large angles (steep ramp). From a pattern formation perspective, participants were instructed (i.e., restructured/brain state shifted) such that under the impact of the visual forces exerted by the ramp, X0/X2 systems brain and sensory

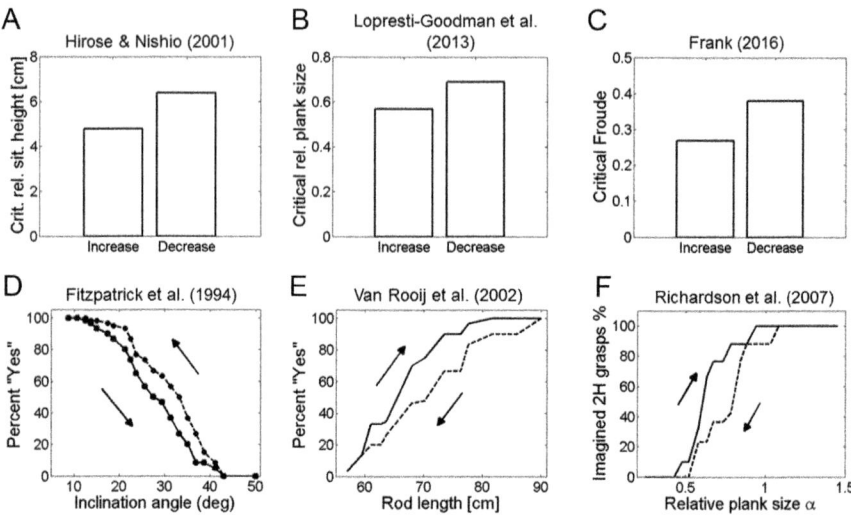

Fig. 7.21 Experimental results reported in experiments on negative hysteresis. Panels (**a**)–(**c**) report mean values of critical bifurcation parameters in the respective increasing and decreasing conditions of studies by Hirose and Nishio [164], Lopresti-Goodman et al. [216], and Frank [109]. Panels (**d**)–(**f**) report percentage functions of observed states (assumed to correspond to X0/X2 systems BBA patterns) for increasing and decreasing conditions of studies by Fitzpatrick et al. [83], van Rooij et al. [331] and Richardson et al. [269]. Solid and dotted lines refer to increasing and decreasing conditions, respectively. See also the arrows

body activity patterns emerged in certain parts of the brain and sensory system about standing on the ramp. The patterns did not involve a movement or force production component with respect to moving and standing on the ramp. That is, participants were instructed to "think" about or to "judge" (see Sect. 6.5.2) whether they could stand on the ramp or not in the usual way. The inclination angle was considered as bifurcation parameter and was increased or decreased. The graphs in panel (d) of Fig. 7.21 reported in Fitzpatrick et al. [83] show the percentage of observed standing-on-the-tilted-ramp BBA patterns associated with the verbal "Yes, I could stand on it" reactions. The graphs corresponds qualitatively to the graphs shown in panel (c) of Fig. 7.20. That is, Fitzpatrick et al. found negative hysteresis.

Panel (e) is taken from a study by van Rooij et al. [331] about human reactions with respect to a target, which was a 6 cm high cylinder placed on a table. The experiment involved tools that changed participants' body properties. The tools were rods that differed in length. Participants were placed in front of the table. Participants were instructed such that they formed X0/X2 BBA patterns about reaching to the target with the rods. That is, (in line with the definition of X0/X2 BBA patterns in Sect. 6.5.2) participants were not allowed to perform the reaching pattern. From a pattern formation perspective, it is assumed that participants formed BBA patterns similar to BA patterns of the isolated human system about reaching with a rod to a target. Rod length was either increased or decreased in steps.

Participants reported their emerging X0/X2 reaching or being-to-short-for-reaching BBA patterns in terms of verbal YES or NO reactions, respectively. Panel (e) of Fig. 7.21 shows the percentage of YES reactions (i.e., "Yes, I could reach" reactions) for the increasing and decreasing conditions. The graphs in panel (e) reported by van Rooij et al. illustrate the phenomenon of negative hysteresis and correspond qualitatively to the graphs shown in panel (d) of Fig. 7.20.

In Sect. 6.5.4 experiments on grasping transitions conducted by Richardson et al. [269] and Lopresti-Goodman et al. [214] have been reviewed. In those experiments, participants grasped consecutively planks while plank size was increased and decreased. Participants grasped planks either with one hand or with two hands. Variants of those experiments were conducted in which participants were only exposed visually to those planks [216, 269]. They were not allowed to grasp the planks. However, by appropriate instructions, participants were restructured or brain state shifted such that they reacted to the planks by producing brain activity patterns similar to the brain and body activity patterns of one-hand and two-hands grasping but without the force production component (i.e., they formed BBA patterns under the constraint of being X0/X2 systems). In those experiments, negative hysteresis was found. Panel (b) of Fig. 7.21 shows mean values of the critical bifurcation parameters as measured in relative plank size observed in a negative hysteresis study by Lopresti-Goodman et al. [216] for the increasing and decreasing conditions. The critical bifurcation parameter was on average smaller in the increasing condition as compared to the decreasing condition indicating that the phenomenon of negative hysteresis was observed (compare also with panels (c) or (d) of Fig. 7.19; the mean values of the critical bifurcation parameters would be located on the horizontal axis and patterns or states A and B would correspond to one-hand and two-hands grasping X0/X2 systems BA attractor patterns, respectively). The experiment has been replicated by Kim and Frank and qualitatively similar results for the sample means values of critical bifurcation parameters have been found [186]. Panel (f) of Fig. 7.21 shows percentage functions reported in Richardson et al. [269] for the percentage values of brain activity patterns that are about two-hands grasping but do not induce grasping movements. The percentage functions shown in Fig. 7.21f of the increasing and decreasing conditions reveal the characteristic features of negative hysteresis percentage curves shown in Fig. 7.20d.

While in Sect. 6.5.4 we reported from experiments showing walk-run gait transitions with positive hysteresis, variants of gait transition experiments involving treadmills have been conducted that showed negative hysteresis. For example, in Frank [109] a study is reported in which participants did not physically move on a treadmill. Rather, from a pattern formation perspective, by instruction participants formed brain activity patterns about walking or running on the treadmill without moving on the treadmill. Importantly, the treadmill speed was manipulated. The Froude number related to the treadmill speed (see Sect. 6.5.4) was considered as bifurcation parameter and was increased and decreased gradually. Participants were asked to verbally report for every given treadmill speed/Froude number whether they would walk or run. Participant reactions on average showed negative hysteresis. That is, the critical Froude numbers at which participants switched from "I would walk" to "I would run" in the increasing condition were on average lower as compared to critical Froude numbers of the "I would run" to "I would walk" switch in the decreasing condition. Panel (c) shows the sample mean values reported in Frank [109] of the critical values separated for the increasing and decreasing conditions.

Negative Hysteresis in Walk-Run Gait Transitions

In addition to the aforementioned acoustic world [325] and visual world [83, 109, 164, 186, 216, 269, 331] experiments on negative hysteresis, negative hysteresis has also been found in walk-run gait transitions when participants moved on a treadmill. Just as in the gait transition experiments discussed in Sect. 6.5.4, in the study by Li [211] participants moved on a treadmill while the treadmill speed was gradually increased or decreased. Walk-to-run transitions in the increasing condition and run-to-walk transitions in the decreasing condition were observed. Unlike the gait transition experiments reviewed in Sect. 6.5.4, in the study by Li [211] the acceleration and de-acceleration of the treadmill speed was considered as an experimental parameter and varied systematically. That is, Li [211] conducted experimental trials in which the treadmill speed changed slowly over time and other trials in which the treadmill speed changed quickly. When treadmill speed was changed quickly (greater magnitude of acceleration and de-acceleration parameters) positive hysteresis was found as in the studies mentioned in Sect. 6.5.4. That is, the treadmill speed scores at which walk-to-run transitions in the increasing condition were observed were larger on average than the treadmill speed scores at which run-to-walk transitions in the decreasing condition were observed. However, when the treadmill speed was changed slowly (acceleration and de-acceleration small in magnitude) then the relationship was reversed. For example, for treadmill acceleration and de-acceleration of 0.04 m/s^2, walk-to-run transitions occurred on average at a treadmill speed of 2.25 m/s, while run-to-walk transitions occurred on average at a treadmill speed of 2.35 m/s. That is, in the decreasing condition the running BBA pattern "disappeared early". Negative hysteresis was observed.

7.6.3   The Human Pattern Formation Reaction Model of Negative Hysteresis

Failure of the Positive Hysteresis Model

In Sect. 6.5.3 the human pattern formation model for positive hysteresis was introduced. As argued there, the positive hysteresis phenomenon observed in various experiments with humans is consistent with a bistable pattern formation system whose structure depends on the bifurcation parameter (that in turn is affected by internal forces produced by emerging BA and BBA attractor patterns or external forces exerted by objects and events in the environment). In contrast, negative

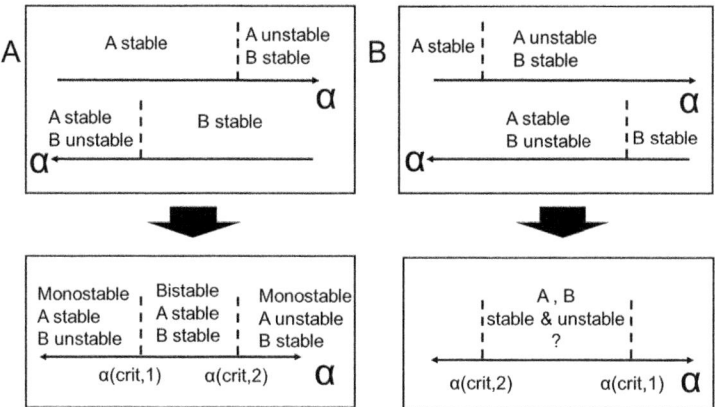

Fig. 7.22 The bistable model involving a single bifurcation parameter explains positive hysteresis (panel (**a**)) but leads to a contradiction when it is applied to explain negative hysteresis (panel (**b**))

hysteresis cannot be explained by means of a bistable dynamical model involving a single bifurcation parameter [216]. Figure 7.22 demonstrates this issue.

Let us consider a n-dimensional vector describing the system under consideration. The vector may describe the state $\mathbf{X} = (x_1, \ldots, x_n)$ of a discrete system (e.g., brain activity recordings measured at n different locations) or correspond to the unstable pattern amplitudes like $\mathbf{X} = (A_1, \ldots, A_n)$ of a pattern formation system in its reduced amplitude space. The argument that will be formulated in what follows does not require to specify the interpretation of the vector. The vector components are assumed to evolve in time. More precisely, it is assumed that the evolution of the vector and consequently of the system is determined by a set of differential equations. That is, the system under consideration is a dynamical system. Furthermore, it is assumed that the system depends on a single bifurcation parameter α. That is, the structure of the system is fixed except for one parameter: the bifurcation parameter. From those assumptions it follows that for any (initial) state given at a certain time point the system evolves in a unique way. However, the way the system evolves in general depends on the bifurcation parameter. The dynamical system is assumed to exhibit two fixed points A and B at locations $\mathbf{X}_A$ and $\mathbf{X}_B$, respectively. The fixed points correspond to stable or unstable fixed points. If both fixed points are stable, the system is bistable.

Let us first demonstrate that this general definition of a bistable system is consistent with the phenomenon of positive hysteresis. To this end, let us consider panel (a) of Fig. 7.22. The top part of panel (a) describes a hypothetical experiment exhibiting positive hysteresis. Accordingly, when increasing the bifurcation parameter α the fixed point A is assumed to be stable for small values of α up to a relatively large critical value $\alpha(crit, 2)$. Due to this property (and assuming that the system initially is prepared in state A), the system is in state A of the fixed point A for small bifurcation parameters values up to the aforementioned, relatively large critical value $\alpha(crit, 2)$. Beyond that critical value, the fixed point A becomes unstable. As seen from panel (a), at such large bifurcation parameter values, the fixed point B is assumed to be stable. Therefore, the state converges to fixed point B and the system is in state B for bifurcation parameters values larger than $\alpha(crit, 2)$. When decreasing the bifurcation parameter from large values it is assumed that fixed point B is stable up to a relatively small value $\alpha(crit, 1)$ of the bifurcation parameter. Since the system is prepared in state B at large values of the bifurcation parameter α, when decreasing α the system remains in the state B for all bifurcation parameter values down to $\alpha(crit, 1)$. At $\alpha(crit, 1)$ the fixed point B becomes unstable. At that value fixed point A is stable. Therefore, the system switches from B to A. Taking the considerations about increasing and decreasing the bifurcation parameter α together, we conclude that the dynamical system is monostable (with fixed point A stable) for $\alpha < \alpha(crit, 1)$, bistable (with both fixed points A and B stable) for $\alpha \in [\alpha(crit, 1), \alpha(crit, 2)]$, and monostable again (with fixed point B stable) for $\alpha > \alpha(crit, 2)$. The switch from A to B takes place at $\alpha(crit, 2)$, while the switch from B to A takes place at $\alpha(crit, 1)$. That is, the switches take place at the boundaries of the bistability domain. Due to this bistability domain, the system exhibit positive hysteresis. The lower part of Fig. 7.22a reproduces the bistable hypothesis for hysteresis that has been presented earlier in Sect. 6.5.3 in panel (d) of Fig. 6.17.

Let us turn next to a negative hysteresis experiment as illustrated schematically in the top part of panel (b) of Fig. 7.22. Let us show that the phenomenon of negative hysteresis leads to a contradiction with respect to the basic bistable model outlined above. First of all, as indicated in panel (b) of Fig. 7.22 (top half), when increasing the bifurcation parameter α, at a relatively low value $\alpha(crit, 2)$ the state of the system is assumed to switch from A to B. In line with the assumption that the system is a dynamical system, this implies that fixed point A becomes unstable at $\alpha(crit, 2)$, while fixed point B is stable at $\alpha(crit, 2)$. Furthermore, since state B is observed for all values of α larger than $\alpha(crit, 2)$ it follows that fixed point B is stable for all bifurcation parameter values larger than $\alpha(crit, 2)$. Decreasing the bifurcation parameter α from large values down to small values, the system switches from B to A at a relatively large value $\alpha(crit, 1)$. Moreover, state A is observed for all values α smaller than $\alpha(crit, 1)$. In view of our assumption that the underlying system is an dynamical system with fixed structure except for the bifurcation parameter, the switch from B to A means that fixed point B becomes unstable at $\alpha(crit, 1)$ and fixed point A is stable at $\alpha(crit, 1)$. Moreover, fixed point A is stable for all values smaller than $\alpha(crit, 1)$. If we combine the arguments made separately for the increasing and decreasing conditions, we arrive at the bottom part in panel (b) of Fig. 7.22. In the interval of the bifurcation parameter from $\alpha(crit, 2)$ to $\alpha(crit, 1)$ fixed point A must be stable (increasing condition) and unstable (decreasing condition). This is a contradiction. As mentioned above, for any given bifurcation parameter the system as a dynamical system evolves in a unique way. This implies that a fixed point cannot be stable and unstable at the same time for a given parameter α. Likewise, in the interval from $\alpha(crit, 2)$ to $\alpha(crit, 1)$ fixed point B must be stable (decreasing condition) and unstable (increasing condition). Again, both statements can not be true at the same time. We arrive at another contradiction.

Negative Hysteresis: Heuristic Explanation

In line with our considerations throughout this chapter on pattern formation that involves structural changes (i.e., pattern formation of grand states), it has been shown that the negative hysteresis phenomenon is consistent with the assumption that the structure of human pattern formation systems is a dynamical system [122, 216, 325]. Following Refs. [122, 216] let us show that it is sufficient to assume that emerging attractor patterns affect the structure of their systems by reducing the eigenvalues of their corresponding basis patterns. Note that this assumption has also been used in several applications discussed earlier in this chapter (see Sects. 7.3.1, 7.3.3, 7.4.2, and 7.4.3).

Panels (a) and (b) of Fig. 7.23 illustrate the components of human pattern formation systems exhibiting negative hysteresis. According to panel (a), we are dealing with E1 or E3 systems. As illustrated in panel (b), the brain state of an individual is described in the reduced amplitude space by means of the pattern amplitudes A_1 and A_2 of the attractor patterns 1 and 2 involved in the phenomenon of negative hysteresis under consideration. The patterns $k = 1$ and $k = 2$ correspond to the patterns that emerge (or can be observed in experiments) for relatively small and relatively large bifurcation parameters, respectively. For example, in the context of the grasping experiments by Richardson et al. and Lopresti-Goodman et al. the amplitudes A_1 and A_2 correspond to one-hand and two-hands grasping BA attractor patterns that are similar to isolated human system BA patterns and do not involve any force production components. The dynamics of the amplitudes A_1 and A_2 is determined by the eigenvalues λ_1 and λ_2, on the one hand, and by the

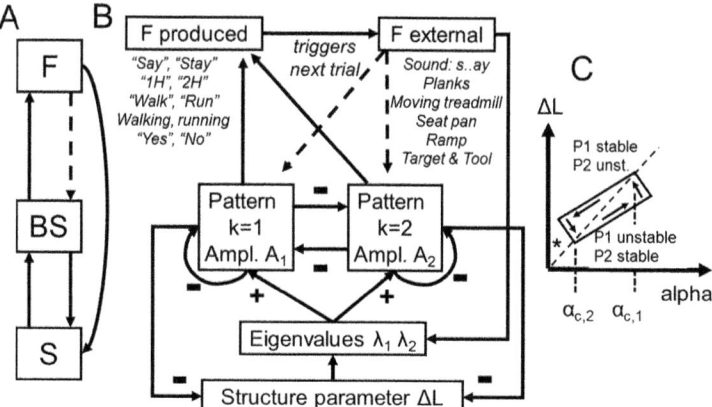

Fig. 7.23 Components and key mechanisms of the human pattern formation model for negative hysteresis. Panel (**a**): The system is assumed to be a E1 or E3 system. Panel (**b**): As compared to the bistable model for positive hysteresis, the negative hysteresis model features a feedback loop from the brain state to the structure via the parameter ΔL. Panel (**c**): The system evolves along a loop in a two-dimensional space spanned by the bifurcation parameter and the auxiliary parameter ΔL

competitive interaction between the amplitudes as measured in terms of the coupling parameter g. As indicated in Fig. 7.23b, the eigenvalues (assumed to be positive) result in an increase of the pattern amplitudes (whence the plus signs), whereas the competitive interaction results in a damping of the pattern amplitudes (whence the minus signs). Moreover, the amplitude dynamics is subjected to nonlinear self-damping mechanisms as discussed in Chap. 4 for general amplitude equations of the LVH type. Negative hysteresis arises due to an additional self-damping loop that acts on a slower time scale than the amplitude dynamics and involves the brain structure as captured in terms of the eigenvalues λ_1 and λ_2. Accordingly, emerging patterns reduce their own eigenvalues, which implies that they reduce their pumping sources. Just as in positive hysteresis systems, the eigenvalues are also subjected to external forces (e.g., exerted by the planks, seat pans, moving treadmill belts, ramps, and target objects to be reached in the experiments reviewed above). In summary, the eigenvalues as part of the structure are assumed to evolve in time under the impact of external forces and under the impact of internal forces generated by emerging BA and BBA patterns.

In order to model the impact of the emerging attractor patterns, an auxiliary structure parameter ΔL is introduced, see bottom part of panel (b). ΔL is related in a particular way to the eigenvalues λ_1 and λ_2. To simplify the presentation, we will assume—as indicated in panel (b)—that ΔL is affected by the emerging BBA patterns and ΔL in turn affects the eigenvalues.[5] If ΔL is large it increases λ_1 as compared to λ_2 such that λ_2 comes closer to the lower boundary of the stability band. In contrast, if ΔL is small it increases λ_2 over λ_1 such that λ_1 comes closer to the lower boundary of the stability band. In the "parameter"[6] domain spanned by ΔL and α, see panel (c), a critical boundary line (given by the dashed line) can be drawn such that there are two domains. In one domain (left-top), the pattern formation system exhibits the attractor $k = 1$ and is monostable for sufficiently large parameters ΔL. That is, the fixed point 1 is stable, whereas the fixed point 2 is unstable. In the other domain (right-bottom), the opposite is true. The pattern formation exhibits the attractor $k = 2$ and is monostable for sufficiently low parameters ΔL. That is, the fixed point 2 is stable, whereas the fixed point 1 is unstable. In the case of negative hysteresis pattern formation systems, the systems do not vary along a one-dimensional parameter space spanned by the bifurcation parameter α as in Fig. 7.22. Rather, negative hysteresis pattern formation systems undergo a loop [216, 331] in a two-dimensional "parameter" space spanned by α and ΔL as indicated in panel (c) of Fig. 7.23.

Let us begin the round trip assuming a low value of ΔL for a small bifurcation parameter α. This starting point is indicated by a star "*" in panel (c). In this case,

---

[5]In fact, from a mathematical point of view, the BBA patterns affect the eigenvalues and, in doing so, the parameter ΔL. However, in some sense both point of views are equivalent, see mathematical notes in Sect. 7.8.6 and Ref. [122].

[6]In fact, ΔL is not a parameter that is fixed by external forces like α. Nevertheless, we will refer to ΔL as parameter.

assuming a sufficiently small value for α the pattern formation system is monostable (due to the impact of the external force α on the eigenvalue spectrum) and exhibits only the attractor $k = 1$. Consequently, the attractor pattern 1 emerges. Increasing the bifurcation parameter decreases the eigenvalue λ_1 (due to the direct impact of the external force/plank size/seat pan height/etc. on the eigenvalue) and brings it closer to the lower boundary of the stability band. At a certain critical value λ_1 drops out of the stability band. The attractor $k = 1$ turns into a repellor. At this critical value, the system exhibits a stable attractor $k = 2$. The system makes a transition from pattern 1 to pattern 2. The bifurcation takes place in panel (b) when the loop hits the critical line in the lower half of the two-dimensional "parameter" space. The corresponding critical value is denoted by $\alpha(crit, 2)$. When increasing the bifurcation parameter beyond $\alpha(crit, 2)$ the emerging pattern 2 damps its own eigenvalue λ_2 via ΔL (whence the minus sign in panel (b) at the feedback loops from the patterns $k = 1$ and $k = 2$ to ΔL). That is, pattern $k = 2$ results in an increase of ΔL and, in doing so, λ_1 is increased over λ_2 (or λ_2 is decreased with respect to λ_1). The increase of ΔL is shown in panel (c). When increasing the bifurcation parameter α further (e.g. making planks longer/increasing the height of the seat pan/etc.), λ_2 is increased due to the increase of the bifurcation parameter (i.e., as a result of the external force). At the same time λ_2 is decreased due to the self-damping of the attractor pattern 2 via ΔL. When the bifurcation parameter α reaches its maximum value (e.g., largest plank/highest seat pan/etc.), ΔL assumes a relatively large number. When the bifurcation parameter is subsequently decreased ΔL increases even further (because of the BBA pattern $k = 2$ self-damping λ_2 by increasing ΔL). In addition, in the decreasing condition, λ_2 is decreased due to the fact the bifurcation parameter α becomes smaller (i.e., as an effect of the changing external forces such as plank size becomes smaller). Taken together, the system approaches the critical line shown in panel (c) at a relatively large value of ΔL and at a relatively large value for the bifurcation parameter α. The loop hits the critical line at a relatively large critical bifurcation parameter $\alpha(crit, 1)$ (at which λ_2 drops out of the stability band). Consequently, $\alpha(crit, 1)$, the bifurcation parameter of the pattern-2-to-1 transition of the decreasing sequence, is relatively large as compared to $\alpha(crit, 2)$, the bifurcation parameter of the pattern-1-to-2 transition in the increasing condition.

Subsequent to the transition from the attractor pattern 2 to the attractor pattern 1, ΔL decreases because of the self-damping mechanism of emerging BBA pattern 1. The BBA pattern 1 causes the eigenvalue λ_1 to decay by decreasing ΔL. However, since the bifurcation parameter becomes smaller, λ_1 increases. When the bifurcation parameter reaches its minimal value, ΔL is relatively small, see panel (c). Let us assume the increasing condition is conducted again, that is, α is increased again. When increasing the bifurcation parameter α while the pattern formation system exhibits the attractor pattern 1, λ_1 decreases as a result of ΔL decaying and as a result of the increasing bifurcation parameter α. Eventually, for sufficiently large bifurcation parameters α the eigenvalue λ_1 drops out of the stability band. In the loop shown in panel (c) of Fig. 7.23 we arrive at the intersection point of the critical line and the loop in the lower half. From this intersection point a new cycle would start if we would continue with the experiment.

In panels (a) and (b) of Fig. 7.23 the possibility is indicated that the external forces affect directly the brain state as in the standard object identification model (see Chap. 6). In fact, it is plausible to assume that to some extent the external forces push the brain state into the basin of attraction of certain attractors that describe whole classes of external forces or objects and events exerting those forces. For example, when participants are confronted with planks the forces exerted by the planks are assumed to shift the brain state into the basin of attraction of some attractor related to objects that can be grasped. The forces do not shift the brain state to the basin of attraction of things on which one can stand. In contrast, when confronted with a ramp that exhibits a particular inclination angle, the forces of the ramp are assumed to shift the brain state into the basin of attraction of an attractor related to that kind of objects that allow for standing. These brain shifts may in particular be important at the very beginning of the experiment to define the initial values of the pattern amplitudes A_1 and A_2. However, once the amplitudes have approached their stationary values for a given value of α, the initial values are assumed to be set primarily by the occurring pattern (in line with the standard bistable model for hysteresis, see Fig. 6.17). Therefore, brain state shifts due to external forces do not play a crucial role—as indicated by the dashed lines in panels (a) and (b) of Fig. 7.23. However, this issue may be debated more thoroughly in future studies.

Negative Hysteresis: Simulation via LVH Model

The human pattern formation reaction model for negative hysteresis described above can be mathematized by means of the LVH model supplements with an appropriate dynamics for the structural components ΔL, λ_1, and λ_2 [122, 186, 216] (for mathematical details see also Sect. 7.8.6). Figure 7.24 shows simulation results obtained from the LVH amplitude equation model for negative hysteresis (using the model parameters from Ref. [216]).

In the simulation, the bifurcation parameter α is increased and decreased in the interval from 0 to 1. Panel (a) shows the fixed point values (i.e., stationary values) of the amplitudes A_1 and A_2 as functions of the bifurcation parameter α. Panel (b) presents the corresponding categorical states. Panels (c) and (d) present the eigenvalues λ_1 and λ_2. Finally, panel (e) shows the ΔL versus α as obtained from the numerical simulation. Panels (a) and (b) indicate that the transitions from patterns 1 to 2 and patterns 2 to 1, respectively, take place at critical values $\alpha(crit, 2)$ and $\alpha(crit, 1)$ with $\alpha(crit, 2)$ smaller than $\alpha(crit, 1)$. From the simulation we read off $\alpha(crit, 2) \approx 0.57$ (1-to-2 transition) and $\alpha(crit, 1) \approx 0.69$ (2-to-1 transition). That is, the model reproduces the phenomenon of negative hysteresis. Panels (c) and (d) show that both in the increasing and decreasing conditions, the eigenvalues λ_1 and λ_2 correspond to monotonically increasing and decreasing functions, respectively, of the bifurcation parameter (when reading the bifurcation parameter on the horizontal axis from left to right). This property reflects the impact of the external circumstances on the structure of the human pattern formation

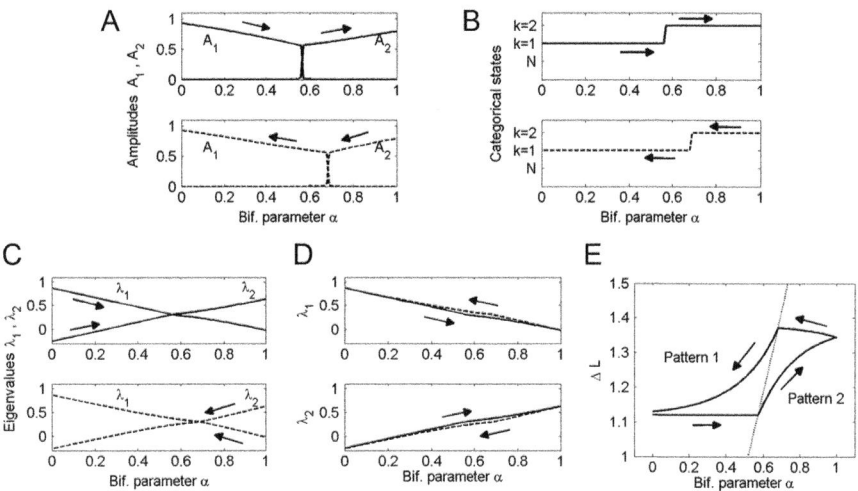

Fig. 7.24 Simulation of the negative hysteresis LVH model. Panel (**a**): Stationary pattern amplitudes A_1 and A_2 as functions of the bifurcation parameter α for increasing (top subpanel) and decreasing (bottom subpanel) conditions. Panel (**b**): Categorical brain states (or ordinary states) obtained from the stationary amplitudes shown in panel (**a**). Panels (**c**) and (**d**): Dependency of eigenvalues λ_1 and λ_2 on the bifurcation parameter and brain state. The λ_1 and λ_2 functions shown in panels (**c**) and (**d**) are the same; however, the way they are presented differs. The presentation in panel (**d**) reveals the impact of the brain state. The split of the curves in panel (**d**) is caused by the brain state and turns positive hysteresis into negative hysteresis. Panel (**e**): Loop dynamics of the negative hysteresis model

system. For example, long planks and fast moving treadmills make the eigenvalues of the one-hand grasping basis pattern and the walking basis pattern small but they make the eigenvalues of the two-hands grasping basis pattern and the running basis pattern large. In fact this dependency of the structure (eigenvalues) on the external forces (plank size, treadmill speed, etc) is the same as for the positive hysteresis model, see Sect. 6.5.3.

In the positive hysteresis model the structure only depends on the external forces, which implies that the λ_1 and λ_2 curves are identical in the increasing and decreasing conditions, see panel (a) of Fig. 6.18. In contrast, for the negative hysteresis model the structure also depends on the brain state. Therefore, as shown in panel (c) of Fig. 7.24 for the increasing (top subpanel) and decreasing (bottom subpanel) conditions the λ_1 and λ_2 curves are not identical to each other—which is indicated by the fact that the λ_1 and λ_2 graphs intersect each other at different values for α (compare the top and bottom subpanels of Fig. 7.24c). This issue becomes more clear from the presentation of the λ_1 and λ_2 functions in panel (d). The top subpanel of panel (d) shows λ_1 as function of α for the increasing condition (solid line) and decreasing condition (dashed line). Likewise, the bottom subpanel shows λ_2 as function of α for the two presentation order conditions. Following the solid lines in panel (d), we see that in the α-increasing conditions, λ_1 decreases linearly and

λ_2 increases linearly from $\alpha = 0$ to $\alpha = \alpha(crit, 2) = 0.59$. For $\alpha > \alpha(crit, 2)$ the eigenvalue λ_1 continues to decay but the decay is less dramatic due to the impact of the brain state. The brain state results in an increase of ΔL (see panel e and Fig. 7.23b), which increases λ_1 such that overall the decay of λ_1 becomes less dramatic. Likewise, λ_2 for $\alpha > \alpha(crit, 2)$ continues to increase but the increase is less dramatic due to the impact of the brain state. The increase of ΔL due to the brain state results in a damping of the increase of λ_2. Reading the subpanels in panel (d) from the right to the left, that is, following the dashed lines for the α-decreasing condition in the direction of the left-pointing arrows, we see that λ_1 increases stronger than it decreased in the α-increasing condition (compare solid and dashed lines). Likewise, λ_2 decays stronger in the α-decreasing condition than it increased in the α-increasing condition (compare solid and dashed lines). This again reflects the impact of the brain state, which is the impact of pattern $k = 2$. As long as pattern $k = 2$ is present it increases ΔL (see panel e and Fig. 7.23b), which results in increasing λ_1 and decreasing λ_2. The impact of pattern 2 stops at the transition point $\alpha(crit, 1) = 0.68$ of the 2-to-1 transition. For $\alpha < 0.68$ we see that the increase of λ_1 (see dashed line, top subpanel) as a function of α slows down slightly. Likewise, the decay of λ_2 (see dashed line, bottom subpanel) as function of α becomes less dramatic. This effect is caused by the brain state, now given by pattern $k = 1$. The pattern $k = 1$ makes that ΔL decays (see panel e and Fig. 7.23b), which makes that (1) λ_1 is affected by an self-damping force that slows down the increase of λ_1 and (2) λ_2 is no longer affected by the self-damping force and, consequently, is affected by an increasing force that makes the decay (due to the changing environmental conditions, i.e., due to the external force) less dramatic.

Panel (e) explicitly illustrates by means of numerical simulations the loop that was drawn in panel (c) of Fig. 7.23 on the basis of theoretical considerations. The loop construction makes clear that the critical bifurcation parameter $\alpha(crit, 2)$ of the 1-to-2 bifurcation is smaller than the critical bifurcation parameter $\alpha(crit, 1)$ of the 2-to-1 bifurcation.

Finally, note that as mentioned above, from the numerical simulation we obtain the critical values of $\alpha(crit, 2) \approx 0.57$ and $\alpha(crit, 1) \approx 0.69$. In fact, the critical bifurcation parameter values $\alpha(crit, 1)$ and $\alpha(crit, 2)$ can be computed from analytical expressions for the negative hysteresis LVH model, see Sect. 7.8.6. If we do so, we obtain $\alpha(crit, 2) = 0.566$ and $\alpha(crit, 1) = 0.684$, which is consistent with the results of the numerical simulation.

7.7   Summary: LVH Model and Structural Dynamics

While in Chaps. 6 and 7 applications of the human pattern formation reaction model have been discussed systematically with respect to the system classes A, B, C, D, and E, as far as the more explicit LVH model is concerned applications may be summarized with respect to characteristic features of the eigenvalue spectrum. Table 7.2 summarizes some applications of the LVH model from this point of

Table 7.2 Eigenvalue spectrum conditions of the LVH model and applications of the model

Conditions	Applications	Section
Multistable due to symmetry (homog. λ spectrum) or close to symmetry (several λ values in stability band)	Formation of BBA patterns	4.4/ 5.4.1
	(1) body movements (e.g. shoe example)	4.6.1
	(2) visual world BBA patterns (e.g., letters)	5.3/ 6.3.1
	(3) acoustic world BBA patterns (e.g., "say"/"stay")	5.3
	Cocktail party example	6.3.3
	"Selective attention"	6.3.4
Bistable (special case of multistability)	House example	6.3.3
Symmetry breaking due to external forces	"Judgments"	6.5.2
	Positive hysteresis	6.5.3
	Grasping transitions	
	Gait transitions	
	Switches of visual world BA patterns	
Symmetry breaking via state-induced restructuring	"Priming", "Retrieval-induced forgetting"	7.4.1
Symmetry breaking due to self-damping	Pattern formation sequences (e.g., child play)	7.3.3
	Pattern formation cycles (oscillatory reactions)	7.4.3
	Scene decomposition	7.4.2
Symmetry breaking due to external forces and self-damping	Negative hysteresis	7.6
Initial bias and self-damping	Overcoming "functional fixedness"	7.4.2

view. Note also that the pattern formation reaction model for negative hysteresis (Sect. 7.6.3) includes the pattern formation reaction model for positive hysteresis (Sect. 6.5.3) as special case (for details see Sect. 7.8.7).

7.8 Mathematical Notes

7.8.1 Mathematics of Type 2 Pattern Sequences in C1 Systems

Let us sketch the LVH models for the lonely speaker (Sect. 7.3.1) and child play (Sect. 7.3.3). Our starting point are the LVH amplitude equations (4.38), which we copy here as

$$\frac{d}{dt}A_k = \lambda_k A_k - g A_k \sum_{j \neq k, j=1}^{m} A_j^{q-1} - A_k^q \qquad (7.1)$$

for m unstable pattern amplitudes $k = 1, \ldots, m$ and $g \geq 1$, $q > 1$. Moreover, the structure given in terms of eigenvalues λ_k satisfies Eq. (5.11), which we copy here as

$$\frac{d}{dt}\lambda_k = f_k(A_1, \ldots, A_n) . \qquad (7.2)$$

Lonely Speaker Dynamics

For the simulations presented in Sect. 7.3.1 about the lonely speaker, we used $m = 2$ and $q = 3$ (note that any other exponent $q > 1$ could be used and that the model could be generalized to any number $m > 1$ of topics). The structure dynamics (7.2) was given by

$$\text{if } \lambda_k > 0 : \left\{ \text{if } A_k > \theta A_{k,st} \ \wedge \ A_k > S(noise) : \frac{d}{dt}\lambda_k = \gamma(\lambda_k - \lambda_0) \right\}$$

$$(7.3)$$

and

$$\text{if } 1 \ \rightarrow \ 2 \text{ transition} : \lambda_1 = \lambda_{out} < 0 . \qquad (7.4)$$

Equation (7.3) states that if the pattern k has emerged (which is described by the two "if" cases), then the corresponding eigenvalue decays towards the value λ_0 with a decay rate of γ. Note that $\theta \in [0, 1]$ is a threshold value that defines together with $A_{k,st} = \sqrt{\lambda_k}$ (for $q = 3$) when the pattern k has emerged and $S(noise) > 0$ makes sure that the pattern is above a certain noise level (where the noise is assumed to reflect deterministic perturbations provided that quantum noise can be neglected, see Sect. 2.4.4). Equation (7.4) simply states that when the transition from topic 1 to 2 takes place the fixed point for topic 1 disappears (i.e., λ_1 is put to a negative value). If the condition in Eq. (7.3) was not satisfied for a pattern k then we used

$$\frac{d}{dt}\lambda_k = 0 . \qquad (7.5)$$

In the simulation shown in Fig. 7.4 the following parameters were used: $g = 1.5$, $\lambda_1(t = 0) = \lambda_2(t = 0) = 10$, $\theta = 0.9$, $S(noise) = 0.5$, $\lambda_0 = -5$, $\lambda_{out} = -1$. The dotted lines in panel (a) of Fig. 7.4 correspond to $S(noise)$. Equations (7.1), (7.3), (7.4), (7.5) were solved numerically using an Euler forward scheme with single time step $\Delta t = 0.001$. A noise term was added to the time-discrete Euler forward scheme of Eq. (7.1) at every time step using a uniformly distributed random variable in $[0, Q]$ with $Q = 0.0001\sqrt{\Delta t}$. Note that the noise term added to the numerical simulation is not related in any way to the noise threshold $S(noise)$.

Child Play Dynamics

For the child play example discussed in Sect. 7.3.3 we used $m = 3$ and $q = 2$. The structure dynamics (7.2) was defined by [103, 104, 112]

$$\text{if } A_k > \theta A_{k,st} \; : \; \frac{d}{dt}\lambda_k = -\gamma_{down}(\lambda_k - \lambda_0) \, ,$$

$$\text{otherwise} : \frac{d}{dt}\lambda_k = -\gamma_{up}(\lambda_k - \lambda_{baseline}) \tag{7.6}$$

with $\lambda_0 < \lambda_{baseline}$. Equation (7.6) states that if the pattern k has emerged (i.e., A_k is larger by θ percent of its stationary value $A_{k,st} = \lambda_k$, note that $q = 2$), then the corresponding eigenvalue decays exponentially towards λ_0 with the decay rate $\gamma_{down} > 0$. Otherwise, the eigenvalue relaxes back to its baseline value $\lambda_{baseline}$ with a rate $\gamma_{up} > 0$. In the simulation leading to Fig. 7.6 the following parameters were used: $g = 1.5$, $\theta = 0.8$, $\lambda_{baseline} = B$, $\lambda_0 = B - \Delta$, $B = 6$, $\Delta = 3$, $\gamma_{down} = 1/\tau_{down}$, $\gamma_{up} = 1/\tau_{up}$, $\tau_{down} = 8$, $\tau_{up} = 5$. Equations (7.1) and (7.6) were solved using an Euler forward method with single time step $\Delta t = 0.001$. A noise term with a uniformly distributed random variable in $[0, Q]$ with $Q = 0.00005\sqrt{\Delta t}$ was added to the time-discrete variant of Eq. (7.1) used by the Euler forward method.

7.8.2   Scene Decomposition in C2 Systems

The pattern formation reaction model describing scene decomposition discussed in Sect. 7.4.2 is described in the reduced amplitude space by the amplitude equations (7.1) again. The external force α exerted by a given scene is assumed to set the initial brain state X at $t = t_{start}$ in terms of the pattern amplitudes A_k with $k = 1, \ldots, m$ like

$$A_k(t_{start}) = \Phi^{A_k} \, . \tag{7.7}$$

If a pattern k has emerged, then it remains for a certain period τ. Subsequently, the pattern causes its own eigenvalue to become negative (or zero), which induces a bifurcation such that another pattern k' emerges [127, 157]. Therefore, the structural dynamics is given by

$$\text{if } A_k > \theta A_{k,st} : \left\{ \text{if } c_k < 1 : \frac{\mathrm{d}}{\mathrm{d}t} c_k = \frac{1}{\tau} , \text{ otherwise } \lambda_k(t) = \lambda_{out} \right\} . \tag{7.8}$$

That is, if the pattern emerges (i.e., $A_k > \theta A_{k,st}$) then it increases a clock variable c_k such that after a period τ the clock variable becomes $c_k = 1$. At that time point the eigenvalue λ_k is shifted to a negative value λ_{out}. For all other times λ_k remains constant, that is, we have

$$\frac{\mathrm{d}}{\mathrm{d}t} \lambda_k = 0 . \tag{7.9}$$

Note that the clock variables c_k are initial all set equal to zero. Figure 7.8 was computed by solving Eqs. (7.1), (7.8), and (7.9) numerical by means of an Euler forward method (with single time step 0.01) for $m = 2$ with $k = 1$ and $k = 2$ corresponding to the house BBA pattern and the number-1 BBA pattern. We used: $\lambda_1(0) = \lambda_2(0) = 3$, $t_{start} = 1$, $\Phi^{A_1} = 0.9$, $\Phi^{A_1} = 0.2$, $g = 1.5$, $\tau = 1$, $\theta = 0.9$, $\lambda_{out} = -1$, and $q = 3 \Rightarrow A_{k,st} = \sqrt{\lambda_k}$.

7.8.3 LVH Limit Cycle Oscillators in A2, Cx, D1, D3, and Ex Systems

The oscillator dynamics of human pattern formation systems addressed in Sect. 7.4.3 is described by the same set of equations as used in the child play example. That is, the LVH limit cycle oscillator model is defined by the amplitude equations (7.1) and the structure dynamics evolution equations (7.6) for $m = 2$, which we write here for sake of completeness like

$$\frac{\mathrm{d}}{\mathrm{d}t} A_k = \lambda_k A_k - g A_k A_j^{q-1} - A_k^q \tag{7.10}$$

for $k = 1, 2$ and $k = 1 \Rightarrow j = 2$, $k = 2 \Rightarrow j = 1$ with

$$\text{if } A_k > \theta A_{k,st} : \frac{\mathrm{d}}{\mathrm{d}t} \lambda_k = -\gamma_{down}(\lambda_k - \lambda_0) ,$$

$$\text{otherwise} : \frac{\mathrm{d}}{\mathrm{d}t} \lambda_k = -\gamma_{up}(\lambda_k - \lambda_{baseline}) \tag{7.11}$$

and $\lambda_0 < \lambda_{baseline}$. Equations (7.10) and (7.11) describe a pattern formation system that bifurcates repetitively between two attractor patterns $k = 1$ and $k = 2$. In doing so, the model describes a limit cycle oscillator. Figure 7.13 was computed for $q = 3$ (note that here the cubic case was used while for child play the quadratic case was used) and $A_{k,st} = \sqrt{\lambda_k}$ with parameters $g = 1.5$, $\theta = 0.8$, $\lambda_{baseline} = B$, $\lambda_0 = B - \Delta$, $B = 6$, $\Delta = 3$, $\gamma_{down} = 1/\tau_{down}$, $\gamma_{up} = 1/\tau_{up}$, $\tau_{down} = \tau_{up} = 5$. Equations (7.10) and (7.11) were solved using an Euler forward method with single time step $\Delta t = 0.01$. A noise term with a uniformly distributed random variable in $[0, Q]$ with $Q = 0.00005\sqrt{\Delta t}$ was added to the time-discrete variant of Eq. (7.10) used by the Euler forward method.

The limit cycle oscillator defined by Eqs. (7.10) and (7.11) was introduced for C2 systems. However, it holds for any system in which a BBA or BA pattern emerging in a system acts back on the structure of its system. Therefore, the limit cycle oscillator model holds for A2 systems as well and for all C and E systems. Those systems exhibit the aforementioned required BS-to-S interaction (see Fig. 5.5) such that structural changes that lead to type 2 pattern sequences can be caused by a direct path (see Fig. 7.1). Moreover, the oscillator model holds for systems producing type 2 pattern sequences in which structure is changed by means of an indirect path (see Fig. 7.1 again). Therefore, the oscillator model given by Eqs. (7.10) and (7.11) also applies to certain D and E systems in which the structural changes leading to an oscillatory instability are not established by a direct path (i.e., BBA pattern acts on structure directly) but by means of an indirect path. In particular, for D1 and D3 systems a direct path is not possible. As far as E1 and E3 systems are concerned, there might be oscillatory E1 and E3 systems for which the BS-to-S interaction (direct path) does not contribute to the oscillatory dynamics of the systems (but is involved in something else), while the oscillatory dynamics is established by means of the BS-to-F-to-S interaction chain (i.e., an indirect path).

7.8.4 Mathematics of Further C2 and C3 Systems

The mathematical details of the pattern formation reaction model for state-induced restructuring (i.e., "priming") discussed in Sect. 7.4.1 are described in detail in Refs. [90, 98]. Mathematical details of the model for "retrieval-induced forgetting" (see Sect. 7.4.1 again) can be found in Ref. [98]. For the mathematical details to describe "functional fixedness" as in Sect. 7.4.2 the reader is referred to Ref. [106].

7.8.5 Mathematical Considerations on D Systems

Searching for Something to Drink

The type 2 pattern sequence shown in Fig. 7.17 about searching for something to drink was computed from the standard LVH model defined by Eq. (7.1) for the cubic case ($q = 3$). It was assumed that thirstiness β (as measured by a relative variable in the range from 0 to 100%) increases according to a Lotka-Volterra law like

$$\text{if pattern } k = 2 \text{ not present} : \quad \frac{\mathrm{d}}{\mathrm{d}t}\beta = \gamma_\beta \beta (1 - \beta) ,$$

$$\text{if pattern } k = 2 \text{ present for 1 time unit: } \beta = 0 , \lambda_2 = \lambda_{out} < 0 . \quad (7.12)$$

That is, as long as the person does not drink something, the degree of thirstiness increases towards the fixed point value of 100%. If the person takes a drink then after 1 time unit, the person stops being thirsty. Consequently, the eigenvalue λ_2 is put to a negative value λ_{out} such that the drinking fixed point pattern disappears. Furthermore, it was assumed that the eigenvalues λ_1 and λ_2 of the searching BBA attractor pattern and the drinking pattern satisfy the following structural dynamics

$$\text{if drink not found} : \quad \lambda_1(t) = 100\beta(t) ,$$

$$\text{if drink found} : \quad \lambda_1 = \lambda_{out} < 0 , \quad \lambda_2 = \lambda_{priority} , \quad (7.13)$$

where $\lambda_{priority}$ is a relatively large positive value such that the system becomes monostable. Equations (7.1), (7.12), and (7.13) have been solved numerically using an Euler forward scheme with time step $\Delta t = 0.001$. The following parameters and initial conditions were used: $g = 1.5$, $\lambda_3 = 50$ (constant), $\beta(0) = 0.1$, $\lambda_{priority} = 100$, $\lambda_{out} = -1$, $A_1(0) = A_2(0) = 0.1$ and $A_3(0) = 0.9A_{3,st}$. In order to obtain categorical states (for the if conditions in Eqs. (7.12) and (7.13)) a threshold of $\theta = 0.9$ was used. A noise term was added with uniformly distributed random numbers in $[0, Q]$ and $Q = 0.0001\sqrt{\Delta t}$.

7.8.6 Mathematics of the Human Pattern Formation Reaction Model for Negative Hysteresis

The model for negative hysteresis involves a time-continuous scale t and a time-discrete time scale. Pattern amplitudes evolve as defined by the LVH amplitude equations (7.1) on the continuous time scale. In contrast, it is assumed that humans encounter with the environment in discrete events. For example, in laboratory experiment, participants are exposed in discrete steps to planks [269] or sounds [325]. This discrete time scale of events will be denoted by n. The amplitude dynamics is assumed to evolve much faster than the discrete events evolve in time. Therefore, at every time step n the amplitude dynamics converges to a fixed point. The structure is assumed to evolve on the slow time scale n on which the discrete events enfold. As discussed in Sect. 7.6, negative hysteresis can be understood as phenomenon resulting from the impact of (1) the external force—just as in positive hysteresis paradigms and (2) a self-damping—just as in the oscillatory human pattern formation model. In order to take both aspects (1) and (2) into account it has been suggested that the eigenvalues λ_1 and λ_2 evolve like [122, 216]

$$\lambda_1(n) = L_1(n) - \alpha(n) , \quad \lambda_2(n) = L_2(n) + \alpha(n) \tag{7.14}$$

where L_1 and L_2 are offset parameters that evolve like

$$L_k(n+1) = L_k(n) - \frac{1}{T}\left(L_k(n) - [L_{k,0} - s_k(n)]\right) ,$$

$$s_k(n) = 0 \text{ if } A_k(t_n) \approx 0 \text{ i.e., pattern is absent} ,$$

$$s_k(n) = h > 0 \text{ if } A_k(t_n) \approx A_{k,st} \text{ i.e., pattern has emerged} . \tag{7.15}$$

In Eq. (7.15) we typically put $L_{1,0} = 1$, while $L_{2,0}$ is a free parameter. Equation (7.14) accounts for the impact of external forces as described by the scalar variable α that also acts as bifurcation parameter. If the bifurcation parameter α is sufficiently small, then λ_1 is large with respect to λ_2 such that for appropriate conditions the system is monostable and features only the attractor $k = 1$. In contrast, if α is sufficiently large, we have $\lambda_2 \gg \lambda_1$ and the system is monostable and exhibits only the attractor $k = 2$. That is, for fixed parameters L_1 and L_2, Eq. (7.14) just describe the basic positive hysteresis model discussed in Chap. 6. However, Eq. (7.15) states that the offset parameters L_1 and L_2 are not fixed. Rather, they evolve in time from event to event. Over a sequence of events they converge to the term $L_{k,0} - s_k$, when assuming s_k would be fixed. s_k in turn depends on the pattern amplitude A_k. If A_k is sufficiently large such that we can say the pattern k has emerged then s_k becomes large. In doing so, the fixed point for L_k is reduced, which in turn means that λ_k is relative small as compared to λ_j with $j \neq k$. This corresponds to the aforementioned self-damping mechanism of patterns. The critical values $\alpha(crit, 1)$ and $\alpha(crit, 2)$ for the decreasing and increasing conditions can be computed from the model parameters g, $L_{2,0}$ and h (assuming $L_{1,0} = 1$). We obtain [216]

$$\alpha(crit, 1) = \frac{1 - g(L_{2,0} - h)}{1 + g} , \quad \alpha(crit, 2) = \frac{g(1 - h) - L_{2,0}}{1 + g} . \tag{7.16}$$

Equation (7.1) with $m = 2$ together with the structure dynamics given by Eqs. (7.14) and (7.15) define the human pattern formation model for negative hysteresis. However, for didactical purposes it is convenient to define $\Delta L(n) = L_1(n) - L_2(n)$. From Eq. (7.15) it follows that ΔL evolves like [122]

$$\Delta L(n+1) - V = (1 - 1/T)(\Delta L(n) - (L_{1,0} - L_{2,0}) - V) ,$$

$$V = h \text{ if } k = 1 \text{ is present} ; \text{ otherwise } V = -h . \tag{7.17}$$

Therefore, ΔL increases when the pattern $k = 1$ is present and decreases when the pattern $k = 2$ is present. This property was used in the main text above, see Sect. 7.6.

Figure 7.24 was obtained by increasing and decreasing the bifurcation parameter α in discrete steps of $\Delta \alpha = 0.01$ at the discrete presentation events $n = 1, 2, 3, \ldots$. For any given $\alpha(n)$ the offset parameters $L_1(n)$ and $L_2(n)$ and the auxiliary variable $\Delta L(n)$ were computed. From $L_1(n)$ and $L_2(n)$ the eigenvalues $\lambda_1(n)$ and $\lambda_2(n)$ were obtained. Subsequently, the amplitudes $A_1(t)$ and $A_2(t)$ were computed using the LVH amplitude equations (7.1). The following parameters taken from Ref. [216] were used: $g = 1.04$, $L_{2,0} = -0.25$ (and $L_{1,0} = 1$ as mentioned above), $h = 0.13$, $T = 22$, and $q = 3$. Note that a better characterization of the time constant T is to express the value of T in terms of $\Delta \alpha$ [216]. In this sense, a value of $T = 0.22$ in units of α increments was used in the simulation.[7] Initial conditions for $n = 1$: $A_1(0) = 0.1, A_2(0) = 0.1, L_1(1) = 1 - h, L_2(1) = L_{2,0}$. Equation (7.1) was solved using an Euler forward method with single time step $\Delta t = 0.01$ on an interval from $t = 0$ to $t = 600$ time units. The usual noise term at every time step was added with uniformly distributed random noise in $[0, Q]$ with $Q = 0.001\sqrt{\Delta t}$. As far as Eq. (7.15) was concerned, a pattern k was assumed to be present if its amplitude A_k reached at least θ percent of its finite fixed point value $A_{k,st} = \sqrt{\lambda_k}$. A value $\theta = 0.9$ was used.

7.8.7   Positive and Negative Hysteresis Models

The LVH model for negative hysteresis defined in Sect. 7.8.6 contains a special case the LVH model described in Sect. 6.6.4 for positive hysteresis. Recall, the positive hysteresis LVH model is defined by Eqs. (6.16) and (6.17). The negative hysteresis model is defined by Eqs. (7.1) with $m = 2$ and Eqs. (7.14), (7.15). If we put $h = 0$ into Eq. (7.15), then this implies that $s_k(n) = 0$ for $k = 1, 2$ and any $n \geq 1$, which in turn implies that the structural dynamics defined by Eq. (7.15) becomes stationary and independent of the brain state. We obtain $L_1 = 1$ (or in more general case $L_1 = L_{1,0}$) and $L_2 = L_{2,0}$. Substituting these time-independent values into Eq. (7.14) we see that Eqs. (7.1) for $m = 2$ and $q = 3$ and (7.14) become identical to Eqs. (6.16) and (6.17).

In particular, the critical values defined by Eq. (7.16) of the negative hysteresis model become identical to the critical values defined by Eq. (6.19) for the positive hysteresis model if we put $h = 0$.

---

[7]E.g., if we would change the α increments from 0.01 to 0.001 then we should increase T from 22 to 220 in order to obtain the same dynamics.

Chapter 8
Pattern Formation and Continuous Reactions

In Chaps. 6 and 7 primarily qualitative aspects about the emergence of patterns in human pattern formation systems were discussed. For example, individuals performing different activities in succession have been described in terms of attractor pattern sequences. In contrast, in this chapter quantitative aspects of pattern formation will be discussed and will be related to continuous reactions of humans. That is, considerations will be made about the properties of attractor patterns measured in terms of continuous variables. Examples for continuous variables are reading times, reaction times, magnitude "estimates", biophysical quantities such as oxygen consumption, and ratings on Likert scales.[1] Research conducted on humans (and animals) frequently uses observables measured on continuous (or discrete numerical) scales. While in Sect. 7.4.1 on state-induced restructuring, we actually explained a continuous observable, the reading time, on the basis of a quantitative aspect of pattern formation, namely, the time it takes for a pattern to emerge, in this chapter a broader point of view will be taken. In this chapter, it will be discussed how observations made in general in studies involving continuous observables can be explained from a physics perspective within the framework of pattern formation.

8.1 Experimental Designs in Human Subjects Research: A Physics Perspective

8.1.1 Experimental Designs

There are various types of experimental designs to test human beings [142, 183]. Let us briefly introduce three of them. Figure 8.1 provides an overview about the kind

---

[1] Although strictly speaking, Likert scales are discrete scales, what is at issue here is that variables on Likert scales do not correspond to categorical variables but to numerical variables.

© Springer Nature Switzerland AG 2019
T. Frank, *Determinism and Self-Organization of Human Perception and Performance*, Springer Series in Synergetics,
https://doi.org/10.1007/978-3-030-28821-1_8

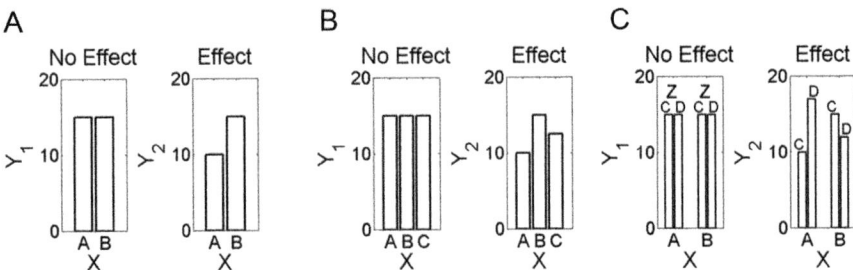

Fig. 8.1 Outcomes that can be observed in three different experimental designs. Panel (**a**): Two-levels single factor design. Panel (**b**): Three-levels single factor design. Panel (**c**): Two-by-two factorial design. For each design on the left-hand-side a situation is shown, in which there is no cause-and-effect relationship between the manipulation X and the observable Y_1. On the right-hand-side an example is given for which there is a cause-and-effect relationship between X and Y_2

of experimental results that can be obtained in those experimental designs. First, let us consider an experimental design that involves two conditions A and B. They are described by a variable: the independent variable X. The independent variable is also called factor. The independent variable or factor assumes two values $X = A$ and $X = B$ describing the two conditions. For example, in Sect. 7.4.1 on state-induced restructuring (i.e., "priming") participants were tested on words that they have previously read (condition A) and new words (condition B). We may denote the independent variable X as the novelty character of words and the two levels are $X =$ "previously read" and $X =$ "new". We will refer to the design involving the conditions A and B as single factor two-levels experimental design because the conditions are described by a single factor (independent variable) and the factor involves two levels.

Let Y denote the observable that is measured. Y is also called the dependent variable because researchers typically hypothesize that it depends on the conditions tested in their experiments. Participants are tested under conditions A and B and the observable Y is measured. Typically, mean values are reported. Panel (a) of Fig. 8.1 exemplifies for two cases mean values obtained in such a design. In the first case shown on the left the observable Y_1 is measured. Y_1 in both conditions assumes the same mean value. This outcome is consistent with the hypothesis that the variable X does not affect the observable. The second case is shown on the right and involves an observable Y_2. The mean values of Y_2 differ across the two conditions. This suggests that the conditions A and B have an effect on the observable. In fact, mathematical tests can be performed (e.g., so-called t tests) to support the hypothesis that the change in Y_2 is due to the difference between the conditions. In particular, if the variability among the scores entering in the mean values is negligibly small, then the hypothesis that the independent variable X affects the dependent variable Y_2 is supported. The variable X is considered as cause. The variable Y_2 describes the effect on the human system caused by X.

The single factor two-levels experimental design is a special case of a single factor N-levels experimental design. In the N-levels design N different conditions are test. Accordingly, in a single factor three-levels experimental design the conditions are described by the independent variable X that assumes three values: $X = A, B, C$. Participants are tested under those conditions and a variable (dependent variable) Y is measured. Possible outcomes with respect to the mean values thus obtained are shown in panel (b) of Fig. 8.1 for two exemplary cases. In the first case shown on the left about the observable Y_1 all mean values of Y_1 assume the same value. This outcome is consistent with the hypothesis that the variable X does not affect the observable Y_1. In the second case shown on the right about the observable Y_2 the mean values of Y_2 differ across the conditions. This suggests that the variations in Y_2 are due to the impact of the conditions described by X. In particular, if the variability among the scores within the three samples of scores is negligibly small, then—on the basis of standard hypothesis testing procedures—the conclusion can be drawn that the variable X affects the human system such that variations in X cause variations in the human reactions measured in terms of Y_2.

Finally, individuals can be tested under different types of conditions that can be varied more or less simultaneously. For example, as mentioned in Sect. 1.8.1, psychophysics experiments have tested human reactions to the weight of objects that can be picked up. A researcher may ask the question whether the size and color of objects affects the reactions about weight. For example, the following question may be asked: Do humans react to bigger and smaller objects of the same weight in the same way or do humans react to bigger objects as if they were heavier than smaller objects? Accordingly, participants pick up objects that all have the same weight but vary in size and report (e.g., verbally) the weight of the objects. That is, participants form BBA attractor patterns of picking up objects. The emerging BBA attractor patterns come with supplementary components that reflect the reports or trigger the emergence of other BBA attractor patterns reflecting the reporting activities. Object size is one condition varied in the experiment. Another condition would be color. Let us consider only gray shades. Let us assume that the objects come in two gray shades: in light gray and dark black. The color condition can be combined with the size condition. In doing so, four conditions are obtained. That is, each object comes in four variants as a small light gray object, a small dark black object, a large light gray object, and a large dark black object. A design with two separate types of manipulations (i.e., two factors or two separate sets of conditions), where each manipulation involves two conditions (i.e., two levels), is called a 2-by-2 experimental design. Let X and Z denote the two factors (e.g. size and color). Let $X = A, B$ denote the two conditions described by the first factor (e.g., being small or large). Likewise, let $Z = C, D$ denote the two conditions described by the second factor (e.g., being shaded in light gray or in dark black). Let Y denote the observable (e.g., reported weight), which from a physics perspective is a part or component of an emerging BBA attractor pattern (e.g., the BBA attractor pattern of lifting an object). Panel (c) describes possible outcomes that can be observed in such 2-by-2 experimental designs. Two cases are considered. In the case shown on the left, all mean values of the observable Y_1 assume the same value. This observation is consistent with the hypothesis that there is no effect of the factors X and Z on the human system with respect to the observable Y_1. On the right hypothetical mean values observed for a measure Y_2 are shown. The mean values do not assume all the same value. This suggests that the reactions of the humans are somehow affected by the variables X and Z. In particular, in statistics, one can distinguish between main and interaction effects of the variables X and Z on the measured variable Y_2. (Depending on the reader's brain state and structure and the way this book interacts with the reader, the reader will or will not consult standard textbooks on statistics and hypothesis testing in this regard; the reader will do so without having a "choice" or making a "decision".)

8.1.2   Physics Perspective: Illustration for Stretching Rubber Strings

From a physics perspective, there are laws that determine the relationships between dependent variables being measured and the independent variables given in terms of external forces acting on the test persons and animals, that is, on the variables describing the structure and the brain states of the humans and animals under consideration. These laws state whether or not a dependent variable Y depends on an independent variable X and, if Y does depend on X, in which way Y depends on X. This physics perspective holds for systems of the inanimate world just as for systems of the animate world. Therefore, let us illustrate the physics perspective of outcomes as described in panels (a)–(c) of Fig. 8.1 for a non-living system: a rubber string.

Let us consider two rubber materials with different structures. A soft rubber that can easily be stretched and a hard rubber that requires a lot of forces to be stretched. Panel (a) of Fig. 8.2 illustrates schematically the laws that determine the two rubber materials. More precisely, panel (a) of Fig. 8.2 shows the relative elongation of two pieces of rubber as functions of the applied pulling force. If we stretch or pull with a certain force at a piece of rubber then the length of the piece changes. The relative elongation is the increase of length divided by the length of the piece for zero forces. That is, it is the elongation in percent. For zero forces, the relative elongation (i.e., the elongation in percent) is zero. In panel (a), for sake of simplicity, it is assumed that for both rubber materials linear relationships hold between the applied force and the resulting relative elongations. Such linear relationships are called the Hooke law of elastic materials. For the soft rubber material the length and, consequently, the relative elongation increases rapidly when the force is increased, see panel (a). In contrast, for the hard rubber material the increase as a function of the applied force is moderate. Panel (a) also illustrates the basic physics perspective of all systems (belonging to the animate or inanimate worlds). Accordingly, systems are subjected to forces, and depending on the structure of the systems, systems react to those forces in certain ways. In our example, the applied force to the rubber is the external

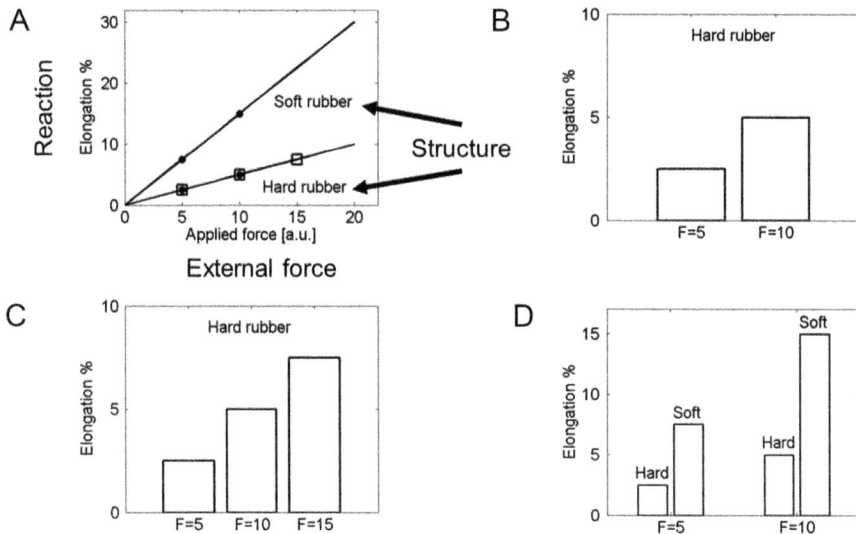

Fig. 8.2 Laws (panel (**a**)) determining how rubber reacts to applied forces and experiments (panels (**b**)–(**d**)) designed to uncover those laws

force. The rubber material is the structure. The elongations of the pieces of rubber are the reactions of the systems.

From the laws demonstrated in panel (a) we can derive the outcomes of experiments that feature certain designs. Panels (b)–(d) of Fig. 8.2 demonstrate outcomes that would be obtained from the laws shown in panel (a) in experimental designs as described in panels (a)–(c) of Fig. 8.1. For example, let us assume the hard rubber material is tested in a single factor two-levels design. The applied force is the factor or independent variable. The force levels 5 and 10 are used. From the lower graph in panel (a) we can read off the elongation of the hard rubber material for those force levels. The experiment is repeated with several pieces of the same type of hard rubber. Assuming that the rubber was produced with a high quality, then all pieces satisfy the same graph shown in panel (a). Consequently, the variation among the scores obtained from the pieces can be neglected and the sample mean values obtained for the two force levels correspond to the values that can be read off from the graph in panel (a). These values are shown in panel (b). Since we assuming that for the high quality rubber pieces under consideration the variability across the pieces can be neglected, from the difference between the mean values shown in panel (b) we would conclude that the applied force has an effect on the length of the rubber. This conclusion is consistent with the law determining the reaction of the hard rubber to stretching forces. Rather than using only two force levels, we could consider an arbitrary number of force levels. Panel (c) demonstrate an single factor three-levels experimental design to test properties of the hard rubber. In this experiment, the force levels 5, 10 and 15 are tested. From the lower graph in panel (a) the relative elongation can be determined for those force levels. The values (as argued above) are assumed to correspond to the mean values that one would obtain for a sample of pieces of the same type of hard rubber. On the basis of the observed differences between the mean values shown in panel (c) (and under the aforementioned assumption of an idealized situation in which variations among the scores within each force level can be neglected) we would conclude that there is a cause-and-effect relationship: the applied force affects the length of the hard rubber material.

Finally, the soft and the hard rubber materials may be tested in a 2-by-2 design. One factor is the applied force. The force levels 5 and 10 are used. The other factor is the rubber material. The soft and hard rubber materials are used. From the graphs shown in panel (a), the sample mean values obtained in such a 2-by-2 design can be obtained and are shown in panel (d) of Fig. 8.2. The mean values differ from each other suggesting that there are some effects related to the two factors (external force and structure) of the experimental design. Given the (assumed) negligible variability among scores obtained from several pieces of the same types of hard and soft rubber tested under the same conditions, the mean values shown in panel (d) would be taken as empirical support for the hypothesis that applied forces and the rubber structure determine the elastic reactions of rubber materials.

8.2 Continuous Observables of Pattern Formation Systems

8.2.1 Continuous Observables

From a physics perspective, there are laws that describe how continuous observables of systems depend on the conditions and properties of the systems. Let us assume that the conditions are described in terms of continuous variables. In particular, let us assume that we are dealing with experiments on human subjects and animals that involve both continuous observables (i.e., continuous dependent variables) and continuous independent variables.[2] The values that the observables assume under varying conditions are described graphically as functions. These functions can either be continuous functions or exhibit jumps and kinks. Figure 8.3 illustrates some of those functional relationships between observables and conditions. As discussed in Sect. 3.2, jumps and kinks are characteristic features of bifurcations and phase transitions. Panels (a) and (b) (which have been shown earlier in Fig. 3.5) provide two examples of two functions that exhibiting kinks related to bifurcations of the systems. The first example shows the roll velocity of rolls emerging in the Benard experiment when a fluid layer is heated from below as a function of the temperature difference between the bottom and top of the layer. At a temperature difference of about 4.5 °C the roll velocity becomes different from zero. The reason for this is

---

[2]Much of what will be presented in what follows also applies to experiments that involve categorical independent variables (for an example see Sect. 10.6.2).

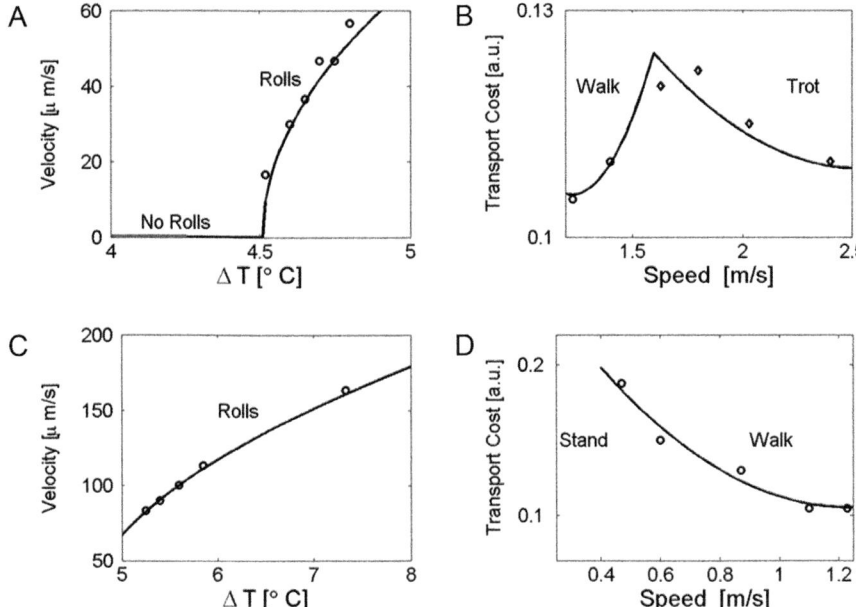

Fig. 8.3 Examples of non-smooth (panels (**a**) and (**b**)) and smooth (panels (**c**) and (**d**)) relationships between experimental conditions and observables

that at that value roll patterns emerge. That is, a bifurcation takes place from the spatially homogeneous layer to a layer organized in rotating rolls. At that critical temperature difference the roll velocity as function of the temperature difference exhibits a kink. The second example shown in panel (b) shows the transportation costs (as measured in oxygen consumption) of miniature horses as function of the locomotion speed of the horses. The solid lines are interpolation lines, while the circles are experimental data. The interpolation lines suggest that there is a critical locomotion speed at which the transportation costs graph exhibits a kink in terms of a sharp peak. The kink indicates that there is a transition between two gait patterns: walk and trot. Panels (c) and (d) show that within the stability domain of an attractor pattern, observables change in a continuous fashion as functions of the varying conditions. Panel (c) shows the increase of the roll velocity in the Benard experiment in the parameter domain of temperature differences for which roll patterns are stable. Roll velocity is a smooth, continuous function of the temperature difference. Panel (d) shows the decrease of the transportation costs when walking speed of miniature horses is increased in the range of relatively low speeds. Again, the graph is a smooth function.

In panels (c) and (d) monotonically increasing and decreasing graphs are shown. Observables not necessarily yield monotonically functions within the parameter domains for which patterns or states are stable. Observables may show a non-monotonic dependency on the conditions. For example, the density of water as a function of the temperature is a non-monotonic graph. This graph was earlier discussed in Sect. 3.2.3 (see Fig. 3.5a). The graph corresponds to an inverted U shaped function in the temperature range from 0 to 100 °C with a peak at about 4 °C. Consequently, let us assume we put a piece of ice in a room with a temperature slightly above 0 °C (e.g., 0.1 °C) and wait until all ice has turned into water. Subsequently, we increase the temperature by a few degrees Celsius. Then what happens is that in the temperature range from 0 to 4 °C the water molecules become more densely packed when temperature is increased (see the part to the right of the 0 °C mark in panel a of Fig. 3.5). Therefore, the density increases. At about 4 °C water has the highest density. If the temperature is increased above 4 °C, then in the interval from 4 to 100 °C water molecules become less and less densely packed. That is, the density decays monotonically as a function of temperature. As a result, the density of water assumes an inverted U-shaped function in the temperature range from 0 to 100 °C with a peak at 4 °C.[3] Another example is given by the graphs of transportation costs drawn in panels (b) and (d) of Fig. 8.3. If we put panels (b) and (d) together we obtain the original figure shown in Sect. 3.2.3 (see Fig. 3.5d). When horses walk at relatively low speeds and walking speed is increased then transportation costs decrease until they reach a minimum. Subsequently, the costs increase. The minimum does not describe a kink. There is no bifurcation taking place. Likewise, the peak of the density of water at 4 °C does not correspond to a phase transition point.

8.2.2   Pattern Amplitudes

As mentioned in the introduction to this chapter, a key feature of the framework of pattern formation and synergetics is that pattern amplitudes provide sufficient descriptions of pattern formation systems and in general determine all aspects of pattern formation systems. Consequently, observables of pattern formation systems are functions of pattern amplitudes. Therefore, let us briefly recapitulate how pattern amplitudes depend on system conditions (e.g., external forces) as described by bifurcation parameters.

Figure 8.4 exemplifies some fundamental relationships. In the examples shown in Fig. 8.4 fixed point values and stationary pattern amplitudes either increase or decrease monotonically as functions of bifurcation parameters. For a generic class of dynamical systems, close to bifurcation points at which attractor patterns or stable fixed points emerge the amplitudes of the emerging patterns or the fixed point values of the emerging fixed points increase as functions of the respective bifurcation parameters. Panel (a) shows as an example the pitchfork bifurcation discussed in Sect. 3.5.2. When the bifurcation parameter α is increased beyond its critical value

---

[3] Note that this inverted U-shaped character is not visible in Fig. 3.5a due to the relative large scale of the vertical axis.

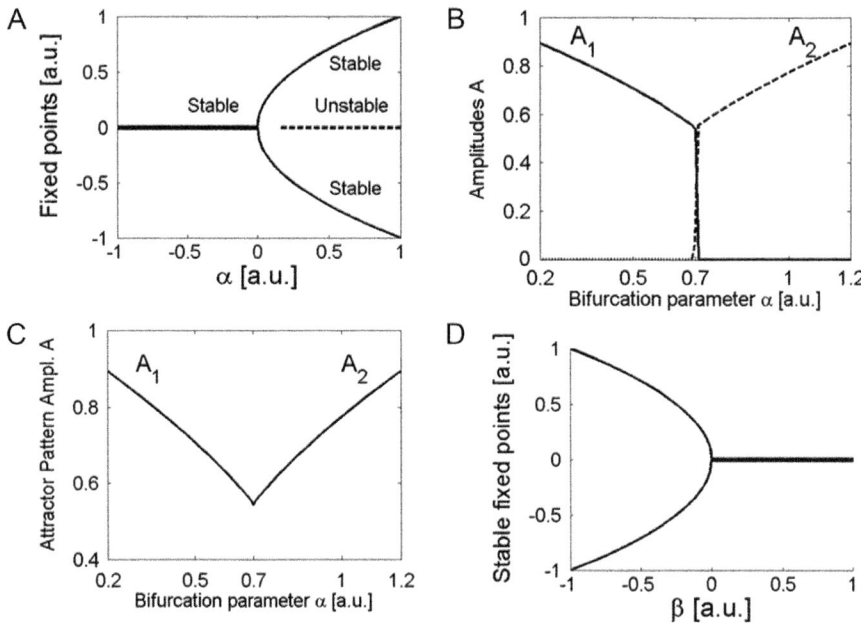

Fig. 8.4 Fixed points and pattern amplitudes as monotonically increasing and decreasing functions of bifurcation parameters. Panel (**a**): Pitchfork bifurcation fixed points increasing in magnitude. Panels (**b**) and (**c**): Amplitudes A_1 and A_2 of the LVH model decreasing (A_1) and increasing (A_2) when the bifurcation parameter α is increased. Panel (**d**): Fixed points of the inverse pitchfork bifurcation decay in magnitude as function of β (only stable fixed points are shown)

of $\alpha(crit) = 0$, then the fixed point values of the emerging fixed points increase in magnitude monotonically as functions of α. Panel (b) shows the amplitude functions computed from the LVH amplitude equations model for two amplitudes A_1 and A_2 and $g = 1$ (as discussed in Sect. 6.5.1 in the context of the experiments by Warren [343] on human reactions to stair steps). At a critical value of $\alpha(crit) = 0.7$, when increasing α beyond 0.7, the fixed point of the first unstable basis pattern becomes unstable, while the fixed point of the second unstable basis pattern becomes stable. Consequently, the amplitude A_1 of the first unstable basis pattern drops to zero, while the amplitude A_2 of the second unstable basis pattern jumps from zero to a finite fixed point value. When we plot only the amplitudes of the stable attractor patterns, we arrive at the graph shown in panel (c). Importantly, as shown in panels (b) and (c) the amplitude A_2 increases monotonically as a function of α. In general, that is, for all kind of amplitude equations (including those of the LVH model), when the fixed point values of pattern amplitudes are increasing functions of eigenvalues and eigenvalues increase as functions of bifurcation parameters, then pattern amplitudes are monotonically increasing functions of bifurcation parameters.

What has been said above with respect to amplitudes that monotonically increase as functions of bifurcation parameters can be reversed. For the pitchfork bifurcation shown in panel (a) we see that finite fixed point values for α larger than zero decay in magnitude if the bifurcation parameter α is decreased. Therefore, let us define the parameter β as the negative of the parameter α like $\beta = -\alpha$ and let us consider β as bifurcation parameter. In doing so, the bifurcation diagram shown in panel (d) of an inverse pitchfork bifurcation [220] is obtained. Increasing β from negative values towards zero means that the underlying dynamical system approaches the bifurcation point at which the stable fixed points with finite values disappear. In this interval of the bifurcation parameter (i.e., in the interval $(-\infty, 0]$), the fixed point values decay monotonically in magnitude. Likewise, the amplitude A_1 shown in panels (b) and (c) describes a monotonically decaying function of the bifurcation parameter α. This is due to the fact that when the bifurcation parameter α is increased, then the pattern formation system under consideration approaches the bifurcation point at which the attractor $k = 1$ becomes unstable. In general, if (1) amplitudes of patterns XYZ are monotonically increasing functions of eigenvalues and (2) eigenvalues approach lower boundaries of appropriately defined stability bands when a bifurcation parameter α is increased because the bifurcation parameter α decreases the eigenvalues,[4] then the XYZ pattern amplitudes decay monotonically as functions of the bifurcation parameter α when α approaches the critical value at with the XYZ patterns become unstable.

8.3   Pattern Formation Perspective of Experiments in Human Subjects Research

In this section, the considerations made in Sects. 8.1 and 8.2 are combined. In doing so, the human pattern formation reaction model can be used to explain outcomes observed in experiments involving human subjects and animals.

8.3.1   Limit Cycle Oscillator Example

The relationship between outcomes of experimental research designs and physical laws has been illustrated by means of the rubber stretching example in Sect. 8.1.2, see Fig. 8.2. Let us introduce another example in this regard that makes closer contact to the notion of a pattern amplitude. Let us consider a limit cycle oscillator. Explicitly, we will consider the limit cycle oscillator addressed in Sect. 3.5.2. This oscillator is a so-called canonical-dissipative oscillator [71, 77, 92]. However, any limit cycle oscillator satisfying the so-called Hopf bifurcation form (with a real-valued parameter of the nonlinear term of this Hopf bifurcation form) could be

---

[4] And not for other reasons; e.g., the lower boundary values are increased as in the retrieval-induced "forgetting" model, see Sect. 7.4.1.

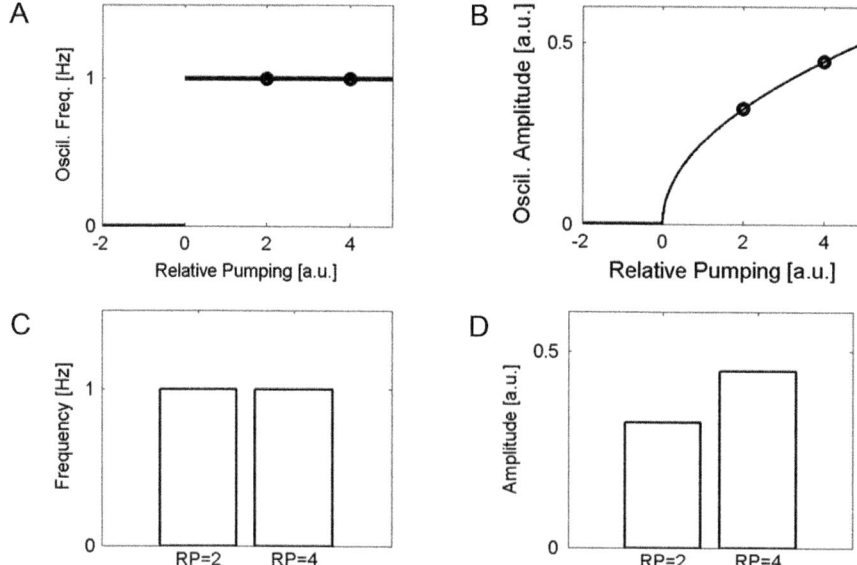

Fig. 8.5 Limit cycle oscillator example to illustrate the connection between outcomes of experimental designs and physical laws. Panels (**a**) and (**b**): Physical laws of the so-called canonical-dissipative oscillator describing oscillation frequency (panel (**a**)) and amplitude (panel (**b**)) as functions of relative pumping. Panels (**c**) and (**d**): Hypothetical outcomes of single factor two-levels experiments testing oscillation frequency (panel (**c**)) and oscillation amplitude (panel (**d**)) for two different pumping conditions (RP: relative pumping)

used as well. Figure 8.5 demonstrates properties of the oscillator as functions of the bifurcation parameter, which is the relative pumping.

As discussed in Sect. 3.5.2, when the relative pumping of the canonical-dissipative oscillator becomes positive, a limit cycle attractor emerges. The oscillator starts to oscillate with a finite amplitude and a particular oscillation frequency. The oscillation frequency as function of the relative pumping parameter is shown in panel (a) of Fig. 8.5. The frequency does not vary with the pumping parameter. This is a special feature of the canonical-dissipative limit cycle oscillator and a general feature of all Hopf bifurcation limit cycle oscillators with real-valued nonlinearity parameter. As shown in panel (b) of Fig. 8.5 (see also Fig. 3.15 in Sect. 3.5.2), the oscillation amplitude increases monotonically as a function of relative pumping. The oscillation amplitude can be regarded as attractor pattern amplitude. Let us consider an experiment in which the oscillator is tested for two conditions that differ in the degree of relative pumping. In one condition, the pumping is at a level of 2 units. In the other condition, the oscillator is pumped at the level of 4 units. Repeating the experiment several times with the same oscillator (or with different devices of the same type) and assuming the variability between repetitions (between the devices) is negligibly small, we would obtain mean values that correspond to the values obtained from a single trial. In particular, these mean values correspond to the function values highlighted in panels (a) and (b). For the oscillator frequency we would obtain the mean values shown in panel (c). Not surprisingly, they assume the same value. For the oscillator amplitude we would obtain the mean values shown in panel (d). The oscillator amplitude on average is larger if the oscillator is pumped at a higher value. This example illustrates how outcomes that differ (panel d) or do not differ (panel c) from a physics perspective are related to the fundamental laws determining the dynamics and reactions of physical systems.

8.3.2   Differences Within and Between Pattern Categories

When experimental conditions are varied (i.e., several values or levels of indepen-dent variables are tested) and the variations affect an observable (i.e., a dependent variable), we can distinguish between two basic scenarios. In the the first scenario, the switches between the conditions do not induce bifurcations. The same kind of pattern (e.g., walking pattern of a human) is present for all conditions. However, the amplitude of the pattern varies across the experimental conditions and those amplitude variations lead to variations of the measured observable. The experimen-tal conditions result in quantitative differences with respect to the same kind of attractor pattern. Therefore, *according to the first scenario, observed differences reflect differences within a pattern category* of a pattern formation system under consideration.

   In the second scenario, the switches between the conditions induce bifurcations. For sake of simplicity, let us consider only a single factor two-levels design. There are two conditions. Switching from one condition to the other induces a bifurcation such that the pattern formation system exhibits different attractor patterns in the two conditions under consideration. In general, the amplitudes of the attractor patterns differ in magnitude. In addition, the relationships between those amplitudes and the observable at hand may not be identical. According to the second scenario, the experiment involves different patterns categories. Again, the emerging patterns of the two categories and their corresponding amplitudes are assumed to determine all characteristic features of the system under consideration and, consequently, the observable (dependent variable) measured under the two conditions. *According to the second scenario, observed differences reflect differences between different pattern categories.*

8.3.3   On a Fundamental D0/D1 System Mechanism: IV-DV Relationships

Within the framework of the theory of pattern formation the relationship between dependent and independent variables is determined by three mapping. The first

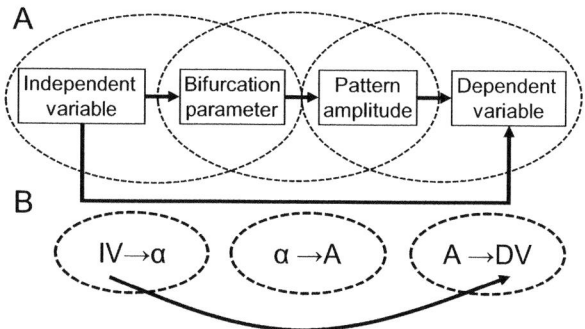

Fig. 8.6 Three fundamental mappings determining how dependent variables measured in experiments on humans and animals (when considered as pattern formation systems) depend on independent variables (IV: independent variable; α: bifurcation parameter; A: pattern amplitude; DV: dependent variable)

mapping describes how the bifurcation parameter α depends on the independent variable (IV): IV $\rightarrow \alpha$. The second mapping describes how the pattern amplitude A (here: the fixed point value A_{st}) of the relevant BBA attractor pattern[5] depends on the bifurcation parameter α: $\alpha \rightarrow A$. The third mapping describes how the observable (i.e., the dependent variable DV) depends on the BBA attractor pattern. Simplifying the third mapping, we will only consider how the observable depends on the pattern amplitude A of the BBA attractor pattern: $A \rightarrow$ DV.

Figure 8.6 illustrates these three mappings. Panel (a) shows the assumed causal chain: the independent variable affects the bifurcation parameter that in turn affects the relevant pattern amplitude that in turn affects the dependent variable. Panel (b) breaks down the three mapping using the aforementioned symbols. Panels (a) and (b) both point out that effectively this cause-and-effect chain describes how the dependent variable (i.e., the observable) depends on the independent variable. Explicit applications of the scheme displayed in Fig. 8.6 will be presented in Sects. 8.4 and 8.5.

Note that in Fig. 8.6 and in the discussion above only a single independent variable (i.e., a single factor) and a single bifurcation parameter are considered. The scheme can be generalized to take several factors into account (e.g., to describe outcomes obtained from a 2-by-2 factor design, see Sect. 8.1.1). Likewise, several bifurcation parameters can be taken into consideration. For example, with respect to the LVH model several parameters may affect the eigenvalues of basis patterns that in turn affect the stability of the fixed points of the basis pattern amplitudes and determine the emergence of attractor patterns. In this regard, in a study by Frank et al. [121] the factors age and object size have been taken into account in order to explain different grasping types during infant development using the LVH model.

---

[5]More precisely, it is the amplitude of the unstable basis pattern giving rise to the BBA pattern of interest.

As anticipated in Sect. 1.8.1, reactions of humans and animals with different magnitudes (i.e., scaled reactions) can be explained within the pattern formation framework. In particular, D0 and D1 pattern formation systems (see Fig. 5.5) and related systems (see below) exhibit a fundamental mechanism to produce continuous reactions to continuously varied forces. Figure 8.7, which was presented in a simplified form as Fig. 1.9 in Sect. 1.8.1, illustrates this point. Accordingly, objects, humans and animals, and events of the surrounding of a person or animal exert forces. In laboratory experiments these forces are described by the independent variable. The independent variable in turn affects the bifurcation parameter α. The bifurcation parameter can be considered as reflecting external forces (as in previous chapters) or as part of the structure of the human and animal (as in Fig. 8.7). The structure determines the formation of patterns, in general, and the magnitude of the amplitude A of the emerging pattern in particular. This is indicated by the mapping $\alpha \to A$.

Note that if we consider α as measure for external forces then the impact of α on the structure could be described by means of the eigenvalue spectrum. That is, we would replace α by the set of eigenvalue $\lambda_1, \lambda_2, \ldots$. Likewise, we would replace the mapping $\alpha \to A$ by $\alpha \to \lambda_1, \lambda_2, \cdots \to A$.

Finally, the dependent variable may be measured on the level of brain activity or in terms of observables related to body reactions. In the former case, we deal with D0 systems. In the latter case, the humans and animals tested correspond to D1 systems. Therefore, the mapping $A \to \mathrm{DV}$ from the emerging pattern in terms of its amplitude A to the observable, DV, is located, in general, in the ordinary state composed of the brain state and the variables describing force production and limb positions and limb movements.

Importantly, the loop in clockwise direction shown in Fig. 8.7 that takes us from the independent variable via structure and brain state (and produced forces) to the dependent variable can formally read in the opposite, counter-clockwise direction.

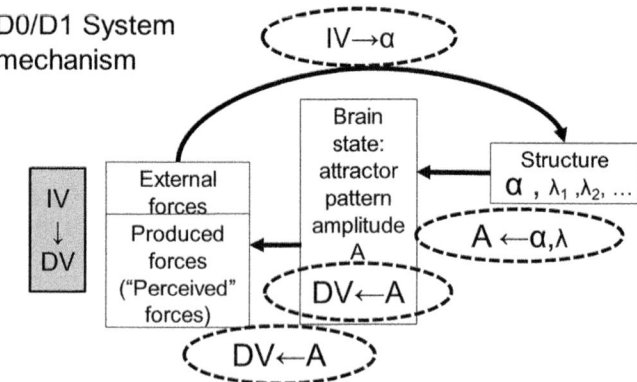

Fig. 8.7 Cause-and-effect relationship between independent variable and dependent variable established in several steps in D0 and D1 pattern formation systems or related systems (see text)

As highlighted by the gray box in Fig. 8.7 the loop establishes a cause-and-effect relationship between independent variable (cause) and dependent variable (effect).

If the dependent variable is measured on the brain activity level, we are dealing with D0 systems or D2, E0, and E2 systems (see Fig. 5.5 for the system classification scheme). In contrast, if the observable is related to a force produced by the human being or animal under consideration, then we deal with a D1 system or a D3, E1, and E3 system. That is, the D0/D1 system scheme shown in Fig. 8.7 holds for related D and E systems as well.

First, let us discuss an important qualitative aspect of the causal chain composed of the aforementioned three mappings as illustrated in Figs. 8.6 and 8.7. The following question arises: Under what circumstances does the independent variable affect the dependent variable? Since the causal chain involves three steps, within each step there must be an effect of one variable onto the other variable. That is, only if the independent variable affects (i.e., varies) the bifurcation parameter and variations of the bifurcation parameter lead to changes in the attractor pattern amplitude and the attractor pattern amplitude affects the dependent variable, then the independent variable effectively affects the dependent variable.

Second, let us assume that in fact in each step there is an effect of the one variable onto the other variable. Then the questions arises how does the relationship between the dependent and the independent variable look like quantitatively? For the sake of simplicity, in what follows, it is assumed that all three mappings are described by either monotonically increasing or decreasing functions. Furthermore, again to simplify the presentation, it is assumed that there is a linear relationship with a positive slope between the bifurcation parameter α and the independent variable. In this special case, the bifurcation parameter can be used as proxy of the independent variable and vice versa the independent variable can be regarded as bifurcation parameter. Consequently, only two mappings need to be considered: the mapping from the bifurcation parameter to the pattern amplitude and the mapping from the pattern amplitude to the dependent variable. In Fig. 8.8 possible mappings are schematically illustrated. The increasing straight lines stand for all monotonically increasing functions. Likewise, the decreasing straight lines stand for all monotonically decreasing functions. As shown in panels (a) and (d), the dependent variable increases monotonically as a function of the independent variable if the mappings $\alpha \to A$ and $A \to DV$ are both monotonically increasing functions or monotonically decreasing functions. In contrast, as illustrated in panels (b) and (c), if the mapping $\alpha \to A$ is an increasing function while the mapping $A \to DV$ is a decreasing function or vice versa the mapping $\alpha \to A$ is a decreasing function while the mapping $A \to DV$ is an increasing function, then the dependent variable decreases as a function of the independent variable.

As discussed in Sect. 8.2.2 the cases A and B that assume that pattern amplitudes increase as functions of bifurcation parameters are likely to occur close to bifurcation points when attractor patterns just have emerged. In contrast, the cases C and D that assumes that pattern amplitudes decreases as functions of bifurcation parameters are likely to occur when attractor patterns are about to become unstable.

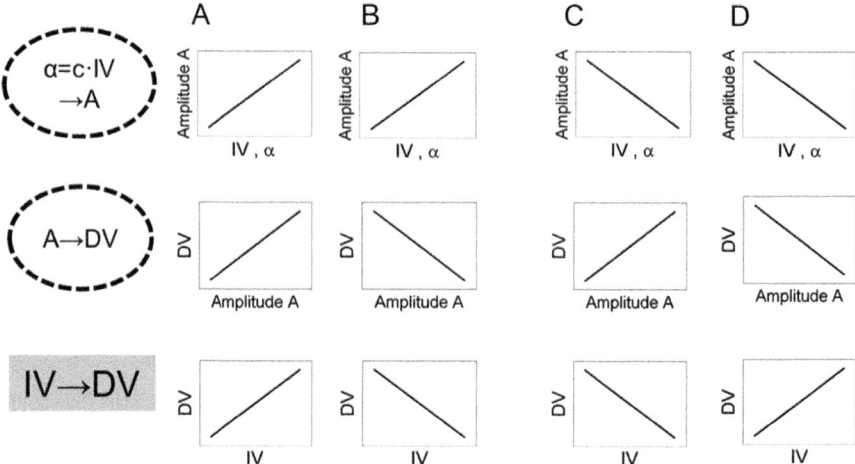

Fig. 8.8 Illustration of how the "local" mappings (or cause-and-effect relationships) in terms of $\alpha \to A$ and $A \to$ DV contribute to the "total" cause-and-effect relationships between independent variables (IV) and dependent variables (DV) studied in experiments on human and animals

8.4   Stevens' Psychophysical Power Laws

Under certain circumstances, humans produce scaled reactions (or reactions that vary in magnitude) to continuous properties of the world (see also Sect. 1.8.1). A box on the floor that a person is about to lift up (i.e., the person has not yet grasped the box) makes that the person produces certain forces that vary in magnitude depending on properties of the box (e.g., box size). For example, in psychophysics experiments (as reviewed in Sect. 1.8.1) participants do lift up boxes of different weight and then produce for each box a measure for the weight of the box on a certain continuous scale. In various cases the magnitude of the reactions depends on the continuous properties of the world in terms of a power law. This is Steven's psychophysical power law [55, 60, 310, 311].

Let us discuss two scenarios within the framework of the human pattern formation reaction model for continuous reactions that lead to power law scaled reactions. In order to discuss this topic quantitatively, we will make use of the LVH amplitude equations. For sake of simplicity, it is assumed—just as in the context of Fig. 8.8—that the bifurcation parameter α is a linearly increasing function of the independent variable: $\alpha = c_1$IV with $c_1 > 0$ constant. Moreover, let λ denote the eigenvalue of the pattern that describes the reaction of interest. Then, it is assumed that the eigenvalue λ increases linearly with the bifurcation parameter (as it is also the case for various models that we have discussed in Chap. 7) like $\lambda = L + c_2\alpha$ with $c_2 > 0$. Without loss of generality, the additive constant (offset constant) L can be neglected. Likewise, the proportionality factors c_1 and c_2 can be put equal to one. In short, we put the bifurcation parameter α equal to the independent variable IV and put the eigenvalue λ of the basis pattern of the emerging attractor pattern equal to α, which implies that $\lambda = \alpha =$ IV.

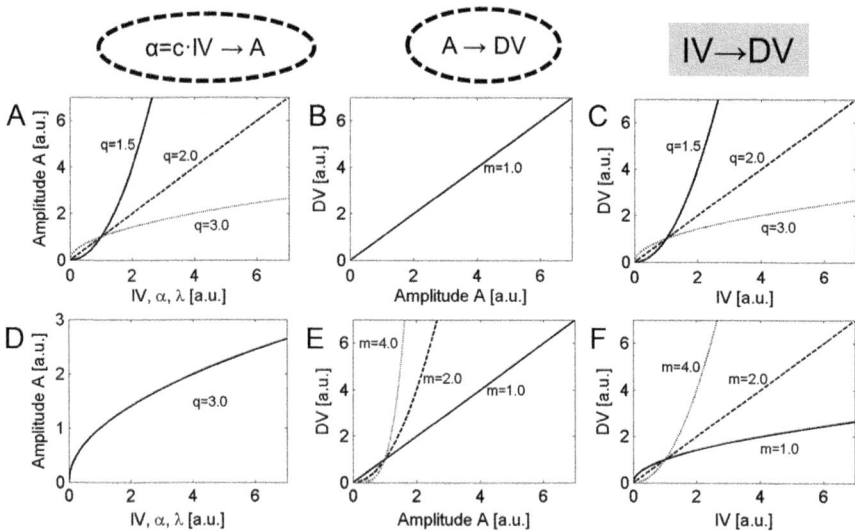

Fig. 8.9 Psychophysical power laws are an in-built property of the human pattern formation reaction model. Illustration of two scenarios (A-B-C versus D-E-F) leading to cause-and-effect power law relationships. See text for details

The first scenario is illustrated in the upper panels (a)–(c) of Fig. 8.9. Accordingly, it is assumed that continuous reactions are determined by LVH amplitude equations that in general differ with respect to the nonlinearity exponent q. Differences in the nonlinearity exponent q lead to power laws with different power law exponents. In order to see this, we follow the upper row panels from panel (a) to panel (c), see Fig. 8.9. As discussed in Sect. 4.6.6, the stationary amplitude A of the emerging attractor pattern depends on the corresponding eigenvalue λ in form of a power law. The power law exponent of the function $\lambda \rightarrow A$ depends on the nonlinearity parameter q. Some examples are illustrated in panel (a). For $q = 3$ the amplitude increases like a square root function. For $q = 2$ the amplitude increases linearly. For $q = 1.5$ the amplitude increases like a quadratic function. The dependent variable, DV, is assumed to be a linear function of the amplitude α, see panel (b). In other words, it is assumed that the dependent variable as a function of the pattern amplitude A satisfies a power law with exponent one. Taking panels (a) and (b) together, we obtain relationships between the independent variable IV and the dependent variable DV that vary with q as shown in panel (c). Since the mapping between DV and A is linear, in fact, the relationships shown in panel (c) are just the same as the relationships shown in panel (a). The key difference is on the vertical axis. While in panel (a) on the vertical axis we have the pattern amplitude, in panel (c), on the vertical axis we have the dependent variable. In fact, the graphs shown in panel (c) exemplify two power laws that have been identified by experimental work. The graphs describe scaled reactions of humans to object weight and to the brightness of light sources (compare with Fig. 1.9 in Sect. 1.8.1).

The second scenario is described in the lower panels (d)–(f) of Fig. 8.9. It is assumed that the reactions to continuous properties are determine by amplitude equations that more or less exhibit the same nonlinearity exponent. Without loss of generality, in panel (b), the case of a cubic nonlinearity (i.e., $q = 3$) is considered. Any other nonlinearity (e.g., the Lotka-Volterra nonlinearity $q = 2$) could be used as well. Panel (d) illustrates the increase of the relevant pattern amplitude as a function of the eigenvalue, where the eigenvalue as mentioned above is assumed to be identical to the bifurcation parameter and the independent variable. Different physical properties are assumed to exhibit different power law relationships between the dependent variable and the pattern amplitude, see panel (e). The power law exponent of those relationships is denoted by m. For $m = 1$ the dependent variable increases linearly with the amplitude. For $m = 2$ we obtain a parabolic increase. For $m = 4$ we obtain an increase that is even faster than the parabolic one. Taking the mappings IV $\rightarrow$ A and $A \rightarrow$ DV shown in panels (d) and (e) together, we obtain the functional dependencies between the dependent variable and the independent variable shown in panel (f). For the linear relationship (i.e., $m = 1$) between DV and A it follows that the dependent variable increases like a square root function of the independent variable. That is we obtain a psychophysical power law with exponent 0.5. For the case $m = 2$ we obtain a linear relationship (a power law with exponent 1). For the case $m = 4$ we obtain a parabolic relationship between the dependent variable and the independent variable (i.e., a power law with exponent 2). In fact, the power law graphs shown in panel (f) are identical with those shown in panel (c). That is, both scenarios can yield the same kind of power law functions. Importantly, both scenarios can produce accelerating (power law exponent larger than 1) as well as de-accelerating (power law exponent small than one) graphs as observed in psychophysical experiments.

8.5 Further Applications

8.5.1 Confidence Ratings

Several studies have been examined secondary continuous reactions on primary categorical reactions to physical properties (forces) of the world. From a pattern formation perspective, participants were producing BA attractor patterns similar to brain activity and body activity patterns without performing the body activity. That is, they produced brain activity patterns as X0/X2 systems (see Sect. 6.5.2). In the study by Warren [343] participants reacted to stair steps of different height by producing X0/X2 systems climbing-up BA patterns. In the study by Fitzpatrick et al. [83] and Wagman and Hajnal [336, 337] participants were exposed to the forces of ramps that varied in ramp inclination and reacted to the ramps by forming as X0/X2 systems standing-on-the-ramp BA patterns.[6] The secondary continuous

---

[6]That is, participants produced BA patterns about standing on the ramps without performing the posture body pattern. In doing so, participants made "judgments" about whether or not they were able to stand on ramps given certain inclination angles.

reactions were about how much "confidence" participants had in their emerging patterns. From the pattern formation perspective, these "confidence ratings" are consistent with the amplitudes of the emerging patterns or measures related to those pattern amplitudes. Since the fixed point values of amplitudes depend on the eigenvalues and eigenvalues may be interpreted as pumping parameters (see Sect. 4.6.4), roughly speaking, "confidence ratings" are consistent with the reporting of the degree to which emerging patterns are pumped. If a pattern is pumped strongly, the pattern amplitude is relative large and, consequently, the participant has a lot of "confidence" in that pattern. In contrast, if a pattern is only weakly pumped, the pattern amplitude is relative small and, consequently, "confidence" is low.

Figure 8.10 illustrates the experimental "confidence ratings" reported in the three aforementioned studies by Warren [343], Fitzpatrick et al. [83], and and Wagman and Hajnal [336, 337]. In addition, the observed graphs are explained in terms of the pattern formation reaction model discussed above. Panel (a) shows the "confidence

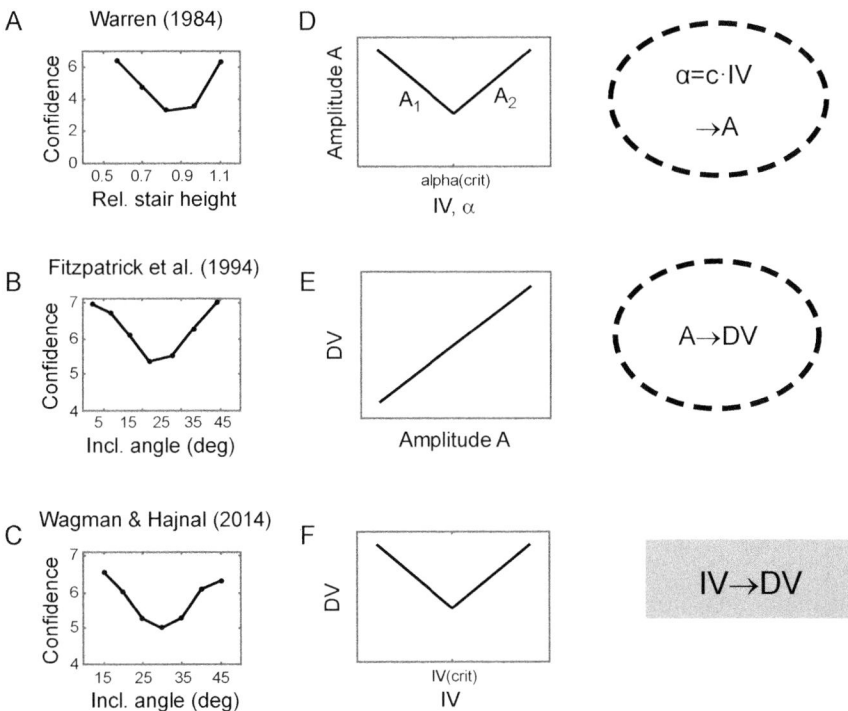

Fig. 8.10 Experimental results of experiments on "confidence" and their explanation using the human pattern formation reaction model. Panels (**a**)–(**c**) shows "confidence" ratings as functions of certain bifurcation parameters of experiments in which participants switched from a "Yes" to a "No" reaction. U-shaped functions were observed with minima at the bifurcation points. Panels (**d**)–(**f**) show how the U-shaped functions can be explained using the human pattern formation reaction model, in general, and properties of the LVH model, in particular

ratings" observed in the study by Warren [343] on stair climbing. The ratings are taken from the tall group. The ratings drop to a minimum when the BA pattern switches from a "Yes, I can" pattern to a "No, I can not" pattern. Panels (b) and (c) show the "confidence ratings" observed in the studies by Fitzpatrick et al. [83] and Wagman and Hajnal [337] about standing on ramps. The graphs are U-shaped just as the graph reported by Warren [343]. In particular, the "confidence" of the participants was lowest when they switched from a "Yes, I can" reaction pattern to a "No, I cannot" reaction pattern. The confidence ratings shown in panel (b) are taken from the visual condition used in Fitzpatrick et al. [83]. The ratings shown in panel (c) are taken from the haptic one-hand condition used by Wagman and Hajnal [337].

Panels (d)–(f) explain the experimentally observed graphs from a pattern formation perspective. As discussed in Sect. 8.2.2 above, according to the LVH model, pattern amplitudes decay when they are about to become unstable and increase when they just have emerged. This leads to the U-shaped pattern shown in panel (c) of Fig. 8.4. This U-shaped pattern is schematically depicted in panel (d) of Fig. 8.10. The U-shaped graph describes the amplitudes of the relevant attractor patterns as a single function of the bifurcation parameter α and as function of the independent variable, when assuming that the bifurcation parameter increases linearly with the independent variable (IV) like $\alpha = c_1 IV$. The mapping $A \rightarrow DV$, which is the mapping of the amplitude A of the attractor pattern that is present at a given value for α or IV to the feeling of "confidence", is assumed to be a monotonically increasing function. For the sake of simplicity, it is assumed that $A \rightarrow DV$ relationship is linear. As a result, the cause-and-effect relationship between the independent variable (stair step height or inclination angle) and the dependent variable ("confidence") is given in terms of the U-shaped function shown in panel (f).

The studies by Warren [343], Fitzpatrick et al. [83], and Wagman and Hajnal [336, 337] did not report from hysteresis. If hysteresis is taken into consideration, the proposed model may be revised as suggested in Ref. [187].

8.5.2   Oxygen Consumption in Gait Transitions Close to Walk-Trot Transitions

Gait transitions in horses have been examined in various studies, see Sect. 3.2.3. As mentioned in Sect. 3.2.3, transitions from walk to trot exhibit the characteristic feature of phase transitions and bifurcations, namely, they are associated with an observable that changes in a non-smooth fashion at the gait transition point. In the study by Griffin et al. [143] transport costs of horses in terms of their oxygen consumption was measured. Panel (a) of Fig. 8.11 (which is a copy of panel (b) of Fig. 8.3) shows a detail of the observed graph. The full graph is shown in Fig. 3.5d. The animals are moving on a treadmill. Therefore, the treadmill speed can be regarded as locomotion speed of the horses. The speed is shown on the horizontal axis of panel (a). The transport costs are given on the vertical axis. The graph of

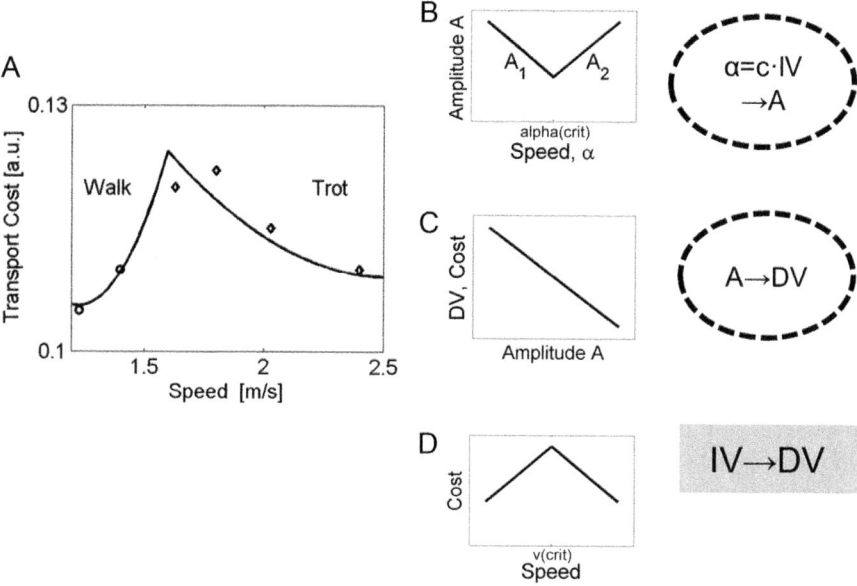

Fig. 8.11 Experimental results (panel (**a**)) and modeling (panels (**b**)–(**d**)) of the dependency of transport costs on locomotion speed before and after the walk-trot gait transition

the transport costs versus speed increases for locomotion speed in the interval from 1.2 m/s to 1.6 m/s. Subsequently, in the interval from 1.6 m/s to 2.5 m/s the transport costs decay as a function of speed. Around the critical locomotion speed of 1.6 m/s the horses switch from walk to trot. Overall the graph corresponds to an inverted U-shaped function with the maximum indicating a bifurcation point. In analogy to the previous example on "confidence ratings" constituting a U-shaped graph, the inverted U-shaped graph can be explained by means of the pattern formation reaction model as shown in panels (b)–(d) of Fig. 8.11.

Panels (b)–(d) of Fig. 8.11 illustrate assumed functional dependencies of the pattern formation systems of horses that are consistent with the inverted U-shaped graph of transport costs. First, as far as the IV= $\alpha \rightarrow A$ relationship between the pattern amplitude A and the locomotion speed (IV) is concerned, it is assumed that the relationship is of the same kind as in the model about "confidence rating" discussed in the previous section, see Sect. 8.5.1. The bifurcation parameter α increases linearly with the independent variable, which is speed. The eigenvalues λ_{walk} and λ_{trot} of the basis patterns of walking and trot are assumed to increase and decrease, respectively, as linear functions of the bifurcation parameter (see the application of the pattern formation reaction model to gait transitions in humans in Sects. 6.5.3 and 6.5.4). The coupling parameter g of the unstable amplitudes is put equal to 1. As a result, the amplitude of the walking attractor pattern decays monotonically as shown schematically in panel (b) of Fig. 8.11 when walking speed is increased until the bifurcation parameter (i.e., the locomotion speed) reaches a critical value. At the bifurcation point, the walking BBA attractor pattern disappears and the trot BBA attractor pattern emerges. The amplitude of the latter pattern increases monotonically with the bifurcation parameter (i.e., locomotion speed). A U-shaped graph for the amplitudes of the occurring BBA attractor patterns is obtained. Second, the animal pattern formation system is described in terms of a state that accounts not only for brain activity and muscle activity but also involves the oxygen consumption as a variable. That is, the emerging gait BBA patterns are assumed to feature a supplementary component that describes oxygen consumption. It is assumed that the oxygen consumption component of the walking and trot gait attractor patterns is inversely related to the amplitudes of the patterns, see panel (c). That is, if the pattern amplitude is large, the oxygen consumption is low. If the pattern amplitude is low, the oxygen consumption is relatively high. The two dependencies illustrated in panels (b) and (c) lead to the overall functional dependency of the transport costs (i.e., oxygen consumption) on locomotion speed shown in panel (d). The U-shaped graph for the amplitudes (panel b) of the animal pattern formation system turns into an *inverted* U-shape graph for the transport costs (panel d) that is consistent with experimental observations.

8.5.3   Example from "Social" Psychology: Malicious Pleasure (Schadenfreude)

Schadenfreude is a noun (taken from the German language) that refers to the phenomenon that under certain conditions human beings can feel some kind of pleasure when they see someone else suffer or fail. That is, individuals experience some kind of satisfaction when they see somebody else being in trouble. Schadenfreude is some kind of malicious pleasure. From a pattern formation perspective, feelings like joy, happiness, pleasure, satisfaction, and schadenfreude are certain brain activity and body activity patterns that also involve components from the human hormone system. The verb "to feel" in this context means that a BBA pattern related to feelings emerges.

Among the many studies on schadenfreude, Leach et al. [207] investigated factors affecting the degree of schadenfreude. To this end, the failure of the German soccer team in the 1998 world soccer cup was used as an event that may trigger schadenfreude. Leach et al. [207] tested Dutch students with respect to the degree of schadenfreude that the students experienced from the misfortune of the German team. In what follows only a part of the study will be reported. The reader should consult the original work for the complete study.[7] Students were tested under four conditions described by two independent variables. The first independent variable was the degree of which students had an interest in soccer. Students with low and

---

[7] And will or will not do so depending on the reader's brain structure and brain state without making any "decision".

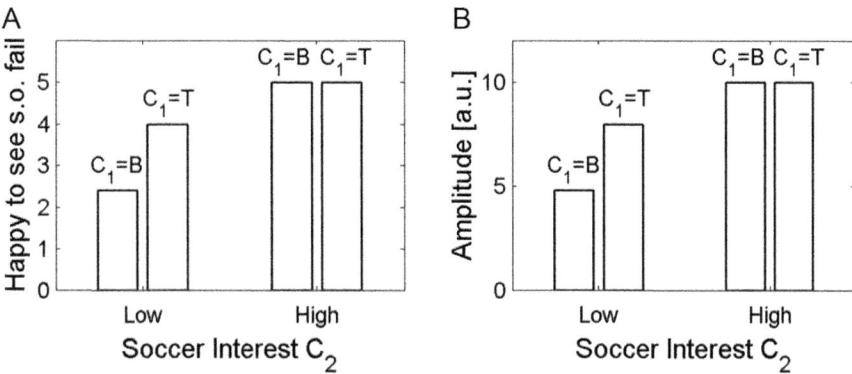

Fig. 8.12 Panel (**a**): Experimental results in an experiments on conditions affecting the emergence of schadenfreude patterns in the human emotional system. The conditions tested were (1) being exposed to some kind of threat (C_1: B = baseline, no threat versus T = mild form of threat) and (2) soccer interest (C_2: low versus high). Panel (**b**): Explanation of experimental results in terms of the human pattern formation reaction model applied to patterns of the emotional system

high interest in soccer were identified. In doing so, two groups of students were obtained. Within each group participants were assumed to have a similar brain structure as far as soccer interest was concerned. In contrast, between the groups students were assumed to differ with respect to the brain structure. The second independent variable was the condition of threat. Some students were tested under baseline conditions. Other students were tested under some mild form of threat. The researchers hypothesized that the students under the threatening condition would experience a higher degree of schadenfreude.

For this 2-by-2 experimental design, the results are summarized in panel (a) of Fig. 8.12. The degree of schadenfreude was measured on a Likert scale with 7 points. The number 1 on that scale meant that a student did not experience any form of malicious pleasure. In contrast, the number 7 meant that a student experienced a large degree of malicious pleasure. As shown in panel (a), the low soccer interest group showed a low degree of malicious pleasure when tested under baseline conditions, while there was a relatively high degree of malicious pleasure when those students were tested under the threatening condition. As far as the students with a strong interest in soccer were concerned, the students of the two subgroups (i.e., the baseline and threatening subgroups) showed approximately the same degree of malicious pleasure.

The scores observed in the four groups can be explained within the framework of the pattern formation reaction model by assuming that the two independent variables C1 (degree of threat) and C2 (interest in soccer) affect in combination the eigenvalue λ of the BBA attractor pattern associated with malicious pleasure. The eigenvalue λ in turn affects the amplitude A that determines the reaction of the participants. Consequently, in line with Figs. 8.6 and 8.7 the causal chain reads like $C1, C2 \rightarrow \lambda \rightarrow A \rightarrow DV$.

At this stage, for the sake of brevity, only a qualitative discussion about the impact of C1 and C2 on the eigenvalue and amplitude of the malicious pleasure BBA pattern will be given. In line with the considerations presented in Sect. 8.3, it is assumed that the amplitude A of a BBA pattern associated with malicious pleasure that emerges in a person eventually determines the degree to which malicious pleasure is experienced and reported by that person. More explicitly, the magnitude of the experience as measured on the 7 point Likert scale is assumed to be a monotonically increasing function of the pattern amplitude. Consequently, the scores observed in the four conditions as shown in panel (a) reflect qualitatively the amplitudes of the BBA patterns that emerged in the participants in the four conditions. Therefore, the human pattern formation reaction model states that the observed scores shown in panel (a) result from pattern amplitudes A that at least qualitative exhibit an order among each other as shown in panel (b).

In particular, as illustrated in panel (b), the amplitude A of the malicious-pleasure BBA pattern is assumed to be at a medium level for individuals whose structure is characteristic for a low interest in soccer ($C2 =$ low) and whose structure or brain state (i.e. grand state) is characteristic for persons under threat ($C1 = T$). If the structure is characteristic for low interest in soccer ($C2 =$ low) but the grand state is otherwise consistent with the baseline ($C1 = B$) grand state of everyday people, then the amplitude of the schadenfreude BBA attractor pattern is at a relatively low level. Finally, if individuals exhibit a brain structure that is characteristic for people with a strong interest in soccer ($C2 =$ high) then exposing those individuals to forces related to the threatening situation does not affect the eigenvalue λ of the malicious-pleasure BBA pattern and, consequently, does not affect the magnitude of the pattern amplitude A and the DV. It is assumed that due to the impact of the soccer-interest factor C2 the eigenvalue λ and (as a result of that) the amplitude A and the DV assume relatively large values.

The aforementioned discussion about amplitudes and eigenvalues illustrates how the experimental observations shown in panel (a) can be explained in terms of the attractor pattern amplitudes shown in panel (b). The discussion is limited to qualitative statements. However, as it has been exemplified in an earlier study by Frank et al. [121], in principle, the way in which two factors (in our example the factors C1 and C2) affect eigenvalues of pattern formation systems can be discussed quantitatively. In doing so, experimental data can be fitted.

8.6   Mathematical Notes

8.6.1   *Shape of IV-DV Cause-and-Effect Relationships*

Figure 8.8 illustrates that the shape of IV-DV cause-and-effect relationships is determined by the two local cause-and-effect relationships $\alpha \rightarrow A$ and $A \rightarrow$ DV.

Let us formulate the local relationships like

$$A = f(\alpha) , \quad DV = g(A) \tag{8.1}$$

where f and g are functions that either increase or decrease monotonically. From Eq. (8.1) it follows that

$$DV = g(f(\alpha)) = g(f(cIV))) , \tag{8.2}$$

using $\alpha = cIV$ with $c > 0$. At issue is to determine whether the dependent variable increases or decreases as function of the independent variable. Differentiating Eq. (8.2) with respect to IV and using the chain rule, we obtain

$$\frac{d\,DV}{d\,IV} = c\beta_g\beta_f , \tag{8.3}$$

where β_g and β_f are the slopes of the functions g and f, respectively. Consequently, we obtain the four cases

$$A : \beta_f > 0 , \beta_g > 0 \Rightarrow \frac{d\,DV}{d\,IV} > 0 ,$$

$$B : \beta_f > 0 , \beta_g < 0 \Rightarrow \frac{d\,DV}{d\,IV} < 0 ,$$

$$C : \beta_f < 0 , \beta_g > 0 \Rightarrow \frac{d\,DV}{d\,IV} < 0 ,$$

$$D : \beta_f < 0 , \beta_g < 0 \Rightarrow \frac{d\,DV}{d\,IV} > 0 , \tag{8.4}$$

which correspond to the four cases shown in Fig. 8.8. Note that while Fig. 8.8 illustrates the case in which f and g are linear functions, the four cases hold in general provided that f and g are functions that either increase or decrease monotonically. That is, our considerations also hold for nonlinear local cause-and-effect relationships provided they correspond to monotonic functions.

8.6.2   Psychophysic Power Laws of Human Pattern Formation Systems

Let us briefly sketch the mathematical details underlying Fig. 8.9. For both schemes (A-B-C and D-E-F) we have

$$\alpha = c_1 IV , \lambda = L + c_2\alpha , c_1 = c_2 = 1 , L = 0 \Rightarrow \lambda = \alpha = IV . \tag{8.5}$$

First, let us consider the A-B-C scheme. In order to describe the $\alpha = \lambda = \text{IV} \rightarrow A$ relationship, we use Eq. (4.51), which reads $A = \lambda^{1/(q-1)}$, which implies

$$A = \text{IV}^{1/(q-1)} . \tag{8.6}$$

Panel (a) of Fig. 8.9 displays the following cases

$$q = 1.5 : A = \lambda^2 = \text{IV}^2 ,$$
$$q = 2 : A = \lambda = \text{IV} ,$$
$$q = 3 : A = \sqrt{\lambda} = \sqrt{\text{IV}} . \tag{8.7}$$

As far as the mapping $A \rightarrow \text{DV}$ is concerned, it is assumed that the cause-and-effect relationship between A and DV is linear like

$$DV = aA + b , \tag{8.8}$$

where a is the slope coefficient and b is the intercept coefficient. It is sufficient to consider the special case $a = 1, b = 0$, which reads

$$DV = A \tag{8.9}$$

and is shown in panel (b) of Fig. 8.9. Substituting Eq. (8.9) into Eq. (8.6), we obtain

$$\text{DV} = \text{IV}^z \tag{8.10}$$

with power law exponent $z = 1/(q - 1)$. The special cases

$$q = 1.5 : \text{DV} = \text{IV}^2 ,$$
$$q = 2 : \text{DV} = \text{IV} ,$$
$$q = 3 : \text{DV} = \sqrt{\text{IV}} \tag{8.11}$$

are shown in panel (c) of Fig. 8.9. Recall that in the derivation above we have put $c_1 = c_2 = a = 1$ and $L = b = 0$. In the general case for $c_1, c_2, a \geq 0$, b arbitrary, and L satisfying $\lambda = L + c_1 c_2 \text{IV} > 0 \Rightarrow L > -c_1 c_2 \text{IV}$ for all values of IV, we obtain

$$\text{DV} = a(L + c_{eff} \text{IV})^z + b \tag{8.12}$$

with $c_{eff} = c_1 c_2$. Equation (8.12) generalizes Eq. (8.10).

Next let us consider the D-E-F scheme shown in Fig. 8.9. The nonlinearity is fixed. Without loss of generality, we consider the cubic case $q = 3$. In this case

Eq. (4.51) reads $A = \sqrt{\lambda}$, which implies

$$A = \sqrt{IV} . \tag{8.13}$$

This square root function is shown in panel (d) of Fig. 8.9. Furthermore, it is assumed that

$$DV = A^m \tag{8.14}$$

holds. Panel (e) shows the corresponding curves for $m = 1, 2, 4$. Substituting Eq. (8.13) into Eq. (8.14), we obtain

$$DV = IV^{m/2} \tag{8.15}$$

and the special cases

$$m = 4 : DV = IV^2 ,$$
$$m = 2 : DV = IV ,$$
$$m = 1 : DV = \sqrt{IV} \tag{8.16}$$

shown in panel (f). Note that the two special cases A-B-C and D-E-F can be combined by replacing Eq. (8.9) by Eq. (8.14). Then, Eq. (8.12) reads

$$DV = a(L + c_{eff} IV)^u + b . \tag{8.17}$$

with power law exponent $u = 0.5m/(q-1)$. However, in terms of fitting data to the human pattern formation reaction model this step does not provide an improvement. The reason for this is that any exponent $z = 1/(q-1) > 0$ in Eq. (8.12) can already be perfectly expressed in terms of the nonlinearity parameter $q > 1$. Introducing an additional parameter m is not needed. Having said that, there might be arguments to use a specific nonlinearity exponent (e.g., $q = 3$ which reflects symmetry in the bifurcation diagram, see the pitchfork bifurcation diagram). If we fix q, then the additional parameter m allows us to fit power laws with arbitrary exponents.

Chapter 9
Restructuring Humans and Animals

9.1 Restructuring Brains by Paired Forces and "Classical Conditioning" Experiments

Pavlov's experiments on dogs have been briefly reviewed in Sect. 2.1. In general, experiments as those conducted by Pavlov involve two forces that occur in pairs. The experiments are about how the human and animal brain can be restructured by means of paired forces. In physics a plenitude of systems subjected to two forces have been studied. For example, a ball rolling down on a ramp is a system subjected to two forces. On the one hand, the gravitational force pulls the ball towards the earth. On the other hand, the ramp exerts forces on the ball such that the ball cannot move downwards on a vertical line. As a consequence of the pair of forces (i.e., gravitational force and force of the ramp) acting on the ball, the ball follows the slope of the ramp in a downwards movement. Note that in the literature on the restructuring of human and animal brains by means of two forces the experiments conducted by Pavlov and similar experiments have also been called "classical conditioning" experiments.

Let us consider restructuring of brains by means of two forces from the perspective of pattern formation systems. Accordingly, the structure of human and animal pattern formation systems is changed such that a given BBA pattern that typically emerges as a reaction to a particular kind of force can also emerge as a reaction to another force. While below the pattern formation perspective will be explicitly discussed in the context of experiments involving animals, the procedure has been used for therapeutically purposes with humans to help individuals to overcome undesired habits [332] and addictive behaviors [302]. Therefore, the considerations that will be made in this chapter are not only relevant to explain animal experiments but may be useful to understand better the treatment of patients using paired-forces procedures.

© Springer Nature Switzerland AG 2019   375
T. Frank, *Determinism and Self-Organization of Human Perception and Performance*, Springer Series in Synergetics,
https://doi.org/10.1007/978-3-030-28821-1_9

In paired-forces experiments on animals, animals are exposed to a force that produce a typical reaction. For example, dogs are exposed to chemical driving forces of food that lead to a salivation reaction and rabbits are exposed to pressure forces on their eyes induced by air puffs that lead to an eye-blink reaction. The force related to the reaction X of interest may be referred to as the original force that triggered the reaction X. In addition to the original force the animals are exposed to another type of force that will be referred to as secondary force. For example, animals are exposed to acoustic sound waves. Dogs may be exposed to the sound waves of a ringing bell as in Pavlov's experiments [51, 288] and rabbits may be exposed to computer generated tones [284]. When the two forces (the original and the secondary) act on the animals approximately at the same time, then after several repetitions of producing the animal reaction X using dual forces the original force can be switched off. It is sufficient to expose the animals to the secondary force. The animals react to the secondary force just as they reacted at the beginning of the experiment to the original force. From a physics perspective, such experiments change the structure of animals such that they act in novel ways to the forces imposed on them. Panel (a) of Fig. 9.1 (that was presented earlier in Sect. 2.1, see Fig. 2.1) shows the percentage of eye-blink reactions in rabbits to forces exerted by a tone played to the rabbits as a function of the trials in which the animals were restructured with the help of paired forces. As indicated in panel (b) of Fig. 9.1 (and anticipated already in Fig. 2.1c of Sect. 2.1) such changes in cause-and-effect relationships (i.e., in the the reactions to forces) are consistent with structural changes taking place in C3 pattern formation systems.

Physiological research has identified various neuronal circuits involved in the phenomenon of dual-force brain restructuring [62, 320]. Inspired by these findings from physiological research, network models for Pavlovian "learning" have been suggested [228, 258, 317]. In what follows, let us address paired-forces experiments from a pattern formation perspective in which pattern formation takes place in neuronal networks. To this end, Haken's attractor network [157] that involves the

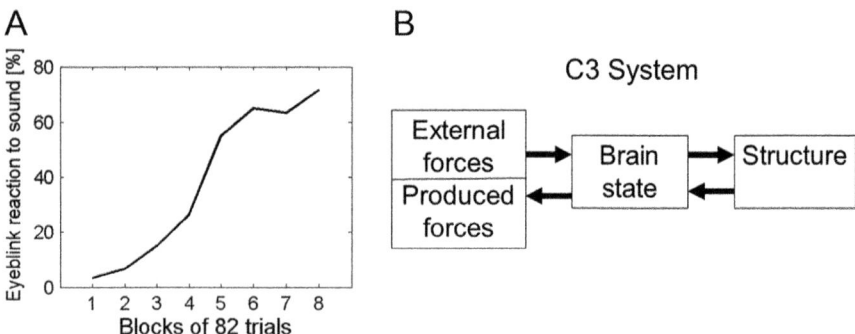

Fig. 9.1 Restructuration-by-paired-forces phenomenon. Panel (**a**): Eyeblink reactions to tones as function of the number of restructuration-by-paired-forces trials. Panel (**b**): Scheme of C3 pattern formation system that can account for restructuration by means of paired forces (cf. also Fig. 5.5)

LVH amplitude equations for the cubic case as special case will be considered. Both the build-up of reactions to additional forces as shown in panel (a) as well as the disappearance of the reactions after an extended periods in which only the secondary forces are presented will be addressed. Two related mechanistic models will be presented: the component-shift model and the eigenvalue-shift model. The first model will be discussed in some details. The latter model will be sketched only.

9.1.1   Component-Shift Model

Let us consider two neuronal networks. One network describes pattern formation with respect to the original force. The other network describes pattern formation related to the secondary force. Let us consider the relatively simple but non-trivial case in which each network is composed of three units. Accordingly, the states of the networks are given by vectors with three components. The network units may be interpreted as neuron populations or single neurons. Accordingly, the variables describing the network units describe the activity of these populations or neurons. The activity may be given in terms of population or single neuron firing rates. In other contexts, as far as neuron populations are concerned, the activity of populations may be measured indirectly via electroencephalography (EEG). To this end, one typically assumes that the electrical potentials measure by means of EEG sensors reflects the neuronal activity of certain neuron populations. Irrespective of the explicit interpretation of the three-components state vectors, the two three-dimensional state vectors are assumed to describe the brain state of the animal under consideration.

In order to explain dual-force brain restructuring by means of the human and animal pattern formation reaction model in terms of C3 systems, it is useful to introduce four components. First, there are certain external forces of interest that act on a human or animal under consideration. These external forces can be electromagnetic forces produced by the conditions in the visual world of a human or animal, mechanical forces produced by the pressure waves of sounds in the acoustic world of a human or animal, chemical driving forces related to tasting and smelling, mechanical forces that act on skin receptors when a human or an animal is in physical contact with other objects, and so on. Second, there are emerging BBA attractor patterns in the neuronal networks. Third, these patterns have original and supplementary components. The original components are those components in the brain state that are directly impacted by the external forces. That is, brain state shifts induced by external forces take place in the original components. In contrast, the supplementary components are not affected by brain state shifts. However, they emerge as part of a given whole BBA attractor pattern [127, 157] (and see Sect. 6.1.3 in this book). In the context of the component-shift model it is assumed that the supplementary components determine subsequent pattern formation processes associated with body reactions (e.g., in terms of salivation or eye-blink reactions). Fourth, the networks exhibit structural parameters that reflect

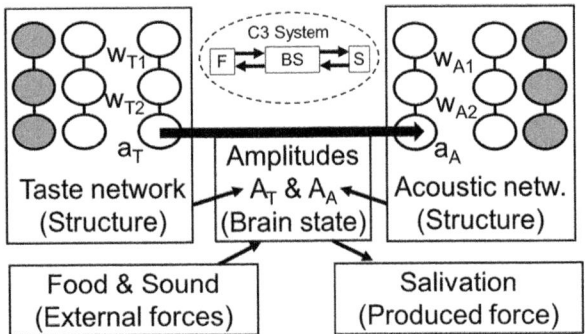

Fig. 9.2 Haken's neuronal network model to describe restructuration of the animal brain by means of paired forces in the context of C3 pattern formation systems. The basic elements of the component-shift model are shown. See text for details

physiological parameters (e.g., synaptic weights) and describe the interactions between the neurons of interest (or between the neuron populations at hand). The structure parameters are assumed to vary as a result of BBA attractor patterns that emerge as reactions to the two types of external forces.

Figure 9.2 describes the model schematically in some detail in the context of Pavlovian experiments on restructuring dogs. The network state spaces are three-dimensional (e.g., activity of three neuron populations are considered). Therefore each network features three basis patterns, where each basis pattern corresponds to a three-dimensional vector that points in a certain direction. The boxes on the left and right describe the structure of the networks for taste and sound in terms of three basis patterns each featuring three components. In line with the distinction between the original and supplementary network components, each network is assumed to involve one basis pattern that is stable. The stable basis pattern is highlighted in grey and does not play a role in what follows. The two remaining basis patterns are assumed to exhibit positive eigenvalues (i.e., correspond to unstable basis patterns). In the network for taste (left box) the middle basis pattern describes non-eatable things, while the basis pattern on the right describes eatable things. That is, chemical driving forces exerted on the taste receptors of dogs by eatable things are assumed to shift the brain state of the taste network in the direction of the basis pattern described schematically on the right. The eatable things basis pattern has components W_{T1}, W_{T2}, and a_T. The component a_T denotes the supplementary component mentioned above and is used to distinguish between eatable and non-eatable goods. Without loss of generality, the basis pattern for eatable goods exhibits a supplementary component $a_T > 0$, whereas the basis pattern for non-eatable goods exhibits a supplementary component $a_T = 0$. Consequently, according to the pattern formation reaction model if the dog tastes something eatable the brain state evolves towards an attractor that is located in state space in the "direction" of the unstable basis pattern of eatable goods. This attractor will be referred to as eatable goods attractor. The BA attractor pattern of the eatable goods attractor describes a brain state with non-vanishing activity of the a_T neuron or neuronal population. In particular, it is assumed that if the a_T neuron or neuronal population is active, a salivation reaction is triggered. Consequently, if the BA attractor pattern for eatable goods emerges which is described by an increase of the corresponding pattern amplitude A_T, then the a_T neuron or neuronal population is active and the dog starts to produce saliva (as indicated in Fig. 9.2).

While the taste network (on the left) is affected by chemical driving forces that are produced when dogs are tasting something, the acoustic network (on the right) is affected by forces of pressure waves that occur when dogs are listening to something. As far as the acoustic network is concerned, in view of the key role of the ringing bell in the dog restructuring experiment under consideration the unstable basis pattern on the left describes BA patterns emerging as reactions to the sound of a ringing bell, while the middle unstable basis pattern describes all kind of BA patterns emerging as reactions to sounds different from a ringing bell. Both unstable basis pattern exhibit supplementary components.

A key element of the component-shift model is the hypothesis that the taste and sound networks exhibit components that have the same consequences for the subsequent formation of patterns. The component a_A considered in the model shown in Fig. 9.2 is that hypothesized component that has the same impact as the component a_T. Consequently, if a_A is active then the dog produces saliva. From a mechanistic point of view, the activities of the neurons or neuronal populations a_T and a_A are assumed to be projected to the same target region in the brain. A different scenario will be discussed below in the context of the eigenvalue-shift model.

Given the hypothesized properties of the component-shift model, the reactions of the animals under baseline and after they have been restructured can be explained as follows. Under baseline, the supplementary component a_A is at zero. If the sound of a ringing bell occurs the ringing bell BA attractor pattern emerges, which is described by an increase of the corresponding amplitude A_A (note that a capital letter "A" is used to denote the amplitude, whereas a lower case letter "a" is used to describe the supplementary component). As just pointed out, under baseline conditions, the basis pattern of the ringing bell is assumed to exhibit a component $a_A = 0$. Therefore, when the ringing bell pattern amplitude A_A converges towards its fixed point value and the ringing bell BA attractor pattern emerges the a_A neuron or neuronal population exhibits only baseline activity (i.e., is inactive). The dog does not produce saliva. As indicated in Fig. 9.2 by the arrow pointing from a_T to a_A, the restructuring is assumed to shift the pattern component a_A to non-finite values. Consequently, after having restructured a dog, when the dog is exposed to the sound of the ringing bell, a ringing-bell BA attractor pattern emerges that involves a non-vanishing component a_A. In other words, after the network structure has been change, when the pattern amplitude A_A increases towards its fixed point and the corresponding BA attractor pattern emerges, then the a_A neuron or neuronal population is active and the dog produces saliva.

Let us consider the process of restructuring the networks. When the experimenter puts food in the mouth of a dog and at the same time rings a bell, then sensory taste receptors of the dog are affected by the chemical driving force of the food and receptors in the dog's ears are effected by the forces of the acoustic pressure waves produced by the ringing bell. In the network for tasting the BA pattern for eatable goods emerges. In terms of amplitude dynamics, this is indicated by the fact that the pattern amplitude A_T reaches a finite fixed point value. Likewise, in the acoustic network for listening the BA pattern of ringing a bell emerges, which is indicated by the fact that the amplitude A_A converges to a finite fixed point value. When both BA patterns have emerged, the synaptic weights in the network for listening change such that the pattern component a_A is shifted in the direction of the pattern component a_T (as indicated by the arrow in Fig. 9.2). Qualitatively, in those event during which both amplitudes A_T and A_A become finite, a_A increases monotonically from the baseline value of zero towards the finite value given by a_T. Any rule describing this monotonic increase would work well for our purposes of modeling. Without loss of generality, the following parameter dynamics (or restructuring dynamics) will be considered:

$$a_A(n' + 1) = a_A(n') - \epsilon_1 (a_A(n'))^r \left(a_A(n') - a_T \right). \tag{9.1}$$

Here, $n' = 1, 2, 3, \ldots$ are the events or trials in which both the BA patterns for eatable goods and ringing bell emerge. That is, two different time-discrete variables are considered. The variable $n = 1, 2, 3, \ldots$ describes instances in which something happens, that is, either food is given to a dog or the bell is ringing or both. In contrast, $n' = 1, 2, 3, \ldots$ describes the aforementioned dual-force events. The parameter $\epsilon_1 > 0$ determines the speed of the structure change. Moreover, the exponent r assumes the values $r = 0$ or $r = 1$. The parameter r determines the shape of the graph describing the increase of a_A as a function of n' and, in doing so, the restructuring dynamics. For $r = 0$ the dynamics can be cast into the form

$$\Delta a_A(n') = \epsilon_1 (a_T - a_A(n')) \tag{9.2}$$

where $\Delta a_A(n')$ denotes the change in neuronal activity: $\Delta a_A(n') = a_A(n' + 1) - a_A(n')$. This case $r = 0$ is consistent with the Rescorla-Wagner (RW) model [46]. The RW-model predicts that the network structure changes by a relatively large amount if there is a relative large mismatch between a_A and a_T, that is, between the activities of the respective neurons or neuron populations of the two networks. For $r = 1$ the curve assumes a more sigmoid shape. Such sigmoid shaped curves have often been found in dual-force brain restructuration experiments [288].

Note that the change of neuronal activity is a consequence of the change of the network structure. In the context of Haken's network the neuronal activity of the a_A neuron or neuron population follows an appropriately defined structure parameter Z associated to a_A. Therefore, a change in the a_T neuron activity or activity of the a_A neuron population can be used as proxy for a change of the network structure component Z.

Typically, in experimental research on brain restructuration the probability is measured that the desired reaction occurs with respect to the secondary force of the force pair. Accordingly, it is observed how the probability of that reaction changes over the restructuration period. The reaction to the secondary force, that is, in our context, the reaction to the ringing bell, has also been called the "conditional response". In what follows, a more neutral phrase will be used: we will refer to the reaction as secondary force reaction (SFR). According to the proposed model, if $a_A = 0$ holds then animals do not exhibit secondary force reactions. In contrast, if $a_A = a_T$ holds then the secondary force reaction occurs in any case, that is, the probability is 1. In the literature there is some agreement that the probability of a secondary force reaction, $p(SFR)$, is a monotonically increasing function of the mismatch between a_A and a_T [288]. Therefore, we consider the ratio a_A/a_T and put

$$p(SFR) = f\left(\frac{a_A}{a_T}\right). \qquad (9.3)$$

with $f(0) = 0$, $f(1) = 1$ and f monotonically increasing. In the simplest case, the function f is the identity $f(z) = z$.

Figure 9.3 present results from simulations of the C3 pattern formation system describing Pavlovian restructuration by means of paired forces. Panel (a) shows the instances in which the experimenter gave food in the dogs mouth (top subpanel) and/or rang the bell (bottom subpanel). On the horizontal axis there is time. On the vertical axis the chemical driving force F_T exerted by food affecting the taste network (top subpanel) and the mechanic force F_A produced by the pressure-waves of the ringing bell affecting the acoustic network (bottom subpanel) are shown schematically. A non-zero magnitude means that the respective force was present. The external forces shifted the brain states in the directions of the eatable goods and ringing-bell basis patterns, respectively, as discussed above. Panel (b) shows how the grand state (brain state A_T and A_A plus structure parameter Z denoted here for sake of simplicity as a_T) of the pattern formation system evolved under the impact of their respective forces. Only the pattern amplitudes A_T (top subpanel) and A_A (middle subpanel) relevant for brain restructuration are shown. The dashed horizontal lines denote the thresholds values indicating that the respective attractor patterns were fully emerged. According to the dynamics of structural changes discussed above, coincidence events n' of neuron activity such that both A_T and A_A were above the threshold values resulted in changes of the structure of the acoustic network: a_T increased at those coincidence events n' (see bottom subpanel).

In panels (a) and (b) only two restructuration events (i.e., the bell was ringing while food was given) were simulated. In contrast, panel (c) shows the results of a simulation for several coincidence trials n'. As can be seen in the top subpanel of panel (c), the pattern component a_A in the acoustic network increased monotonically as function of trials. Using $f(z) = z$, we calculated the probability $p(SFR)$ of a secondary force reaction. Since the function f is the identical function, not surprisingly, the probability increased monotonically in just the same way as the pattern component a_A.

Experimental research on dogs after they have been restructured has shown that ringing the bell without delivering food results in a decay of the probability of a secondary force reaction. That is, the salivation becomes less and eventually the

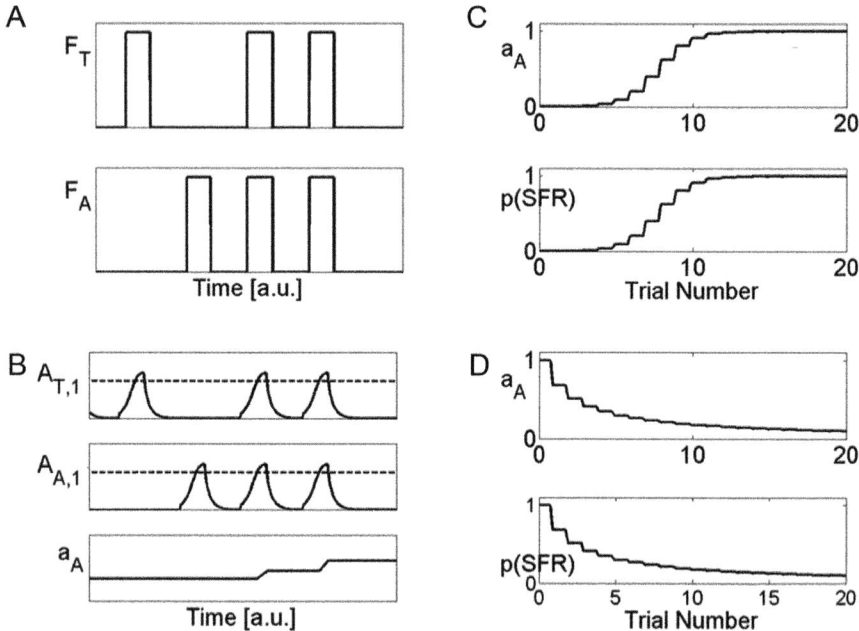

Fig. 9.3 Restructuration in individual paired-forces trials (panels (**a**) and (**b**)) and in a sequences of trials using paired or unpaired forces (panel (**c**) and (**d**)). Panel (**a**): Chemical driving force F_T acting on taste receptors (top subpanel) and sound-induced mechanical force F_A (bottom subpanel) acting on the acoustic system of a dog being restructured. Panel (**b**): Amplitudes $A_{T,1}$ (top) and $A_{A,1}$ (middle) describing the emergence of the eatable-goods BBA pattern and the ringing-bell BBA pattern, respectively. Dashed lines indicate when the BBA patterns are assumed to be fully emerged. The supplementary component a_T of the ringing-bell BBA pattern (more precisely: the structural counterpart variable Z) is shown as well (bottom) as function of time. Panel (**c**): Evolution of the supplementary component a_T (Z) over the course of paired-forces trials (top) and probability of a secondary force reaction (SFR), that is, probability of a salivation reaction to the sound of a ringing bell (bottom). Panel (**d**): As in panel (**c**) but for unpaired force trials, in which animals are exposed to the ringing-bell sound only

salivation reaction to the ringing of the bell stops to occur. In general, experimental research has shown that the change in animal reactions due to restructuring by pairing external forces is temporary in the sense that the reaction to the secondary force disappears after a while when the original force is not presented for a relatively long period. Consistent with this phenomenon, we assume that the structure of the acoustic network decays exponentially to the baseline state, which can be expressed by

$$\Delta a_A(n) = -\epsilon_2 (a_A(n))^{1+r} . \tag{9.4}$$

Here, n denotes the events when the secondary force is presented alone. Note that above the parameter ϵ_2 was introduced in order to account for the fact that in general the speed of the build-up of the new structure is different from the speed of the decay to baseline. Importantly, except for a possible change in the numerical value of the speed coefficient ϵ Eqs. (9.2) and (9.4) actually correspond to a single law. Equation (9.4) describing the decay to the baseline corresponds to Eq. (9.2) when we put the reference value a_T in Eq. (9.2) to zero because there is no coincidence of brain activity A_T and A_A. That is, Eqs. (9.2) and (9.4) can be written as a single restructuration equation like

$$\Delta a_A = -\epsilon (a_A)^r (a_A - \xi) ,$$

if $A_T \wedge A_A$ at finite stationary values $\Rightarrow \xi = a_T$,

otherwise $\Rightarrow \xi = 0$.   (9.5)

Panel (d) presents the results of simulations that describe the return of an animal's brain structure to its baseline structure. The pattern component a_A (top subpanel) and the probability $p(SFR)$ (bottom subpanel) are shown as functions of trials n, in which only the bell was ringing and, consequently, only A_A was finite but $A_T = 0$. As expected, the simulation shows that the pattern component a_A and the probability $p(SFR)$ decayed monotonically over the course of several events that involved the unpaired, secondary external force of a ringing bell.

Note that in the literature the return to the baseline has also been called the "extinction of a learned conditional reaction". However, if we stretch a rubber string, then the length of the string increases, which does not mean that the rubber string has "learned" how to make its size longer. Likewise, if we relax the rubber string, the length of the rubber string will get back to the rest length. This does not mean that an "extinction of the learned length-increase of the rubber string" did take place.

9.1.2   Eigenvalue-Shift Model

The component-shift model assumes that the neurons or neuron populations of the two networks related to the pair of external forces inducing brain restructuration can project their signals to the same target areas in the human or animal brain. Therefore, when those neurons or neuron populations are active then this leads to the same consequences. Let us briefly address a different scenario that comes with a slightly revised version of the component-shift model and is sketched schematically in Fig. 9.4. Accordingly, the network for tasting affects the structure of the acoustic network in the same way as for the component-shift model. However, the a_A neuron or neuron population of the acoustic network does not project signals into the same region as the a_T neuron or neuron population of the taste network. Rather, the neuronal activity a_A is assumed to act back on the network for tasting. More precisely, it acts back on the structure underlying the BA attractor pattern that caused the structural change in the first place and supports its emergence. That

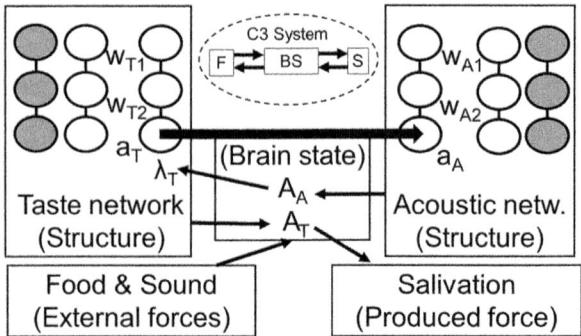

Fig. 9.4 As in Fig. 9.2 but for the eigenvalue-shift model

is, the eigenvalue-shift model describes an amplifier effect. In other words, while the BA pattern for eatable goods with amplitude A_T in combination with the BA pattern for the ringing-bell with amplitude A_A lead to a change of the structure in terms of an increase of the component a_A towards a_T, the a_T neuronal activity acts back on the emergence of the BA pattern for eatable goods by increasing its eigenvalue λ_T (see Fig. 9.4 again). That is, we deal with a pattern sequence in the grand state as discussed in Chap. 7. If the ringing-bell BA pattern emerges after the structure has changed then the A_A neuronal activity is assumed to increase the eigenvalue λ_T such that the taste network becomes monostable and the BA pattern for eatable goods emerges. The BA pattern for eatable goods emerges even if there is no corresponding external force acting on the animal and there is no chemical driving force acting on the taste receptors of the sensory body of the animal. The possibility that BA pattern emerges that typically are reactions to external forces without the presence of the relevant external forces has been discussed in detail in the context of pattern formation in the human isolated system (see Sect. 5.5). BA patterns of eatable goods can form in human individuals in the isolated brain in the awake person (i.e., when people "think" about food) and in individuals in the REM stage (i.e., when people "dream" about food). Once the BA pattern for eatable goods has emerged the a_T neuronal activity leads to the production of saliva.

9.2   Restructuration in E1 Systems, "Operant Conditioning", and "Reinforcement"

Let us consider a typical experiment in which an animal over time changes its reactions to the environment as a result of paired forces acting from the environment on the animal that are in part related to a reward. A rat is put into a box with a lever. If the rat presses the lever down then after a certain period a piece of food (a food pellet) is given to the rat. That is, every time the rat presses the lever it receives a

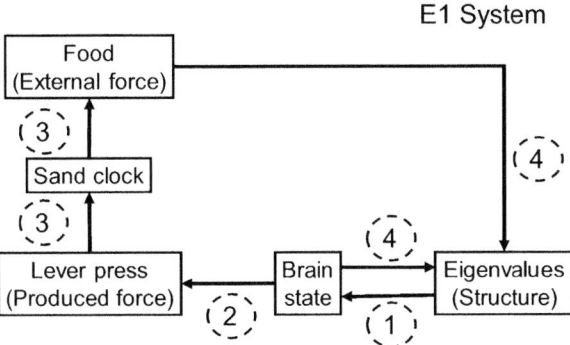

Fig. 9.5 E1 pattern formation perspective of the restructuration of rats leading to an increase in lever pressing. The numbers 1,2,3,4 refer to the mathematical model described in Sect. 9.5.2

reward after a certain waiting period. The fundamental observation in such animal experiments is that the animal presses the lever more and more often over time and, in doing so, it receives more food. This observation has been called "learning". The food reward has been called "reinforcer of behavior". The phenomenon of a change in the reactions of the animal to the environment (i.e., the lever) has been called "reinforcement".

From a physics perspective and in particular from the perspective of the human pattern formation reaction model, such observed changes are consistent with grand state dynamics in E1 pattern formation systems. Figure 9.5 describes the E1 system mechanism leading to brain restructuration in the context of the aforementioned experiment in which rats are restructured to press a lever.

Let us consider the E1 system mechanism sketched in Fig. 9.5 and start with the brain state. Let us assume the ongoing brain activity is in the basin of attraction of the attractor for pressing the lever. The lever pressing BBA pattern emerges and the animal presses the lever. The lever press switches a sand clock that counts time. After the period measured by the sand clock (in the simulation the period is 2 time units) the sand clock is empty and a food pellet is given to the animal. During that waiting period the animal is assumed to do either something else or to press the lever. Accordingly, while the sand clock is running down the brain state evolves and other BBA patterns emerge or the lever pressing BBA pattern is maintained. That is, we take into consideration that attractor pattern emerge and disappear in a more or less continuous fashion. In particular, after the lever pressing BBA pattern has emerged there a two scenarios. The BBA pattern becomes unstable, disappears, and another BBA pattern emerges. In this case, the pattern amplitude of the lever pressing BBA pattern decays. The animal is doing something else. In the second scenario the animal continuous to press the lever or during the waiting period presses the lever again after having done something else in-between. In both cases of the second scenario the lever pressing amplitude increases such that the lever pressing BBA pattern emerges. At issue is that the lever pressing amplitude is at some non-zero finite value at the time point when the sand clock is empty and the food is given to the animal. That is, in comparison to a relatively large number of other pattern amplitudes the lever pressing amplitude is at a relatively large value. It is assumed that the external force in terms of a reward increases the eigenvalues of all basis patterns that have relatively large amplitudes. That is, all eigenvalues are increased that are associated with large pattern amplitudes at the time point of the reward. This implies that for the attractors related to those basis patterns the basins of attraction become larger. However, if the basin of attraction of an attractor is increased then "the chance" that the corresponding pattern emerges increases as well. Therefore, the BBA patterns whose eigenvalues are increased by the external force exerted by the reward are more likely to emerge again. Importantly, if the waiting period is not too long, then the lever pressing pattern is in any case among the basis patterns whose eigenvalues are increased. In this case, since the lever pressing BBA pattern is in any case among the set of patterns with increasing eigenvalue, whereas all other patterns are only sometimes in that special set of patterns, there is a net effect that specifically increases the eigenvalue of the lever pressing basis pattern. The eigenvalues of the remaining basis patterns increase as well but the effect is smaller in magnitude. Consequently, specifically the lever pressing reaction to the lever increases over time, that is, increases over the number of completed cause-and-effect cycles involving lever pressing and structural changes in terms of eigenvalues.

The E1 system described schematically in Fig. 9.5 can be modeled by means of the LVH model introduced in Chap. 4. In particular, the numbers shown in Fig. 9.5 refer to equations determining the relationships between structure and pattern dynamics (1), produced force and brain state (2), produced force and environment (3), and environment and brain state and structural changes (4).

Figure 9.6 shows simulation results obtained on the basis of the LVH model for the general scheme shown in Fig. 9.5 and the relationships (1), (2), (3), and (4) pointed out there (see mathematical notes for details, Sect. 9.5.2). For sake of simplicity, only ten possible BBA patterns have been considered. Moreover, it was assumed that each pattern occurs for one time unit. Subsequently, it disappears or it is maintained for another time unit. The switches between the BBA attractor patterns are assumed to depend on the basins of attraction of the attractors related to the patterns. Panel (a) shows the sequence of simulated BBA patterns and the switches between the patterns. In panel (a) only 100 time units are shown. Panel (b) presents the amplitudes of the BBA patterns. It was assumed that if a pattern emerges the amplitude increases to a magnitude of 3 units. Subsequently, every time unit the amplitude decays by 1 unit provided that the pattern is not maintained or does not re-appear. The ten subpanels in panel (b) show the amplitudes of the ten BBA attractor patterns, where the subpanel on the bottom refers to pattern 1 and the subpanel on the top refers to pattern 10. The BBA pattern $k = 1$ was assumed to describe the lever pressing. That is, when the pattern $k = 1$ emerged, then the simulated animal pressed the lever and the sand clock was started. In the simulation the sand clock took 2 time units to run out. That is, after 2 time units a food reward was given to the animal. In the bottom subpanel of panel (b) the dots indicate the time points when food rewards were given to the simulated animal. At those time points the

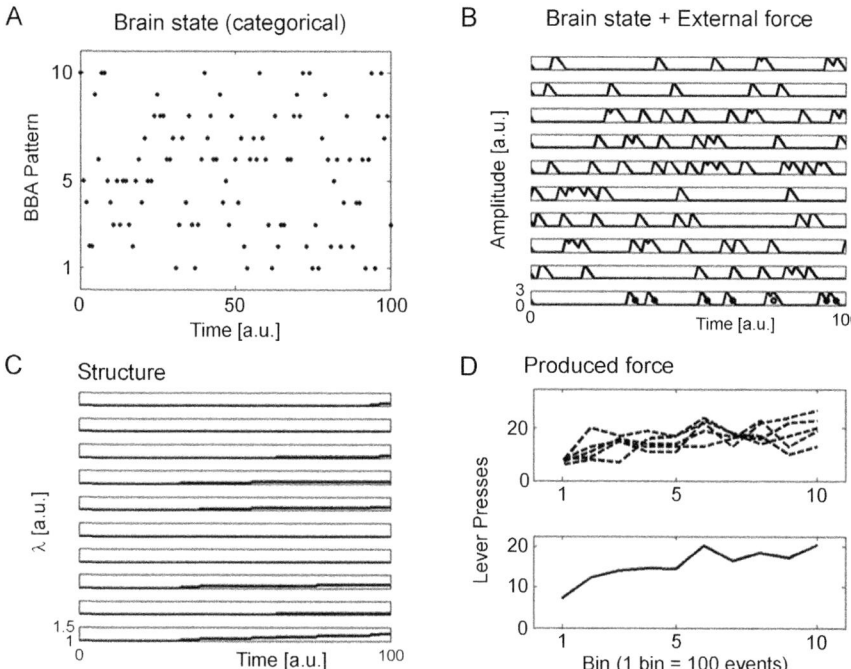

Fig. 9.6 Simulation results obtained for the E1 system described in Fig. 9.5 on the basis of the LVH model. In the simulation 10 BBA patterns are considered. Pattern $k = 1$ describes level pressing body activity and the corresponding brain activity. Patterns $k = 2, \ldots, 9$ are related to other activities. Panels (**a**), (**b**), (**c**) show simulations for a single rat over a time period of 100 time units (panel (**a**): categorical brain state; (**b**): BBA pattern amplitudes $A_1, \ldots, A_{10}$ and external food force; (**c**): eigenvalues $\lambda_1, \ldots, \lambda_{10}$). Bottom subpanel $k = 1$. Top subpanel $k = 10$. The circles in the bottom subpanel for $A_1(t)$ in panel (**b**) indicate time points when food is given to the rat. Panel (**d**): Simulation results for five animals over time periods of 1000 time units. Time is binned into episodes of 100 time units. The number of emerging lever pressing BBA patterns in each bin is shown for individual rats (top subpanel) and averaged across all rats (bottom subpanel)

amplitude of the lever pressing basis pattern was finite in any case (see the solid line in the bottom subpanel). In the simulation, the eigenvalues of all basis patterns with finite (i.e., non-zero) amplitudes were increased by a particular amount. The eigenvalues as functions of time are shown in panel (c). Again, only 100 time units are shown. The eigenvalues of all ten BBA patterns were initially fixed at 1 units. Panel (c) demonstrates that the eigenvalues increased over time (consistent with the theoretical considerations made above). However, the eigenvalue of the lever pressing pattern $k = 1$ increased most (compare the bottom subpanel of panel (c) with all other subpanels). Panel (d) shows for 5 simulated animals the frequency of lever presses over time (top panel) for simulations of 1000 time units. The number of lever presses were counted in bins of 100 time units.

Panel (d) shows that overall, as expected, the lever press frequency increased as a function of time (in fact, the lever press frequency increased as a function of the number of cause-and-effect-cycles involving lever pressing and structural changes). The bottom subpanel of panel (d) shows the average over the five simulated animals. Lever pressing on average increased consistent with what has been reported in the literature in the context of so-called "reinforcement" experiments of animals.

The model described above and sketched in Figs. 9.5 and 9.6 is a simplified model in which eigenvalues only increase (e.g., until they reach some saturation values as in the simulations shown in Fig. 9.6). The model could be generalized to take both increasing and decreasing eigenvalues into account. For example, the reward could increase the eigenvalues of the basis patterns with relatively large amplitudes at the time of the reward and decrease the eigenvalues of all remaining basis patterns. Likewise, eigenvalues could decay to baseline levels explicitly as functions of time (just like a rubber string relaxes back to its rest length when it is no longer stretched; see also the model on child play in Sects. 7.3.3 and 7.8.1 and Eq. (7.6)).

Note also that in the model above based on an E1 system the impact of the external forces on the brain state are ignored. For example, while the force exerted by the lever on the animal is assumed to shift the brain state into the basin of attraction of the attractor of the BBA lever pressing attractor pattern, this impact on the animal is not taken into consideration. Likewise, while the sand clock is running out the animal is assumed to react in various ways to the cage. In this context, external forces exerted by the cage again are assumed to shift brain activity in various ways, which is not captured by the E1 model. If such direct impacts of features of the environment on the brain state are taken into consideration, then we are dealing with an E3 system rather than a E1 system (see the system classification in Fig. 5.5).

9.3   A Third Look at Tolman's Experiments with Animals

In Sects. 6.4.3 and 7.4.4 two models were proposed to address experimental settings that restructure animals such they find food rewards. The experiment conducted in this regard by Tolman [319] involved two doors that differed with respect to some visual properties (e.g., by color). Only one door led to the food reward.

The first model was proposed in Sect. 6.4.3 on basis of considerations about ordinary states and brain states, in particular. The pattern formation model proposed in Sect. 6.4.3 to explain some of the data reported from this experiment assumes that the forces exerted by the doors on the rats result in a shift of the brain states of the rats such that shifted brain states fall into the basins of attraction of attractors associated with movement patterns that take the animals through the food doors. The model will be referred to as brain state shift (BS-shift) model. The degree of the shift (as quantified by the shift parameter μ) is assumed to change during the restructuring period from a zero shift to a dominant shift that makes that in any trial the brain state is shifted into the aforementioned basin of attraction, see Fig. 6.15. In the case of the zero shift there is a 50% "chance" that the animals go through the food door (i.e., the door that leads to the food reward), whereas in the case of a dominant shift the animals walks through the food door in any repetition of the experiment.

Another model was proposed in Sect. 7.4.4 on the basis of considerations on the grand states of animals that are composed of ordinary states (brain states plus "body" states) and structure. The model in Sect. 7.4.4 assumes that the structure in terms of the eigenvalues of the movement patterns towards and away from the doors is changed during the restructuring period. That is, the eigenvalues are shifted in some sense. Therefore, the model will be referred to as eigenvalue-shift (λ-shift) model. In this section, mechanisms will be discussed that are consistent with the experimental results reported by Tolman [319] and lead to the hypothesized changes of the parameters that determine brain state shifts, on the one hand, and eigenvalue shifts, on the other hand.

Figure 9.7 describes the restructuration mechanisms of the BS-shift and λ-shift models schematically. Accordingly, the mechanisms are assumed to be part of E1 pattern formation systems that determine the reactions of the animals, see panel (a). When the animal is confronted with the black or white doors (assuming that the doors are colored black and white), then depending on the door color and the brain structure the brain state converges to the walk-through-the-white-door or walk-through-the-black-door BBA attractor pattern. The animal passes through the corresponding door. Subsequently, it is assumed that it takes some time for the animal to run through the maze and reach the food. During that period the amplitude of the BBA pattern that determined the animals reaction at the doors

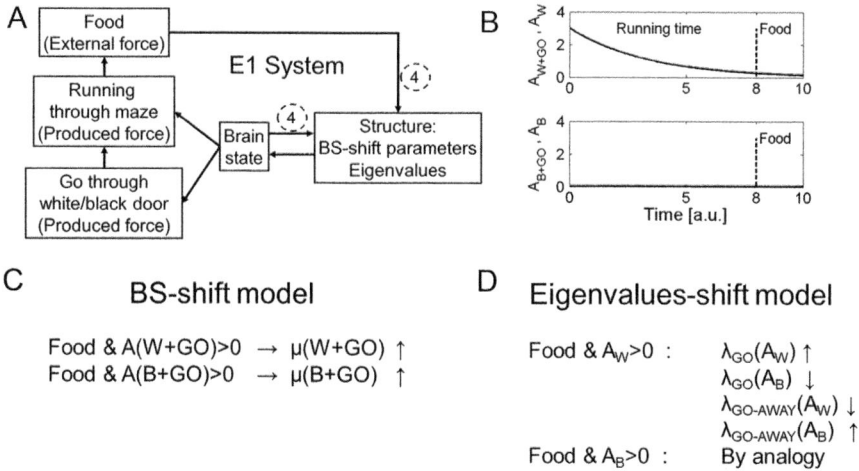

Fig. 9.7 E1 pattern formation model to explain brain restructuring of rats by means of (forces exerted by) food rewards. See text for details

decays. Without loss of generality, let us assume the white door is the food door. Panel (b) illustrates for the BS-shift model the case when the animal indeed runs through the food door and the amplitude $A_{W\text{-}GO}$ of the white-door-and-go-through-the-door BBA pattern decays while the animal runs through the maze (see top subpanel). Importantly, although the amplitude decays the amplitude is larger than the amplitude $A_{B\text{-}GO}$ of the black-door-and-go-through-the-door pattern that is assumed to be at a negligibly small level (see bottom subpanel). Likewise, panel (b) also illustrates for the eigenvalues-shift model the case when the animal goes through the food door. In the context of the λ-shift model, the amplitude A_W of the white door BBA pattern decays as function of the running time but it is still larger than the amplitude of the black door BBA pattern. As a consequence, when the animal reaches the food reward the amplitude ($A_{W\text{-}GO}$ for the BS-shift model and A_W for the λ-shift model, respectively) of the pattern that caused the animal to find the food source is relative large as compared to the amplitude ($A_{B\text{-}GO}$ or A_B) of the pattern that does not allow the animal to find the food source.

Just as in the pervious section (Sect. 9.2), it is assumed that parameters are changed that are related to patterns that exhibit relative large amplitudes at the time point of the reward. Panels (c) and (d) describe the changes schematically for the BS-shift and λ-shift models, respectively. If the food is received and the amplitude of the white-door-and-go pattern is finite (because the white door is the food door) then the shift parameter μ for the white-door-and-go pattern introduced in Sect. 6.4.3 is increased. The parameter μ is gradually increased, see panel (b) of Fig. 6.15. The increase of that parameter μ leads to a change in the reactions of the animal to the doors as discussed in Sect. 6.4.3. In particular, when looking at a sample of animals or several repetitions of trials for the same animal then the percentage of runs through the food door (here the white door) increases, see panel (d) of Fig. 6.15. If the food door is the black door, analogous considerations can be made. Namely, as described in panel (c) of Fig. 9.7, if the food reward is received by the animal and the amplitude of the black-door-and-go pattern is finite (because the black door is the food door) then a shift parameter related to the black door (in analogy to the shift parameter of the white door) is increased. As pointed out in panel (c), pattern amplitudes and external forces (given in terms of or exerted by the food) in combination affect shift parameters that are assumed to be a part of the structure of the animal. In other words, according to the model, structural changes are caused by an interplay between the brain state and external forces. It is a key property of E1 pattern formation systems that such combined effects are possible (e.g., A,B,C, and D systems cannot account for such effects). The label 4 (used in analogy to the example discussed in the previous section) in panel (a) indicates this combined effect of external forces and the brain state on the structure.

Panel (d) schematically describes structural changes that take place according to the eigenvalue-shift model consistent with the experimental data of Tolman's experiment reported in Sect. 6.4.3. Accordingly, if at the time the food is received by the animal the amplitude of the white door BBA pattern is finite (because the white door is the food door and the animal passed through that door), then functional relationships between the eigenvalues $\lambda(go)$ and $\lambda(go\text{-}away)$ (i.e., the eigenvalues of the go-through-the-door and the go-away-from-the-door BBA patterns) and the amplitudes A_W and A_B of the white and black door BBA patterns are changed. The eigenvalue $\lambda(go)$ as a function of A_W is increased and/or as a function of A_B is decreased. Likewise, the eigenvalue $\lambda(go\text{-}away)$ as a function of A_W is decreased and/or as a function of A_B is increased. If the food door is the black door analogous changes of the functional relationships can be defined. In any case, these changes result that the eigenvalues become dependent on the amplitudes of the white and black door BBA patterns as described in Chap. 7 and illustrated in Fig. 7.14.

Figure 9.8 exemplifies structural changes taking place over the restructuration period for an explicit (but very simplified) λ-shift model. The model describes the functional relationships between states (as measured by pattern amplitudes A_W and A_B) and structure (as given by eigenvalues λ) in terms of two linear regression equations. Accordingly, it is assumed that the eigenvalues depend on the amplitudes like $\lambda(go) = L_1 + c_{11}A_W - c_{12}A_B$ and $\lambda(go\text{-}away) = L_2 - c_{21}A_W + c_{22}A_B$, where $L_1, L_2, c_{11}, c_{12}, c_{21}, c_{22}$ are parameters and (without loss of generality) signs in front of the parameters have been used that simplify the presentation. In line with the discussion in Sect. 7.4.4 and panel (b) of Fig. 7.14, in what follows we will consider only the special case in which c_{11} and c_{22} are equal to zero. In this special case, we have $\lambda(go) = L_1 - c_{12}A_B$ such that the eigenvalue $\lambda(go)$ depends only on the amplitude A_B of the black-door BBA pattern (or is constant if $c_{12} = 0$). Likewise, we have $\lambda(go\text{-}away) = L_2 - c_{21}A_W$ such that the eigenvalue $\lambda(go\text{-}away)$ depends only on the amplitude A_W of the white-door BBA pattern (or is constant if $c_{21} = 0$). Figure 9.8 illustrates the linear relationships in form of graphs over the 20 days restructuring period used in Tolman's animal experiment.

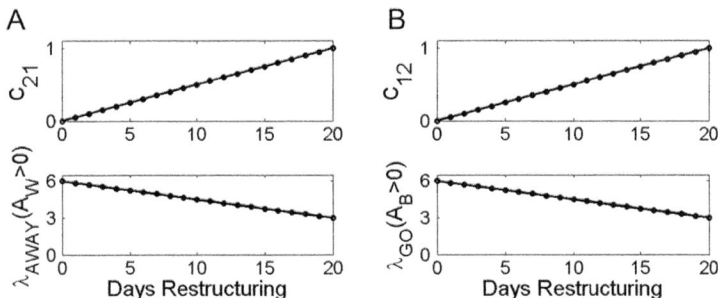

Fig. 9.8 Structural changes as described by the λ-shift model. Panels (**a**) and (**b**) show how the λ-related proportionality factors c_{21} and c_{12} increase (i.e., are shifted) over discrete events of restructuration induced by food reward forces (top subpanels). Panel (**a**): The increase of the structure proportionality factor c_{21} (top subpanel) results in a decay of the eigenvalue $\lambda_{go\text{-}away}$ (bottom subpanel) of the basis patterns for walking away if the BBA pattern of the white door emerges ($A_W > 0$). Eventually, this decay at day 20 makes the system monostable such that the animal walks in any case through the white door. Panel (**b**): The increase of the structure parameter c_{12} (top) makes that λ_{GO} decays (bottom) when the black-door BBA pattern emerges ($A_B > 0$) such that eventually at day 20 the system becomes monostable and the animal goes away from the black door in any case

At the beginning of the restructuring period, the parameters c_{12} and c_{21} are assumed to be zero. L_1 and L_2 are assumed to be equal. In panel (b) of Fig. 7.14 it was assumed that the eigenvalues $\lambda(go)$ and $\lambda(go\text{-}away)$ assume a value of 6 units at the beginning of the experiment. Accordingly, we put L_1 and L_2 equal to 6 units. As mentioned above, the parameters c_{21} and c_{12} are assumed to be zero. As a result, as shown in panels (a) and (b) of Fig. 9.8 for zero parameters c_{21} and c_{12} we obtain eigenvalues $\lambda(go)$ and $\lambda(go\text{-}away)$ of 6 units. The pattern formation system is symmetric. Consequently, in 50% of the runs the animal will go through the white door and in the remaining 50% the animal will go through the black door. Due to the mechanism described above, when the food is received by the animal the structure changes. Assuming that the white door is the food door, as described in panel (d) of Fig. 9.7 the eigenvalue $\lambda(go)$ as a function of A_B decays. Likewise, the eigenvalue $\lambda(go\text{-}away)$ as function of A_W decays. This effect can be modeled by assuming that the coefficients c_{12} and c_{21} become positive and increase in every run in which the animal finds the food by passing through the white door. Assuming that the pattern amplitude A_W and A_B are at constant levels of 3 units when the corresponding patterns emerge, it follows that increasing the parameters c_{12} results in a decay of $\lambda(go)$ when the animal stands in front of the black door and the black door BBA pattern emerges. Therefore, the animal has a reduced "tendency" to go through the black door. Likewise, increasing c_{21} results in a decay of $\lambda(go\text{-}away)$ when the animal stands in front of the white door and the white door BBA pattern emerges. The animal has a reduced "tendency" to go away from the door and there is a higher "chance" that is goes through the door. After the 20 days of restructuration it is assumed that the parameters c_{12} and c_{21} are at a level of 1 unit. This implies that the corresponding eigenvalues drop to a level of 3 units and, in doing so, drop out of the stability band. This case was considered in Sect. 7.4.4 and is illustrated in panel (b) of Fig. 7.14. In this case, the pattern formation system is monostable and the animal goes through the food door in any case. In summary, the λ-shift E1 pattern formation system sketched in Fig. 9.7(panels a, b and d) can explain the changes in the animal reactions observed in Tolman's experiment when the system is supplemented with an explicit mechanism as sketched in Fig. 9.8 that describes how structural changes (i.e., eigenvalue shifts) are caused by an appropriate combination of external forces (related to food rewards) and brain states.

Note that eventually the structural changes are caused by an appropriate pairing of external forces. On the one hand, the forces exerted by the door colors induce the emergence of certain door-color BBA patterns. On the other hand, the food rewards exert chemical forces on the animals. The chemical driving forces show up explicitly in the E1 scheme. The external forces of the doors are not taken into account. In analogy to the discussion at the end of the previous section (Sect. 9.2), the E1 model may be generalized to a E3 model in which it is explicitly considered that the door colors exert forces and shift brain states such that those shifted brain states fall in the basins of attraction of the appropriate door-color attractors.

As suggested in Sect. 7.4.4, models that explain the animal reactions (and the changes of those reactions) to the white and black doors of the rudimentary maze used in Tolman's study can be generalized to describe the animals reactions while

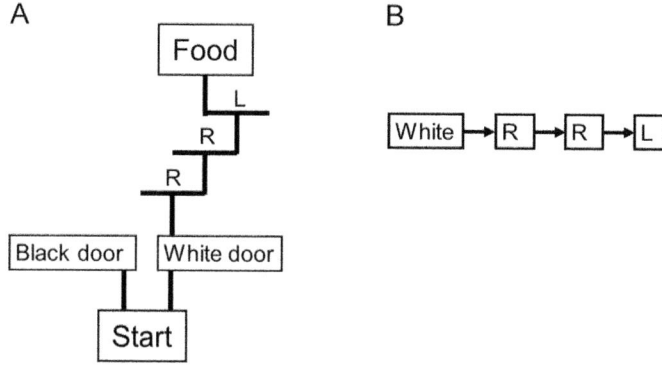

Fig. 9.9 Maze with two differently colored doors and T-junctions (panel (**a**)) and solution sequence of BBA attractor patterns that takes the animal to the food reward (panel (**b**))

the animal is running through an ordinary maze composed of several T-junctions. For example, for mazes involving several T-junctions, pattern formation system could be describe separately for every T-junction and, subsequently, the individual pattern formation systems could be merged together to an overall pattern formation system. In this context, an even more general point of view can be taken. Let us consider panel (a) of Fig. 9.9. Panel (a) shows part of a hypothetical maze involving T-junctions. For this maze, in order to get to the food the animal must pass through the white door, turn right at the first T-junction, turn right again at the following T-junction, and turn left at the final T-junction. If so, the animal produces an elementary sequence of BBA attractor patterns that is shown in panel (b): there is the white door BBA attractor pattern, a turn right BBA pattern, another turn right BBA pattern, and a turn left BBA pattern. The sequence is referred to as elementary because in fact the animal while walking through the maze would produce various additional BBA patterns that are not listed in the elementary sequence. The elementary pattern sequence can be put into the context of the pattern sequences discussed in Chaps. 6 and 7. In particular, the BBA pattern sequences of child play discussed in Sect. 7.3.3 and the sequences induced by oscillatory human reactions studied in Sect. 7.4.3 can be regarded as patterns on a higher hierarchical level. These higher hierarchical patterns (by definition) are dynamic in nature, that is, exhibit a temporal aspect. In short, we may interpret sequences of BBA patterns that emerge and disappear over time as spatio-temporal BBA attractor patterns. Importantly, they may emerge as a whole on reference points that become unstable just as individual BBA patterns emerge at reference point. If so, the amplitude equation description applies to those higher hierarchical patterns. Accordingly, those higher hierarchical BBA patterns exhibit amplitudes that describe the emergence of pattern sequences as a whole.

Let us dwell on this issue. As argued above the sequence of left- and right-turn patterns shown in panel (b) of Fig. 9.9 correspond to a higher hierarchical spatio-temporal pattern that comes with its own amplitude. At reference points that become unstable (i.e., when animals are placed at starting locations of a maze) it is assumed that several higher hierarchical BBA patterns exist in terms of unstable basis patterns. Some of them have eigenvalues in the stability band, others not. By definition, when an animal reaches a food source and receives a food reward, then the amplitude of the corresponding spatio-temporal BBA (sequence) attractor reaction pattern (that made that the animal received the food) is finite and large compared to other spatio-temporal BBA (sequence) patterns that do not lead the animals to the respective food source. In line with our previous considerations, the food reward is assumed to increase the eigenvalue of that higher hierarchical BBA pattern that comes with a finite amplitude at the time of the food reward. In doing so, the "chance" that the corresponding sequence (i.e., higher hierarchical pattern) occurs again increases over time (i.e., as a function of the completed maze runs that involve a food reward).

9.4 Why "Learning", "Forgetting", and "Knowledge" Are Unnecessary Concepts

9.4.1 The "Learning" Seesaw

Changes in the reactions of animals (and humans) to external forces have been considered in the literature as "learning". For example, the graph shown in panel (a) of Fig. 9.10 that shows the percentage of runs through the food door as function of the "training" days observed in the aforementioned experiment by Tolman [319] has been interpreted as "learning curve". In fact, from a physics perspective all kind of system (self-organizing or not self-organizing systems, living or non-living systems) exhibit structural changes and as a result of such changes those system change the way they interact with other systems. As an example, a guitar string changes the way it reacts to the forces exerted by the guitar player when the string is stretched (see Chap. 1). Explicitly, when the string is stretched, the string produces sounds with a higher frequency. The guitar string does not "learn" to produce higher frequency sounds. Likewise, humans and animals do not learn anything. Rather, they change their structure.

Let us illustrate this point with anther example: a special kind of seesaw with a basket sketched in panels (b) and (c) of Fig. 9.10. Let us consider a seesaw that does not have a straight bar but a bar that is shaped like an inverted V, as shown in panel (b). That is, an inverted V-shaped bar is attached to a straight pole. If there is no load then the two levers of the bar point down and form 30° angles with the horizontal, see panel (b) again. If we load one side then the other side goes up. For example, if we push the right lever down by 31°, then the left lever goes up by 31°. The left lever points slightly upwards at an angle of 1° with respect to the horizontal, see also panel (c). It is assumed that there is a torsion spring attached (not shown in the graphs) that holds the inverted V-shaped bar in a horizontal orientation (i.e., the

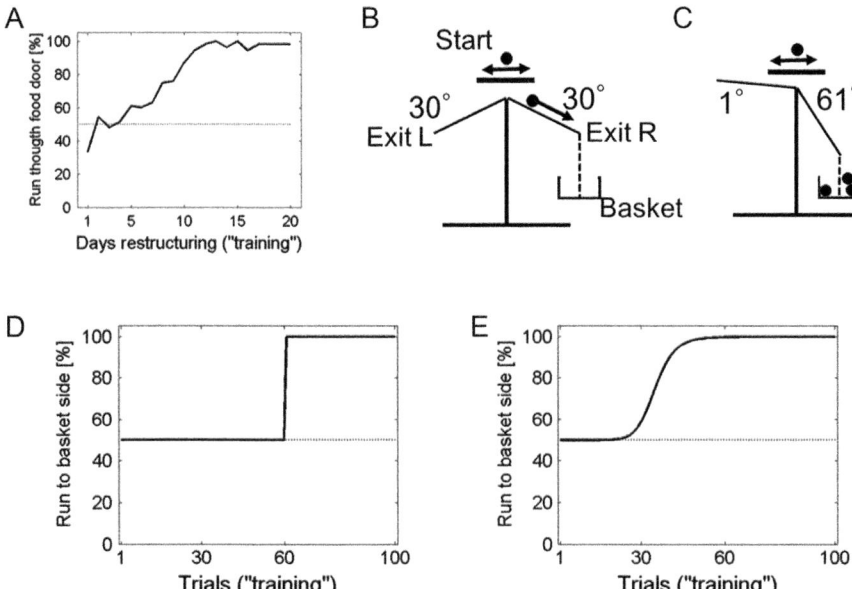

Fig. 9.10 Panel (**a**): Change in the reactions of animals in Tolman's experiment [319] due to restructuration. Panels (**b**) and (**c**): Set-up of a seesaw that changes its structure when balls are released at the top. Panels (**d**) and (**e**): Change in the performance of the seesaw due to structural changes for the idealized case of a sharp peak of the inverted V-shaped bar (panel (**d**)) or a rounded peak (panel (**e**))

orientation with the 30° angles on both sides). From a non-scientific point of view, the aim is to "train the seesaw to deliver metal balls to one specific exit". That is, we want that the seesaw "learns" something. To this end, we attach a basket that has negligible weight on the lever with the exit of interest. In Fig. 9.10 the exit is the right one. From a physics perspective we modify the seesaw and, in doing so, change it properties. Given the modification with the basket, metal balls are released one after another at the top of the seesaw. The balls are released not exactly over the top of the seesaw, see panels (b) and (c). Rather, they are released at positions that fall into a small interval to the left and right of the top of the seesaw. The position within that interval is determined by a "random generator". Let us assume that the inverted V-shaped bar has a sharp peak at the point where the right and left lever are merged together. In this case, if a ball is released to the left of the straight pole then it rolls down the left lever as long as this lever points down. If it is released to the right of the pole then it rolls down the right lever as long as that lever points down. In the example the basket is attached to the right side. Any time a ball rolls down to the right side the ball falls in the basket. This increases the load on the right side. Let us assume the torsion springs and the weight of the balls is such that each ball in the basket rotates the inverted V-shaped bar by 1° clockwise. Consequently, when 61 metal balls have been released and 30 have been released to the left hand side, while 31 have been released to the right hand side, then there are 31 balls in the basket on the right and the right lever points down with an angle of $61°$ while the left lever points up with an angle of $1°$ with respect to the horizontal. This situation is shown in panel (c). Any further metal ball rolls down to the right irrespective of the release point. That is, even if a ball is released to the left of the straight pole, the ball rolls down to the right side. The reason for this is that the left lever points up. If the experiment is repeated several times and we calculate the percentage of ball runs to the basket given a number a "training" runs, we obtain the graph in panel (d). At about 60 runs the basket contains about 30 balls and the critical condition is reached. Therefore, the percentage of ball runs to the right increases sharply from 50% to 100%.

So far we have considered an inverted V-shaped bar with a sharp peak. Let us assume that the two levers are merged by means of a round connection piece. In this case, as soon as the inverted V-shaped bar is slightly rotated clockwise, there is a higher chance that balls roll to the right (because some balls that a released to the left but sufficiently close to the pole roll to the right). As a result, the percentage of ball runs to the basket as function of the "training" runs shows a smooth transition from the 50% to the 100% level as indicted in panel (e). From a non-scientific perspective, the graphs in panel (d) and (e) may be interpreted as "learning curves". The seesaw has "learned" to deliver the balls to the basket side. Catching a ball in a basket might be interpreted as "reward" that "reinforces" the seesaw to rotate clockwise in order to get more rewards. From a physics perspective we are dealing with a system that changes its structure when it interacts with the environment and due to these changes in structure the reaction of the system changes.

In fact, the change of structure due to forces acting on system is a general feature of many system. For example, the author of this book has shoes. When they are new they have a certain structure. When the author has used his shoes for half a year, they look different and have changed their structure. The author of this book also has T-shirts that change there shape and elasticity when the author uses them and puts them into the washing machine. The author of this book speculates that the reader also has shoes and T-shirts that change their structure when they are used. Shoes and T-shirts are not supposed to "learn" anything. Like the guitar string, they do not "learn". In a similar vain, from a science point of view, the seesaw described above does not "learn" anything and the animals in Tolman's experiment do not "learn" anything. In general, humans and animals do not "learn".

9.4.2 Who Says That the Air over the Atlantic Ocean Is "Learning" and "Forgetting"?

The concept of "learning" seems to be a key concept in developmental psychology. Experimental studies have identified various developmental milestones of children such as sitting and walking [27, 42, 94, 350]. Some further milestone are listed in

Table 9.1 Examples of gross motor milestones

Age [month]	Motor BBA pattern	References
4.5	Rolling over	[5, 18, 123]
5	Sitting with support	[5]
6	Sitting without support	[5, 123]
7.5	Manages to sit up	[5, 18]
8	Crawling	[5]
14	Walk	[5, 18]
24	Running	[18]
30	Walking on tiptoe	[18]
35	Balancing on a foot	[123]

Table 9.1. Accordingly, at certain ages infants begin to exhibit certain motor skills. From a dynamical systems perspective [250] and pattern formation perspective [94, 103, 121] this means that certain BBA patterns begin to emerge at characteristic time points in the lives of human beings. In this context, the question arises whether from a physics perspective infants actually "learn" to produce those BBA patterns. Do they "learn" to roll over, to sit up without support, and to walk? From the considerations made above, the answer is that they do not "learn" anything at all.

In order to clarify this issue let us have a look at infant-mother face-to-face visual world interactions. During the first few months of an infant's life, an infant interacts with his/her mother or caregiver among other things by means of facial expressions and gestures [169]. In particular, at the end of the second month there is a relatively dramatic increase in the use of facial gestures of the infant that are aimed towards the mother [163]. For example, infants begins to smile in social contexts [353]. That is, infants react to the forces of the visual or acoustic world (or mechanic forces related to touch) exerted by their caregivers by producing a smiling BBA pattern. Panel (a) of Fig. 9.11 shows how much time infants spend on performing face-to-face interactions during 3 min observation periods. That is, the durations during which infants produce face-to-face interaction BBA attractor patterns are shown for several infant ages [103, 206]. The developmental milestones listed above in Table 9.1, the increase in the emergence of face-to-face interaction BBA patterns shown in Fig. 9.11a, and developmental transitions in grasping styles have been explained within the framework of the pattern formation reaction model in previous studies [94, 103, 121].

In this section, the goal is to point out that motor milestones from a pattern formation perspective are seen in analogy to seasonal effects that involve the emergence of large scaled atmospheric patterns such as monsoon rainfalls, winter snow, lake effect snow, and hurricanes. That is, the step from infancy to childhood should be seen in analogy to the step from the non-hurricane season to the hurricane season.

For example, summer monsoon rainfalls and winds occurs in India over the summer months (June, July, August) [173], see panel (c) of Fig. 9.11. These rainfalls and winds are part of dynamical, cyclic patterns that take up water over the oceans

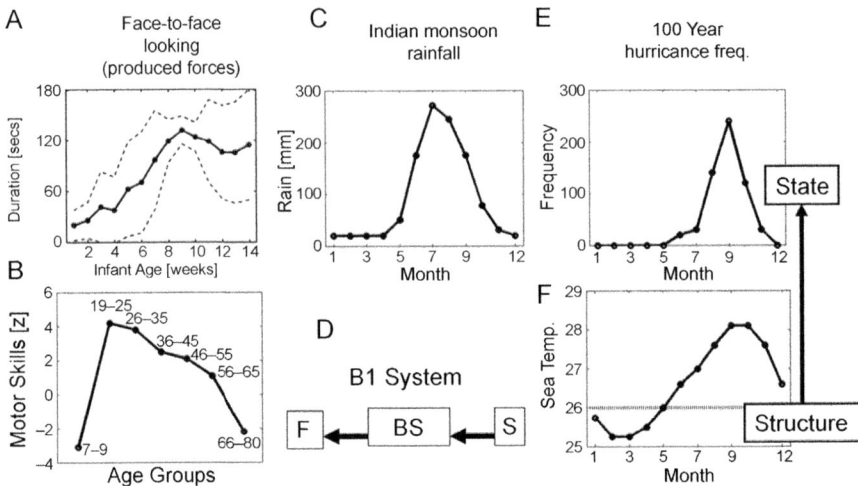

Fig. 9.11 Structural changes leading to emergence of attractor patterns in the animate and inanimate worlds. Panels (**a**): The duration of infant-mother face-to-face interactions shows a dramatic increase in the second month of life of infants consistent with structural changes in B1 pattern formation system (panel (**d**)). Figure taken from Frank [103] under open access CC license. Panel (**b**): Index of how well motor skills are performed in certain age groups [210]. The index corresponds to a so-called z-score. Positive and negative values indicate above- and below-average performance, respectively. Performance is poor in the young and elderly age groups and is best in the age group of 19–25 years old. In the context of the pattern formation approach, the poor performance is interpreted as the emergence (age group 7–9) and disappearance (age group 66–80) of attractors due to changes in the human structure (see panel (**d**)) in analogy to panels (**a**), (**c**), and (**e**). Panel (**c**): Indian monsoon rainfall data [173] showing seasonal emergence of the monsoon atmospheric patterns. Panels (**e**) and (**f**): Frequency of hurricanes [328] (average number of hurricanes, by month, observed in a 100 year period computed from observations during 1851–2015) and sea surface temperature in degrees Celsius [318] for the months of the year. The dotted line in panel (**d**) corresponds to the threshold temperature above which hurricanes can emerge via atmospheric bifurcations

adjacent to India and release the water in form of rainfall over parts of India and some neighboring countries. The atmospheric Indian summer monsoon patterns emerge only under appropriate conditions, which are given in the aforementioned summer months. That is, the earth climate system over India must exhibit an appropriate structure such that the monsoon patterns can emerge. Likewise, lake effect snow (see Chap. 1) and winter snow are atmospheric hydrological cycle patterns that emerge when certain climate condition are satisfied. Those conditions are typically satisfied during the winter months. Finally, hurricanes over the Atlantic basin (which includes that Atlantic ocean, the Caribbean sea, and the Gulf of Mexico) occur typically during the summer season from June to November [328], see panel (e) of Fig. 9.11. One variable among the many variables that determine the emergence of hurricane patterns is the sea surface temperature. As shown schematically in panel (f), the sea surface temperature increases above a critical threshold of 26 °C [328] during the period from June to November, which is the reason why during those months hurricane patterns can emerge. Panels (e) and (f) highlight the relationship between structure and state, which is a key element of the human pattern formation model, see panel (d) (see also Sect. 5.2.1 and Fig. 5.4a). The structure of pattern formation systems of the animate and inanimate worlds determines (in combination with other factors) whether or not the states of the systems form certain patterns.

Let us put the seasonal emergence of atmospheric patterns into the context of motor milestones during infant development. Seasonal atmospheric patterns emerge when the relevant climate systems exhibit an appropriate structure. In general, structure related parameters of climate systems vary periodically. When the structure of a given climate system under consideration happens to exhibit appropriate properties, then a particular atmospheric pattern (e.g., hurricane) emerges provided that an appropriately defined atmospheric reference point (e.g., clear sky) becomes unstable. Likewise, when the brain structure of infants and in general the body structure of infants has developed sufficiently, BBA attractor patterns such as sitting and walking can emerge at bifurcation points. From a science point of view, self-organizing pattern formation system exhibit structural changes that lead to conditions under which patterns can emerge. To refer to such change as "learning" in the context of humans and animals is unnecessary because this kind of changes can happen in all kind of pattern formation systems of the animate and inanimate worlds. The climate systems over India and the Atlantic basin change their structures periodically. As a result monsoon patterns and hurricane patterns occur. Those climate systems do not "learn" anything. They neither "learn" how to produce monsoon rainfalls nor "learn" how to make hurricanes. What matters for pattern formation systems of the inanimate world are structural changes and not "learning". Likewise, what matters for humans and animals are the changes of their structures and not "learning". To call those changes "learning" would introduce a difference between the animate and inanimate worlds that does not exist from a physics perspective. Having said that in the context of non-scientific perspectives (e.g., religious perspectives) the concept of "learning" may be a useful concept.

Panel (d) allows for an even stronger illustration of this line of argumentation. Panel (d) shows how good certain motor skills are performed by various age groups according to a study by Leversen et al. [210]. In the study a number of motor skills were tested such as throwing bean bags to a target and walking on a straight line performing a heel-to-toe walk. That is, participants produced a walking pattern in which they put one foot in front of the other such that at every step the heel of the one foot touched the toes of the other. Taking all tested motor skills together, a total performance score was constructed, a z-score. Positive (negative) values indicated above (below) average performance.[1] Leversen et al. found that performance was poor in the young and elderly age groups and was best in the

---

[1] Actually, Leversen et al. [210] presented the data in a slightly different way (see the original study).

age group of 19–25 years old. In the context of the pattern formation approach and consistent with our considerations in Chap. 8, let us assume that poor performance indicates either that motor pattern attractors just have emerged or that they are about to disappear. Accordingly, it is plausible to interprete the performance score graph displayed in panel (b) as a measure that shows that in the age group 7–9 certain motor attractors just have emerged while in the age group 66–80 those attractors are about to disappear. The emergence and disappearance of attractors is assumed to be related to changes in the structure of the human system over the life-span involving not only the peripheral system (e.g., muscle, joints) but also the central nervous system (see Fig. 5.7 in Sect. 5.2.3). For example, it has been shown that the volume of gray matter and white matter in the human brain is reduced in the elderly population [294]. Moreover, there is evidence that the gray and white matter volume is positively correlated with motor performances [294] such that motor performance becomes worse when gray and white matter volume is reduced. Therefore, just as hurricanes emergence and disappear due to structural changes of the sea-air-system (see panels e and f), motor skills emerge, improve, decline, and disappear due to structural changes in the human system (panel b). Both phenomena can be treated on an equal footing without the need to mystify the human system.

In Sect. 7.4.1 it has been shown that the phenomenon of "retrieval-induced forgetting" is consistent with structural changes that are such that an eigenvalue of an emerging BBA pattern becomes sufficiently large such that eigenvalues of other BBA patterns drop out of the stability band. Those patterns can no longer emerge and in this sense are "forgotten" patterns. At the end of the monsoon, winter, and hurricane seasons mentioned above, in the respective climate systems monsoon patterns, water cycle patterns with winter snow and winter lake snow, and hurricanes do no longer emerge. Using the terminology of the state-of-the-art research on humans, we could (but we should not do so) say that the respective climate systems have "forgotten" how to make monsoon rainfall, winter snow, winter lake snow, and hurricanes. Such a terminology is not needed and in fact is not used in the context of seasonally emerging climate patterns. In the experiments on "retrieval-induced forgetting" it has been shown that when participants produce repeatedly word-pair BBA patterns such as Fruit-Banana, Fruit-Grape, Fruit-Lemon, then in some of the participants related word-pair BBA patterns such as Fruit-Orange, Fruit-Pineapple, and Fruit-Strawberry do not longer emerge (see Sect. 7.4.1). As mentioned above, this observation is consistent with the human pattern formation reaction model in which the structure changes for those participants such that the eigenvalues of the Fruit-Orange, Fruit-Pineapple, and Fruit-Strawberry word-pair BBA patterns drop out of the stability band. Therefore, from a physics perspective the non-emergence of certain attractor patterns in humans (and animals) is due to structural changes just as the disappearance of certain seasonal atmospheric patterns at the end of the respective seasons is due to structural changes in the respective climate systems. Again, what matters for pattern formation systems both of the animate and inanimate worlds are mechanisms that lead to structural changes. The concept of "forgetting" is an unnecessary concept in this regard.

As mentioned above, the shoes of the author of this book change their shapes when they are used by the author. Also the T-shirts of the author change their shapes. This does not mean that the shoes or T-shirts of the author "forget" their original shapes. A guitar string that is stretched produces a higher sound when it is picked by a guitar player. When the additional stress is released, then the guitar string produces the lower, original sound again. That does not mean that the guitar string has "forgotten" how to produce a high sound.

9.4.3 Why Brain Activity Patterns Are Not Neural Correlates of "Knowledge"

The number of brain activity patterns than can emerge in the isolated brain for a given structure is determined by the degree of multistability of the brain as a pattern formation system. According to the LVH model, the degree of multistability is given by the number of patterns that have eigenvalues in the stability band (see also the discussion in Sect. 4.6.4 about the degree of multistability).

Frequently, it is assumed that humans and animals have "knowledge". However, the notion of "knowledge" is inconsistent with the pattern formation perspective of humans and animals because self-organizing, pattern formation systems in general do not "know" about anything. Pattern formation is not about "knowing" anything.

Self-organizing pattern formation systems in general exhibit patterns. For example, certain lasers when they are pumped appropriately are bistable and can exhibit two different self-organized states with differently polarized laser light (see Sect. 6.3.4 for polarization bistability). Roll patterns in fluid and gas layers heated from below are bistable: they can rotate in one direction or rotate just in the opposite direction. The chemical pattern formation system introduced in Sect. 4.4.1 and sketched in Fig. 4.12 can produce vertical or horizontal stripe patterns.

In general, the self-organized states and patterns of lasers, chemical systems, and other self-organizing systems reflect that the components of the systems can settle down into different relationships with each other. The states and patterns do not reflect "knowledge". Likewise, when the skin of a zebra shows a stripe pattern, then this does not mean that the skin "knows" how to produce stripes. As mentioned in the previous section (Sect. 9.4.2), under appropriate conditions, hurricanes emerge in the Atlantic basin or in different places. Again, this does not mean that the oceans or the respective climate systems "know" anything. By analogy, when humans and animals are considered as self-organizing, pattern formation systems, for any given structure of a given individual various brain activity patterns in the isolated brain can emerge (and other BBA patterns can emerge when those human and animal individuals react to and/or produce forces). It does not make much sense to postulate that these repertoire of patterns reflect a special property, called "knowledge", given the fact that for inanimate self-organizing systems the pattern repertoires that they can form are not related to that property. In physics, the concept of "knowledge" is not needed. Introducing it for humans and animals just contributes to mystify living things over non-living things.[2]

The argument given in this section echos what has been discussed earlier in Sect. 7.5.3. In Sect. 7.5.3 it has been argued that some patterns cannot emerge in A,B, or C systems. They only emerge in D and E systems. It is sufficient to state and explain that there is a fundamental difference between A,B,C systems, on the one hand, and D and E systems, on the other hand. It is not necessary to introduce the concept of "tacit knowledge" in this context.

9.5   Mathematical Notes

9.5.1   Mathematics of Dual-Forces C3 System Restructuration

We consider here the brain state shift model discussed in Sect. 9.1. The neuronal network for taste in the reduced amplitude space is described by two amplitudes $A_{T,1}$ and $A_{T,2}$ related to eatable goods and non-eatable goods, respectively. As such, the amplitudes are assumed to satisfy the cubic LVH model like

$$\frac{d}{dt} A_{T,p} = \lambda A_{T,p} - g A_{T,p} A_{T,m}^2 - A_{T,p}^3 \tag{9.6}$$

for $p = 1, 2$ and $m = 2$ if $p = 1$, while $m = 1$ if $p = 2$. Likewise, the acoustic neuronal network is described in the two-dimensional reduced amplitude space by the amplitudes $A_{A,1}$ and $A_{A,2}$ associated with the sound of a ringing bell and other sounds, respectively. The amplitudes satisfy the LVH model

$$\frac{d}{dt} A_{A,p} = \lambda A_{A,p} - g A_{A,p} A_{A,m}^2 - A_{A,p}^3 \tag{9.7}$$

for $p = 1, 2$ and $m = 2$ if $p = 1$, while $m = 1$ if $p = 2$. However, since we are only interested in the dynamics of the amplitudes $A_{T,1}$ and $A_{A,1}$ we assume only those variables satisfy the LVH amplitude equations, while the variables $A_{T,2}$ and $A_{A,2}$

---

[2]For sake of simplicity, "knowledge" has been discussed in the context of the brain state of humans. A similar argument can be made for the brain structure or in general for the grand state of humans. If certain brain structures that give rise to attractors or certain grand states that give rise to attractors and attractor patterns would reflect "knowledge", then in order to avoid to put the human system "above" other non-living physical systems (e.g. as vitalism supporter do), we would need to acknowledge that certain structures and grand states of other non-living self-organizing systems such as hurricanes and lasers reflect "knowledge" as well.

are completely determined by the forces acting on the animals. More precisely, it is assumed that $A_{T,1}$ and $A_{A,1}$ satisfy the equations

$$\frac{d}{dt}A_{T,1} = \lambda A_{T,1} - g A_{T,1} A_{T,2}^2(t) - A_{T,1}^3 \qquad (9.8)$$

and

$$\frac{d}{dt}A_{A,1} = \lambda A_{A,1} - g A_{A,1} A_{A,2}^2(t) - A_{A,1}^3 , \qquad (9.9)$$

where $A_{T,2}(t)$ and $A_{A,2}(t)$ are considered as given time-dependent functions that will be defined in a moment. We assume that if the food force acts on the animals it shifts the brain state into the basin of attraction of the eatable goods attractor like

$$\text{If } F_T \text{ switch on } \Rightarrow A_{T,1}(t) = \Phi^{A_{T,1}} = \delta > 0 , A_{T,2} = \Phi^{A_{T,2}} = 0 \qquad (9.10)$$

with $\delta = A_{T,1,st}/10$ and $A_{T,1,st} = \sqrt{\lambda}$. Likewise, the forces exerted by the ringing bell are assumed to shift the brain state into the basin of attraction of the ringing-bell attractor like

$$\text{If } F_A \text{ switch on } \Rightarrow A_{A,1}(t) = \Phi^{A_{A,1}} = \delta > 0 , A_{A,2} = \Phi^{A_{A,2}} = 0 . \qquad (9.11)$$

Here, the phrase "switch on" means a jump from 0 to 1 as shown in panel (a) of Fig. 9.3. However, when there is no food and likewise when the bell is not ringing it is assumed that other forces act on the animals that shift the brain states in both networks to states in the basins of attraction of the non-eatable food attractor and the other sounds attractor. In this context, note that the non-eatable BBA attractor pattern is assumed to describe also situations in which there is nothing to eat. Likewise, the other sound BBA attractor pattern is assumed to describe also situations in which only "background noise" acts on the animals. Therefore, we defined $A_{T,2}(t)$ like

$$\text{If } F_T(t) = 0 \Rightarrow A_{T,2}(t) = \Phi^{A_{T,2}} = A_{T,2,st} ,$$
$$\text{If } F_T(t) = 1 \Rightarrow A_{T,2}(t) = 0 , \qquad (9.12)$$

where the second relation for $F_T(t) = 1$ actually reflects the brain state shift described in Eq. (9.10). Likewise, $A_{A,2}(t)$ is defined by

$$\text{If } F_A(t) = 0 \Rightarrow A_{A,2}(t) = \Phi^{A_{A,2}} = A_{A,2,st} ,$$
$$\text{If } F_A(t) = 1 \Rightarrow A_{A,2}(t) = 0 , \qquad (9.13)$$

The stationary values are given by $A_{T,2,st} = A_{A,2,st} = \sqrt{\lambda}$. The structure dynamics is defined by Eq. (9.5). More precisely, in order to describe that the eatable-goods

BBA pattern and the ringing-bell BBA pattern have emerged, we use (as in various previous applications) the threshold parameter $\theta \in [0, 1]$ that corresponds to the dashed lines in panel (a) of Fig. 9.3. Then, Eq. (9.5) using $r = 1$ reads

$$A_{A,1}(t_n) > \theta A_{A,1,st} \Rightarrow a_A(n+1) = a_A(n) - \epsilon a_A(n)(a_A(n) - \xi) ,$$

$$\text{if} A_{T,1}(t_n) > \theta A_{T,1,st} \Rightarrow \xi = a_T ,$$

$$\text{otherwise} \Rightarrow \xi = 0 . \tag{9.14}$$

Note also that in Sect. 9.1 the subindices "1" in the amplitude variables $A_{T,1}$ and $A_{A,1}$ has been neglected to simplify the presentation. That is, the variables A_T and A_A in Sect. 9.1 refer to the amplitudes $A_{T,1}$ and $A_{A,1}$ described in this paragraph. The simulation results presented in Fig. 9.3 have been computed by solving the grand state dynamics defined by the amplitude equations (9.8) and (9.9) and the structure dynamics defined by Eq. (9.14) using the brain state (or ordinary state) shift mechanism Φ described by Eqs. (9.10), (9.11), (9.12), and (9.13). Parameters: $\lambda = 0.5$, $g = 2$, $\theta = 0.8$, $\epsilon = 0.1$ (panel b), $\epsilon_1 = 0.3$ (panel c), $\epsilon_2 = 0.15$ (panel d), and $a_T = 1$.

9.5.2 Mathematics of Reward-Induced E1 System Restructuration

Let us provide some mathematical details of the model discussed in Sect. 9.2. For sake of simplicity, the LVH amplitude dynamics will not be modeled explicitly. Rather, the characteristic properties of the LVH model will be used in the simulation. Let us describe the four relationships 1,2,3,4 illustrated in Fig. 9.5.

Relation 1: Structure Affects Ordinary State Dynamics
Structure is given in terms of the eigenvalues $\lambda_1(t), \ldots, \lambda_n(t)$ with $n = 10$. Here, t is a discrete time step. For example, in panel (a) of Fig. 9.6 we have $t = 1, \ldots, 100$. The ordinary state (i.e., brain activity and force production/body movement) is described in terms of the amplitudes $A_1, \ldots, A_n$ of the $n = 10$ BBA patterns or activities under consideration. The BBA pattern $k = 1$ corresponds to pressing the lever. At any given discrete time step t the transition probability $T(i \rightarrow j)$ from state $i \in \{1, \ldots, n\}$ to $j \in \{1, \ldots, n\}$ is assumed to depend on the size of the basins of attraction of the attractors. The size of the basin of attraction of the attractor k is approximately given in terms of the corresponding eigenvalue λ_k. Consequently, we put

$$T(i, t \rightarrow j, t+1) = \frac{\lambda_j(t)}{\sum_{m=1}^{n} \lambda_m(t)} . \tag{9.15}$$

Note that T is independent of the current state i. Therefore, we deal with a simple random walk model. At every time step t, with the help of $T(i, t \to j, t+1)$ and a random number generator we can determine the state at time $t+1$. Let $X(t) \in \{1, \ldots, n\}$ denote the state at time t. If $X(t+1) \neq X(t)$ then a new attractor pattern emerges at time step $t+1$. If $X(t+1) = X(t)$ then the current attractor pattern is maintained. In any case, if a new attractor pattern k emerges or if the pattern k is maintained we put the corresponding amplitude to a fixed point value $A_{st} = 3$. Otherwise, pattern amplitudes decay to zero. We have

$$X(t+1) = k \Rightarrow \left\{ A_k(t+1) = 3 , \right.$$
$$\forall j \neq k : \text{if } A_j(t) > 0 : A_j(t+1) = A_j(t) - 1 ,$$
$$\left. \text{if } A_j(t) = 0 \Rightarrow A_j(t+1) = 0 \right\} . \tag{9.16}$$

Step 2: Lever Pressing
The BBA pattern $k = 1$ describes that the animal presses the lever. In order to describe the force produced by the animal, we introduce the variable $F_{lever\ press}(t)$ that takes on the values 0 and 1. If $F_{lever\ press}(t) = 1$ then the animal presses the lever at time t. Otherwise, it does not press the lever. The dynamics of $F_{lever\ press}(t)$ is determined by

$$\text{If } A_1(t) = 3 \Rightarrow F_{lever\ press}(t) = 1 ,$$
$$\text{Otherwise} \Rightarrow F_{lever\ press}(t) = 0 . \tag{9.17}$$

Step 3: Reaction of the Environment to the Forces Produced by the Animal
If the clock is not running, then a lever press produced by the animal starts the sand clock. If the sand clock is running already, then the lever press is ignored. In order to capture, whether or not the sand clock is running, we introduce the categorical variable $S_{clock,cat}(t)$ that assumes the values 0 and 1. $S_{clock,cat}(t) = 0$ means the clock is switched off, while $S_{clock,cat}(t) = 1$ means the clock is running. In order to describe the sand clock quantitatively, we introduce $S_{clock,num}(t)$ that assumes the values 0,1,2 and counts time back from 2 to 1 to 0. The sand clock dynamics is described by

$$\text{If } S_{clock,cat}(t) = 0 \wedge F_{lever\ press}(t) = 1$$
$$\Rightarrow S_{clock,cat}(t+1) = 1 , S_{clock,num}(t+1) = 2.$$
$$\text{If } S_{clock,cat}(t) = 1 \wedge S_{clock,num}(t) = 2$$
$$\Rightarrow S_{clock,cat}(t+1) = 1 , S_{clock,num}(t+1) = 1 . \tag{9.18}$$

When the sand clock is empty (i.e., has counted down 2 time units), then a food reward is given and the clock is turned off. In order to describe the external force related to the food reward, we introduce the variable $F_{food}(t)$ that assumes the values 0 and 1, where $F_{food}(t) = 0$ means that no food is given and $F_{food}(t) = 1$ means that food is given to the animal. We have

$$\text{If } S_{clock,cat}(t) = 1 \ \wedge \ S_{clock,num}(t) = 1$$
$$\Rightarrow \ S_{clock,cat}(t + 1) = 0, \ F_{food}(t + 1) = 1,$$
$$\text{Otherwise } F_{food}(t + 1) = 0. \tag{9.19}$$

Step 4: Restructuration

The restructuration step takes place when a food reward is given. The restructuration affects all basis patterns k whose amplitudes A_k are larger than zero at the time t at which the food is received. All basis patterns k with zero amplitudes are not affected by the restructuration. Accordingly, the restructuration dynamics is described by

$$\text{If } F_{food}(t) = 1 \ \Rightarrow \ \forall k \ : \ \big\{ \text{if } A_k(t) > 0 \ \Rightarrow \lambda_k(t + 1) = \lambda_k(t) + \gamma(\lambda_k(t) - C),$$
$$\text{if } A_k(t) = 0 \ \Rightarrow \lambda_k(t + 1) = \lambda_k(t) \big\},$$
$$\text{If } F_{food}(t) = 0 \ \Rightarrow \ \forall k \ : \lambda_k(t + 1) = \lambda_k(t). \tag{9.20}$$

The graphs presented in Fig. 9.6 have been computed by solving the equations listed above iteratively for $\gamma = -0.01$, $C = 5$ with grand state initial conditions $A_k(t = 1) = 1$ and $\lambda_k(t = 1) = 1$ for $k = 1, \ldots, n$ and $S_{clock,cat} = 0$.

Chapter 10
Applications in Clinical Psychology

10.1 Dynamical Diseases

10.1.1 Disease Development, Therapy, and Relapse

As pointed out in Sects. 1.5, 3.1, and 3.2 the pattern formation perspective is part of a broad perspective that addresses how system of the animate and inanimate worlds can form qualitatively new states and how transitions between qualitatively different states take place. In this context, aggregate phase transitions (e.g., transitions from ice to water), phase transitions in non-equilibrium systems of the inanimate world (e.g., emergence of roll patterns in fluids and gases heated from below), and transitions between qualitatively different movement patterns of humans and animals (e.g., walk to trot gait transitions in horses) have been considered on an equal footing (see Fig. 3.5 in Sect. 3.2.3). It has been pointed out that when an individual stands up from a chair and starts to walk, then the sit-to-stand and stand-to-walk transitions from a physics perspective are considered as counterparts to aggregate phase transitions such as ice-to-water and water-to-gas transitions (see Fig. 1.4). Mackey and Glass [135, 223] proposed to take this analogy to medicine and clinical psychology, see Fig. 10.1. Accordingly, just as ice turns into water via a phase transition (panel a) and humans turn from sitting BBA patterns into standing BBA patterns via bifurcations (panel b), a disease state emerges via a bifurcation from a healthy state (panel c). Diseases emerging from a bifurcation have been called dynamical diseases. In clinical psychology, for example, mood disorders [11, 99, 287, 307, 321] and schizophrenia [52] have been discussed as dynamical diseases. However, not only the emergence of diseases has been addressed from a dynamical systems perspective. As suggested by Schiepek and co-workers [161, 277–280] (see also Tschacher et al. [323]) therapy and therapy success may be addressed from the pattern formation perspective presented in this book. In line with the aforementioned studies, it has also been proposed to consider

© Springer Nature Switzerland AG 2019

T. Frank, *Determinism and Self-Organization of Human Perception and Performance*, Springer Series in Synergetics, https://doi.org/10.1007/978-3-030-28821-1_10

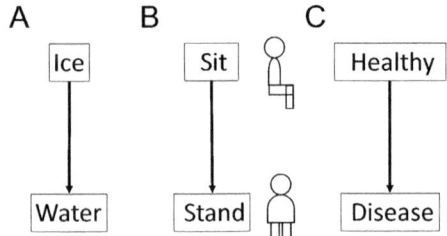

Fig. 10.1 Disease as a pattern formation phenomenon as suggested by Mackey and Glass. Phase transitions (panel (**a**)), bifurcations between (brain activity and) body patterns (panel (**b**)), and the emergence of diseases (panel (**c**)) are considered as analogous phenomena

relapse (i.e., the return to the disease state after successful therapy) as another phase transition or bifurcation phenomenon [101].

10.1.2   Risk Factors

Since dynamical diseases emerge via bifurcations and bifurcations involve bifurcation parameters, from a clinical point of view, bifurcation parameters may be considered as risk factors.

10.1.3   Dynamical Disease Scenarios

There are various possible scenarios related to the emergence of a disease as a dynamical disease, the treatment of a disease, and relapse. Figure 10.2 summarizes some of them schematically.

First of all, the states of a healthy individual is in general considered as a grand state. That is, both the ordinary state (brain state and produced forces) and the structure of the individual matters. Likewise, the disease state of an individual is considered as a grand state. As mentioned above, a dynamical disease is a disease that emerges by means of a bifurcation. That is, there is a disease related bifurcation (involving a disease related bifurcation parameter) that makes the healthy state unstable or makes that the corresponding attractor of the healthy state disappears entirely and leads to the emergence of a disease grand state. This fundamental scenario of the emergence of a dynamical disease is illustrated in panel (a) of Fig. 10.2. Treatment and therapy of a disease can be understood in a similar way as a bifurcation that makes the disease state unstable or causes the disease-related attractor to disappear. In the ideal case, the bifurcation brings patients back to the state of healthy individuals, as illustrated in panel (b). However, the symptoms free state after successful treatment and therapy does not necessarily correspond

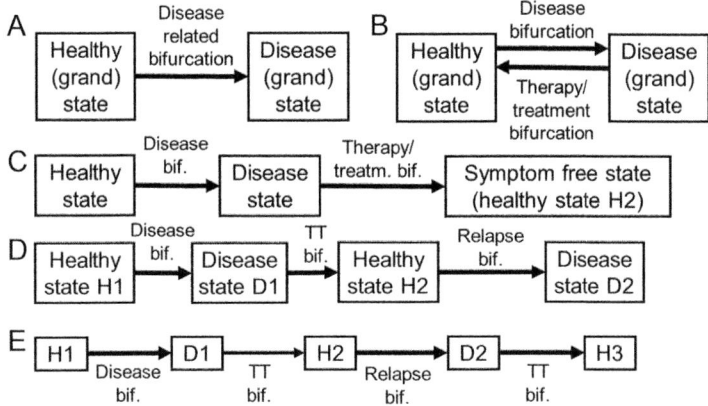

Fig. 10.2 Bifurcations between healthy and disease states. Various possible scenarios are shown. See text for details. (TT: abbreviation for treatment/therapy)

to the state of likewise healthy individuals. As illustrated in panel (c), the treatment-induced or therapy-induced bifurcation may lead to a symptoms free state that differs qualitatively from the original healthy state. For example, the individual under consider has to take daily medications that healthy adults do not need to take. Let us denote the original healthy state by H1 and the disease state emerging from H1 by D1. Let us denote the healthy state after successful treatment and therapy by H2. Then we obtain the first three steps in the sequences shown in panels (c)–(e).

If the healthy state H2 requires the in-take of drugs sometimes individuals stop for various reasons to take the prescribed drugs. As a result the disease re-emerges. There is a relapse. Relapse can be considered as a bifurcation in analogy to the bifurcations that lead to the emergence of dynamical diseases in the first place and the bifurcations related to treatment and therapy. A relapse related bifurcation may take an individual from state H2 back to the disease state D1. However, given the fact that the individual has restructured already due to treatment and therapy, it is also possible that the emerging disease grand state qualitatively differs from the previous disease state. This case is illustrated in panel (d). The relapse related bifurcation takes the individual to a disease state D2. Panel (e) finally illustrates the impact of treatment and therapy on a disease state D2 that emerged due to relapse. For example, if the relapse was due to medication non-adherence (i.e., the individual did not take the drugs as prescribed) then the individual may start to take the drugs again. The individual may end up in the state H2 (scenario not shown) or in a new healthy state H3. In particular, the experience that the individual has made, namely, that stopping taking drugs makes the disease to re-appear, can change the structure of the individual such that the treatment after relapse does the individual not bring back to the healthy state H2 but to a different healthy state H3. It is plausible to assume that another relapse is less likely for H3 state individuals as compared to H2 state individuals.

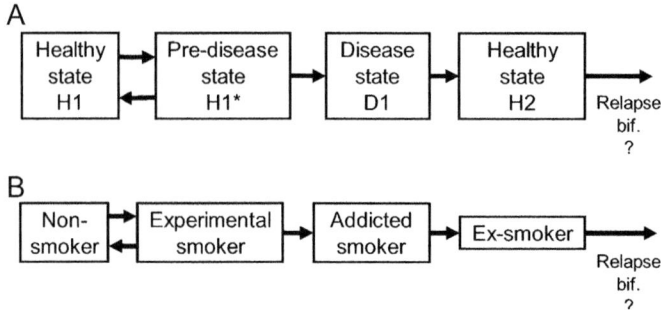

Fig. 10.3 Grand state pattern formation sequences (i.e., type 2 pattern sequences) that involve pre-disease states

The schemes illustrated in Fig. 10.2 are some examples of possible scenarios centered around the notion that the emergence of a disease, the success of therapy and treatment, and the disease relapse are bifurcation phenomena. In particular, subclasses that describe again qualitatively different states may be introduced. For example, a healthy state may turn into a particular pre-disease state that might be considered either as healthy state or disease state. Panel (a) of Fig. 10.3 illustrated schematically such a transition to a pre-disease state. The healthy state is denoted by H1, while the pre-disease state is denoted by H1*. A pre-disease states may be defined as a state that is qualitatively different from a healthy state but corresponds to a grand state with a bifurcation parameter α such that there is a bifurcation that takes the individual back to the original H1 state. That is, increasing α such that $\alpha > \alpha(crit)$, where $\alpha(crit)$ is a critical value, the individual makes a transition to the pre-disease state, while decreasing α such that $\alpha < \alpha(crit)$ the individual makes a transition back to the healthy state.[1] In particular, an individual that returns from a pre-disease state to a healthy state does not need to take medications to maintain the healthy state. Let us return to panel (a) of Fig. 10.3. The emergence of the disease state D1 is considered as a transition from the pre-disease state H1* to the state D1 (rather than a transition from the original healthy state H1 to D1). Once the individual is in the disease state D1, treatment and therapy can take the individual to a new healthy state H2 that differs from H1.

Panel (b) of Fig. 10.3 gives an example from the field of substance abuse. Individuals naturally start as non-smokers. Non-smokers may become experimental smokers when the conditions are such that the non-smoker state disappears in a bifurcation or becomes unstable. Experimental smokers are smokers that can stop smoking without side-effects. That is, when the conditions are appropriate (e.g., external force in terms of interactions with other people who are all non-smokers and urge the individual to stop smoking, or the individual is female and becomes pregnant, or the individual has a new girl-friend/boy-friend who does not like that

---

[1]Note that there might be hysteresis such that there are actually two different critical values.

the individual is smoking, or the individual is a student and there is a discussion in class about the effects of smoking on the human brain, etc.) the experimental smoker state may become unstable or the corresponding attractor may disappear entirely and a transition to the non-smoker state occurs. The question is open for debate whether the state that emerges in this way can indeed be considered to be qualitatively similar to the state of non-smokers who have never consumed tobacco. However, we will not dwell on this issue. As indicated in panel (b), not only a bifurcation from the experimental smoker state to the non-smoker state may occur. It is also possible that the individual under consideration bifurcates from the experimental smoker state to the state of an addicted smoker. Finally, due to therapy and treatment the conditions of the individual of interest may change such that the state of the individual as an addicted smoker becomes unstable or the corresponding attractor disappears. The person under consideration stops smoking and becomes an ex-smoker. Being an ex-smoker is frequently considered to be different from being a non-smoker. This distinction between non-smokers and ex-smokers and the sequence shown in panel (b) of Fig. 10.3 provides another example for disease progression sequences illustrated in panels (c)–(e) of Fig. 10.2 that do not return back to the original healthy states at hand.

10.1.4   Smoking Addiction and the Mystic Concepts of "Decision Making" and "Vitalism"

As stated in the context of Fig. 10.3, in a first step non-smokers may become experimental smokers when the conditions are such that the non-smoker state becomes unstable or disappears entirely. Accordingly, this step takes place because the non-smoker happens to be in conditions that lead to a bifurcation just as ice turns into water when the conditions are appropriate or a horse changes from walk to trot when the conditions are appropriate. Note that from the pattern formation perspective it is not necessary to bring up concepts like "decision" or "choice" in this context. A non-smoker does not make a "decision" to start smoking. A non-smoker does not "choose" to smoke. This is just like ice does not make a "decision" to turn into water and ice does not "choose" to become water.

Advocates of the concept "decision" may be compared with advocates of "vitalism". "Vitalism" supporters assume that human beings and living things in general exhibit a "vital force" that goes beyond the forces of physics discussed in Sect. 1.7. Just as advocates of "decision theory" assume that there exist something like a "decision event" to explain certain observations, advocates of "vitalism" assume that there exists a "vital force" to explain certain observations in biology. However, as it has been shown in the past, physics and chemistry can explain those observations in biology without the need to introduce the concept of a "vital force". Likewise, in this book it is shown that what humans do can be explained without the need to introduce the concept of "decision making".

10.2   Diseases as Categories and Disease Definition: Physics Perspective

As discussed in Sects. 3.2.4 and 3.2.5, from a physics perspective qualitatively different states, i.e., categories of states, are defined by means of phase transitions and bifurcations. Although it is tempting to consider the states of an electric bulb or an electric fan that are switched on as qualitatively different from the states of the electric devices, when they are switched off, in fact, the on and off states differ just quantitatively from each other. They are two extreme states within a single category of states (see Sect. 3.2.4). Consequently, if (and the author of this book does not say if we should do so or not) disease states are defined as states that are qualitatively different from the states of healthy individuals, then from a physics perspective, the disease states and diseases must emerge from bifurcations (strong category case, see Sect. 3.2.5) or there must be at least "routes" towards the disease states that involve bifurcations (weak category case, see Sect. 3.2.5). In other words, if (see comment above) diseases describe humans, who are in qualitatively different conditions as compared to healthy humans, then diseases are dynamical diseases as proposed by Mackey and Glass [135, 223]. Importantly, diseases that have been defined in medicine in terms of specific symptoms but that do not emerge via bifurcations are not diseases in a strict sense but should be considered as something else.

10.3   Bipolar Disorder (BD)

10.3.1   BD as a Dynamical Disease

Bipolar disorder affects a relatively large proportion of the world's population. It is estimated that 1–3% of the US population is affected by bipolar disorders [146, 181] and similar estimates have been reported for European countries [81]. Typical signs of bipolar disorder are experiences of episodes of mania and depression that are more intense than the mood fluctuations that people of the general population experience. Mood variations of bipolar disorder patients exhibit different types of temporal, cyclic patterns. Some patients experience repeated episodes of the same polarity (e.g., mania) that are separated by episodes of "normal mood". Other patients experience alternating episodes of opposite polarity (mania, depression, mania, depression, etc). These two types of temporal patterns are both referred to as cycling bipolar disorder patterns [17]. A further distinction is made into rapid and non-rapid cyclers. Rapid cyclers experience more than four episodes per year, whereas non-rapid cyclers experiences fewer episodes [34, 73, 201, 224]. The distinction between rapid and non-rapid cycler is helpful because in general the disease condition of rapid cyclers is more severe and more difficult to treat [34, 73, 201, 224]. Bipolar disorder seriously affects the quality of life. In particular, suicide rates are higher than in the general population.

It has been suggested that bipolar disorder is a dynamical disease. That is, it emerges via a bifurcation. Let us briefly mentioned three approaches in this regard. First, Steinacher and Wright have suggested that the disease state corresponds to a multistable dynamical system [307]. Accordingly, the severe mood variations are stochastic jumps between the multiple stable states of the systems. Bipolar disorder emerges as a bifurcation from monostability (healthy state) to multistability (bipolar disorder). Second, Tretter et al. have suggested that bipolar disorder is an oscillatory phenomenon that emerges via a delay-induced bifurcation [321]. Accordingly, the bifurcation parameter is a delay in a feedback loop of the human cortisol system. If the delay is below a critical value, the system is non-oscillatory and describes the healthy state. In contrast, if the delay increases beyond a critical value, the system becomes oscillatory. The model applies in particular to variations from mania to depression. The episodes of mania and depression corresponds to situations in which the physiological variables of the cortisol assume the positive and negative extreme values of the oscillation cycle. Delays reflect propagation times of forces and input-output relationships of reaction chains [85, 87]. Delays do not reflect memory in the sense that humans or physical systems could access the past. The presence completely determines any physical system (including humans and animals).

Third, Schreiber and Avissar [11, 287] and Frank [99, 101] have suggested that bipolar disorder is an oscillatory disease just as suggested by Tretter et al. [321]. Unlike the suggestion by Tretter et al. [321], it is assumed that the protein kinase A and C pathway in certain brain cells (e.g., neurons or supporting brain cells) becomes oscillatory. That is, bipolar disorder emerges from an oscillatory instability that does not involve a delay.

10.3.2 Structural Oscillation Hypothesis

In what follows, we review the medical hypothesis that bipolar disorder is an oscillatory instability on the cellular level that involves the protein kinase A and C pathway [11, 99, 101, 287]. Accordingly, Schreiber and Avissar [11, 287] suggested that cyclic bipolar disorder patterns emerge on the cellular level due a malfunction of the signaling pathways of protein kinase A and protein kinase C. In fact, several authors have suggested that the kinase A and C pathways are relevant for understanding mood disorders [56, 80, 178]. Schreiber and Avissar proposed a linearized model of the protein kinase A and C pathway that describes how an oscillatory instability can occur on the cellular level. Frank [99] proposed a nonlinear version of such a interacting protein kinase A and protein kinase C network that can describe the approach towards the limit cycle.

Figure 10.4 depicts the basic elements of the model. First of all, the oscillation takes place on the level of the brain structure, see panel (a). That is, the structure exhibits an oscillatory dynamics. This implies that the brain state and the hormone system becomes oscillatory as well. Consequently, the basic model describes an A1 pattern formation system. However, it is assumed that cells are exposed to unspecific

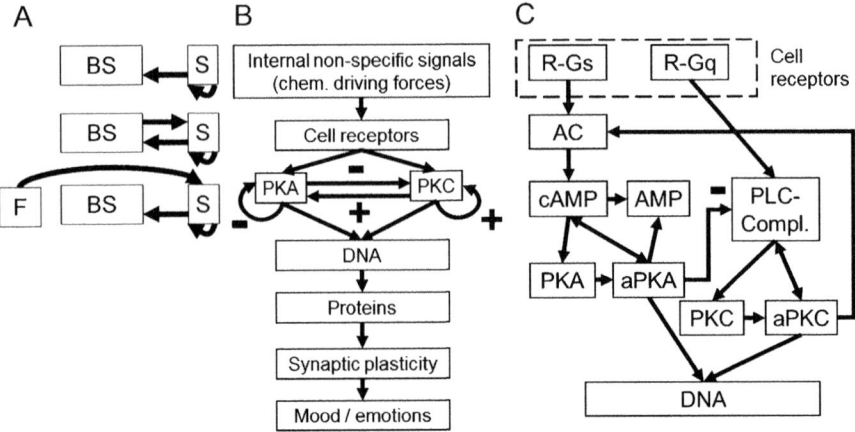

Fig. 10.4 Three steps zoom (panels (**a**), (**b**), (**c**)) into the cellular BD oscillator model. Panel (**a**): The oscillation is assumed to take place on the level of human A1, A2, or D0 pattern formation systems (from top to bottom). Panel (**b**): Levels of the cellular network in which unspecific signals induce an oscillatory bifurcation of protein kinase A (PKA) and protein kinase C (PKC) concentrations. The bifurcation, in turn, leads to oscillatory variations of mood and emotions. The essential oscillator components are given by the couplings between the PKA and PKC pathways in terms of damping (PKA to PKC) and pumping (PKC to PKA). Panel (**c**): Explicit modeling of the cellular level signaling network. Connections describe biochemical reactions. Importantly, active PKA causes a reduction of the PLC-complex, which describes the damping mechanism from PKA to PKC. The network includes the following species: G-protein receptors R-Gs and R-Gq; AC = adenylate cyclase; inactive and activated protein kinase A (PKA and aPKA); the sub-network composed of phospholipase C (PLC), phophotidylinositol, and diacylglycerol denoted as PLC-complex; inactive and activated protein kinase C (PKC and aPKC). For details see [99]

forces that provide some kind of background pumping. If they correspond to brain activity, we would deal with an A2 system. If they are related to external forces, we would deal with a D0 system. Since the background pumping is assumed to be constant and unspecific it is sufficient to consider the most parsimony model, which is the model of an A1 system. In Fig. 10.4a there is a curved arrow pointing from the structure symbol to the structure symbol. This indicates that the structure itself is considered to be a dynamical system. Like in Figs. 1.5 and 1.11 curved arrows point from the brain state to the brain state in order to indicate that the brain state evolves in time as dynamical system.

Panel (b) sketches the overall biochemical cellular model for mood variations experienced by bipolar disorder patients. Accordingly, protein kinase A (PKA) and protein kinase C (PKC) react to certain signals exerted by external chemical driving forces on cell receptors. These signals are external to the receptors. However, for A1 and A2 systems they are internal to the human individual, which is the reason why in panel (b) they are referred to as internal non-specific signals. As reviewed in Ref. [99] there is experimental evidence that the PKA pathway effectively leads to damping of (i.e., "inhibits") the PKC pathway. Likewise, it can be argued that the PKC pathway pumps (i.e., supports "activation of") the PKA pathway [99]. Kinase

A and kinase C affect the production of certain biochemicals that are transported outside the cell and are known to affect, in turn, the plasticity of the nervous systems, that is, lead to synaptic changes [56, 72, 80, 178, 202, 315]. In addition, it is also known that PKA and PKC affect mood in general [56, 80, 141, 178, 218, 221]. Therefore, as indicated in panel (b) of Fig. 10.4 it is assumed that the following cause-and-effect chain holds: variations on the level of PKA and PKC concentrations within cells cause changes in between-cells synaptic transmission parameters that, in turn, result in mood variations.

The aforementioned interplay between pumping and damping (i.e., "inhibitory and excitatory couplings") of the PKA and PKC pathways are crucially important for the proposed limit cycle oscillator model. It has been hypothesized that these pumping and damping couplings effectively give rise to a negatively damped cellular oscillator [11, 287]. That is, in healthy adults, we deal with a stable fixed point in terms of a stable focus, see Chap. 3. Accordingly, bipolar disorder occurs when the stable focus becomes unstable such that a limit limit cycle attractor emerges via a so-called Hopf bifurcation, which implies that bipolar disorder belongs to the class of dynamical diseases introduced in Sect. 10.1.

Panel (c) of Fig. 10.4 illustrates schematically the relevant biochemical reactions. They are described in detail in Ref. [99]. Under certain assumptions the high-dimensional biochemical network depicted in Fig. 10.4c can be reduced to a set of two coupled first order differential equations that describe the evolution of the PKA and PKC concentrations. Let a and c denote appropriately rescaled, dimensionless concentration variables for PKA and PKC molecules. Then, the evolution equation of the cellular brain structure reads [99]

$$\frac{d}{dt}a = b_1(1-a)Z(a,c) - b_2 a \tag{10.1}$$

$$\frac{d}{dt}c = d_1(1-c)Y(a,c) - d_2 c \tag{10.2}$$

where t denotes time and

$$Z(a,c) = \frac{S + g_{C \to A}\, c}{k_1 + W(a)}\ , \tag{10.3}$$

$$Y(a,c) = \frac{Q + \gamma\, c}{k_2 + g_{A \to C}\, a} \tag{10.4}$$

with

$$W(a) = \Omega\, a \exp\{-Ha\}\ . \tag{10.5}$$

All model variables and parameters and their descriptions (i.e., short names) are listed in Table 10.1. All parameters $b_1, b_2, d_1, d_2, S, g_{C \to A}, Q, \gamma, k_1, k_2, g_{A \to C}, \Omega, H$ are positive.

Table 10.1 Variables and parameters and their short descriptions of the bipolar disorder cellular oscillator model

Symbol	Description	Variable	Parameter
a	Relative concentration of activated protein kinase A	x	
c	Relative concentration of activated protein kinase C	x	
Z	cAMP concentration	x	
Y	Concentration of PLC-PI-DAG complex	x	
b_1, b_2	Protein kinase A pathway rate constants		x
d_1, d_2	Protein kinase C pathway rate constants		x
S	Time-invariant activation strength of AC by Gs-protein receptors		x
$g_{C \to A}$	Constant of excitatory coupling from protein kinase C to protein kinase A pathway		x
k_1	Constant degradation rate of cAMP		x
Q	Time-invariant production rate of Y due to inputs to Gq-protein receptors		x
γ	Rate constant of positive feedback loop in protein kinase C pathway		x
k_2	Degradation rate of Y		x
$g_{A \to C}$	Constant of inhibitory coupling from protein kinase A to protein kinase C pathway		x
W	Negative feedback function for reduction of cAMP to AMP due to protein kinase A	x	
Ω	Rate constant of cAMP negative feedback loop		x
H	Hypothesized pathological parameter for breakdown of monotonic behavior of W		x

AC = adenylate cyclase, Gs-protein receptor = G-protein receptor of sub-type "s", Gq-protein receptor = G-protein receptor of sub-type "q"

Special attention should be directed to the parameter H. The PKA pathway involves a sub-network that describes the production and reduction of cyclic AMP (cAMP), see Fig. 10.4c. That sub-network exhibits a negative feedback loop mediated by PKA. In the case $H = 0$ the reduction rate of cAMP to AMP increases linearly with the concentration of PKA. In this case, the structure dynamics defined by Eqs. (10.1)–(10.5) describes a stable fixed point (either node or focus). This case is assumed to reflect the cellular dynamics of healthy adults. It has been hypothesized that for bipolar disorder patients the cAMP reduction mechanism fails at higher PKA concentrations. That is, the self-damping mechanism breaks down at higher PKA concentrations. In this case, the feedback function W depends on the relative concentration a of PKA in a non-monotonic way. The parameter H measures the degree of the breakdown of the function W from a simple linear relationship . For $H = 0$ the function W is linear. For $H > 0$ but $H \approx 0$ the "malfunction" (i.e., deviation from linearity) is weak. In contrast, if H is large, the "malfunction" is severe. In particular, it has been shown [99] that if all other model parameters are fixed, then there exists a critical value $H(crit)$ such that for $H < H(crit)$ the model exhibits a stable fixed point (either node or focus), whereas for $H > H(crit)$ the biochemical model shows self-oscillations. That is, at $H = H(crit)$ a Hopf bifurcation to the oscillatory case occurs and the disease state emerges as a limit cycle attractor. The parameter H is considered as bifurcation parameter. From a clinical point of view, H may be considered as risk factor.

10.3.3 Disease Development and Therapy: Cellular Level

The cellular model defined by Eqs. (10.1)–(10.5) can be used to model disease development and drug treatment. As mentioned previously, the parameter H is considered as bifurcation parameter. Disease development can be model by increasing H from $H = 0$ gradually such that H eventually exceeds the critical value $H(crit)$. Likewise, taking a simplified perspective, we may assume that drug treatment such a Lithium therapy has the opposite affect on H. Accordingly, drug treatment is assumed to reduced the severity of the hypothesized "malfunction" measured by H and, consequently, reduces H.

In order to illustrate these two aspects of the model, Eqs. (10.1)–(10.5) were solved numerically. As model parameters we used the parameters reported in Figs. 11 and 12 of Ref. [99]. For those parameters it has been shown [99] that $H(crit) \approx 9.10$. Panel (a) of Fig. 10.5 shows the results of the simulation. In the

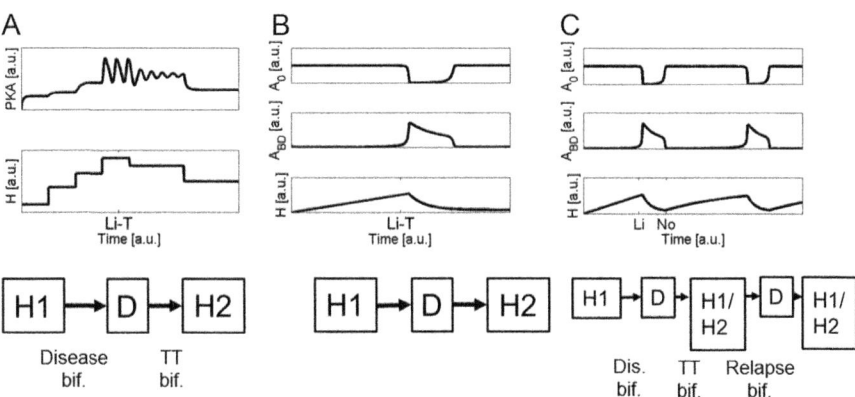

Fig. 10.5 Disease-induced bifurcation, treatment bifurcation, and relapse bifurcation. Panel (**a**): Numerical solution of the cellular signaling network described by Eqs. (10.1)–(10.5). Only the variable $a(t)$ is shown (top subpanel). The risk factor H was gradually increased and decreased (bottom subpanel) to mimic disease development and the impact of drug (e.g. Lithium) medication, respectively. Panel (**b**): Simulation of the signaling network by means of the reduced description in terms of the LVH amplitude equations defined by Eqs. (10.14)–(10.17). Panel (**c**): Simulation of Eqs. (10.14)–(10.16) with Eq. (10.18) of the LVH model for relapse due to medication non-adherence

top subpanel the evolution of $a(t)$ is shown. In the bottom subpanel the bifurcation parameter H as a function of time is shown.

In the simulation, the parameter H was increased in steps as shown in the bottom panel. This first part of the simulation mimicked disease development. Note that any monotonically increasing function could be used for that purpose. When H reached the critical value $H(crit)$, the Hopf bifurcation occurred. The PKA concentration a started to oscillate. After onset of the self-oscillation, the parameter H was decreased in steps as shown in Fig. 10.5a (bottom subpanel). This part of the simulation mimicked the impact of drug treatment such as a Lithium-based drug therapy (see the marker "Li-T"). Again, the explicit shape of the graph $H(t)$ in the decaying branch is not of primary interest. At issue is that at a particular point in time the bifurcation parameter H reached the critical value $H(crit)$ again. At that bifurcation point, the self-oscillations disappeared.

The model describes disease development and therapy as a type 2 pattern sequence that satisfies the schemes C or D shown in Fig. 10.2. The healthy grand state H1 is described by a structure parameter $H = 0$ (or $H < H(crit)$) and a PKA and PKC dynamics given in terms of a stable focus. The brain state is determined by the stable focus on the structural level and, consequently, does not exhibit mood oscillations. The disease related bifurcation is the bifurcation induced by the bifurcation parameter H, when H exceeds $H(crit)$. The bifurcation point is an oscillatory instability. The bifurcation is a Hopf bifurcation. The disease state is characterized by the structure parameter $H > H(crit)$ and limit cycle oscillations of the PKA and PKC concentrations. The brain state (not shown) follows the structure oscillations and is assumed to show oscillatory mood variations taking individuals from episodes of mania to episodes of depression. Lithium therapy reduces H and induces a treatment-induced bifurcation when H drops below $H(crit)$. While the pattern formation system as such is back at the state H1 in the sense that $H < H(crit)$ holds and that PKA and PKC exhibit a stable focus, the system actually is in a state different from H1. The risk factor H only assumes values below $H(crit)$ as long as the individual takes the prescribed Lithium-based drug. That is, the system is in a symptoms free state H2.

10.3.4   BD Pattern Formation Model: Amplitude Equations

Amplitude Equation for the Bipolar Disease Pattern
As pointed out in Sect. 3.5.2 a limit cycle oscillator may be described in terms of an oscillation amplitude. Let us take advantage of this amplitude description. In line with earlier work [101], it will be shown next that the description of the oscillator by means of its oscillation amplitude allows us to make contact to the LVH amplitude equations used throughout this book.

In order to derive amplitude equations from the biochemical cellular oscillator model, the rescaled PKA concentration $a(t)$ is decomposed into two parts like

$$a(t) \approx A_{BD}(t)u_{BD}(t) , \tag{10.6}$$

where $u_{BD}(t)$ denotes an oscillatory component with a fixed amplitude and A_{BD} denotes the non-oscillatory amplitude of the oscillation in $a(t)$. Such a decomposition has frequently been used to study the oscillatory nature of laser light [154, 155]. In principle, using the so-called rotating wave approximation [154, 155, 239], by substituting Eq. (10.6) into Eqs. (10.1)–(10.5) the evolution equation for A_{BD} can be derived and assumes the form

$$\frac{d}{dt}A_{BD}(t) = \lambda_{BD}\, A_{BD} + \text{nonlinear terms} . \tag{10.7}$$

At this stage, the explicit form of the nonlinear terms is not known. However, since A_{BD} corresponds to the amplitude of an oscillatory pattern, it is plausible to assume that Eq. (10.7) is invariant under the transformation: $A_{BD} \leftrightarrow -A_{BD}$. Therefore, the nonlinear terms will only involve odd power law terms. The lowest nonlinear, odd power law term is the cubic term. Therefore, in lowest order, Eq. (10.7) reads

$$\frac{d}{dt}A_{BD}(t) = \lambda_{BD}\, A_{BD} - A_{BD}^3 , \tag{10.8}$$

where for sake of conveniency the factor in front of the cubic term has been put equal to 1 (in fact, without loss of generality it can be shown that the factor can be eliminated by appropriate variable transformations [215]). Equation (10.8) describes the pitchfork bifurcation studied in Chap. 3 (see Sect. 3.5.2 in particular).

The parameter λ_{BD} in the linear terms of Eqs. (10.7) and (10.8) can assume positive and negative values. For $\lambda_{BD} < 0$, Eq. (10.8) has only one fixed point given by $A_{BD} = 0$ and λ_{BD} describes the exponential decay rate of the amplitude $A_{BD}(t)$ when it is close to zero. In contrast, for $\lambda_{BD} > 0$, Eq. (10.8) has the fixed point $A_{BD} = 0$ and another fixed point with $A_{BD} > 0$ (namely, $A_{BD} = \sqrt{\lambda_{BD}}$). The fixed point $A_{BD} = 0$ is unstable and λ_{BD} describes the exponential growth rate of A_{BD} when the variable is close to zero. In the context of the amplitude equation dynamics defined by Eq. (10.8) the parameter λ_{BD} is the real part of the type 2 eigenvectors (see Chap. 4) that describe the unstable focus giving rise to the limit cycle attractor of interest. For sake of simplicity, λ_{BD} will be refer to as eigenvalue associated with the emerging bipolar disorder pattern. The eigenvalue λ_{BD} is assumed to depend on the properties of the biochemical cellular oscillator model and on the bifurcation parameter H, in particular. When H exceeds the critical value $H(crit)$, then λ_{BD} switches from a negative to a positive value. Formally, this implies that

$$\lambda_{BD}(H) = \beta \ \text{with} \ \begin{cases} \beta > 0 \ \text{if} \ H > H(crit) \\ \beta = 0 \ \text{if} \ H = H(crit) \\ \beta < 0 \ \text{if} \ H < H(crit) \end{cases} \tag{10.9}$$

Amplitude Equations for Disease-Related and Disease-Free Structural Dynamics
So far, the focus was on cyclic variations on the cellular level that are assumed to drive the cyclic mood variations of bipolar disorder patients. In this context, the state of healthy adults was described by $\lambda_{BD} < 0$. Let us assume that on the structural level the dynamics can show variations in time that are not disease related. For example, as mentioned in Sect. 10.3.1 rapid cyclers have more than four episodes per year with mood variations that are relatively strong in magnitude. In contrast, let us consider variations of structural properties that take place on even slower time scales (e.g., one cycle per year) and that might come with less pronounced oscillation amplitudes. The network shown in Fig. 10.4c is only a detail of a much more comprehensive biochemical network. That is, in general, the evolution of concentrations a and c will be determined by evolution equations of the form

$$\frac{d}{dt}a = b_1(1 - a)Z - b_2 a + f_1(a, c, \dots) , \tag{10.10}$$

$$\frac{d}{dt}c = d_1(1 - c)Y - d_2 c + f_2(a, c, \dots) , \tag{10.11}$$

where f_1 and f_2 are unknown functions. The aforementioned different dynamical patterns on the structural level will be related to different temporal patterns of variations in a and c. Focusing only on the evolution of a, we may replace Eq. (10.6) by the more general form

$$a(t) \approx A_{BD}(t)u_{BD}(t) + \sum_k A_k(t)u_k(t) . \tag{10.12}$$

Here, u_k are temporal patterns of structural variations with a fixed amplitude. The amplitudes of those dynamic variations are given by the variables A_k that are referred to as amplitudes. The decomposition shown in Eq. (10.12) is typically used in lasers physics [154, 155]. For our purposes it is sufficient to consider only the disease-related variations $u_{BD}(t)$ and the one of the disease-free structural variation u_k. Without loss of generality, let us use $k = 0$ and consider $u_0(t)$ as a disease-free structural dynamics related to day-to-day mood variations as experience by people in the general population. In total, we will distinguish between two modes $k = 0$ and $k =$"BP" and the corresponding amplitudes A_k.

Without coupling between these temporal "patterns" or structure variations u_{BD} and u_0, the amplitudes A_k satisfy Eq. (10.6) for $k =$"BD" and likewise

$$\frac{d}{dt}A_0(t) = \lambda_0 A_0 - A_0^3 , \tag{10.13}$$

for $k = 0$, where λ_0 is the relevant eigenvalue. However, taking interactions between the disease-related and disease-free structural variations into account, and restricting ourselves only to competitive interactions, we may replace the eigenvalues λ_{BD} and λ_0 in Eqs. (10.7) and (10.13) like $\lambda_{BD} \rightarrow \lambda_{BD} - g A_0^2$ and $\lambda_0 \rightarrow \lambda_0 - g A_{BD}^2$.

That is, using the argument made in the derivation of the Lotka-Volterra model for competitive populations, we assume that each dynamic structure damps the other structure by reducing its pumping parameter (i.e., eigenvalue). That is, the eigenvalue λ_{BD} is negatively affected by the presence of the disease-free variations as measured in terms of the amplitude A_0. Vice versa, the eigenvalue λ_0 is negatively affected by the presence of the disease-related variations as measured in terms of the amplitude A_{BD}. Here, g is the coupling parameter introduced in the context of the LVH model. For multistability and winners-takes-all dynamics we require $g > 1$. In total, Eqs. (10.7) and (10.13) become

$$\frac{d}{dt} A_0(t) = \lambda_0\, A_0 - g\, A_0 A_{BD}^2 - A_0^3\,,$$

$$\frac{d}{dt} A_{BD}(t) = \lambda_{BD}\, A_{BD} - g A_{BD} A_0^2 - A_{BD}^3 \qquad (10.14)$$

and correspond to the standard LVH amplitude equations with cubic nonlinearities and $n = 2$ amplitudes.

10.3.5 Disease Development and Therapy: Amplitude Equations

In order to illustrate the order parameter model defined by Eq. (10.14), let us put

$$\lambda_0 = 1\,,\ \lambda_{BD} = H - H(crit) \qquad (10.15)$$

such that λ_{BD} switches signs at $H = H(crit)$ as discussed in the context of Eq. (10.9). For $g = 0$ we would observe the emergence of bipolar disorder in terms of a transition from $A_{BD} = 0$ to $A_{BD} = \sqrt{\lambda_{BD}}$ when H is increased beyond the critical value $H(crit)$. However, we consider competing structural variations (while ignoring variations that may co-exist), which is mathematically expressed by the condition $g \geq 1$. Therefore, the transition from the healthy condition H1 with $A_0 > 0 \wedge A_{BD} = 0$ to the bipolar disorder state D1 with $A_0 = 0 \wedge A_{BD} > 0$ occurs at a critical value $H^*(crit) > H(crit)$. Let us illustrate this point by means of a numerical simulation that also takes drug treatment into account.

In order to model disease development and the impact of drug treatment within the LVH model (10.14), we postulate (without loss of generality) that the risk factor H increases linearly in time during disease progression (rather than in steps as assumed in Fig. 10.5a) like $H = r_H t$ with $r_H > 0$. Furthermore, it is assumed that the Lithium-based drug treatment results in an exponential decay of H with a decay rate of $\gamma_H > 0$. In total, we obtain

$$\frac{d}{dt} H = r_H - \gamma_H\, H\,. \qquad (10.16)$$

Drug treatment is assumed to take place when the severity of the disease pattern measured in terms of the amplitude A_{BP} exceeds a diagnostic threshold $A_{diagnose} > 0$. That is, if $A_{BD} > A_{diagnose}$ holds then symptoms become clinical and treatment is given to the individual under consideration. At that point in time γ_H is switched from $\gamma_H = 0$ to a non-vanishing value. This non-vanishing value corresponds to the drug dose recommended to the patient provided that the patient takes the drug as recommended. Let $\gamma_{H,required}$ denote the prescribed drug dose. Then, the parameter γ_H can be described as follows:

$$\gamma_H(A_{BD}) = \begin{cases} 0 & \text{if } A_{BD} < A_{diagnose} \\ \gamma_{H,required} & \text{if } A_{BD} \geq A_{diagnose} \end{cases} \tag{10.17}$$

Equations (10.14)–(10.17) were solved numerically for the following parameters and initial conditions: $g = 2$, $H(crit) = 0$, $r_H = 0.02$, $\gamma_{H,required} = 0.07$, $A_0(0) = 1$, $A_{BD}(0) = 0.1$, and $H(0) = 0$. Panel (b) of Fig. 10.5 summarizes the results.

Panel (b) in Fig. 10.5 shows the evolution of the amplitudes A_0 (top panel) and A_{BD} (middle panel) and the risk factor (i.e., bifurcation parameter) H (bottom panel) over time. As expected, at a sufficiently high level of H a transition from H1 with $A_0 > 0 \wedge A_{BD} = 0$ to D1 with $A_0 = 0 \wedge A_{BD} > 0$ occurred. At that time point, drug treatment was started (see marker "Li-T" on the horizontal axis in the bottom panel). Consequently, H started to decay. At a sufficiently low level of H the disease pattern disappeared. That is, a transition from D1 with $A_0 = 0 \wedge A_{BD} > 0$ to H2 with $A_0 > 0 \wedge A_{BD} = 0$ occurred.

Note that the LVH model defined by Eqs. (10.14)–(10.17) exhibits hysteretic transitions for $g > 1$, see Sect. 6.5.3. Consequently, the transition to the symptoms-free state H2 required that H decayed beyond the critical value at which the disease emerged in the first place. This is evident from the graph $H(t)$ shown in Fig. 10.5b (bottom panel) and the values of H at the two aforementioned transition points. Roughly speaking, "it is much easier to catch the disease than to get rid of it."

10.3.6   Relapse Due to Medication Non-adherence

Non-adherence to medication is a major problem in the drug treatment of bipolar disease [238]. Here, medication non-adherence means that patients do not take their drugs as prescribed. Poor adherence typically increases the risk of relapse. That is, a key factor leading to relapse to the disease state after a primary therapy has been successfully completed is medication non-adherence of bipolar disorder patients [238].

The bipolar disorder LVH model described above can capture relapse due to poor adherence of medication. In order to illustrate this, we consider bipolar patients who stop drug treatment or take a lower dose than prescribed as soon as they become symptoms free. That is, we assume that if the structure dynamics u_0 related to the symptoms-free healthy state H2 is re-established, then medication adherence falls below the recommended level. Accordingly, the decay parameter γ_H defined by Eq. (10.17) becomes

$$\gamma_H(A_0, A_{BD}) = \begin{cases} 0 & \text{if } A_{BD} < A_{diag} \\ \gamma_{H,req} & \text{if } A_{BD} > A_{diag} \\ \gamma_{H,req}(1-\Delta) & \text{if } A_0(t) \approx A_{0,st} \wedge \exists \tau < t : A_{BD}(\tau) > A_{diag} \end{cases}$$
(10.18)

The parameter $\Delta \in [0, 1]$ in Eq. (10.18) measures the degree of non-adherence as a proportion between 0 and 100%. For $\Delta = 0$ the patients follows the drug prescription, while for $\Delta = 100\%$ the patients stops taking any medication.

Panel (c) of Fig. 10.5 shows a worked-out example of a relapse due to non-adherence as described by the LVH model that takes Eq. (10.18) into account. Eqs. (10.14)–(10.16) together with Eq. (10.18) were solved numerically. Just as in the simulation presented in Fig. 10.5b, drug treatment started when the disease pattern emerged (A_{BD} jumps to a finite value, that is, when the oscillatory structural dynamics settled in (and related to that the bipolar disease state emerged). In Fig. 10.5c drug treatment begins at the time point indicated by the marker "Li" (see bottom panel). Just as in the example shown in panel (b), in panel (c) the bifurcation parameter and risk factor H decayed and eventually the disease-related fixed point attractor disappeared and A_{BD} switched to zero indicating that the computer-simulated patient became symptom free. At that time point, the amplitude A_0 of the symptoms free state increased towards its finite fixed point value. According to Eq. (10.18), at that bifurcation point (indicated on the horizontal axis by "No") the decay rate γ_H switched from $\gamma_{H,required}$ to $\gamma_{H,required}(1-\Delta)$. In the simulation, the degree of non-adherence (i.e., the non-adherence "probability") was $\Delta = 90\%$. As a result, H increased and eventually the structural oscillations re-appeared triggering bipolar disorder episodes of mania and depression. The numerical simulation of Eqs. (10.14)–(10.16) with Eq. (10.18) presented in panel (c) of Fig. 10.5 used the same parameters as in panel (b) of Fig. 10.5 except for Δ: $\Delta = 0.9$ was used.

10.4   Obsessive Compulsive Disorder (OCD)

Obsessive compulsive disorder (OCD) is a disorder that is about obsessive "thoughts" (i.e, various types of BA patterns of the human isolated brain, see Tables 6.2 and 6.3) and rituals. The "thoughts" are undesired and irrational [54]. For example, a OCD patient may fear dropping something that will hurt somebody. To deal with obsessive "thoughts" of this kind, patients engage themselves in rituals, that is, they form certain sequences of BBA patterns. For example, OCD patients use washing rituals to deal with obsessive "thoughts" about contamination. They perform checking rituals to deal with "thoughts" about hurting others [170]. These rituals are most likely preformed because they reduce the feelings of distress, discomfort, and anxiety that are triggered by the obsessive "thoughts" [84, 235, 273]. Typically, the rituals are time consuming. OCD patients do not derive pleasure from the rituals and would rather not perform them.

10.4.1   Pattern Formation Perspective of OCD

As mentioned above, the obsessive "thoughts" are irrational. They do not make sense even to the OCD patients. Nevertheless, patients cannot "stop their thoughts" to occur. In that context, note that humans in general cannot "stop their thoughts". The notion of "control" violates physics. Therefore, what is meant by "patients cannot stop their thoughts" is that OCD patients exhibit different types of BBA pattern sequences as compared to healthy adults. Likewise, the rituals of OCD patients are often referred to as "compulsive behaviors". Again, the notion of something "compulsive" is not part of the physics perspective of the animate and inanimate worlds because everything in this universe is determined by laws and in this sense is "compulsive". That is, the moon spinning around the earth performs a dynamics that is as "compulsive" as any body dynamics of any human and animal living in this world and as "compulsive" as any emergence of any brain activity pattern in any human and animal. There are no "free" actions.[2]

There is some experimental evidence that the emotional distress states and the ritual patterns of OCD patients correspond to certain brain activity patterns [235, 278, 279]. We will consider these patterns as BBA attractor patterns and in particular as BBA pattern formation sequences within the framework of the human pattern formation reaction model.

---

[2]For example, when the author of this book is happily drinking a cup of coffee while typing these lines the actions performed by the author are as compulsive as a the actions performed by a OCD patient during a ritual. The classifications of actions in terms of voluntary and involuntary actions, automated and non-automated actions, compulsive actions, reflex-like actions, self-aware actions, conscious and non-conscious actions, all involve an element or the negation of an element of "choice" or "free will". However, when taken a physics-based scientific approach automated just as non-automated actions are parts of cause-and-effect chains. In non-automated actions the cause-and-effect relationships are not "less strong" or "less dominant" as compared to automated actions. Likewise, compulsive actions and reflex-like actions just as self-aware actions follows cause-and-effect chains and follow in every step and moment in time the laws of physics just as every tick of a mechanical pendulum clock follows from the laws of physics. Self-aware actions are not "less bounded" by physics as compared to compulsive actions or reflex-like actions. The amount of "freedom" involved in self-aware actions is just the same as the amount of "freedom" involved in compulsive and reflex-like actions, namely, zero. The same applies to the distinction of conscious and non-conscious actions (in addition see Sect. 7.2.2). In this sense, from a physics-based scientific perspective the phrases voluntary and involuntary actions, automated and non-automated actions, compulsive actions, reflex-like actions, self-aware actions, conscious and non-conscious actions, are meaningless. Such phrases might be useful when understanding humans and animals from spiritualistic, religious perspectives.

In summary, OCD is not about being "unable to stop something". OCD is not about something being "compulsive". From a pattern formation perspective, OCD is about certain BBA pattern sequences that do not appear in healthy adults.

10.4.2   Type 2 Pattern Sequences with "Directional" and "Non-directional" Transitions and Rituals

It has been suggested to consider the rituals as type 2 pattern sequences [100, 104] similar to type 2 pattern sequences discussed in Chap. 7 (see Sects. 7.3.3 and 7.4.3 in particular). Figure 10.6 illustrates schematically the underlying mechanism proposed in Refs. [100, 104] leading to a pattern sequence that describes a ritual. The pattern sequence model sketched in Fig. 10.6 is similar to the pattern sequence model describing child play, see Sect. 7.3.3. However, in the sequence model for child play it is only considered that a BBA pattern becomes unstable because the eigenvalue of its corresponding basis pattern drops out of the stability band (for the stability band see Sect. 4.6.3). When patterns that emerged under certain circumstances change their circumstances such that they become unstable, then transitions between BBA patterns necessarily occur. Which BBA pattern emerges at a given bifurcation point depends on the initial conditions of the system at the bifurcation point. In contrast, in the context of the eigenvalue-shift model of restructuring rats by means of mazes and food rewards, the possibility of increasing specific eigenvalue has been discussed such that a specific BBA pattern emerges (see Sects. 7.4.4 and 9.3). In general, a specific transition from a particular BBA pattern k to another BBA pattern j irrespective of the brain activity (or ordinary state) at the bifurcation point occurs when the eigenvalue λ_j of pattern j is sufficiently increased

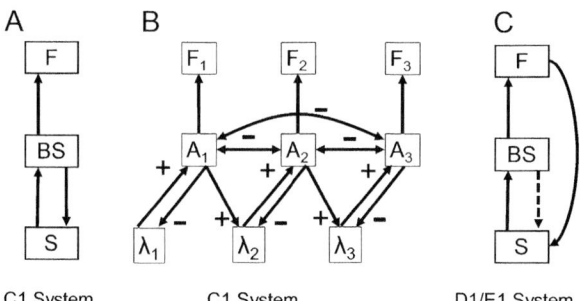

Fig. 10.6 Pattern formation reaction model applied to rituals or type 2 pattern formation sequences with "directional" transitions. Panel (**a**): C1 system scheme as in Fig. 5.5. Panel (**b**): Mechanism leading to specific (i.e., "directional") transitions between patterns in the context of C1 systems. Panel (**c**): Rituals may emerge in D1 or E1 systems as well, where changes in the external forces (caused by the human beings, when producing BBA patterns) act back on the structure in a similar way as in panel (**b**)

Table 10.2 Different kinds of type 2 sequences (i.e., subtypes of type 2 sequences)

Sequences	Self-damping (pattern k reduces its eigenvalue λ_k)	Pumping (pattern k increases eigenvalue λ_j of pattern j)	Additional feature
Pattern sequences without specific order	Yes	No	–
Pattern sequences with specific order (rituals)	Yes	Yes	–
OCD rituals	Yes	Yes	Linked to unwanted obsessions

such that the system becomes monostable. Figure 10.6 illustrates this point. To this end, the system under consideration is considered as a C1 system as shown in panel (a) of Fig. 10.6. More specifically, as shown in panel (b), a pattern $k = 1$ is assumed to emerge because its associated basis pattern exhibits a sufficiently large eigenvalue λ_1 and the initial state of the system is in the basin of attraction of the pattern 1. Subsequently, pattern 1 reduces its own eigenvalue λ_1 and, in doing so, the pattern becomes unstable after a certain period. At the same time, pattern 1 increases the eigenvalue λ_2 of a specific pattern $k = 2$ such that the system becomes monostable. The transition takes place from 1 to 2 (rather than from pattern $k = 1$ to pattern $k = 3$). Then the process is repeated with pattern $k = 2$. That is, pattern $k = 2$ results in damping of λ_2 but increases λ_3 such that eventually a bifurcation from pattern $k = 2$ to $k = 3$ occurs. Finally, panel (c) illustrates that such sequences with a particular order can also emerge in D1 or E1 systems. Accordingly, for rituals emerging in D1 systems the changes in the environment that are due to the produced forces of the BBA pattern k act back on the structure and decrease λ_k and increase λ_{k+1}. In E1 systems structural changes of λ_k and λ_{k+1} are due to the brain state and the changes in the environment. In this context, there are various possibilities how λ_k and λ_{k+1} are explicitly affected.[3]

Table 10.2 points out differences between pattern sequences without a specific order and pattern sequences that exhibit a specific order. Roughly speaking, in pattern sequences without a specific order transitions are "non-directional", while in rituals transitions have a "direction" in the sense that they "point" from a specific BBA pattern to another specific BBA pattern. Table 10.2 also points out a key difference between everyday rituals and rituals of OCD patients. BBA pattern sequences of OCD patients are related to or induced by unwanted obsession BA attractor patterns of the human isolated system.

---

[3]If—after reading these lines—the brain state and brain structure of the reader is in a certain condition, then he/she will produce human isolated brain BA patterns about those possibilities.

10.4.3 OCD Rituals

Let us discuss the example of an OCD ritual described in Frank [100, 104] of a person performing a checking ritual in order to deal with obsessive "thoughts". We consider a ritual involving three actions: checking water taps, checking the stove, and checking doors. The activities are performed in this order (i.e., checking the water taps first, subsequently, checking the stove, after that checking doors). The BBA attractor patterns of the three activities are described by three amplitudes that evolve in time. In addition, a catch-all class of BBA patterns that are not disease related is considered. More precisely, as can be motivated by mathematical arguments [104], an amplitude of a pattern is considered that reflects the amplitude of any pattern not related to the ritual.

In addition to the evolution of brain activity patterns involving muscle force production and body movements, the emotional dynamics is considered. On the one hand, there is the distress feeling that is related with the obsessive "thought" of the person under consideration. On the other hand, emotions are considered that are not related to the obsessive "thought". They are taken together as a class of emotions. For sake of simplicity, emotions are considered as patterns of their own kind. That is, two pattern formation systems are considered: (1) the pattern formation system related to the unstable basis patterns describing the four aforementioned body movements and their associated brain activity patterns and (2) the pattern formation system of emotions featuring two unstable basis patterns. The two pattern formation systems are explicitly modeled as two coupled LVH amplitude equation models [100, 104]. A similar approach based on a coupling between emotional dynamics and body movement dynamics has been proposed in Rabinovich et al. [265].

The eigenvalues of the body movement related patterns evolve in line with the scheme shown in Fig. 10.6. On the one hand, the eigenvalue λ_k of pattern k increases due to the impact of the BBA pattern $k - 1$ that comes earlier in the sequence. On the other hand, the eigenvalue λ_k of pattern k decreases when BBA pattern k has emerged. With respect to the emotional dynamics, a damping mechanism is assumed. While the ritual is performed, that is, while the ritual related BBA patterns are present, the magnitude of the eigenvalue of the distress emotion pattern decays. This implies that the amplitude of the distress emotion pattern decays over time and that the pattern itself becomes unstable at a certain point of time because its eigenvalue drops out of the stability band. Finally, when the distress pattern emerges, the emerging pattern affects the body movement related pattern formation system by increasing the eigenvalue of the first BBA pattern of the ritual such that the ritual is initiated.

10.4.4 OCD Ritual Simulations

In order to illustrate explicitly the application of the human pattern formation reaction model to OCD rituals, we will take three steps. In the first step, the emotional pattern formation system will be ignored. In the second step, both the emotional pattern formation system and the body movement pattern formation system will be considered and the interactions between them. In the third step, it will be considered that the emotional pattern formation may not only initiate rituals but may also terminate the ongoing performance of rituals.

Figure 10.7 presents results from computer simulation of the pattern formation model involved in body dynamics ignoring the emotional dynamics. Panel (a) presents the amplitudes of the four movement patterns as functions of time. Panel (b) shows the corresponding categorical states. Panel (c) presents the eigenvalues of the basis patterns related to the three BBA patterns of the ritual sequence and the BBA pattern of the class of baseline (not ritual related) activities.

According to Fig. 10.7, at 5 time units the obsessive "thought" as BA pattern emerged (not simulated here) and initiated the ritual. The ritual BBA pattern sequence was initiated by putting the eigenvalue of the first activity (checking the water taps) to a sufficiently high value such that the pattern formation system became monostable, see Fig. 10.7 (panel c, top row). That is, the fixed point of

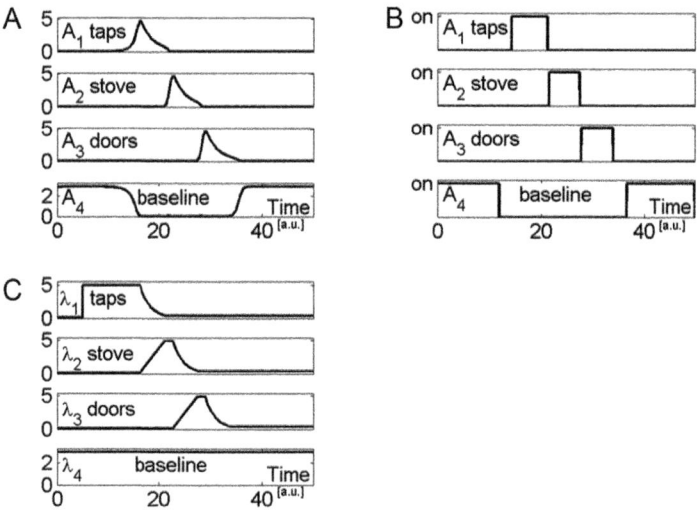

Fig. 10.7 Evolution of the grand state of a simulated OCD patient as defined by the scheme shown in Fig. 10.6b and the LVH model. Panels (**a**) and (**b**) describe the evolution of the ordinary state in terms of the dynamics of the pattern amplitudes (panel (**a**)) for taps checking, stove checking, and doors checking and in terms of the corresponding categorical states (panel (**b**)). Panel (**c**) shows the evolution of the corresponding eigenvalues. Figures are taken from the study by Frank [100] under open access CC license

the not disease-related activities became an unstable fixed point and the fixed point of the first activity of the OCD ritual was the only stable fixed point. As a result, the amplitude of the non-disease related BBA pattern decayed to zero (see row 4 in panel (a)). In contrast, the amplitude of the water-tap-checking BBA pattern increased towards a finite fixed point value (see row 1 in panel (a)). According to the model, the person under consideration was checking the water taps for a certain period. Note that during this period the eigenvalues of the second and third patterns were relatively low.

At this stage, the water-tap-checking BBA pattern caused the eigenvalue of the follow-up activity (i.e., the stove-checking pattern) to increase and damped its own eigenvalue (see panel (c), rows 1 and 2). Panel (a) (top row) shows that the amplitude of the taps-checking BBA pattern decayed. This was due to a drift of the fixed point of the taps-checking attractor that was due to the decaying eigenvalue λ_1.

Since λ_1 decayed and λ_2 increased, at a certain point in time, the eigenvalue λ_1 dropped out of the stability band. That is, the water-tap-checking BBA pattern became unstable. In contrast, at that stage, the stove-checking BBA pattern was stable and was the only pattern with an attractor (i.e, stable fixed point) in the reduced amplitude space. Consequently, the pattern formation system showed a bifurcation from the first to the second ritual activity (see panels (a) and (b) of Fig. 10.7, rows 1 and 2). Accordingly, the computer-simulated OCD patient started checking the stove and other kitchen gadgets. Once the stove-checking BBA pattern emerged, the pattern caused self-damping of its own eigenvalue and caused the eigenvalue of the follow-up activity (checking doors and windows) to increase (see panel (c), rows 2 and 3). At a certain point in time, the eigenvalue λ_2 of the stove-checking BBA pattern dropped out of the stability band and the stove-checking pattern became unstable. At that point, the pattern formation system was monostability and exhibited only the attractor of the checking doors BBA attractor pattern. Consequently, the system exhibited a bifurcation and the ordinary state made a transition from the stove-checking BBA pattern to the checking doors BBA pattern. The computer-simulated person stopped checking the stove and started checking doors. The checking doors BBA pattern was the final BBA pattern in the ritual sequence. Consequently, the pattern only damped its own eigenvalue but did not cause another eigenvalue to increase. Since it was assumed that the eigenvalue λ_4 of the basis pattern related to baseline BBA patterns was constant over the whole period of the ritual, at a certain time point (more precisely after having performed the doors-checking activity for a certain period), the eigenvalue λ_3 dropped out of the stability band whose lower boundary at that time point was determined by λ_4 and the coupling parameter between the amplitudes (see Sect. 4.6.3). The fixed point of the doors-checking activity became unstable. The doors-checking attractor turned into a repellor. The pattern formation system exhibited a bifurcation towards the fixed point that was stable under these conditions, which was the fixed point of the non-disease related baseline patterns (see panels (a) and (b), rows 3 and 4). The ritual was completed.

10.4.5 OCD and the Human Emotional System

Let us turn next to the coupling between pattern formation in the emotional system and the brain activity and body movement system. In this step, LVH amplitude equations for both pattern formation systems were solved simultaneously. The results are presented in Fig. 10.8. Panel (a) shows the amplitude dynamics of the emotional pattern formation system, while panels (b)–(d) refer to amplitudes, categorical states, and eigenvalues of the movement related pattern formation system (and show the same quantities as panels (a)–(c) in the previous Fig. 10.7).

In the second computer experiment, the obsessive "thought" (not simulated) triggered the distress feeling. More precisely, the eigenvalue L_1 of the distress emotion was put to a sufficiently high value such that the emotional pattern formation system became monostable. Here, we use L_1 as symbol for an eigenvalue to point out that the eigenvalue is about the emotional system. The emotional dynamics exhibited a bifurcation towards the distress BA pattern (see panel a in Fig. 10.8). The experience of distress caused the eigenvalue of the first activity of the OCD ritual to jump to a sufficiently high level such that the OCD ritual was initialized as discussed above in Sect. 10.4.4. The dynamics in the movement related pattern formation system evolved as discussed above in the context of Fig. 10.7.

The movement related BBA patterns of the OCD ritual were assumed to lead to a decay of the eigenvalue L_1 of the distress emotion pattern. The eigenvalue of the distress BA pattern decayed monotonically (not shown). At a certain point in

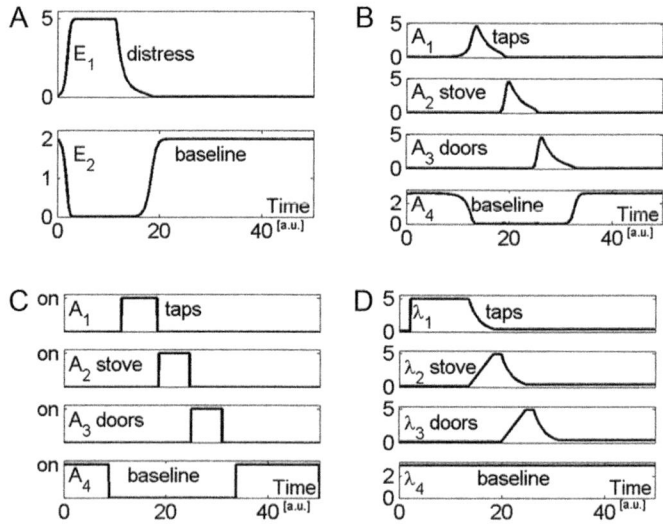

Fig. 10.8 Panel (**a**) shows the dynamics of the amplitude E_1 of the distress emotional pattern and the amplitude E_2 of baseline emotional patterns. Panels (**b**), (**c**), and (**d**) correspond to panels (**a**), (**b**), and (**c**) in Fig. 10.7. Figures are taken from the study by Frank [100] under open access CC license

time, the eigenvalue dropped out of the stability band. The distress emotion BA pattern disappeared and the emotional system made a bifurcation to the baseline state (see panel (a) of Fig. 10.8). Accordingly, the computer-simulated OCD patient stopped experiencing the distress feeling. For the parameters chosen in the computer experiment shown in Fig. 10.8, this emotional bifurcation happened before the ritual was completed (compare panel (a) with panels (b)–(d)). Importantly, the individual continued to perform the ritual although the individual did not experience any distress. This corresponds to a situation in which there is no feedback from the emotional level to the body dynamics level with regard when to terminate the ritual. For this reason, the computer-simulated person continued performing the ritual even in the absence of a distress feeling. In general, Fig. 10.8 illustrates a situation such that once an OCD ritual is initiated then is will be completed irrespective of the state of the emotional system.

In a third computer experiment, the situation was considered in which the ritual is terminated once the distress emotion BBA attractor patterns disappears. The results of the third computer experiment are shown in Fig. 10.9. As it is clear from Fig. 10.9, the emotional dynamics evolved just as in the second computer experiment shown in Fig. 10.8 (compare panels (a) of Figs. 10.8 and 10.9). However, the third activity of the ritual was not performed—as expected—because of the assumed interaction between the emotional dynamics and the brain and body activity dynamics. Taking a mechanistic point of view, in the third simulation it

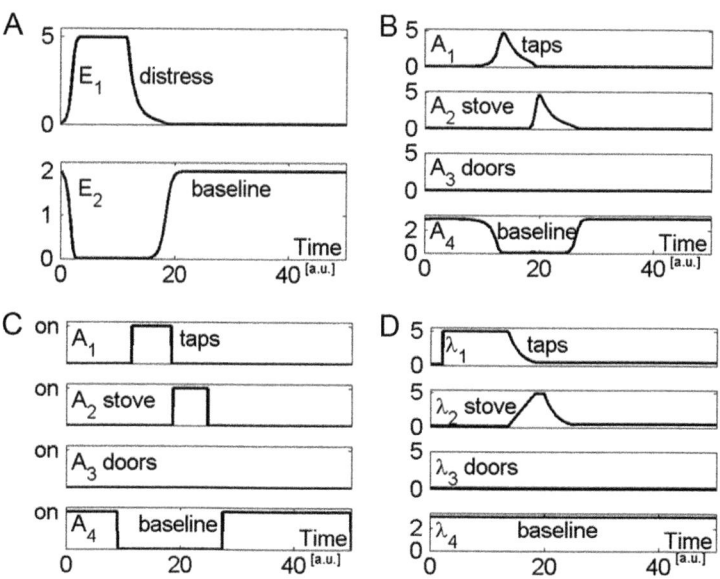

Fig. 10.9 Simulation as in Fig. 10.8 but assuming that the disappearance of the distress emotional pattern (indicated by amplitude E_1 goes to zero) results in an immediate termination of the ritual. See text for details. Figures are taken from the study by Frank [100] under open access CC license

was assumed that when the distress emotion BBA disappeared the breakdown of this pattern caused the eigenvalues of the ritual-related activities 2 and 3 (stove checking and doors checking, respectively) to go to zero as well. While the second activity (stove checking) was performed already, the third activity (doors checking) was not yet performed. Since the eigenvalue λ_3 of the doors-checking activity was put to zero (and was not increased by the BBA pattern of checking the stove), the door-checking BBA pattern did not emerge and the corresponding activity was not performed.

10.4.6   OCD as Dynamical Disease

Support for the hypothesis of OCD as a dynamical disease may come from the studies by Schiepek and co-workers on the treatment of OCD patients [277, 278]. In those studies it has been argued that the successful treatment of OCD patients involves a bifurcation from a disease state to a symptom-free (healthy) state. In particular, in Refs. [277, 278] a case study is presented of a female OCD patient, who developed OCD of moderate severity. The severity of her disease was determined on the Yale-Brown obsessive compulsive scale (Y-BOCS). Y-BOCS scores were measured once a week during a 8 weeks period. Around the third week the Y-BOCS scores decayed dramatically from the aforementioned moderate severity level to scores that are characteristic for symptom-free individuals (see Fig. 1 in Ref. [278] and Fig. 5 in Ref. [277]). Importantly, daily scores of the therapy process were determined as well. Those scores showed intensive fluctuations before the onset of the dramatic decay of the symptom severity [277, 278]. It has been suggested [277] to consider these fluctuations as critical fluctuations that are known to occur at phase transitions (see Sect. 3.2.6). Likewise, the dramatic reduction of the Y-BOCS scores may be considered as non-equilibrium phase transition [277, 278]. In fact, in line with our discussion on continuous human reactions in Chap. 8, the decay of the Y-BOCS scores resembles the decay of an amplitude A for which the corresponding fixed point becomes unstable (i.e., the fixed point attractor turns into a repellor) or disappears entirely. That is, let DV denote the decaying Y-BOCS scores shown in Figs. 1 and 5 of Refs. [277, 278]. Then the variable DV as function of time may be fitted to the LVH amplitude dynamics $dA/dt = \lambda A - A^q$ with $\lambda < 0$ and $DV = cA$ and $c > 0$.

   In line with these considerations on the treatment success of obsessive-compulsive-disorder as a bifurcation in which a disease pattern disappears, it is plausible to assume that the disease pattern of OCD emerges in the first place via a bifurcation. If so, OCD would belong to the class of dynamical diseases. The disease pattern, in turn, is then assumed to change neurophysiological parameters of OCD patients such that OCD patients form type 2 pattern formation sequences that do not occur in healthy adults, which are the OCD rituals characteristic for OCD patients as described above (and distinguished from rituals of healthy adults in Table 10.2).

10.5 Schizophrenia

10.5.1 Medical Hypothesis

Patients with schizophrenia react in various ways differently to the visual world as compared to healthy adults [41]. Such deviating reactions have also been called "perceptual disturbances". In some instances differences in the reactions of schizophrenic patients and healthy adults are subtle [184, 203] (e.g. the "failure" to react to certain structures in dot patterns [203]). In other instances the reactions of schizophrenic patients and the dynamics of the isolated schizophrenic human brain exhibit dramatically different properties as compared to healthy adults. For example, visual and auditory hallucinatory BA patterns are characteristic for many schizophrenic patients, while they are not for healthy adults [196, 344]. Although it is difficult to estimate the prevalence of schizophrenia [275], there is some consensus that schizophrenia affects between 0.5 and 1% of the world population [222, 275]. In what follows it will be shown that the human pattern formation reaction model allows to address various examples of deviating reactions of schizophrenic patients from a unified perspective. Four experimental findings will be addressed: "perceptual inflexibility" under schizophrenia (i.e., reduced multistability of schizophrenic human pattern formation systems), motion-induced blindness under schizophrenia, "priming-induced recognition failures" under schizophrenia, and the reduction of the impact of the proximity principle due to schizophrenia.

The main medical hypothesis is illustrated in Fig. 10.10. In line with the consideration made in Sect. 10.1 on dynamical diseases and diseases in general, it is assumed that schizophrenia is a disease that emerges via a bifurcation, see panel (a) in Fig. 10.10. In healthy individuals (just as in schizophrenic patients) the structure determines among other things the evolution of the brain state (see Chap. 5). Schizophrenia as a dynamical disease is assumed to shift certain primary disease parameters (i.e., risk factors) that act as bifurcation parameters (or contribute to a single bifurcation parameter) such that the dynamics of the brain state makes a transition from the baseline state of healthy individuals to an intermediate brain state of schizophrenia. In this intermediate condition, pattern formation of brain activity, sensory body activity, and muscle activity is determined by the values of the disease-related bifurcation parameters. However, as discussed in various examples on pattern formation of grand states (see Chaps. 7 and 9), brain activity in general is assumed to change structure. In line with this general assumption it is assumed that the intermediate state acts on the structure and affects certain structure parameters. They, in turn, determine the brain activity dynamics that, in turn, acts back on the structure parameters. Eventually, this kind of grand state pattern formation is assumed to produce a set of relatively constant abnormal structure parameters, on the one hand, and BBA attractor patterns characteristic for schizophrenia, on the other. Accordingly, the development of schizophrenia is a cyclic process of changes in structure that result in changes of BBA attractor patterns that induce

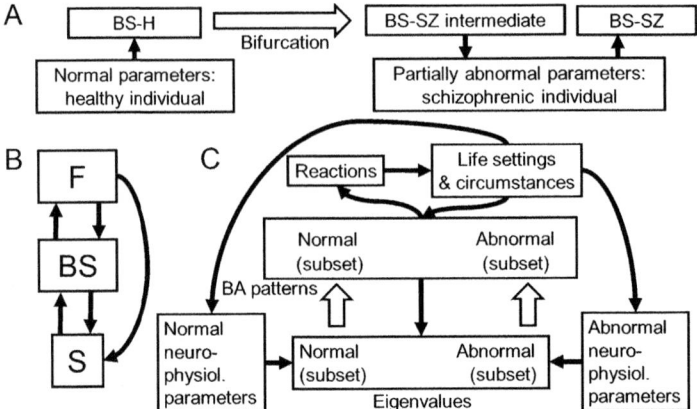

Fig. 10.10 Panel (**a**): Schizophrenia as a dynamical disease. Schizophrenia affects structure, in general. Panel (**b**): General scheme of E3 pattern formation systems (the most general pattern formation system considered in this book). According to the dynamical disease hypothesis, the structure component of the E3 system is affected by schizophrenia. Panel (**c**): Schizophrenia, in particular, affects the eigenvalue spectrums at reference points. Abnormal neurophysiological parameters under schizophrenia are assumed to result in abnormal eigenvalues in a variety of settings. The abnormal eigenvalues, in turn, lead to the emergence of abnormal BBA patterns

changes in structure that result in further changes of BBA attractor patterns and so on. Consequently, from a pattern formation perspective, according to the medical hypothesis sketched in panel (a), schizophrenia is a higher-hierarchical (see also Sect. 9.3) attractor pattern in the grand state of human individuals.

The schizophrenic intermediate brain state (BS-SZ intermediate) may or may not correspond to a pre-disease state as discussed in Sect. 10.1.3. If it corresponds to a pre-disease state then a treatment approach or therapy that could catch patients in this intermediate state would—if successful—reverse the disease-related bifurcation.

As illustrated in panels (b) and (c) of Fig. 10.10, pattern formation in schizophrenia in the most general case takes place in E3 systems (panel b) and is affected by the aforementioned hypothesized set of abnormal parameters (panel c). In particular, abnormal parameters are assumed to affect eigenvalues of basis patterns at unstable reference points such that some of the eigenvalues differ in a characteristic way between healthy adults and schizophrenic patients. These abnormal eigenvalues in turn result in abnormal BA and BBA patterns that determine reactions. As pointed out in panel (c), in general, external forces exerted by the life settings and life circumstances of schizophrenic patients affect the emerging BA/BBA patterns as well such that the evolution of the brain state is both determined by the external forces acting on the individuals and the structure of the individuals. Not all parameters are assumed to be abnormal (in the sense of being different with respect of healthy adults, where we would need to define what is meant by different). Consequently, it is plausible to assume that there are eigenvalues at unstable reference points that are primarily determined by normal parameters and, in doing so, are comparable with the respective eigenvalues of pattern formation systems of healthy adults. The attractor patterns related to the basis patterns of those normal eigenvalues are assumed to be similar to the attractor patterns of healthy adults. Note that panel (c) just describes details of the E3 system sketched in panel (b). Therefore, panel (c) may be regarded as the component and interaction map of E3 systems of schizophrenic patients.

The hypothesis of schizophrenia as a dynamical disease is consistent with suggestion made in the literature that schizophrenia emerges via a bifurcation that is triggered by emotional experiences [52]. In line with this hypothesis, the bifurcation illustrated in panel (a) of Fig. 10.10 leading to schizophrenia is induced by external forces leading to certain BBA patterns of the emotional system. Consequently, according to the hypothesis of emotion-induced bifurcations, similar to our considerations on OCD (see Sect. 10.4 and Figs. 10.8 and 10.9), certain emotional BBA patterns are assumed to shift structure parameters, in general, and eigenvalues of reference points relevant for human reactions and human isolated brain dynamics, in particular.

In the following applications the hypothesis will be investigated that schizophrenia (eventually, i.e., after the system has settled down) affects the spectrum of eigenvalues at bifurcation points. Schizophrenia is assumed to bring eigenvalue spectrums out of their natural "balance". That is, as discussed above, schizophrenia is assumed to be characterized by abnormal neurophysiological parameters that, in turn, determine the magnitude of eigenvalues relevant for the formation of BBA attractor patterns. In doing so, schizophrenia increases or decreases eigenvalues that determine the human isolated brain activity dynamics and the reactions of schizophrenic patients to external forces of their surroundings. The schizophrenia-induced shifts (increases or decreases) of eigenvalues are the cause of differences in reactions (and isolated brain dynamics) between healthy adults and schizophrenic patients.

10.5.2 *"Perceptual Inflexibility" Under Schizophrenia*

As part of a comprehensive study, Martin et al. [234] tested participants with respect to their ability to detect screams in short, 600 ms long audio recordings. The screams were nonverbal, vocal expressions of fear produced by two female native British English speakers. Each scream was considered as being a signal. The screams were concealed by a random noise sound to different degrees. Signal to noise ratios from -30 dB to -16 dB were used and presented in sequences with increasing and decreasing intensities as shown in Fig. 10.11 (panels a, b). Participants were asked to "judge" the audio recording with respect to the presence or absence of a scream. That is, participants were set up to form brain activity patterns of screams or ordinary noise as reactions to the audio recordings. Both schizophrenic patients and healthy adults were tested. For both groups, Martin et al. [234] found hysteresis

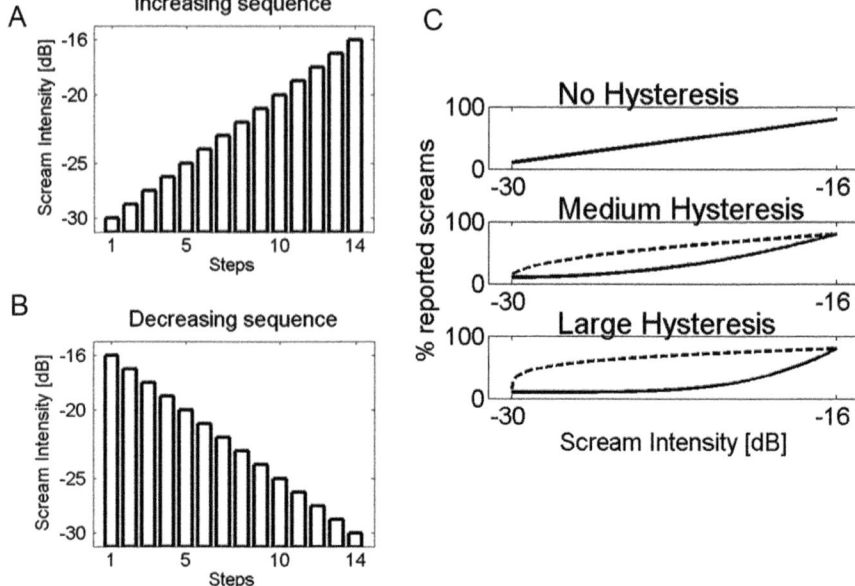

Fig. 10.11 Experimental details and schematic results of the study by Martin et al. [234]. The scream intensity was changed systematically in steps in two different experimental conditions referred to as increasing (panel (**a**)) and decreasing (panel (**b**)) conditions. Panel (**c**): The percentage of screams reported from a sample of test persons is shown schematically for three different scenarios that involve no hysteresis, medium hysteresis, and large hysteresis (from top to bottom). The solid and dashed lines describe the percentage scores of reported screams for the increasing and decreasing presentation conditions, respectively. In the case of no hysteresis both lines fall on top of each other. Martin et al. [234] found an increase in the degree of hysteresis under schizophrenia similar to the increase from medium to large hysteresis depicted in panel (**c**)

in the sense that in the decreasing condition participants reported more frequently that they have heard screams as compared to the increasing condition (see panel (c) and compare with Fig. 6.19e). From a pattern formation perspective, this result suggests that the "Yes-there-is-a-scream" or scream BBA attractor pattern emerged more frequently in the decreasing intensity condition. The splitting between the decreasing and increasing presentation conditions was larger for the schizophrenic patients, meaning that hysteresis was stronger under schizophrenia. Martin et al. [234] interpreted "perceptual" hysteresis as a measure for "perceptual inflexibility" (the "failure" to update experiences) and accordingly concluded that schizophrenia increases the degree of "perceptual inflexibility".

In line with the medical hypothesis given above, it is proposed that schizophrenia increases the eigenvalues of BBA patterns associated with experiences of existence/presence (e.g., there is a scream) relative to the eigenvalue of BBA patterns of experiences of absence (e.g., there is no scream). To this end, hysteresis observed on the level of samples of participants in the study by Martin et al. [234] is understood as hysteresis on the level of individual participants given in terms of different critical bifurcation parameters (see Sects. 6.5.3 and 6.5.4). Accordingly, while in the increasing condition the scream BBA pattern emerged for the first time at a particular critical value $\alpha(crit, 2)$ of the signal to noise ratio, in the decreasing condition this BBA pattern associated with a scream became unstable and disappeared at a different critical intensity value $\alpha(crit, 1)$. This critical value of the decreasing condition was lower than the critical value of the increasing condition: $\alpha(crit, 1) < \alpha(crit, 2)$. Moreover, hysteresis size $\Delta\alpha = \alpha(crit, 2) - \alpha(crit, 1)$ (see Sect. 6.4.3) typically is proportional to the degree of the splitting shown in panel (c) between the two presentation conditions. That is, if the splitting shown in panel (c) is large (small), then hysteresis size on average is large (small) as well. Therefore, according to our model-based interpretation of the data by Martin et al. [234], schizophrenia is assumed to change structural parameters such that hysteresis size becomes larger. Vice versa, the increased magnitude of the hysteresis phenomenon under schizophrenia can be taken as evidence for a disease-induced shift of structural parameters (e.g., eigenvalues).

Following our analysis on hysteresis and bistability in human pattern formation systems (see Sect. 6.5.3), it is assumed that the experimentally observed hysteretic reactions result from the competition between the two BBA patterns related to the absence and presence of a scream. Figure 10.12 displays the phase portraits (panels b–d) of the LVH amplitude equations describing this kind of pattern competition. In panels (b)–(d) the amplitude of the BBA pattern of the presence of a scream (i.e., abbreviated as "scream amplitude") is shown on the horizontal axis and the amplitude of the BBA pattern of the absence of a scream (abbreviated as "no scream amplitude") is shown on the vertical axis. Panel (a) is an auxiliary panel that demonstrates the construction of the phase portraits in line with our general considerations in Sect. 3.3.3 (compare also Figs. 3.8 and 10.12). Panel (a) of Fig. 10.12 shows the two amplitudes as functions of time as computed from the LVH amplitude equations for particular initial conditions. The dotted line shown in panel (b) was obtained from the two functions shown in panel (a). In particular, for the time points of 0.02, 0.05, 0.10, and 0.20 time units, the values of the two amplitudes have been indicated by markers in panel (a) and the same markers have been used in panel (b) to describe the corresponding locations. All solid lines in panels (b)–(d) have been obtained by means of this procedure by varying the initial conditions. In our context, the key parameter that determines the overall layout of the phase portraits is the scream intensity. Panels (b)–(d) show phase portraits for relatively low, medium, and relatively high scream intensities, respectively, consistent with the experimental data reported by Martin et al. [234]. For relatively low or relatively high scream intensities the pattern formation system is assumed to be monostable as illustrated by the phase portraits in panels (b) and (d) such that irrespective of the initial conditions only the BBA pattern of the absence (panel b) or presence (panel d) of a scream emerges. In contrast, for medium scream intensities the system is assumed to be bistable as illustrated by the phase portrait shown in panel (c). Accordingly both patterns can emerge. Which pattern emerges depends on the initial conditions, that is, the ongoing brain state. As discussed earlier in Sect. 6.5.3 in the

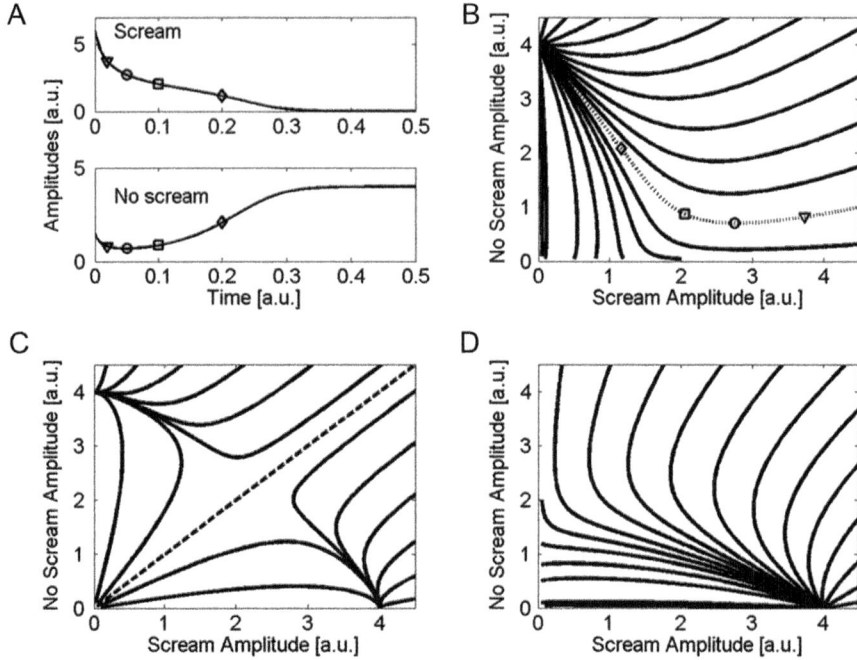

Fig. 10.12 Demonstration of bistable case (relevant for understanding hysteresis) of the LVH model in the context of the study by Martin et al. [234]. Panel (**a**): Auxiliary panel presenting the two relevant amplitudes as functions of time as obtained by solving the LVH model numerically under specific initial values for the amplitudes. The two functions of time give rise to the dotted line shown in panel (**b**). All lines in panels (**b**), (**c**), (**d**) have been constructed in a similar way. Panels (**b**), (**c**), (**d**) show the phase portraits of the LVH model for two monostable ((**b**) and (**d**)) conditions and a bistable condition (panel (**c**)). Panels (**b**) and (**d**): For arbitrary initial values the amplitude dynamics converges to the unique fixed points at 4 units shown on the vertical (panel (**b**)) and horizontal (panel (**d**)) axes, respectively. Panel (**c**): Depending on the initial brain state (i.e., the initial values of the amplitudes), the amplitude dynamics converges to either of the two fixed points shown in panels (**b**) and (**d**)

context of the hysteretic human reactions, BBA or BA attractor patterns that have emerged can serve as initial states for the formation of patterns in the next moment of time. Consequently, the BBA pattern of a scream for a certain scream intensity makes that it emerges again (or is maintained) even when the scream intensity is lowered as long as it does not become unstable. This leads to the hysteresis phenomenon (see Sect. 6.5.3).

In order to see how schizophrenia affects the acoustic world reaction task, let us model explicitly the impact of the forces exerted by the acoustic signal on the eigenvalue spectrum, see panel (a) of Fig. 10.13. Let us label the BBA patterns associated with the absence and presence of a scream by 0 and 1, respectively, and the corresponding eigenvalues by λ_0 and λ_1. In line with the human pattern formation reaction model of hysteresis (see Sect. 6.5.3), it is assumed that the

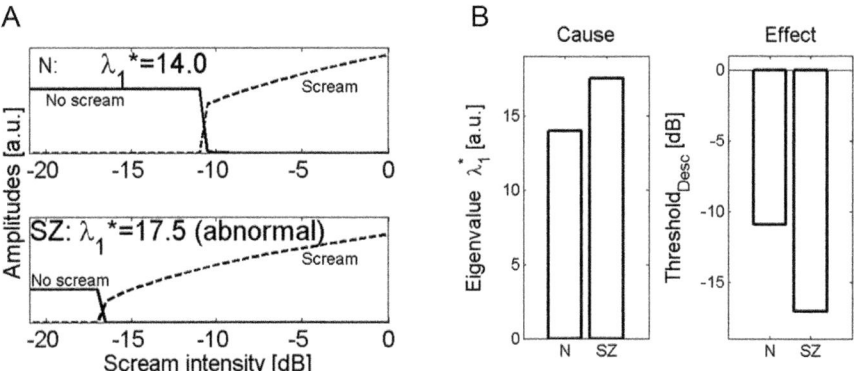

Fig. 10.13 Disease-induced increase of eigenvalues of BBA patterns associated with experiences of the presence of something can lead to a decrease of the corresponding acoustic threshold. Panel (**a**): Simulation results obtained for a hypothetical healthy subject with normal eigenvalue (top subpanel) and a hypothetical schizophrenic individual with abnormal eigenvalue (bottom subpanel). Panel (**b**), left: Eigenvalue λ_1^* of the scream BBA pattern for healthy adults (N) and patients with schizophrenia (SZ). Panel (**b**), right: Acoustic thresholds $\alpha(1, crit)$ measured in decibel predicted by the LVH amplitude equations for healthy adults (N) and schizophrenic individuals (SZ). They have been computed from the values of λ_1^* shown on the left by means of analytical expressions presented in Sect. 10.7.2 and are consistent with the bifurcation points obtained from the numerical simulations shown in panel (**a**)

eigenvalues λ_0 and λ_1 of the no scream and scream BBA patterns remain constant (λ_0) and increase (λ_1), respectively, when the signal to noise ratio is increased. The signal to noise ratio is considered as bifurcation parameter α. Using linearized model equations as first order approximations (see [120] and the models in Chap. 7), let us put $\lambda_0 = L - \delta/2$ and $\lambda_1 = L + \delta/2 + C\alpha$. We will explain these two expressions in three steps. First, for $\delta = 0$ and $\alpha = 0$ we have $\lambda_0 = \lambda_1 = L$. Accordingly, L is a constant offset parameter that is independent of scream intensity. Second, in general, the eigenvalues λ_0 and λ_1 exhibit different constant offset parameters. In order to account for this feature, we could use two different L variables (such as L_1 and L_2). From a mathematical point of view, however, it is more convenient to introduce a bias parameter δ. Accordingly, for $\alpha = 0$ from $\lambda_0 = L - \delta/2$ and $\lambda_1 = L + \delta/2 + C\alpha$ it follows that $\lambda_0 = L - \delta/2$ and $\lambda_1 = L + \delta/2$. To re-iterate, this means that BBA patterns associated with the absence (0) and presence (1) of a signal in general exhibit different offset parameters. Let us consider two participants A and B. If δ is larger for A as compared to B then for A there is a stronger bias that pattern 1 emerges which may be interpreted as a stronger bias to react to an acoustic signal (irrespective of the actual signal intensity). Third, in order to model the impact of the scream intensity, it is assumed that the eigenvalue λ_1 increases linearly[4] with α. In this context, $C > 0$ is a gain factor.

In line with these considerations and in view of the experimental observations reported by Martin et al. [234], it is hypothesized that schizophrenia increases the bias parameter δ such that δ for schizophrenic patients on average is larger than δ for healthy adults. This implies that the constant component $\lambda_1^* = L + \delta/2$ of the eigenvalue λ_1 associated with the reaction to signals is larger under schizophrenia. The increased bias towards the emergence of pattern $k = 1$ leads under appropriate circumstances to a lower threshold $\alpha(crit, 1)$ in the decreasing condition. This can be shown by detailed calculations similar to the one presented in Ref. [104].

A worked out example is presented in Fig. 10.13. Panel (a) shows simulation results. Panel (b) shows analytical results as obtained in Sect. 10.7.2. Panel a shows the stationary values of the scream amplitudes (dashed lines) and the no scream amplitudes (solid lines) as obtained by numerical simulation of the LVH amplitude equations for scream intensities ranging from 0 to -20 dB. The top subpanel illustrates the situation for a healthy subject with a normal eigenvalue component λ_1^* (here: $\lambda_1^* = 14$ units). In contrast, the bottom subpanel refers to a hypothetical schizophrenic individual with an abnormal eigenvalue component λ_1^* (here: $\lambda_1^* = 17.5$ units). In order to interpret panel (a) correctly, we need to read the horizontal axis from the right to the left. Following the horizontal axis from the right to the left, scream intensity decreases. By visual inspection of both panels, we see that for a scream intensity of 0 dB the amplitudes of the BBA patterns related to the presence of a scream assumed finite values in both simulations, whereas the amplitudes of the BBA patterns associated with the absence of a scream were zero. Consequently, the simulated healthy and schizophrenic subjects formed initially scream BBA attractor patterns. In contrast, for very low scream intensities of -20 dB the opposite situation was true. For the two simulated individuals the no scream amplitudes assumed a finite value, whereas the scream amplitudes were zero. Accordingly, at the end of the simulated decreasing sequences the subjects reacted to the acoustic signals in a similar way as they would react to signals without screams. For the healthy adult, the switch at which the BBA pattern indicating a scream became unstable and disappeared took place between -10.5 and -11.0 dB. For the schizophrenic individual the switch took place between -16.5 and -17.0 dB. Consequently, the switching threshold (critical bifurcation parameter $\alpha(crit, 1)$) was lower for the computer simulated schizophrenic individual. In this sense, the individual reacted to the acoustic signals "longer" by forming the scream BBA pattern $k = 1$. In other words, according to our model simulations, the schizophrenic patient "perceived" screams for weaker signal to noise ratios as compared to the simulated healthy adult, which is consistent with the experimental findings mentioned above.

---

[4]In addition, we could assume that λ_0 decreases with α as in Sect. 6.5.3. In the context of grasping transitions and gait transitions such a decrease can be motivated explicitly. However, here we just assume that λ_0 remains constant.

For a sample of healthy adults, we may assume that the parameter δ is distributed around a mean value in a particular way. Likewise, in order to simulate a sample of schizophrenic individuals, we would take δ from a distribution of values with a certain mean. An example of a simulation in which variability across participants has been taken into account has been presented in Sect. 7.4.1, see Fig. 7.7. Another example will be presented in Sect. 10.5.4 below. However, in order to illustrate the main idea, in what follows, variability among individuals will be neglected. In this case, the values obtained for the two simulated subjects can be interpreted as sample mean values. This perspective is taken in panel (b) of Fig. 10.13. Panel (b) summarizes the eigenvalue components λ_1^* (left subpanel) and the critical bifurcation parameters $\alpha(crit, 1)$ (right subpanel) for a sample of healthy adults (i.e., normal adults, N) and a sample of schizophrenic patients (SZ). Panel (b) highlights that according to the LVH amplitude equation model the difference between the eigenvalue components causes the difference in the switching thresholds (critical bifurcation parameters). Note that in panel (b) the switching thresholds have not been taken from the simulations shown in panel (a). Rather, they have been calculated from the analytical expressions derived in Sect. 10.7.2 below. From those analytical expressions the threshold values of $\alpha(crit, 1) = -10.9$ dB for healthy adults and $\alpha(crit, 1) = -17.1$ dB for schizophrenic subjects were obtained. Those analytical results are in excellent agreement with the numerical values.

Taking panels (a) and (b) of Fig. 10.13 together, the "perceptual inflexibility" postulated by Martin et al. [234] is explained as the result of an increased (abnormal) eigenvalue of the BBA pattern emerging under experiences of existence.[5]

Finally note that the experimental results reported by Martin et al. [234] suggest that when the scream intensity was increasing then the switching thresholds were approximately the same for the simulated healthy subjects and the schizophrenic individuals. However, the switching threshold was lower in the decreasing condition for the SZ-individuals, which implies that the hysteresis was larger in magnitude for those individuals. In the simulations discussed above all remaining parameters of the LVH model have been chosen such that the switching thresholds were the same in the increasing condition across the simulated SZ- and N-individuals (see mathematical notes in Sect. 10.7.2 for more details).

---

[5]More precisely, as discussed in Sect. 5.4.1 it is assumed that forces exerted by objects and events shift under idealized symmetric conditions the brain state of humans into the basin of attraction of a certain attractor. The attractor may be labeled as the attractor related to the BBA pattern of "there is something" but any other label could be used as well. The attractor is associated with an unstable basis pattern (which makes that the attractor exists at all). The eigenvalue of that basis pattern is assumed to be increased under schizophrenia.

10.5.3 Motion-Induced Blindness Under Schizophrenia

Tschacher et al. [324] investigated differences between healthy adults and schizophrenic patients with regard to the motion-induced blindness (MIB) phenomenon that was briefly reviewed in Sect. 7.4.3. Recall that the MIB phenomenon refers to an oscillatory instability in which features of a stationary foreground pattern "get lost" for short time periods due to the impact of a moving background pattern. That is, participants typically report that foreground pattern features disappear and re-appear in an alternating (i.e., oscillatory) fashion. The background pattern used by Tschacher et al. [324] was a rotating array composed of 49 blue crosses that were arranged on the grid points of a 7 by 7 grid. The background pattern rotated clockwise at a rotation speed of 1 rotation in 3 s. The foreground pattern was composed of three yellow dots that formed the corners of a triangle that pointed downwards. Three one-minute runs were administered to the participants. Patients with schizophrenia and healthy adults were tested. The number of disappearances of the foreground pattern as a whole or parts of it and the durations of the disappearance episodes were measured and accumulated across the three one-minute runs. On average, healthy individuals experienced 42 disappearances for a duration of 42 s, whereas schizophrenic patients experienced 29 experiences for a duration of 33 s. This implies that the single MIB periods had approximately the same duration for healthy individuals (1.0 s) and patients (1.1 s). However, the MIB free periods (i.e., periods during which the participants reported to see the full foreground pattern) were longer for schizophrenia patients (4.5 s) as compared to healthy adults (3.3 s) [102].

Consequently, for schizophrenic patients the BBA attractor pattern emerging as reaction to the foreground features of the motion-induced blindness movie was less affected by the external forces exerted by the background features of the movie.

Within the framework of Gestalt psychology, it is assumed that the "human perceptual system" "reorganizes", integrates, and groups features of visual "stimuli" to obtain a "holistic", "perceptual" experience. Taking this perspective it has been suggested that this property of "reorganizing", integrating, and grouping is impaired under schizophrenia [324].

In line with the medical hypothesis stated above that schizophrenia affects eigenvalues of basis patterns related to BBA patterns and, in doing so, the formation of BBA patterns, in what follows, the experimental findings by Tschacher et al. [324] will be explained by assuming that schizophrenia increases eigenvalues of the visual world BBA pattern associated with the triangular foreground pattern. Accordingly, if the eigenvalue increases in magnitude, then the BBA pattern emerges for a longer time before it disappears due to the interaction with other, competing BBA patterns. These other BBA patterns are the background pattern itself and patterns in which parts of the foreground pattern are missing [102]. A detailed mathematical and numerical analysis of the LVH amplitude equations confirms this argument (see Ref. [102] and the mathematical notes below in Sect. 10.7.2).

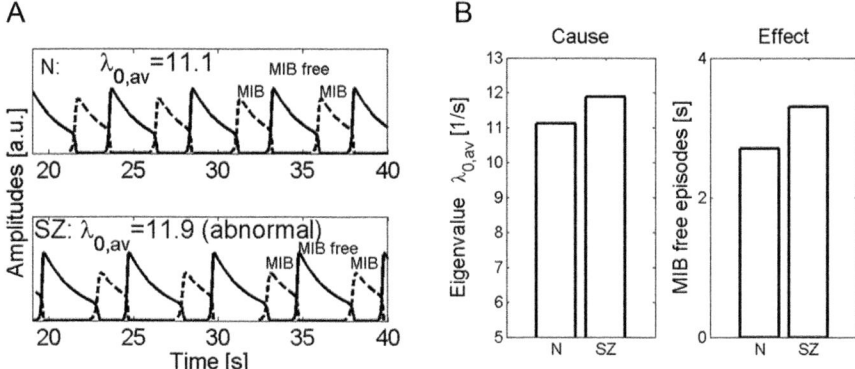

Fig. 10.14 Abnormally increased eigenvalue associated with the foreground BBA attractor pattern of a MIB movie leads to longer periods without MIB episodes. Panel (**a**): Time course of the relevant pattern amplitudes obtained from numerical simulations of the LVH amplitude equations for a healthy subject (top subpanel) and a schizophrenic individual (bottom subpanel) with normal and abnormal eigenvalues $\lambda_{0,av}$. MIB free episodes are those episodes during which the amplitudes indicated by the solid lines assume finite values. Likewise, episodes of partial or complete "blindness" towards the foreground dots are those episodes during which the amplitudes indicated by the dashed lines assume finite values. Panel (**b**), left: The eigenvalue $\lambda_{0,av}$ of the triangle, foreground BBA pattern is shown for healthy adults (N) and schizophrenic individuals (SZ). Panel (**b**), right: The duration of single MIB free episodes is shown for healthy adults and schizophrenic individuals as determined from the numerical results shown in panel (**a**). Note that Fig. 10.14 illustrates the qualitative relationship between the magnitude of the parameter $\lambda_{0,av}$ and the MIB free episode durations and is not a fit to experimental data

Figure 10.14 illustrates the relationship between the eigenvalue of the visual world triangular foreground BBA pattern and the duration of MIB free episodes. In order to produce Fig. 10.14, the LVH model discussed in Sect. 7.4.3 for oscillatory phenomena has been applied to the MIB paradigm. Panel (a) shows simulation results obtained by solving numerically the LVH model. Panel (b) summarizes the relevant key measures obtained from the graphs shown in panel (a). Accordingly, in the computer experiment, it was assumed that on average the eigenvalue $\lambda_{0,av}$ of the foreground BBA pattern (labeled $k = 0$) is larger for schizophrenic patients (SZ) as compared to healthy adults (i.e., normal adults, N). Panel (a) shows the amplitudes of the foreground BBA patterns indicated by solid lines for a simulated healthy adult and a simulated schizophrenic subject. Likewise, the amplitudes of the BBA patterns related to foreground dot "disappearances" as indicated by dashed lines are shown. For the schizophrenic individual an (abnormal) larger value of $\lambda_{0,av}$ was used (here: 11.9 units) as compared to the healthy subject (here: 11.1 units). As a result of this difference, by visual inspection of the amplitude functions presented in Fig. 10.14a we found that the MIB free periods were longer for the simulated schizophrenic individual as compared to the simulated healthy individual. From the numerical simulations shown in panel (a), the durations of single MIB free episodes were also calculated numerically for the two simulated subjects. Consistent with the aforementioned theoretical considerations and in line with our visual impression, we found that the MIB free episodes were longer for the patient with schizophrenia (3.3 s) as compared to the healthy (2.7 s). Panel (b) presents those results in form of sample averages as we would obtained them in experimental experiments. In panel (b) variability among individuals is neglected, which is the reason why the sample means simply correspond to the values used for the two simulated subjects (i.e., the health subjects, on the one hand, and the schizophrenic patient, on the other hand). As such, the simulations presented above may be conducted a more comprehensive way in order to take variability among participants in account (see Sect. 7.4.1 and Fig. 7.7; see also the subsequent section, Sect. 10.5.4).

10.5.4   Instruction-Induced "Recognition Failures"

Schizophrenic patients are known to have "deficits" in the interpretation of emotions of facial expressions [231], which might be a negative symptom specific to emotions or reflects that in general "perception" is "impaired"[6] under schizophrenia [185]. Barbalat et al. [15] examined "recognition failures" in "perception" of emotions under schizophrenia induced by a subtle "priming" cue. In the experiment, images were used that showed the faces of male and female actors, who demonstrated a neutral emotion along with three non-neutral emotions: fear, anger, and happiness. Sequences were shown in four runs to each participant. Each run had a "target" emotion[7] (either neutral, anger, fear, or happiness). Within a run the task was to indicate whether or not a given face showed the "target" emotion (i.e., a so-called "two-alternatives" "forced-choice" task was used[8]). Participant indicated their "decisions"[9] by pressing either of two buttons. The experiment was conducted with schizophrenic patients and healthy adults.

All faces showed clearly one of the four emotions. That is, taken out of the context of the experimental task, each face could "easily" be assigned to one of the emotions. However, within the experimental paradigm that involved "target"

---

[6]That is, taking a scientific perspective, schizophrenic patients react to facial expressions of emotions *differently* as compared to healthy adults and this difference might be a specific symptom for schizophrenia or it might reflect that in general schizophrenic patients react *differently* to the forces of the world acting on them. Note that from a scientific perspective there are no such things as "deficits" or "failures" or "impairments". Such a terminology might be useful within spiritualistic, e.g., religious, approaches to understand humans. E.g., key elements of Christianity are that humans *fail*, on the one hand, and that God is a forgiving entity, on the other hand.

[7]The concept of a "target" emotion is not a scientific one, just as the concept of "attention" is not a scientific concept. However, we will not dwell on this issue here.

[8]Of course, taking a science point of view, at no point in time the participants in the experiment had a "choice". Neither were there any "alternatives".

[9]Of course, saying that participants in this experiment made "decisions" is like saying that water makes "decisions" when it turns into ice. From a science perspective, participants did not make "decisions" at all.

emotions, participants reacted in various instances differently as they would react to the faces taken out of context. For sake of brevity, in what follows only the impact of the fearful "target" condition on the reactions to angry faces and, vice versa, the impact of the anger "target" condition on the reactions to fearful faces will be reviewed (for more details the reader is referred to the original study [15]). When the task was to distinguish between angry faces and faces that showed different emotional expressions, the correct identification of angry faces was above 95% both for patients and healthy adults. However, for both participant groups the "correct" identification of the fearful faces was reduced to about 90% for healthy adults and 75% for schizophrenic patients. Likewise, when the task was to distinguish between fearful faces and faces that showed emotional content different from fear, then the "correct" identification of fearful faces was relatively high but the "correct" identification of the angry faces was poor. For healthy adults "correct" identification of fearful versus angry faces was at about 95% versus 90%. For schizophrenic patients "correct" identification was at about 90% versus 77%. Again, the effect of "priming" (i.e., restructuration by forces, which happens to be external forces exerted by the instructions; see also Sect. 7.4.1 for "priming" as restructuration by internal forces) was stronger for schizophrenic patients as compared to healthy adults.

Taken together, it was found that schizophrenia increases the frequency of instruction-induced "recognition failures". Barbalat et al. [15] interpreted their finding in line with related studies that showed that schizophrenic patients as compared to healthy adults were less willing to re-evaluate and—if necessary— change their beliefs when confronted with novel "pieces of information" that challenge their current beliefs [355]. Accordingly, the instructions that were given to the participants and involved "target" emotions created the expectations to be exposed to the "target" emotions. These expectations in turn negatively impacted the identification of "non-target" emotions.

In view of our medical hypothesis (see above), the observation by Barbalat et al. [15] will be explained by assuming that the "priming cue" (i.e., the restructuring) strengthens the BBA pattern associated to the "prime" by increasing its eigenvalue. This assumption is in line with the human pattern formation model applied to "priming" in general and "retrieval-induced forgetting" as discussed in Sect. 7.4.1. If restructuring increases the eigenvalue of the "primed" BBA pattern then due to the competition between patterns other BBA patterns related to "non-primed" emotional expressions can no longer emerge even if an human individual is exposed to a face showing one of those "non-primed" emotions. From a mechanistic point of view, it is assumed that the instruction-induced restructuring (i.e., the "priming") increases the eigenvalue of the "prime-related" BBA pattern [90] such that BBA patterns of "non-primed percepts" drop out of the stability band (e.g. see "retrieval-induced forgetting"). In other words, we exploit that a sufficiently strong increase of a particular eigenvalue can make that eigenvalues of other BBA patterns drop out of the stability band such that the corresponding patterns can no longer emerge even when an individual is exposed to the forces typically leading to the emergence of that patterns [106]. The individual becomes "blind" towards those patterns with eigenvalues out of the stability band. The reactions to the "non-primed" external circumstances change and in this sense the "non-primed" external circumstances are identified "incorrectly". The LVH amplitude equations allow to demonstrate this "priming effect" (i.e., effect of restructuring) explicitly. By means of detailed analytical calculations (sketched in the mathematical notes given in Sect. 10.7.2), the impact of increasing the eigenvalue related to a "primed" emotion basis pattern (e.g., anger) on the reactions to "non-primed" facial emotions (e.g., fear) can be determined.

In order to model the experimental observations, let us consider populations rather than individuals. That is, let us consider the populations of schizophrenic patients and healthy subjects. Without loss of generality, for both populations it is assumed that the (increased) "primed" eigenvalue, here denoted by λ_1, is normally distributed around a mean value $\lambda_{1,av}$. It is assumed that under schizophrenia the mean value $\lambda_{1,av}$ is abnormal in the sense that it is larger than for healthy subjects.

Panel (a) of Fig. 10.15 shows hypothesized distributions of λ_1 for schizophrenic patients (dotted graph) and healthy adults (solid line). The locations of the mean values ($\lambda_{1,av} = 4.0$ units for healthy individuals versus $\lambda_{1,av} = 5.5$ units for schizophrenic individuals) of those distributions are indicated on the horizontal axis as well. The mean value locations correspond to the locations of the peaks of the distributions. The vertical bar in panel (a) indicates a critical value $\lambda(crit, 1)$ (here: $\lambda(crit, 1) = 4.5$ units). From the LVH amplitude equations, it follows that only if λ_1 is smaller than $\lambda(crit, 1)$, then the eigenvalues of the "non-primed" BBA patterns belong to the stability band and the corresponding patterns can emerge if the corresponding emotional expressions are presented. Consequently, the percentage values of "correct" identifications of "non-primed" emotional expressions correspond to the areas under the normal curves to the left of the vertical line. From the graphical construction shown in panel (a), it follows that this particular area under the curve is larger for healthy subjects (i.e., the solid line distribution) as compared to schizophrenic individuals (i.e., the dotted-line distribution). Therefore, the LVH model predicts that the percentage of "correct" identifications is larger for healthy adults than for the schizophrenic patients. In fact, the area to the left of the vertical line and consequently, the probability (or percentage) of "correct" reactions or "correct" identifications can be computed analytically for both population distributions (see Sect. 10.7.2). For the parameters used in panel (a), we obtain 73% "correct" identifications for healthy adults and only 11% "correct" identifications for schizophrenic patients. Panel (b) summarizes the key measures of modeling the impact of schizophrenia on "recognition failures" due to instruction-induced restructuration as observed by Barbalat et al. [15]. The left subpanel displays the normal and abnormal mean values $\lambda_{1,av}(N)$ and $\lambda_{1,av}(SZ)$ for the population of healthy adults and schizophrenia patients, respectively. The right subpanel present the aforementioned predicted percentage values (73% versus 11%) of "correct" identifications of the "non-primed emotion" (here: the emotion fear). The larger (abnormal) mean value $\lambda_{1,av}(SZ)$ for the population of schizophrenic patients results in a lower percentage value of "correct" identifications.

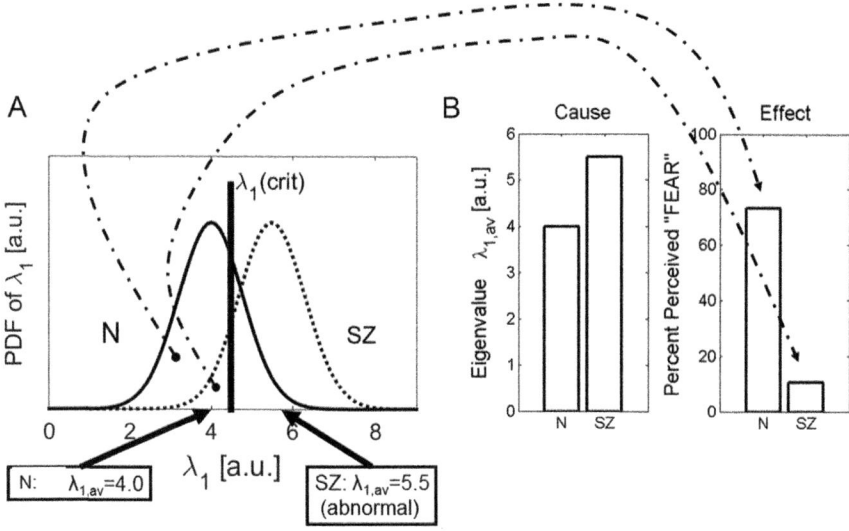

Fig. 10.15 Abnormally increased eigenvalue due to restructuration-by-instruction ("priming") under schizophrenia and its effect on the percentage of "correctly" identified facial expressions not mentioned in the instruction (i.e., "non-primed" expressions). Panel (**a**): The probability density functions (PDFs) of eigenvalues λ_1 for the population of healthy individuals (solid line marked N) and the population of schizophrenic individuals (dashed line marked SZ) are shown. The vertical line indicates the critical threshold $\lambda(crit, 1)$. The areas under the curves to the left of the vertical critical line corresponds to the percentage of population members (Ns and SZs) who can "correctly" identify the "non-primed" emotion (here: fear). All other population members are "blind" towards the forces exerted by "non-primed" emotional facial expression. The mean eigenvalue $\lambda_{1,av}$ is assumed to be larger under schizophrenia, which implies that the percentage of individuals that can solve the identification task "correctly" is smaller under schizophrenia. Panel (**b**), left: The mean eigenvalue $\lambda_{1,av}$ of the PDFs are shown for healthy (N) and schizophrenic (SZ) individuals. Panel (**b**), right: The percentage values of "correct" identifications (corresponding to the areas indicated in panel (**a**)) are shown for N- and SZ-individuals. They have been computed analytically, see Sect. 10.7.2

10.5.5 Proximity Principle Under Schizophrenia

There is an extensive literature about to what extent schizophrenia affects phenomena addressed by Gestalt psychology (for reviews see, e.g., Refs. [322, 327]). As such, such Gestalt phenomena are concerned with the main proposition of Gestalt psychology, which states that human beings have the tendency to reorganize details of their surroundings to a whole. Accordingly, the human system is constructed such that it re-organize the forces that act on it and are exerted by various sources such that those forces give rise to a "meaningful" whole [189, 348]. This process of re-organization is assumed to follow several principles, the so-called Gestalt principles. The proximity principle is a frequently studied Gestalt principle that states that the "process of perceiving" (i.e., the reaction to external forces by forming BBA patterns) binds features of a figure together to a whole if those features are relatively close to each other. That is, features of a figure are grouped together to a whole if they are in close proximity to each other. A benchmark example are dots presented in a line next to each other to which humans react in a similar way as humans react to lines (i.e., dot in close proximity to each other are "perceived" as lines).

As part of a comprehensive study (that will be reviewed here in parts only), Kurylo et al. [203] presented arrays of white dots on a black background to participants. The arrays extended in two dimensions (vertical and horizontal) and formed squares. In all arrays the distances between the dots in the vertical direction were constant and likewise the distances in the horizontal direction were constants. However, in general, the vertical and horizontal between-dot distances were not the same. In the so-called horizontal condition, the horizontal between-dot distance was varied while the vertical between-dot distance was kept constant. As a result, for small horizontal between-dot distances participants reacted to the arrays in a similar way as they typically react to patterns composed of horizontal lines. By analogy, in the so-called vertical condition, for relative short vertical between-dot distances participants reacted to the arrays as if the arrays showed vertical lines.[10] Kurylo et al. [203] introduced a measure of relative proximity that was defined in such a way that the 100% mark on the measure indicated that proximity in the "target" direction (e.g., horizontal) relative to the "non-target" direction (e.g., vertical) was maximal. That is, dots were placed next to each other with no gaps between them. In contrast, the 0% mark indicated that proximity in the one direction relative to the other direction was minimal. More precisely, the distances in the "target" direction (e.g., horizontal) were the same as in the "non-target" direction (e.g., vertical). Consequently, for arrays with zero relative proximity the patterns exhibit perfect spatial homogeneity. Kurylo et al. [203] determined the threshold scores (measured in terms of relative proximity) at which participants stopped to react to the dots as if they were lines. Consequently, the threshold score thus obtained for a given participant indicated that for relative proximity values larger than the threshold the participant reacted to the dots in a similar way as if there were lines, whereas for relative proximity values lower than the threshold this was no longer the case. Averaged across horizontal and vertical conditions, Kurylo et al. [203] found relative proximity thresholds, $\alpha(crit)$, of approximately $\alpha(crit, SZ) = 20\%$ for schizophrenic patients and approximately $\alpha(crit, N) = 10\%$ for healthy adults. That is, the schizophrenic patients stopped to produce lines-reactions earlier as compared to healthy adults when the distance between dots in the "target" direction was increased and the arrays became more and more spatially homogeneous.

Kurylo et al. [203] argued within the framework of Gestalt psychology that schizophrenic patients demonstrate an "impaired" ability to group (or integrate) parts of images into a whole. In other words, the re-organization in terms of grouping

---

[10]An analogous phenomenon caused by forces of the acoustic world has been discussed in the context of the "say" to "stay" bifurcation of phrases "sa..y" that are equipped with a gap, see Sect. 5.3 with Fig. 5.10 and Sect. 7.6.2.

by proximity was "impaired" under schizophrenia. In particular, Kurylo et al. [203] argued that the assumed "malfunctioning" of the re-organization processes is likely to be caused by "impaired" neural connections in brain circuits relevant for "higher-order perceptual information processing". In doing so, Kurylo et al. [203] proposed a neuroanatomic interpretation of their experimental findings. For example, in word production task experiments it has been found that the brain activity of healthy adults exhibits temporal correlations (which are assumed to hint to functional connections) between the anterior cingulate and superior temporal regions, while for schizophrenic patients those correlations are almost absent [126]. This observation and similar observations have been taken as support for the aforementioned neuronal disconnection hypothesis [126].

Let us address the observed difference between the relative proximity thresholds from a physics perspective using the human pattern formation reaction model and the medical hypothesis introduced above. First of all, it is assumed that humans react to arrays of dots with high relative proximity in one direction in a similar way as they react to line patterns because the forces exerted by dot patterns with high relative proximity shift the brain state into the basin of attraction of attractors of line BBA attractor patterns. That is, it is assumed that when individuals are put into an experimental setting like the one used by Kurylo et al., an appropriately defined reference point becomes unstable (e.g., due to the instruction given to the participants that they should report what they see). The unstable reference point comes with a set of basis patterns. Among those patterns are patterns that emerge as attractor patterns when the participants are exposed to forces exerted by lines. In line with the terminology introduced in Chap. 5.4, we refer to those basis patterns as visual world line basis patterns and to the corresponding BBA attractor patterns as visual world line BBA attractor patterns or simply as line BBA attractor patterns. Irrespective of the presence or absence of external forces exerted by lines, the line BBA attractor patterns can emerge (e.g. as BA patterns of the isolated human system of the awake person or in REM stage) when the eigenvalue of the line basis pattern is in the stability band and the ongoing brain activity is in the basin of attraction of the line attractor. Therefore, the forces exerted by the dots arrays under appropriate conditions (i.e., when relative proximity is sufficiently high in one direction) are assumed to induce the emergence of those visual world line BBA attractor patterns. In this context, relative proximity, denoted here by α, is considered as bifurcation parameter. In addition to the vertical and horizontal line basis patterns and their corresponding line BBA attractor patterns, we consider a basis pattern for dots (or dot arrays) and its corresponding visual world dots BBA attractor pattern. Accordingly, participants exhibit bifurcations between the vertical or horizontal line BBA attractor patterns and dots BBA attractor patterns induced by the bifurcation parameter α given in terms of relative proximity.

Just as for the human pattern formation reaction model of grasping transitions (transitions between one-handed and two-handed grasps) and gait transitions (walk-run transitions) reviewed in Sects. 6.5.3 and 6.5.4, it is assumed that the eigenvalue of the visual world line BBA pattern increases when relative proximity is increased. That is, when the distances between the dots in one direction become smaller (i.e., the proximity measure by Kurylo et al. [203] becomes larger) then the eigenvalue of the line basis pattern and, consequently, the basin of attraction of the line attractor increases such that there is a stronger "tendency" that the line BBA pattern emerges. Mathematically speaking, we put $\lambda_1 = \lambda_1^* + C\alpha$, where λ_1 is the eigenvalue of the line BBA pattern (more precisely: its associated basis pattern), λ_1^* is a constant component (independent of external forces), α is the relative proximity, and $C > 0$ is a constant factor. Consistent with the experimental findings by Kurylo et al., it is assumed that schizophrenia reduces the eigenvalue λ_1. This could be modeled by reducing λ_1^* or the factor C. For sake of brevity, let us consider only on the eigenvalue λ_1^*. Assuming that the parameter C is constant across healthy adults and schizophrenic patients and assuming that $\lambda_1^*(SZ) < \lambda_1^*(N)$, a detailed calculation (see Sect. 10.7.2 below) based on the LVH amplitude equations shows that the threshold $\alpha(crit, SZ)$ under schizophrenia differs from the threshold $\alpha(crit, N)$ for healthy adults. More precisely, it can be shown that $\alpha(crit, SZ) > \alpha(crit, N)$ holds.

Figure 10.16 demonstrates for the two participant groups of interest the relationships between the eigenvalue offset parameters $\lambda_1^*(SZ)$ and $\lambda_1^*(N)$ and the relative proximity thresholds $\alpha(crit, SZ)$ and $\alpha(crit, N)$. Panel (a) shows simulation results of the LVH model, whereas panel (b) shows analytical results as obtained from the model (see Sect. 10.7.2). Panel (a) shows the stationary amplitude values

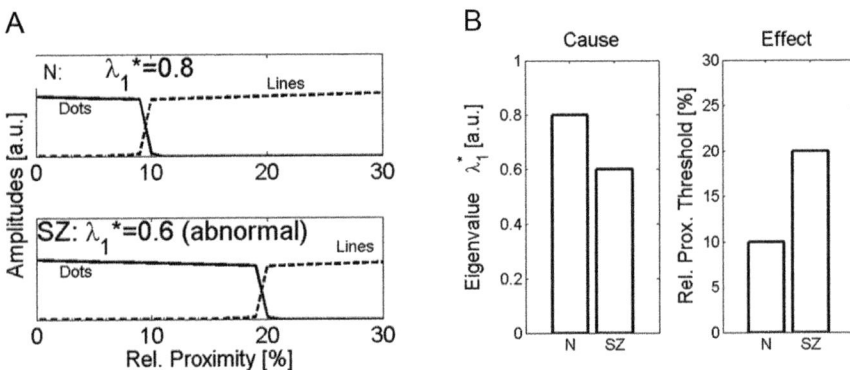

Fig. 10.16 Disease-induced decrease of the eigenvalue of BBA attractor patterns of line structures under schizophrenia can lead to a higher proximity threshold for the emergence of the line BBA patterns. Panel (**a**): Numerical results obtained from the LVH model for two simulated subjects: a healthy subject (top subpanel) and a schizophrenic individual (bottom subpanel) characterized by a normal and abnormal eigenvalue λ_1^*, respectively. The stationary values of the amplitudes of the line BBA pattern (dashed curves) and dots BBA pattern (solid curves) are shown for various relative proximity values. Bifurcations take place at different critical relative proximity values (compare top and bottom subpanels). Panel (**b**), left: Eigenvalues λ_1^* of the line BBA pattern as used in the simulations shown in panel (**a**) for the N- and SZ-individuals. Panel (**b**), right: Threshold values at which the simulated individuals begin to see lines as computed from analytical expressions discussed in Sect. 10.7.2. The threshold values correspond to the bifurcation points shown in panel (**a**)

of the dots BBA pattern (solid lines) and line BBA pattern (dashed lines) for relative proximity values ranging from 0 to 30%. The top and bottom subpanels, respectively, refer to a healthy subject with a normal eigenvalue component λ_1^* (here: $\lambda_1^* = 0.8$ units) and a schizophrenic individual with an abnormal eigenvalue component λ_1^* (here: $\lambda_1^* = 0.6$ units). Panel (a) shows the situation when the relative proximity parameter was increased gradually. However, the same results (not shown) was obtained for decreasing the relative proximity parameter (for the issue of hysteresis see the mathematical notes in Sect. 10.7.2 below). We see that for zero relative proximity the simulated healthy subject and the schizophrenic individual reacted to the array as an array of dots without a line structure. This is indicated by the fact that the amplitude of the dots BBA pattern was finite for both simulated individuals. Furthermore, for the healthy subject the brain state switched from the dots BBA pattern to the line BBA pattern when relative proximity was increased from 9% to 10%. The healthy subject reacted to the dots as lines at a relative proximity value of 10%. For the schizophrenic individual the brain state switched when relative proximity was increased from 19% to 20%. Only for relative proximity values larger than 20% the simulated schizophrenic individual formed line BBA patterns as reactions to the dots arrays. Consequently, the switching threshold was higher (as expected) for the simulated schizophrenic individual.

Panel (b) assumes that we are dealing with two samples of participants rather than two individuals. For sake of simplicity, variability among participants is neglected (for simulations and modeling in which variability among participants is taken into account see Sect. 7.4.1 and Fig. 7.7 and Sect. 10.5.4 above). Neglecting variability in structure parameters, it is assumed that all healthy participants of the sample under consideration follow the simulated healthy participant. Likewise, the schizophrenic individuals in the sample under consideration follow the simulated schizophrenic individual. Panel (b) provides the offset parameters λ_1^* (left subpanel) and the relative proximity thresholds (right subpanel) for the samples of healthy adults (i.e., normal adults, N) and schizophrenic patients (SZ). Panel (b) highlights again that according to the LVH model of pattern formation the difference between the eigenvalue components causes the difference in the switching thresholds. Note also that in panel (b) the thresholds have been calculated from the analytical expressions (presented in the mathematical notes in Sect. 10.7.2 below). From the analytical expressions, we obtained for healthy adults and schizophrenic individuals the threshold values of 10% and 20%, respectively. These analytical results were in excellent agreement with the numerical values.

10.5.6 Concluding Remarks

The hypothesis has been worked out that schizophrenia affects the spectrum of eigenvalues of the brain and body activity (BBA) patterns that are formed as reactions to forces that act on human individuals and are exerted by the acoustic and visual worlds in which they live. According to the physics perspective of thingbeings, in general, and the pattern formation perspective, in particular, the eigenvalue spectrum at unstable reference points constitutes a key set of parameters that determines (among other things) whether or not a given BBA pattern emerges as a reaction to certain external forces and/or as part of a pattern formation sequence of the human isolated system. In general, what matters is not the absolute value of the eigenvalues. Rather, it is the relationship between several eigenvalues that determines the formation of BBA and BA patterns. In line with the concept of dynamical diseases, the medical hypothesis has been formulated that schizophrenia comes with a set of abnormal neurophysiological parameters that affect in a variety of settings the magnitudes of the eigenvalues relative to each other. In doing so, the "natural balance" between eigenvalues that can be found in healthy adults is disturbed under schizophrenia.

Implications of the medical hypothesis have been demonstrated for four studies on reactions of schizophrenic patients to forces of the visual and acoustic worlds. These studies examined "inflexibility in perception" of sounds, motion-induced blindness, instruction-induced "recognition failures" of emotions, and "perceptual" grouping based on proximity. In the context of the first three studies, it has been proposed that schizophrenia increases the eigenvalue related to the BBA pattern associated with the experience of a "target" (e.g., a scream versus random noise, a stationary foreground triangle versus a rotating background pattern, a particular "target" emotion). As a result, schizophrenic patients "perceived" screams more frequently, had longer periods without blindness experiences, and showed more "recognition failures" as compared to healthy adults. However, in the context of "perceptual" grouping it was assumed that the BBA pattern associate with the experience of lines exhibits a reduced eigenvalue under schizophrenia. As illustrated explicitly, a reduced eigenvalue leads to the experimentally observed earlier breakdown of "perceptual" grouping under schizophrenia. Therefore, the questions arises which eigenvalues of the eigenvalue spectrum increase, decrease or remain unaffected under schizophrenia. In this context, a mechanistic modeling of the medical hypothesis sketched in Fig. 10.10 would a promising approach to predict whether eigenvalues are affected by schizophrenia and, if so, the way in which they are affected.

Let us return to a particular characteristic positive symptom of schizophrenia: hallucinations. Schizophrenic patients often experience visual and auditory hallucinations [344]. The emergence of such hallucinations might be explained as an abnormal increase of eigenvalues during the development of the disease (see medical hypothesis). If in the isolated human pattern formation A2 system of a schizophrenic patient an emerging BA pattern changes the structure such that for example the eigenvalue of the scream BA pattern increases and the system becomes monostable, then the scream BA pattern emerges as if the patient would be exposed to a scream. In fact, Bressloff et al. [38, 39] proposed that the emergence of drug-induced hallucination patterns in the visual cortex are due to such a mechanism that involves abnormally (here: drug-induced) increased eigenvalues.

10.6   Further Applications

Let us briefly sketch here two more applications.

10.6.1   Motivation and Depression

First, let us consider the study by Salamone et al. [276] on the dynamics of animals and their reactions to food related circumstances of their environments. The theoretical background to the study (when formulated within the pattern formation perspective) is the observation that healthy humans frequently produce brain and body activity patterns that involve a relatively large amount of effort in terms of time or physical activity provided that those patterns also come with a relatively high reward. In contrast, certain human individuals suffering from certain diseases such as depression no longer produce those kind of patterns. In other words, certain patient groups such as depressive people are less motivated to put effort in getting things done even if this will give them a reward. For example, individuals that used to jog and run, when entering into depressive episodes may do not find the motivation any more to go for a run even running as such is fun for them. That is, high-effort BBA attractor patterns emerge less frequently in patients suffering from depression and in patients suffering from a variety of other diseases.

It has been suggested that the dopamine pathway in the human brain of healthy individuals plays an important role for the emergence of high-effort BBA attractor patterns. Salamone et al. [276] conducted experiments on rats with intact dopamine pathway and rats that had parts of their dopamine pathways destroyed by surgery. In a first phase, rats were restructured over several days to press a lever in a cage in order to obtain food pellets. After that restructuring phase, half of the rats being tested had a surgery that destroyed parts of their dopamine pathways. The other half of the rats being tested had a surgery as well. However, dopamine pathways were not destroyed. The surgery for the second group of rats was a means to avoid a possible placebo effect. In short, after the restructuring phase and the surgery phase, there were two groups of rats. Rats in both groups had been restructured to press a lever to obtain food. Rats in the one group had an intact dopamine pathway. Rats in the other group had a partially destroyed (or damaged) dopamine pathway. After the surgery phase the rats were tested under several conditions that varied with respect to the degree of effort required to obtain a food reward. In the high-effort condition, rats had to press the lever several times in a row in order to obtain food. Panel (a) of Fig. 10.17 present the lever presses observed in the two groups of animals for this high-effort condition (for details see the original study by Salamone et al. [276]). The rats with the damaged dopamine pathway produced only a few lever presses as compared to the rats with the intact pathway. The ratio of the lever presses was approximately 5:1.

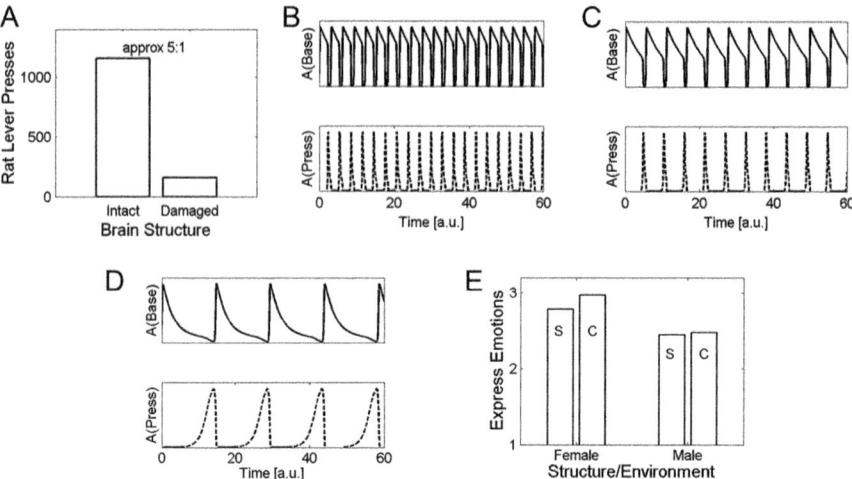

Fig. 10.17 Panel (**a**): Experimental result observed by Salamone et al. [276]: lever presses of rats with intact and damaged dopamine pathways. Panels (**b**), (**c**), (**e**): Explanation of the experimental finding by Salamone et al. in terms of the LVH limit cycle oscillator model. Amplitudes computed from the oscillator model are shown (solid lines/top subpanels: baseline pattern amplitudes; dotted lines/bottom subpanels: lever-pressing pattern amplitude). Panel (**b**): Frequency of the emergence of lever pressing activities of rats with intact dopamine pathway. Panels (**c**), (**d**): Damaging the dopamine pathway is assumed to change certain neurophysiological parameters that, in turn, affect the parameters of the LVH limit cycle oscillator such that the emergence of lever pressing BBA patterns becomes less frequent. Panel (**e**): Gender effect on the importance to express emotions for males and females under stress (bars labeled S) and male and female cancer survivors (bars labeled C) taken from Refs. [50, 306]. The higher importance for females to express emotions may be explained with the LVH limit cycle model by assuming that certain model parameters differ across males and females such that the frequency of the emergence of BBA patterns associated with expressing emotions increases (as in panels (**b**), (**c**), (**d**) when going *backwards* from D to B and replacing the lever-pressing activity with expressing emotions) and/or the amplitude of the oscillations increases such that the patterns emerge with stronger "intensity"

Panels (b)–(d) explains the observation in terms of the pattern formation oscillator introduced in Sect. 7.4.3. For sake of simplicity, we only distinguish between two attractor patterns: the (high-effort) lever pressing BBA pattern and a baseline BBA pattern that describes all other activities not related to lever pressing. Accordingly, rats with intact pathways have a normal eigenvalue of the high-effort BBA attractor pattern. Therefore, the pattern emerges relatively frequently in an alternating sequence composed of lever-pressing BBA attractor pattern and the baseline BBA pattern, see panel (b). Quantitatively, in the simulation shown in panel (b), during the observation interval the simulated animal formed 20 times the lever-pressing BBA pattern. Destroying parts of the dopamine pathway is assumed to decrease the eigenvalue of the lever-pressing BBA pattern. Panel (c) shows a simulation of the pattern formation oscillator model for a reduced value of the eigenvalue. As expected, the lever-pressing BBA pattern occurs less frequently during the observation period. In the simulation, the BBA pattern only occurs 10 times. Let us in addition assume that the overall time scale on which structural changes (caused by emerging BBA patterns) slows down in the damaged brain. Slowing down the dynamics of the structural changes in addition to the aforementioned reduction of the eigenvalue lead to the amplitude dynamics shown in panel (d). The lever-pressing BBA pattern occurs only a few times during the observation internal. In the simulation, there are only 4 events of an emerging lever-pressing BBA pattern. Comparing panels (b) and (d), we see that the lever-pressing BBA patterns for the intact animal simulated in panel (b) occurs 5 times more frequently as compared to the simulated animal in panel (d). While this computer experiment is not an attempt to fit experimental data, it demonstrates that the human pattern formation reaction model when describing an oscillatory instability (see Sect. 7.4.3) leads to results that are consistent with the experimental observation by Salamone et al. [276] on motivation and depression.

10.6.2 Expressing Emotions to Cope with Stress and Cancer

People express emotions for various reasons. Among other things, people express emotions in order to cope with stressful situations and life-threatening situations such as being diagnosed with cancer. In a comprehensive study, Stanton et al. [306] determine how important it is for young adults to express emotions in order deal with stressful situations. In a similar vain, Cho et al. [50] interviewed people that had been diagnosed with cancer and had successfully completed primary cancer treatment. Cho et al. [50] determined to what extent cancer survivors express their emotions in order to deal with the fact that they had been in a life-threatening situation. In both studies, males and females had been interviewed separately. Panel (e) of Fig. 10.17 presents results reported in both studies. On the vertical axis a four-point scale from 1 to 4 is shown, where high scores/low scores mean that it is important/not so important to express emotions. Overall, it was important for both gender groups to express emotions in stressful situations in general and in particular in order to cope with cancer. Females showed higher scores than males in both studies.

From a pattern formation perspective, expressing emotions is considered as a BBA attractor pattern. Again, a simplified point of view is taken in which only a single BBA attractor pattern for expressing emotions is consider (rather than several different BBA attractor patterns). In view of the studies by Stanton et al. [306] and Cho et al. [50], factors like stress and having survived from cancer are assumed to have an impact on the emergence of the emotion-expressing BBA pattern. In particular, it is assumed that such factors increase the eigenvalue of the basis pattern associated with the emotion-expressing BBA pattern.

As a toy model, let us adopt the model of the previous section, Sect. 10.6.1, by replacing the lever-pressing BBA pattern with the emotion-expressing BBA pattern. That is, we are dealing with a model that involves two BBA patterns: the emotion-expressing BBA pattern and a baseline pattern describing all kinds of activities not related to the expression of emotions. If the eigenvalue of the emotion-expressing BBA pattern becomes sufficiently large, the eigenvalue of the baseline pattern drops out of the stability band and the individual under consideration switches from an ongoing activity to an activity in which he/she communicates his/her emotions to somebody. The person under consideration is assumed to be engaged in that activity for a certain amount of time. After that the emotion-expressing BBA pattern is assumed to damp its own eigenvalue such that it drops out of the stability band. A transition to a baseline activity occurs. As long as the person is in the stressful situation and as long as the person is under the impact of the experience of having survived from cancer, it is assumed that the eigenvalue of the expressing-emotions BBA attractor pattern gradually increases over time. Eventually it reaches a critical value at which it dominates the eigenvalue spectrum such that the ongoing activity becomes unstable and a switch occurs again to the emotion-expressing BBA pattern. This sequence repeats itself again and again. In doing so, the human pattern formation reaction model exhibits the oscillatory instability discussed in general in Sect. 7.4.3 and applied in the previous section, Sect. 10.6.1, to the experiment by Salamone et al. [276]. That is, panels (b)–(d) of Fig. 10.17 apply to the experimental findings by Stanton et al. [306] and Cho et al. [50] when re-interpreting the amplitude of the lever-pressing BBA pattern as amplitude of the emotion-expressing BBA pattern. Gender differences as documented in panel (e) can then be explained by differences in the frequencies in which the expressing-emotions BBA attractor pattern occurs. As discussed in the previous section, such differences in the frequencies as illustrated when comparing panels (b) and (c) can be caused by different values of the eigenvalue of the expressing-emotions BBA pattern. In view of the findings by Stanton et al. and Cho et al., it is plausible to assume that the eigenvalue of the expressing-emotions BBA pattern is larger for females than for males. This would explain that females (when under stress or as cancer survivor) more frequently engage in expressing their emotions as compared to males.

10.7   Mathematical Notes

10.7.1   OCD Rituals as Type 2 Sequences with "Directional" Transitions

Let us describe the simulation results discussed in Sects. 10.4.4 and 10.4.5.

Type 2 Sequences with "Directional" Transitions: General Case

In Sect. 10.4.4 type 2 pattern formation sequences with "directional" transitions were described that correspond to rituals, in general, and OCD rituals, in particular.

A sequence involving m activities was considered. One of the m activities actually described a whole class of non-ritual related activities [102]. More precisely, the BBA patterns $k = 1, .., m-1$ described the ritual activities, while the BBA pattern m described all kind of activities not related to the ritual. Without loss of generality, the pattern formation system in which the sequences occurred was modeled by means of the LVH model with a quadratic nonlinearity $q = 2$ for $m = 4$ total activities. In this case, Eq. (4.38) with $k = 1, 2, 3, 4$ reads

$$\frac{d}{dt} A_k = A_k \left(\lambda_k - A_k - g \sum_{j=1, j \neq k}^{4} A_j \right) \tag{10.19}$$

and describes the evolution of the ordinary state (i.e., brain state and force production including limb movements) in terms of amplitudes of the reduced amplitude space. The structural dynamics is assumed to satisfy [100, 104]

$$\frac{d}{dt} \lambda_k = S \gamma_k (A_{k-1}) - \beta_k (A_k) \lambda_k \tag{10.20}$$

for $k = 1, 2, 3$ with $\gamma_1 = 0$ and $\gamma_2, \gamma_3, \beta_1, \beta_2, \beta_3$ defined by

$$\gamma_k = \begin{cases} \gamma_{on} & \text{if } A_{k-1} > \theta A_{k-1, st} \\ 0 & \text{otherwise} \end{cases}, \quad \beta_k = \begin{cases} \beta_{on} & \text{if } A_k > \theta A_{k, st} \\ 0 & \text{otherwise} \end{cases} \tag{10.21}$$

In Eq. (10.20) the parameter $S > 0$ is a constant that can be ignored for a moment. The functions $\gamma_k \geq 0$ and $\beta_k \geq 0$ describe increase and decrease of the eigenvalues (i.e., structural changes) caused by the emerging BBA patterns. The pumping terms $S \gamma_k$ leading to an increase of the eigenvalues describe constant pumping mechanisms, whereas the damping terms $-\beta_k \lambda_k$ leading to a decay of the eigenvalues depend on the magnitudes of the eigenvalues. Equation (10.21) states that an emerging BBA pattern k pumps the eigenvalue λ_{k+1} of the follow-up BBA pattern $k + 1$, whereas any BBA pattern k that is present triggers a self-damping mechanism that leads to the decay of its own eigenvalue λ_k. Equations (10.19), (10.20), (10.21) describe the mechanism illustrated in Fig. 10.6 that leads to the formation of pattern sequences featuring "directional" transitions. Note that $A_{k, st} = \lambda_k$ holds because of $q = 2$. Furthermore, we have $\gamma_{on} > 0$, $\beta_{on} > 0$. The simulation results presented in Fig. 10.7 were obtained by solving Eqs. (10.19), (10.20), (10.21) numerically (by means of an Euler forward method) for the following parameters and initial conditions:

$$\lambda_4 = 3, \ \lambda_{1,2,3}(t = 0) = 0.1, \ \lambda_1(t^*) = D, \ D = g\lambda_4 + 0.5,$$
$$g = 1.5, \ \gamma_{on} = \beta_{on} = 1, \ S = 1, \ \theta = 0.95. \tag{10.22}$$

Emotional Dynamics Coupling and Application to OCD

In Sect. 10.4.5 the coupling between the emotional dynamics and the performed motor actions and related BBA patterns was discussed. Only two BBA patterns of the emotional system were considered: the distress pattern $k = 1$ and a pattern $k = 2$ that describes the class of all kind of disease-free emotional patterns. Accordingly, the evolution of the emotional BBA patterns are described by a LVH model (4.38) with $m = 2$ for $k = 1, 2$ that reads [100, 104]

$$\frac{d}{dt} E_k = E_k \left(L_k - E_k - h E_j \right) \tag{10.23}$$

with $k = 1 \Rightarrow j = 2$ and $k = 2 \Rightarrow j = 1$ and $h \geq 1$, where we have used quadratic nonlinearities again (i.e., $q = 2$).

The parameter assignments and initial conditions presented in Eq. (10.22) do not take the emotional dynamics into account. Rather, according to Eq. (10.22) it is assumed that at the time point t^* the eigenvalue λ_1 is put to a sufficiently large value $\lambda_1(t^*) = D$ such that the system becomes monostable and the first activity of the ritual emerges as BBA pattern. In contrast, by modeling the emotional dynamics explicitly via Eq. (10.23), we can describe that the emotional distress pattern changes the structure of the human system of OCD patients (just like a hurricane that makes landfall changes the atmospheric structural conditions such that a tornado can emerge) and increases λ_1. In particular, for the simulations presented in Sect. 10.4.5 we used

$$E \rightarrow A : t^* \text{ with } \frac{d}{dt} E_1 > 0 \wedge E_1 = \theta E_{1,st} \Rightarrow \lambda_1(t^*) = D, S(t \geq t^*) = 1 \tag{10.24}$$

In Eq. (10.23) the parameter L_2 was constant. In contrast, L_1 decayed like

$$A \rightarrow E : \frac{d}{dt} L_1 = -B L_1 , \ B = \begin{cases} B_{on} & \text{if } A_1 \vee A_2 \vee A_3 \text{ "on"} \\ 0 & \text{otherwise} \end{cases} \tag{10.25}$$

where "on" means $A_k > \theta A_{k,st}$. As indicated, Eqs. (10.24) and (10.25) describe the couplings $A \rightarrow E$ and $E \rightarrow A$ between the emotional and the motor performance pattern formation systems (for analogous consideration see Ref. [265]). Specifically, Eqs. (10.25) describes that the eigenvalue L_1 of the distress pattern decayed in our simulation when the ritual was performed. Note that we had $E_{k,st} = L_k$ because of $q = 2$. The simulation results presented in Fig. 10.8 were obtained by solving Eqs. (10.19), (10.20), (10.21), (10.23), (10.24), (10.25) numerically for the parameters and initial conditions listed in Eq. (10.22) except for λ_1 and the parameters and initial conditions listed here:

$$E_1(0) = 0.05 \ , \ E_2(0) = E_{2,st} = 2 \ , \ L_1(0) = d \ , \ d = h L_2 + 2.0 \ ,$$
$$S(t = 0) = 0 \ , \ L_2 = 2 \ , \ h = 1.5 \ , \ B_{on} = 1 \ . \tag{10.26}$$

Early OCD Ritual Termination

For the simulation presented in Fig. 10.8, according to Eq. (10.24), the parameter S was switched on from $S = 0$ to $S = 1$ when the emotional distress pattern emerged. Subsequently, S stayed on the level 1 even if the distress pattern disappeared. This describes a scenario in which the ritual is completed in any case irrespective whether or not there is a distress feeling. In contrast, Fig. 10.9 shows a scenario, in which the ritual was terminated early as soon as the distress feeling pattern disappeared. In order to capture early ritual termination, we ignored the rule regarding S in Eq. (10.24) and replaced it by

$$S = \begin{cases} 1 \text{ if } E_1 \text{ "on"} \\ 0 \quad \text{otherwise} \end{cases} \tag{10.27}$$

Here, E_1 is "on" means that the distress pattern is stable and present. Note that the stationary amplitude E_1 decayed because L_1 decayed. Nevertheless, in general, the definition that the distress pattern $k = 1$ is "on" by means of $E_1 > \theta E_{1,st}$ works well as long as L_1 changes slowly (see quasi-attractor theory suggested by Haken [157]), which implies that the fixed point $E_{1,st}$ drifts slowly in time to smaller values. Figure 10.9 was obtained by solving Eqs. (10.19), (10.20), (10.21), (10.23), (10.24), (10.25) numerically with Eqs. (10.22) and (10.26) except for λ_1 and Eq. (10.27) replacing the rule for S in Eq. (10.24).

10.7.2 Model-Based Analysis of Human Reactions Under Schizophrenia

The LVH model (4.38) used in the four studies on schizophrenia described in Sect. 10.5 is given by (4.38) with $m = 2$ and reads

$$\frac{d}{dt} A_k = A_k \left(\lambda_k - A_k^2 - g A_j^2 \right) \tag{10.28}$$

with $k = 0 \Rightarrow j = 1$ and $k = 1 \Rightarrow j = 0$. Note that without loss of generality in those applications the cubic nonlinearity $q = 3$ is used.

Qualitative Re-analysis of the Study by Martin et al. [234]

In the application to the study by Martin et al. described in Sect. 10.5.2 the pattern $k = 0$ refers to the no scream BBA pattern, whereas $k = 1$ denotes the scream BBA pattern. As explained in detail in Sect. 10.5.2 the impact of the external acoustic forces (as capture by the bifurcation parameter α) on the structure (as described by λ_0 and λ_1) is given by

$$\lambda_0 = \lambda_0^*, \ \lambda_1 = \lambda_1^* + C\alpha, \ \lambda_0^* = L - \frac{\delta}{2}, \ \lambda_1^* = L + \frac{\delta}{2} \qquad (10.29)$$

with $C > 0$ and $L > \delta/2 \Rightarrow \lambda_0 > 0$. Using the stability conditions defined by the stability band (see Sects. 4.6.3 and 4.7.8) from Eq. (10.29) it follows that the critical bifurcation parameters in the increasing and decreasing conditions are given by

$$\alpha(crit, 1) = -\frac{1}{gC}\left((g-1)L + (g+1)\frac{\delta}{2}\right),$$

$$\alpha(crit, 2) = \frac{1}{C}\left((g-1)L - (g+1)\frac{\delta}{2}\right), \qquad (10.30)$$

which implies $\alpha(crit, 1) \geq \alpha(crit, 2)$ for $g > 1$ and $\alpha(crit, 1) = \alpha(crit, 2)$ for $g = 1$. From Eqs. (10.29) and (10.30) when assuming that L and g are constant across SZ- and N-individuals, we obtain

$$\text{if } \lambda_1^*(SZ) > \lambda_1^*(N) \ \Rightarrow \delta(SZ) > \delta(N) \ \Rightarrow \alpha(crit, 1, SZ) < \alpha(crit, 1, N).$$
$$(10.31)$$

That is, a disease-induced change in the structure given in terms of an increase of λ_1^* results in the experimentally observed decay of the threshold $\alpha(crit, 1)$ in the scream-intensity decreasing condition.

Note that Eq. (10.31) has been derived as mentioned above in part from Eq. (10.30) assuming that the parameter g is constant across SZ- and N-individuals. However, Martin et al. observed that $\alpha(crit, 2, SZ) = \alpha(crit, 2, N)$ holds. In order to model that the thresholds in the increasing condition did not differ across the two populations we cannot keep g constant, that is, we must use $g(SZ) \neq g(N)$. For $g(SZ) \neq g(N)$ Eq. (10.31) does not necessarily hold, while it still is a good hint that an increase of λ_1^* may result in a decrease of $\alpha(crit, 1)$ as long as g does not vary too much. Therefore, for any given parameters $g(SZ), C(SZ), \delta(SZ)$, on the one hand, and $g(N), C(N), \delta(N)$, on the other hand, we must compute the critical bifurcation parameters from Eq. (10.30) and, subsequently, check whether or not the difference in parameters leads to the observed decay of $\alpha(crit, 1)$ under schizophrenia.

Panel (a) of Fig. 10.13 was computed by solving Eq. (10.28) with the help of an Euler forward method for Eq. (10.29) using the following parameters: $L = 10$, $\delta(SZ) = 15, \delta(N) = 8, C = 1, g(SZ) = 6.1$, and $g(N) = 1.95$. The bifurcation parameter α was decreased in step of 0.5 from 0 to -21. For $\alpha = 0$ the initial conditions $A_0(0) = 2$ and $A_1(0) = 1$ were used. For any fixed value α the stationary values $A_{k,st}$ for $k = 0, 1$ were obtained and were used as initial values $A_k(0)$ for the next simulation step with α decreased by 0.5. The stationary values $A_{k,st}$ are shown in Fig. 10.13. A noise term was added (reflecting deterministic perturbations, see Sect. 2.4) to the amplitude dynamics of A_0 in terms of a uniformly distributed random variable in the interval $[0, Q]$ with $Q = 0.01\sqrt{\Delta t}$, where $\Delta t = 0.01$ was the Euler forward single time step. From the numerical simulation we observed the switch to the no scream BBA pattern between $\alpha = -10.5$ and $\alpha = -11$ for the normal individual and between $\alpha = -16.5$ and $\alpha = -17$ for the schizophrenic individual. Panel (b) of Fig. 10.13 shows analytical results obtained from Eq. (10.30). From Eq. (10.30) we obtained $\alpha(crit, 1, N) = -10.92$ versus $\alpha(crit, 1, SZ) = -17.09$ for the decreasing condition (consistent with the numerical results) and $\alpha(crit, 2, N) = -2.30$ versus $\alpha(crit, 2, SZ) = -2.25$ for the increasing condition.

Qualitative Re-analysis of the Study by Tschacher et al. [324]

The human pattern formation reaction model for oscillatory reactions was used as discussed in Sects. 7.4.3 and 7.8.3. In particular, the LVH limit cycle oscillator defined by Eqs. (7.10) and (7.11) and $q = 3$ was used to describe alternations between the foreground BBA pattern $k = 0$ and patterns in which parts of the foreground dots were absent as described by the pattern $k = 1$. Note that the indices $k = 0, 1$ were used rather than $k = 1, 2$ as in Sect. 7.8.3. The upper and lower reference parameters $\lambda_{baseline}$ and λ_0 of the structural dynamics were defined by

$$\lambda_{baseline} = L + \frac{\delta}{2} \, , \; \lambda_0 = L - \frac{\delta}{2} \Rightarrow \lambda_{baseline} = \lambda_0 + \delta \qquad (10.32)$$

with $\delta > 0$. The results shown in panel (a) of Fig. 10.14 were obtained by solving the LVH limit cycle oscillator equations numerically with the help of a Euler forward scheme for the parameters

$$g = \exp(1) \, , \; L = 20 \, , \delta(N) = 8 \, , \; \delta(SZ) = 16 \, , \; \gamma_{down} = \gamma_{up} = 1 \, . \qquad (10.33)$$

Note that Eq. (7.10) of the LVH limit cycle oscillator is identical to Eq. (10.28) mentioned at the beginning of this section, Sect. 10.7.2. That is, all phenomena discussed in this Sect. 10.7.2 are explained mathematically in terms of the same amplitude equations. A noise term was added (to model the impact of deterministic perturbations, see Sect. 2.4) to the Euler forward scheme for solving the amplitude equations (7.10) given in terms of a uniformly distributed noise in $[0, Q]$ with $Q = 0.01\sqrt{\Delta t}$, where $\Delta t = 0.001$ was the single simulation time step.

The averaged eigenvalue $\lambda_{0,av}$ shown in panel (b) (left subpanel) was computed from $\lambda_0(t)$ shown in panel (a) as the average of the maximal and minimal eigenvalue: $\lambda_{0,av} = [\max(\lambda_0) + \min(\lambda_0)]/2$, where the maximal and minimal values were taken after discarding a transient period of 10 time units. The durations of the MIB free episodes shown in panel (b) (right subpanel) were taken directly from peak-to-peak measurements based on the simulated amplitude dynamics shown in panel (a).

Qualitative Re-analysis of the Study by Barbalat et al. [15]

In order to address the experimental results by Barbalat et al. [15] reviewed in Sect. 10.5.4, the LVH model (10.28) with $m = 2$ was used for the interacting BAB patterns $k = 1$ and $k = 2$ associated with the "primed" and "non-primed" emotional content. For example, when participants are "primed" for anger, then $k = 1$ is the visual world anger BBA attractor pattern. More precisely, the BBA pattern $k = 1$ is the BBA pattern of the attractor that has a basin of attraction into which all kind of forces exerted by angry looking faces shift the brain state provided that the pattern formation system of interest shows symmetry, see Sect. 5.4 and Fig. 5.11. In contrast, the BAB pattern $k = 2$ is the fearful expressions BBA attractor pattern. That is, the BBA pattern $k = 2$ is the BBA pattern of the attractor that has—under idealized symmetric conditions—a basin of attraction into which all kind of forces exerted by fearful looking faces shift the brain state. The eigenvalue λ_2 of the "non-primed" BBA pattern drops out of the stability band if the eigenvalues λ_1 and λ_2 satisfy the relation $\lambda_1/g > \lambda_2$, see Sect. 4.7.8. The BBA pattern $k = 2$ of the "non-primed" emotion can no longer emerge in the relevant brain areas of the participants under consideration even if the participants are exposed to a face expressing that emotion. In this case, (in our toy model that involves only two BBA patterns) the "prime"-related BBA pattern emerges as a reaction to the "non-primed" emotion. In other words, if participants are restructured by instruction to see fearful faces and if $\lambda_1/g > \lambda_2$ holds as a result of the instruction-induced restructuration, then participants react to expressions of fear *differently* as they do when their visual world pattern formation system is symmetric.[11]

For sake of simplicity, it is assumed that the variability of λ_2 in the populations of SZ- and N-participants is negligibly small. In contrast, λ_1 is assumed to exhibit for both population groups a normal distributions [120] denoted by $N(\lambda_1, \mu, \sigma)$ with the same standard deviation σ but different mean values μ: $\mu(SZ) \neq \mu(N)$. Note that the mean values μ correspond to the respective averaged λ parameters: $\mu(SZ) = \lambda_{1,av}(SZ)$ and $\mu(N) = \lambda_{1,av}(N)$. From the graphical construction shown in Fig. 10.15 it follows that the percentage of "correct" identifications of the "non-primed" objects (e.g., fearful faces when anger was "primed") can be computed from the integral

$$p(fear) = \int_{-\infty}^{g\lambda_2} N(\lambda_1, \lambda_{1,av}, \sigma) \, d\lambda_1 . \tag{10.34}$$

Equation (10.34) was numerically determined under the assumption discussed in Sect. 10.5.4 that schizophrenia amplifies the instruction-induced shift of $\lambda_{1,av}$ such that $\lambda_{1,av}(SZ) > \lambda_{1,av}(N)$. Panel (b) in Fig. 10.15 shows $\lambda_{1,av}(SZ)$, $\lambda_{1,av}(N)$ and

---

[11] And in this sense participants make "recognition failures" if they are exposed to a fearful face. However, as such the concept of "failures" is a spiritualistic one—like the concepts of the good and the evil—, not a scientific one.

$p(fear, SZ)$, $p(fear, N)$ obtained for the following model parameters: $\sigma = 0.8$, $\lambda_{1,av}(N) = 4$, $\lambda_{1,av}(SZ) = 5.5$, $g = 1.5$, and $\lambda_{2,av}(SZ) = \lambda_{2,av}(N) = 3$. Note that the curves in panel (a) of Fig. 10.15 have also been drawn for the very same parameters.

Qualitative Re-analysis of the Study by Kurylo et al. [203]

Since in the study by Kurylo et al. a possible hysteresis phenomenon was not considered, the LVH model (10.28) was used for the special case $g = 1$ in which the model does not exhibit hysteresis (see Sects. 6.5.1, 6.5.3, and 6.6.4). As explained in detail in Sect. 10.5.5, the impact of the external visual forces exerted by the dot fields (as capture by the bifurcation parameter α) on the structure (as described by λ_0 and λ_1) is given by

$$\lambda_0 = 1 - C\alpha \, , \ \lambda_1 = \lambda_1^* + C\alpha \tag{10.35}$$

with $C > 0$. From Sects. 4.6.3 and 4.7.8 it follows that the critical bifurcation parameter is given by

$$\alpha(crit) = \frac{1 - \lambda_1^*}{2C} \, . \tag{10.36}$$

Accordingly, the proximity threshold $\alpha(crit)$ at which the line BBA pattern disappears or emerges decreases when the offset parameter λ_1^* is increased that describes the baseline pumping of the line BBA pattern. Consequently, from Eq. (10.36) it follows that if $\lambda_1^*(SZ) < \lambda_1^*(N)$ holds (i.e., pumping is reduced under schizophrenia) and C is constant across SZ- and N-individuals, then we have $\alpha(crit, SZ) > \alpha(crit, N)$ – as anticipated in Sect. 10.5.5. Vice versa, the experimental observation $\alpha(crit, SZ) > \alpha(crit, N)$ reported by Kurylo et al. [203] suggests that $\lambda_1^*(SZ) < \lambda_1^*(N)$ holds.

Panel (a) of Fig. 10.16 shows numerical simulation results obtained by solving Eq. (10.28) for Eq. (10.35) with

$$g = 1 \, , \ C = 1 \, , \ \lambda_1^*(SZ) = 0.6 \, , \ \lambda_1^*(N) = 0.8 \, . \tag{10.37}$$

The bifurcation parameter α was increased from 0 to 30% in steps of 1%. For $\alpha = 0$ the initial conditions $A_0(0) = 2$ and $A_1(0) = 1$ were used. The stationary values $A_{k,st}$ obtained for a given value of α were used as initial values $A_k(0)$ for the next simulation step with α increased by 1%. The stationary values $A_{k,st}$ are shown in Fig. 10.16a. A noise term was added (in order to take at least approximately the effect of deterministic perturbations into account, see Sect. 2.4) to the amplitude dynamics of A_1 in terms of a uniformly distributed random variable in the interval $[0, Q]$ with $Q = 0.0001\sqrt{\Delta t}$, where $\Delta t = 0.005$ was the single time step of the Euler forward method to solve the LVH amplitude equations. In the numerical simulations the switch from the dots BBA pattern to the line BBA pattern occurred between $\alpha = 9\%$ and $\alpha = 10\%$ for the computer simulated normal individual and between $\alpha = 19\%$ and $\alpha = 20\%$ for the schizophrenic individual. Panel (b) of Fig. 10.13 shows analytical results obtained from Eq. (10.36). From Eq. (10.36) we obtained (consistent with the numerical results) $\alpha(crit, N) = 10\%$ versus $\alpha(crit, SZ) = 20\%$.

10.7.3   LVH Oscillator Model for Reward-Related Animal Body Dynamics

In Sect. 10.6.1 the LVH oscillator model was applied to describe the animal lever pressing dynamics under various conditions. To this end, a modified version of the LVH limit cycle oscillator model defined by Eqs. (7.10) and (7.11) was used. As such the amplitude and structure dynamics satisfied Eqs. (7.10) and (7.11) with $q = 3$ such that

$$\frac{d}{dt} A_k = \lambda_k A_k - g A_k A_j^2 - A_k^3 \tag{10.38}$$

for $k = 1, 2$ and $k = 1 \Rightarrow j = 2, k = 2 \Rightarrow j = 1$ with

$$\text{if } A_k > \theta A_{k,st} \; : \; \frac{d}{dt} \lambda_k = -\gamma_{down}^{(k)}(\lambda_k - \lambda_0^{(k)}) \,,$$

$$\text{otherwise} \; : \; \frac{d}{dt} \lambda_k = -\gamma_{up}^{(k)}(\lambda_k - \lambda_{baseline}^{(k)}) \tag{10.39}$$

Here $k = 2$ describes the lever presses and $k = 1$ other activities. Note that we considered the case in which the parameters λ_0, $\lambda_{baseline}$, γ_{down}, and γ_{up} differed for $k = 1$ and $k = 2$. Moreover, for the lever pressing activity $k = 2$ the structural equation (10.39) was revised to take into account that the activity was only performed for a fixed amount of time defined by $\tau(press) > 0$. More precisely, we used

$$\text{if } A_2 > \theta A_{2,st} \; : \; \Big\{ \text{first } \tau \text{ time units} \; : \; \frac{d}{dt} \lambda_2 = -\gamma_{down}^{(2)}(\lambda_2 - \lambda_0^{(2)}) \,,$$

$$\text{at } \tau \text{ time units} \; : \; \lambda_2 = \lambda_{off} \Big\}$$

$$\text{otherwise} \; : \; \frac{d}{dt} \lambda_2 = -\gamma_{up}^{(2)}(\lambda_2 - \lambda_{baseline}^{(2)}) \tag{10.40}$$

That is, after a duration of τ during which λ_2 decayed towards $\lambda_0^{(2)}$, the eigenvalue was put to a low value $\lambda_{off} > 0$.

For the simulations shown in all panels (b)–(d) of Fig. 10.17 the following parameters were used:

$$\tau = 1 \, , \; \lambda_0^{(1)} = \lambda_0^{(2)} = 0 \, , \lambda_{off} = 0.1 \, , \theta = 0.95 \, . \tag{10.41}$$

For the simulations shown in panels (b) and (c) the following parameters were used:

$$g = \exp(1) \, , \; \lambda_{baseline}^{(1)} = 21 \, , \; \lambda_{baseline}^{(2)} = 40 - \delta \, ,$$

$$\gamma_{down}^{(1)} = 1/3 \, , \; \gamma_{up}^{(1)} = 10 \, , \; \gamma_{down}^{(2)} = \gamma_{up}^{(2)} = 10 \, . \tag{10.42}$$

Panel (b) shows simulation results for animals with intact dopamine pathway. The simulation results shown in panel (b) were obtained for $\delta = 0$. Therefore, the baseline eigenvalue $\lambda_{baseline}^{(2)}$ of the lever pressing pattern was relatively high (i.e., $\lambda_{baseline}^{(2)} = 40$). In contrast, panel (c) shows simulation results for animals with defect dopamine pathway. The results shown in panel (c) were obtained for $\delta = 20$. The baseline level $\lambda_{baseline}^{(2)}$ of the eigenvalue of the lever pressing pattern was reduced as compared to the healthy animal (i.e., $\lambda_{baseline}^{(2)} = 20$). Comparing panels (b) and (c), we can see that the effect was that the lever pressing activity occurred less frequently. Finally, the simulation results shown in panel (d) were obtained by rescaling model parameters appropriately. The following parameters were used:

$$g = \exp(1)/Z \, , \; \lambda_{baseline}^{(1)} = 21/Z \, , \; \lambda_{baseline}^{(2)} = 40/Z - \delta \, , \delta = 20$$

$$\gamma_{down}^{(1)} = Z/3 \, , \; \gamma_{up}^{(1)} = 10Z \, , \; \gamma_{down}^{(2)} = \gamma_{up}^{(2)} = 10Z \tag{10.43}$$

with $Z = 1.92$.

Chapter 11
Life Trajectories and Pattern Formation Sequences

The formation of sequences of patterns has been discussed in Chaps. 6 and 7 for type 1 and type 2 pattern formation sequences, respectively. Let us apply pattern formation sequences to describe how human individuals and groups of individuals evolve over extended periods such as a day, a week, a year, a life or several generations.

11.1 Pattern Formation Sequences of a Single Day

11.1.1 An Example

In Sect. 7.3.3 the example of a child performing several leisure activities in succession (i.e., forming a sequence of BBA patterns) has been considered, while in Sect. 10.4 rituals have been discussed in general and in the context of rituals performed by OCD patients. The key difference between those two examples of pattern formation sequences has been addressed in Sect. 10.4.2, see Table 10.2. The transitions from one BBA attractor pattern to another pattern when a child is playing is assumed to have no specific "direction". That is, the transition depends on the ongoing brain activity or the current ordinary state. In contrast, rituals typically have a fixed order. Therefore, the transitions are assumed to take place between specific pairs of BBA attractor patterns. From a mechanistic point of view, transitions without a specific "direction" are due to the fact that a BBA attractor pattern becomes unstable and that at that bifurcation point there are multiple attractors A,B,C,D, etc available. Consequently, if the state of the system at the bifurcation point is in the basin of attraction of attractor A, then the corresponding BBA attractor pattern A emerges. If the state is in the basin of attractor B, then the BBA attractor pattern B emerges and so on. In contrast, transitions between specific pairs of BBA

© Springer Nature Switzerland AG 2019
T. Frank, *Determinism and Self-Organization of Human Perception and Performance*, Springer Series in Synergetics,
https://doi.org/10.1007/978-3-030-28821-1_11

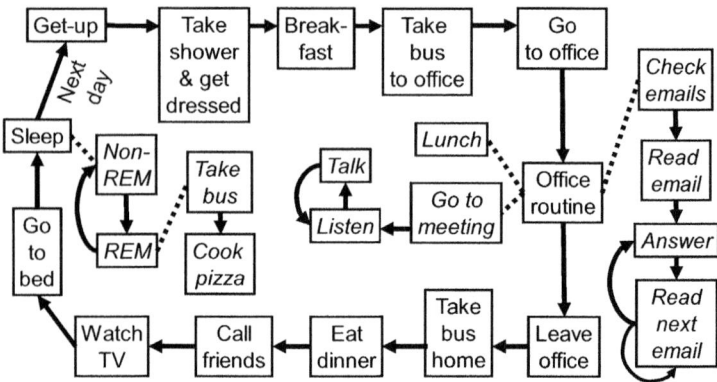

Fig. 11.1 Example of a pattern formation sequence that describes the activities of an individual during a whole working day

attractor patterns are assumed to occur when the system becomes monostable at the bifurcation point. If there is only one attractor then irrespective of the current state of the system the state converges to that attractor and the corresponding BBA attractor pattern emerges. In general, sequences of patterns that emerge and disappear and describe daily activities involve transitions of both types. That is, pattern formation sequences during a whole day of an individual in general exhibit both "directional" and "non-directional" transitions between BBA patterns.

Figure 11.1 illustrates schematically a working day of an individual in terms of a sequence of BBA patterns that emerge and disappear. The individual happens to be single and works in an office. Let us start in the left upper corner. In the morning, the conditions are such that the individual wakes up. Staying in bed becomes unstable. A bifurcation occurs and the individual gets up. That is, the individual forms a getting-up BBA attractor pattern. Standing there in the middle of the bedroom becomes an unstable pattern. A bifurcation occurs. In the example shown in Fig. 11.1, the individual turns into a BBA pattern describing a human taking a shower. Subsequently, another bifurcation takes place and the individual becomes a BBA pattern of a human getting dressed. The conditions in the stomach of the individual (i.e., assumed emptiness) increase the eigenvalue of the food-intake BBA attractor pattern (see the example in Sect. 7.5.4 about being thirsty and getting something to drink). If that eigenvalue becomes sufficiently large a bifurcation takes place. However, even without the impact of the feeling being hungry, the eigenvalue of having breakfast may simply increase as part of the morning ritual of the individual. In this scenario not the feeling of being hungry but the getting-dressed BBA attractor pattern increases the eigenvalue of the having breakfast activity. In both scenarios, the individual has breakfast. Having finished breakfast another bifurcation occurs. The individual gets ready to leave[1] and takes

---

[1] And puts on shoes. See our extensive discussions on this topic in this book.

the bus to the office. This implies that several BBA attractor patterns emerge in succession like walking to the bus, waiting at the bus stop, getting on the bus, standing or sitting in the bus, while the bus is moving. Finally, when the bus stops at the bus stop close to the office the standing or sitting in the bus BBA attractor pattern becomes unstable. A bifurcation occurs. The individual gets out of the bus and walks to the office. In the office the individual forms several sequences of patterns that are part of the office routine. Some examples of sequences that are part of the office routine are indicated in Fig. 11.1 by dotted lines connected to the box labeled office-routine. The checking emails office sub-routine consists of reading the first email. Here it is assumed that the individual answers the email. Subsequently, the individual under consideration is assumed to read the next email. The individual may answer the email or skip to the next email and so on. Another office sub-routine is given by a meeting with others. In this context, taking a bird's eye view, the individual forms part of a sequence of emerging and disappearing patterns that involve several people. We will give an example of how to formulate pattern formation in such human many body systems in Sect. 11.3 below. For sake of simplicity, we describe the meeting from the perspective of the individual of interest. First of all the individual forms several BBA patterns that bring him or her to the meeting like walking down a floor, opening doors, taking the elevator, and so on. At the meeting the individual forms a BBA pattern of listening. The forces exerted by the other people attending the meeting by talking to the individual under consideration are assumed to make the listening BBA pattern unstable at a certain point of time. The individual starts talking. After a while, the talking BBA pattern becomes unstable and disappears (see the lonely speaker example in Sect. 7.3.1). A transition back to the listening BBA pattern takes place. Talking and listening BBA patterns form an alternating sequence similar to the oscillatory sequences discussed in Sect. 7.4.3 but involve several people. This topic will be discussed in more detail in Sect. 11.3 below. Typically, the conditions for working in the office become unstable due to external conditions simply given by time. Let us assume the office closes at 6:00 PM. Close to 6:00 PM all eigenvalues related to office activity BBA patterns decay. There might be a getting-out-of-the office ritual that emerges as a sequence of BBA patterns. In any case, staying in the office becomes an unstable state. A bifurcation occurs. The individual leaves the office and takes the bus home. Being at home and doing nothing is assumed to be an unstable reference point. There might be a coming-home ritual. In this case, there might be a "directional" transition to eat dinner first when coming home. In a different scenario, the individual produces a sequence of patterns similar to the playing child, where the transitions depend on the ongoing ordinary state or brain activity. Given a particular ordinary state X when coming home, the individual may make some phone calls first, subsequently watches TV, and has dinner before going to bed. Given a different ordinary state Y when coming home, the individual may produce the sequence shown in Fig. 11.1. Accordingly, the individual eats dinner first. After a while, eating becomes an unstable activity. A bifurcation happens and makes that the individual calls some friends. After a while, the calling activity becomes unstable (e.g., everything has been said, the individual and the calling partner have run out of topics, etc., see again the lonely speaker example in Sect. 7.3.1). A bifurcation occurs. The calling BBA attractor pattern disappears. The individual watches TV. In this context note that throughout this book stable co-existence fixed points of pattern formation systems are not taken into consideration (although they exist even for the LVH amplitude equation model [104]). Such co-existence fixed points may be used to describe that the individual talks on a phone with a friend while watching TV. However, a more detailed analysis of such situations is needed. As indicated in Sect. 6.3.4 when discussing "selective attention" a promising way to describe the situation when the individual is calling a friend, while watching TV, is to use an oscillator model again. Accordingly, the individual would switch permanently between the BBA patterns of calling and watching TV (this would be similar the "attentional resource theory" [270] or the reactive inhibition theory stating that there are permanent switches between periods of "attention" and "distraction"). Coming back to Fig. 11.1, eventually staying awake becomes unstable. There is a bifurcation and the individual goes to bed. Sleep itself is a state that involves transitions between different sleep stages as indicated by the dotted line in Fig. 11.1 starting at the box labeled sleep. In particular, the sleep cycle is composed of the REM stage and several non-REM stages. The REM stage in turn (as indicated by the dotted line) is assumed to make an appropriately defined reference point unstable such that BA patterns of the human isolated brain emerge. They may emerge as a type 1 or type 2 pattern formation sequence. In Fig. 11.1 for example a BA attractor pattern of taking the bus (i.e., a pattern similar to a visual world and body movement walking-to-the-bus stop BBA pattern that however comes without external forces and produced forces) emerges and disappears and, subsequently, a BA attractor pattern about cooking pizza (similar to a visual world BBA attractor pattern about cooking pizza but without force components) emerges. The pattern sequence describing a day of the individual under consideration is completed with the sleeping activity. At the next day, a new pattern sequence emerges. The pattern sequences of different days might be similar to each other. In particular, if daily sequences involve rituals such as a getting-up ritual then these rituals correspond to sub-sequences that are the same across pattern sequences of different days.

If daily pattern sequences are similar to each other across several days, the sequences may be considered as pattern formation cycles. Accordingly, the individual of interest is regarded as a pattern formation system that evolves on a limit cycle attractor. The patterns formed every day constitute one cycle along the limit cycle (see also Sect. 11.4 below). Taking a different perspective, daily sequences may be put one after another to build sequences that cover several days. This perspective will be discussed next.

11.1.2   Sequences of the Awake, Dreaming, and Hallucinating Individual

The sequence shown in Fig. 11.1 describes pattern formation while the individual under consideration is in an awake state and while he/she is in the REM stage of sleep. In both cases the sequences correspond to cause-and-effect chains. In both cases the emerging BA patterns (REM stage person) and BBA and BA patterns (awake person) emerge from bifurcations and change external and internal conditions such that they become unstable and new patterns emerge.

> On a mechanistic level, there is no difference between pattern formation sequences of the awake, dreaming, and hallucinating human being. On a mechanistic level, there is no difference between pattern formation sequences of healthy people and patients with schizophrenia, bipolar disorder, and obsessive-compulsive disorders.

What we mean by that is that the amount of "control" or the degree of "choice" that is involved in a pattern formation sequence is always the same, namely, zero. This holds for pattern formation sequences that can be observed under any kind of circumstances in humans and animals and in systems of the inanimate world (e.g., hurricanes that produce tornadoes when hurricanes make landfall; for an engineered example see the Belousov-Zhabotinsky hand [357]). The kind of sequences that are produced by human systems during the awake stage and the REM stage are different. Likewise, the kind of patterns and sequences that people experience during drug-induced hallucinations [38] are different from those of healthy people. In a similar vein, the pattern formation sequences of people suffering from certain diseases are different from the pattern formation sequences of healthy people. However, they are "only" different with respect to the "content" of the sequences. The mechanism of self-organization and pattern formation are the same for all kind of self-organizing systems. There is not a physics of healthy people and another physics that holds for people, who suffer from diseases. There is not a physics of human beings and another physics of lasers, hurricanes, and cloud patterns.

11.2   Life Trajectories

The sequence of emerging and disappearing patterns over the day of an individual describes how the grand state of that individual evolves over time during that day. That is, the sequence describes a trajectory in the grand state of the individual. In general, as mentioned in the previous section daily trajectories exhibit some common sub-sequences but also differ from each other. In particular, pattern sequences of weekdays typically differ from pattern sequences of weekend days. Irrespectively, the sequences of several days can be taken together to describe the evolution of an individual over extended durations such as weeks and years. Figure 11.2 illustrates this issue.

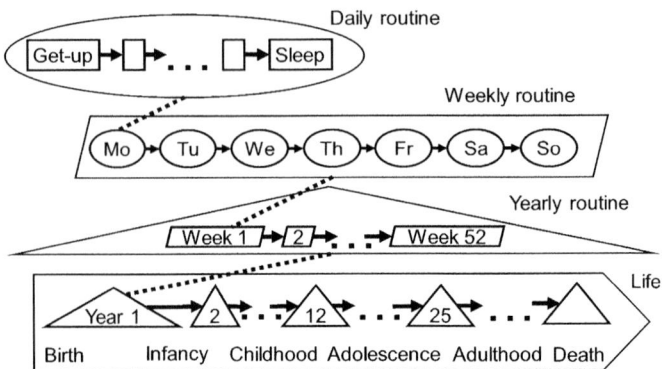

Fig. 11.2 Formation of life trajectories on the basis of single events of emerging and disappearing patterns as in daily routines

Accordingly to Fig. 11.2, the pattern sequences of the 7 days of the week are taken together to form sequences of weekly routine. Again, the weekly pattern sequences are assumed to show similarities and differences. For example, if weeks are holiday weeks of the individual, then the BBA patterns formed during those weeks are likely to be different from the BBA patterns formed during working weeks of that individual. Moreover, external seasonal impacts affect the pattern sequences. For example, which shoes the individual puts on (e.g., in the morning as part of his or her getting-to-work ritual) typically differs across summer, autumn, winter, and spring seasons. A year in turn consists of about 52 weeks. Therefore, weekly sequences of 52 weeks approximately constitute the yearly trajectories of the individual under consideration. These yearly trajectories typically differ from each other across major life stages. That is, yearly trajectories during infancy of the individual differ considerably from yearly trajectories of later life stages such as childhood and adolescence. Likewise, the kind and frequency of emerging BBA attractor patterns that constitute yearly trajectories during childhood differ from the kind and frequency of emerging BBA patterns during later life stages. In this context, recall the developmental milestones mentioned in Sect. 9.4.2 and listed in Table 9.1. Infants at the age of 14 months and older walk. That is, the walking BBA patterns can be found in the yearly trajectories of year 2 and in later years while they do not occur in the trajectory of the first year of life. Other milestones and major life events can be defined for childhood, adolescence, and adulthood. For example, the child is restructured ("learns") to read and the teenager or young adult is restructured ("learns") to drive a car. Marriage, giving birth to a child, and so on, are further examples of major life events. In general, characteristic changes in the type of emerging BBA attractor patterns across different life stages are due to changes in the structure of individuals. According to the human pattern formation reaction model, such changes are caused in general by an interaction between the emerging BBA patterns themselves and other factors such as external forces acting on the individuals under consideration. Finally, taking the yearly trajectories together, we arrive at a description of an human individual over his or her whole lifespan. The corresponding trajectory in the grand state is the life trajectory of that individual and describes the evolution of that person from birth to death.

11.3   Trajectories of Two People Systems

In the previous sections, pattern sequences have been considered from the perspective of individuals. Forces exerted by other people on an individual have been considered as external forces. Taking a bird's eye view, groups of people can be taken together to constitute many body pattern formation systems featuring interacting components. In this case, pattern formation takes place on a scale that is larger than the scale of the individuals, see Sect. 4.2.4. As a result, individuals are parts of emerging or disappearing patterns.

Let us illustrate the emergence of an oscillatory temporal pattern on the level of two people. The example shown in Fig. 11.3 describes a part of a conversation between a teenage girl and her father that could be a verbal conversation between the two of them or could correspond to text messages exchanged on cell phones. The father and the daughter are considered as coupled E1 pattern formation systems as shown in the top part of Fig. 11.3. In particular, the forces produced in terms of verbal utterances or text messages by one of them are assumed to act on the structure and brain state of the other (see dotted and dashed lines). In particular, in line with what has been discussed in earlier chapters (e.g., see Chaps. 5 and 6), brain states are shifted to the basins of attraction of certain attractors such that the father and

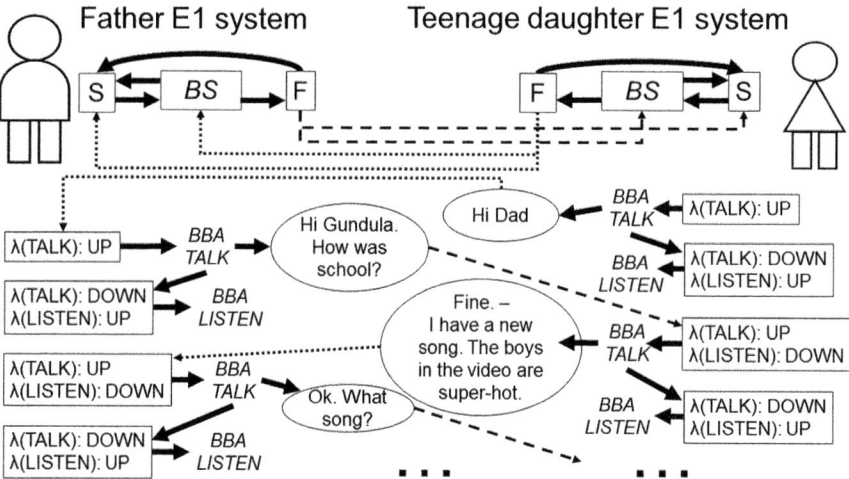

Fig. 11.3 A pattern formation sequence in a father-daughter system. The daughter and the father are parts of an oscillatory pattern formation sequence that forms due to the interactions between them

the daughter react "appropriately" (i.e., in the "usual way") to the produced forces. For sake of simplicity, however, this aspect of the conversation will be ignored. Rather, at issue is to show how pattern formation in one of the person depends on the produced forces of the other person irrespective of the content of the conversation. This aspect can be understood by assuming (as mentioned above) that the external forces produced by a person and exerted on another person change the structure of that other person. The lower part of Fig. 11.3 exemplifies explicitly this kind of mechanism.

Let us start with the teenage daughter at the top right corner of the lower part of Fig. 11.3. It is assumed that an appropriate reference state becomes unstable. For example, the father comes home and enters the living room in which the daughter sits. Ignoring the father is an unstable situation. Therefore, a bifurcation occurs. In a different scenario, the daughter may finish homework. The father is not at home. Doing nothing is an unstable situation and the daughter takes her cell phone (mobile phone) and starts texting the father. In the example, we distinguish between two BBA attractor patterns: talking and listening. On a structural level, the two patterns are related to eigenvalues λ of the corresponding basis patterns. Accordingly, there are two eigenvalues. One for talking and one for listening. When the eigenvalue for talking is sufficiently large compared to the eigenvalue of listening, then the pattern formation system is monostable and the talking BBA attractor pattern emerges. Vice versa, when the eigenvalue for listening is sufficiently large compared to the eigenvalue for talking, then the system is monostable and irrespective of the ordinary state (or brain state) the listening BBA pattern emerges (see Sect. 4.6.3 about eigenvalues and the stability band). At the bifurcation point when the reference state becomes unstable (father comes home or daughter finishes homework) it is assumed that the eigenvalue for talking is sufficiently large. This is indicated in Fig. 11.3 by the phrase $\lambda(TALK) : UP$. Consequently, the teenage daughter E1 pattern formation is monostable. The talking BBA attractor pattern emerges. The daughter says or is texting something. In the example, the daughter says or is texting "Hi Dad". As mentioned above, note that in this example the content of the conversation is ignored. It is more about the fact that the daughter produces a force that acts on the father. The forces exerted by the pressure forces of the verbal utterance "Hi Dad" or by the forces of the visual world "Hi Dad" message on the cell phone display are assumed to act on the structure of the father E1 pattern formation system. They are assumed to increase the eigenvalue for talking. This is indicated on the left hand side of Fig. 11.3 by the phrase $\lambda(TALK) : UP$. At the same time, the talking BBA pattern in the daughter system is assumed to act back on the structure that resulted in the emergence of the pattern in the first place. The eigenvalue $\lambda(TALK)$ is decreased and/or the eigenvalue $\lambda(LISTEN)$ of the basis pattern for listening is increased. This in indicated by $\lambda(TALK) : DOWN$ and $\lambda(LISTEN) : UP$. In fact, from a mathematical point of view, it is sufficient that either the eigenvalue for talking decreases or that the eigenvalue for listening increases (see Sects. 4.6.3 and 4.7.8). If the "balance" between the eigenvalues for talking and listening is such that the eigenvalue for listening is sufficiently large as compared to the eigenvalue or talking then the BBA attractor pattern for talking becomes unstable. The system becomes monostable and only exhibits the attractor for listening. The listening BBA attractor pattern emerges. In summary, the BBA attractor pattern for talking of the daughter and the produced "Hi Dad" force imply that on the side of the father the eigenvalue for talking increases, while on the side of the daughter a bifurcation occurs and the listen BBA attractor pattern emerges. The daughter stops talking or texting.

As indicated on the left hand side of Fig. 11.3 the increased eigenvalue for talking is assumed to make the father pattern formation system monostable. The talking BBA attractor pattern emerges and the father is talking or texting. In the example, the father says or is texting "Hi Gundula. How was school?". This produced force triggers the same kind of events in the daughter as the "Hi Dad" force had triggered in the father. In the daughter pattern formation system the eigenvalue for talking increases, while the eigenvalue for listening decreases. A bifurcation occurs. The talking BBA pattern emerges with "Fine. – I have a new song. The boys in the video are super-hot". On the father side, the talking BBA pattern with "Hi Gundula" decreases the talking eigenvalue and increases the listening eigenvalue. The father stops talking and listens (or stops texting and waits for an incoming message). When the daughter makes the statement "Fine – I have a new song. . .", the statement affects the father system in the same way as the previous "Hi Dad" greeting or text message by the daughter. The changes in the structure and ordinary state (brain state and state describing force production) on the father side can be considered as reaction to the forces exerted by the daughter. Those changes trigger bifurcations in the father system that come with the formation of BBA attractor patterns that exert forces on the daughter system. Those forces lead to changes in the grand state (structure, brain state, and forces) of the daughter pattern formation system and in particular to bifurcations and switches between BBA attractor patterns in that system. This kind of back and forth between the father and daughter pattern formation systems continues during the whole conversation period and drives the conversation forward. The pattern formation systems settle down in an oscillatory or alternating dynamics on the level of the two people.

11.4 Higher Hierarchical Patterns and Haken's Order Parameters

Type 1 and type 2 pattern formation sequences may be considered as higher hierarchical attractor patterns or as transient states of pattern formation systems evolving towards higher hierarchical attractor patterns. Let us address these two cases separately.

First of all, in Sect. 7.4.3 oscillatory pattern formation systems have been introduced. They have been exemplified for oscillatory BBA reaction patterns induced by the spinning dancer movie (Sect. 7.4.3), the MIB phenomenon (i.e., the oscillatory temporary breakdown of foreground BBA attractor patterns induced by forces exerted by moving backgrounds, see also Sect. 10.5.3), and the oscillatory switching between lever-pressing activities and other activities of rats with intact and damaged dopamine pathways (Sect. 10.6.1). The oscillatory dynamics involves changes on the structural and brain state level of the systems under consideration. Consequently, the type 2 pattern formation cycles are described by closed trajectories in the grand state of the humans and animals under consideration. These trajectories evolve on limit cycle attractors. In other words, the pattern formation cycles correspond to attractor patterns of limit cycle attractors. Since these limit cycle attractor patterns are composed of more elementary attractor patterns, we refer to the limit cycle attractor patterns as higher hierarchical attractor patterns.

As discussed in Sect. 4.2.1 (see also Fig. 4.3) limit cycle oscillations typically emerge from unstable foci that come with type 2 eigenvectors. That is, limit cycle attractors—just like fixed point attractors—can be related to certain unstable basis patterns. For those unstable basis patterns, amplitudes can be defined. This has been shown explicitly for the so-called canonical-dissipative limit cycle oscillator, see Sect. 3.6.3. The oscillator amplitude can be considered as the relevant amplitude characterizing the limit cycle attractor pattern, see, for example, panel (c) in Fig. 3.15. Using the terminology of synergetics, the amplitude acts as order parameter amplitude (for the definition of order parameters see Sect. 4.3.2). Therefore, higher hierarchical limit cycle attractor patterns come with appropriately defined amplitudes or order parameter amplitudes. These amplitudes have the outstanding properties discussed in Sect. 4.3.7. They determine the whole pattern formation cycle. That is, they determine the evolution of the pattern formation sequence along the limit cycle attractor.

> Pattern formation cycles as a whole correspond to higher hierarchical limit cycle attractor patterns that come with appropriately defined order parameter amplitudes that, in turn, characterize the cyclic pattern sequences.

Second, when simulating a playing child in Sect. 7.3.3 it was pointed out that the LVH model eventually settles down on a limit cycle attractor such that the child—after the child has performed a cycle composed of a number of activities—repeats the cycle and performs the cycle again and again. It has been pointed out that in reality the child will just perform a single cycle. The main point to be made here is that the sequence of m emerging BBA attractor patterns can be considered as one cycle evolving on a limit cycle attractor. In general, however, the first sequence of m activities will just approach the limit cycle attractor rather than be perfectly on the attractor. This observation allows us to generalize the aforementioned observation that pattern cycles exhibit order parameter amplitudes to more general pattern sequences that at first sight do not correspond to cycles.

In Sect. 9.3 we discussed rats running through a maze towards a food source. In Sects. 11.1 and 11.2 above, we examined daily routines and life trajectories capturing the activities of individuals from birth to death. These sequences may be considered in analogy to the m activity sequence of a playing child. If we would not allow the rat to eat the food and would put it back to the starting position, the animal would run through the maze and perform a second cycle. At least hypothetically, we can think of resetting the rat system again and again such that the pattern formation sequence of running through the maze corresponds to a cycle of a limit cycle attractor. For example, just before the rat reaches the food it would fall through a trap in the floor and would be placed back into the starting position of the maze.[2] If we consider the maze running pattern sequence as one cycle towards the approach of a limit cycle attractor, then the pattern formation system is characterized by the amplitude of a higher hierarchical attractor pattern or by an order parameter amplitude.

Likewise, rather than putting the daily routines one after another to obtain the activities of a person of a week, we may think that when the person goes to sleep and wakes up, we shift time back by 1 day and reset the system.[3] In doing so, the daily activities would converge to or evolve on a limit cycle attractor. In this sense, the daily activities form a higher hierarchical limit cycle attractor pattern and are characterized by an order parameter amplitude of the day. Likewise, life can be considered as an approach to a higher hierarchical limit cycle attractor pattern. If we imagine to get born again after death (just like the rat falling through the trap when it reaches the food and be placed at the beginning of the maze or rebirth in Buddhism), then life can be considered as a cycle on a limit cycle attractor or at least as a piece along the spiral that takes a person towards a limit cycle attractor. If so, life would be characterized by an order parameter amplitude.

---

[2]Of course, only a few cycles could be constructed in that way. The emerging pattern cycles would act back on the animal structure and change it.

[3]See the movie Groundhog Day (1993) with Bill Murray and Andie MacDowell.

Chapter 12
Ethics "Beyond" Physics

In Chap. 11 it has been pointed out that interacting humans can be modeled as coupled pattern formation systems. From this point of view, the evolution of "societies" is an issue of many body physics. In particular, when taking such a many body physics perspective it follows that mass "behavior" as well as opinions and beliefs that emerge in "societies" follow deterministic laws. For example, the coordinated dynamics of human and animal individuals forming groups [79, 289, 290, 313, 352], the formation of common opinions among human individuals [89, 91] as well as opinion polarization [345] or the group buying phenomenon [108] have been considered as synchronization phenomena within the framework of physics. These synchronization phenomena belong to the class of collective phenomena in physics. The class of collective phenomena in turn includes order-disorder equilibrium phase transitions such as can be observed in ferromagnets and liquid crystals as special cases. Therefore, when individuals of "societies" come to an agreement on an issue with broad impact (e.g., health care should be provided to everyone irrespective of status and income) then this common agreement can be understood as a synchronization phenomenon or a stable state or attractor pattern that has emerged via a non-equilibrium phase transition. Likewise, examples in the history about "societies" engaging in colonialism and slavery and suppressing other countries and ethnic and racial groups [268, 283] can be understood on the basis of humans evolving according to the laws of coupled pattern formation systems. Germans during Nazi Germany (Third Reich) holding the belief to be superior to other races [21, 347] are another example in this regard.

In general, the ethical values of "societies" are a consequence of the laws that determine the interactions between individuals. From a physics perspective, ethics is a collective phenomenon just like the formation of a magnetic phase of a ferromagnet. According to physics, ethics is a state of synchronization. In this context it is important that physics does not attach moral values to the states of physical systems. A laser that operates above the laser threshold and produces laser light is not "better" than a laser that operates below the laser threshold and produces

© Springer Nature Switzerland AG 2019
T. Frank, *Determinism and Self-Organization of Human Perception and Performance*, Springer Series in Synergetics,
https://doi.org/10.1007/978-3-030-28821-1_12

ordinary light. When the elementary magnets of a rod-shaped ferromagnet align themselves in one particular direction and produce a magnetic field with a north pole on the one end and the south pole at the other end, then this is not "better" or "worse" as if they would align themselves in just the opposite way such that north and south poles would be switched.

Physics is an all-encompassing perspective of the animate and inanimate worlds. Therefore, it can explain (e.g., in terms of synchronization phenomena) that people come together and do the things that they do. They do "good" things or "bad" things. "Societies" when considered as human many body systems exhibit general beliefs and have ethical standards in terms of synchronized states. Importantly, from a physics perspective, there is no principle difference between fireflies synchronizing their flashes, birds coordinating their movements and flying in flocks, elementary magnets aligning themselves to produce magnetic fields, elongated molecules of liquid crystals aligning themselves to produce a phase with orientational order and humans "society" holding certain ethical values. In other words, a ferromagnet exhibits as much as ethics as human "societies" (human many body systems). What physics can contribute to the discussion about ethical norms is an ethics of synchronization or the ethics of a ferromagnet.

As point out throughout this book, physics is one perspective among many perspectives about the nature of human beings and the universe as a whole. In particular, in the context of ethics the author as a private person (not as a scientist) believes that it is worthwhile to consider other perspectives (e.g., religious perspectives) that address ethical issue. What should be considered to be "good"? What should be considered to be "bad"? Other perspectives are not necessarily inconsistent with physics (see Sect. 2.7). In particular, if we discuss an ethics "different" from the ethics of physics (i.e., the ethics of a ferromagnet/ethics of synchronization), the term "different" should not be used as it is used in physics. If we use physics to define what is "different", then we make the mistake discussed in Sect. 2.7.2, see Fig. 2.7. Physics will tell us that only the concepts that hold in physics are the concepts that hold in physics and everything else that is different (where the phrase different is here understood as defined in physics) does not exist within the framework of physics. In other words, if we are interested in discussing an ethics that goes "beyond" the ethics of physics or is "different" from the ethics of physics then this kind of ethics should be "different" from physics in a way "not defined" by physics. The question would be open for debate what is meant here by "not defined" because (again) if we understand "not defined" as "different from" and use "different from" as it is used in physics we do not get "out of" the framework of physics. (And, again, the question would be what is meant here by "out of"?)

Ethics is just one topic out of several topics that might be also addressed using perspectives "different" from the physics perspective. For example, questions like "What is love?" or "What is the purpose of life?" can be answered from the physics perspective (because physics is an all-encompassing theoretical framework) but it might be worthwhile to address such questions also from "other" perspectives. Let us dwell on the latter question. The question "What is life?" has been answered from a pattern formation perspective on the level of individuals in the previous chapter, see Sects. 11.2 and 11.4. Accordingly, the life of an individual is a local order parameter in an appropriately defined state space. However, physics does not come with the notion that entities of the universe exhibit a "goal" or "purpose" different from their functioning as parts of the whole. A ball that rolls down a ramp due to the gravitational force acting on the ball exhibits no "goal" or "purpose" other than to do what it is doing, namely, to roll down the ramp. Likewise, from a physics perspective and the perspective of pattern formation systems, a human being has no other "goal" or "purpose" than to be a local order parameter in the universe. However, "other" perspectives may provide "different" answers in this regard. Importantly, those answers can be "consistent" (in the sense of a theoretical framework that compares across perspectives and is "different" from physics) with the description of human lives as local order parameters.

The author of this book as a private person (not as a scientist) hopes that this discussion highlights how important it is to consider perspectives "other than" the physics perspective of the world. Using the metaphor mentioned in Sect. 2.7.2, physics is like a baker that bakes bread, other perspectives are like pilots that fly planes. We need both bakers and pilots.

Chapter 13
Appendix

In this appendix some key concepts of the physics perspective to understand humans and animals are listed. In addition, a second list of concepts is given for which it has been shown in previous chapters that they are unnecessary within the physics perspective. The concepts listed in that second list might be useful concepts in approaches that look at humans and animals from various non-scientific perspectives.

13.1   Key Concepts of the Physics Perspective

The human pattern formation reaction model that has been discussed and applied in the previous chapters involves a number of concepts that are listed below. Importantly, the concepts listed below have been sufficient to describe and explain the experiments reviewed in this book involving humans and animals. Note that some of those concepts are shared concepts of physics, applied mathematics, and dynamical systems theory.

To act on
Amplitudes (as pattern amplitudes)
Amplitude space (and reduced amplitude space)
Attractors
Bifurcations
Bifurcation parameters
Body activity (in the sense of muscle force production, limb movements, body movement and posture, etc.)
Brain activity (in the sense of cell signaling in neurons, neurotransmitter biochemical reactions at synapses, electric pulses in axons, dendritic currents,

© Springer Nature Switzerland AG 2019
T. Frank, *Determinism and Self-Organization of Human Perception and Performance*, Springer Series in Synergetics,
https://doi.org/10.1007/978-3-030-28821-1_13

changes in membrane potentials, changes in local electrical potentials, etc.)
Causality (cause and effect)
Cause-and-effect chains
Damping (damping forces)
Determinism
Dynamics
Dynamical diseases
Effect (cause and effect)
Eigenvalues (eigenvalue spectrum)
Emergence (to emerge)
Equilibrium (and non-equilibrium)
Evolution (to evolve)
To exert a force
Fixed points
Focus (as a type of attractor)
Forces (produced, external, damping, pumping, mechanical, electromagnetic, gravitational, chemical driving forces, etc.)
Grand state
Isolated human brain (human isolated brain dynamics; brain activity patterns of the isolated human brain)
Initial conditions (initial brain/ordinary/grand state, initial structure, initial external structure/external forces)
Limit cycle attractors
Movement
Non-equilibrium (and equilibrium)
Oscillatory instability
Order parameter
Ordinary state
Patterns (basis patterns, fixed point patterns, attractor patterns, repellor patterns)
Pattern formation
Pattern sequences (pattern formation sequences)
Perspectives (physics perspective, non-scientific perspectives, religious perspectives, spiritualistic perspectives)
Phase portraits
Phase transitions
Physics-based determinism
Pumping (pumping forces)
Reactions (to react to)
Repellors
Sensory body
Self-organization
Sequences (pattern formation sequences, pattern sequences)
Stability (stable, unstable, monostable, bistable, multistable, stability band)
State (brain state, ordinary state, grand state)
State space

Structure (internal)
Structural change (restructuring, restructuration)
System classes A,B,C,D,E
Trajectories
Transitions (between qualitatively different states)
Winner-takes-all dynamics

13.2 Unnecessary Concepts Within the Physics Perspective

Throughout this book it has been pointed out that in the literature on the nature of human and animal brain and body activities one can find various concepts that are not needed within the physics perspective. Some of them are listed below together with short explanations why there is no need to use them and/or how they are related to concepts of the pattern formation perspective. As pointed out throughout this book, while the concepts below stand for concepts not needed in the physics perspective of humans and animals and, in this sense, are not needed in a physics-based scientific approach to understand humans and animals, those concepts may be considered as useful concepts in non-scientific, spiritualistic approaches to address humans and animals.

Furthermore, note that as pointed out in Sect. 2.7.1 physics does not need to explain non-scientific concepts. Likewise, physics does not need to propose correlates of non-scientific concepts in terms of physical quantities. For example, physics does not need to explain the human soul. However, physics can explain all observations involving humans and animals. If in a non-scientific discipline a group of observations has been called X, then it might be the case that all observations of that group X can be explained in the same way from a physics perspective, in general, and the pattern formation perspective, in particular. For example, various experimental observations that have been reviewed in this book and that in the literature have been called instances of "perception" have been explained from the physics perspective taken in this book as the formation of BBA patterns under the impact of external forces (or as patterns that form and emerge as reactions to external forces). Therefore, phenomena in the field of "perception" may be seen as pattern formation phenomena in humans and animals related to reactions to external forces acting on humans and animals. The explanations given below should be interpreted in this sense. They are *not* an attempt to find physical correlates for things that do not exist in physics (e.g., physics does not attempt to find physical correlates of God, Allah, or the human soul in terms of neuronal firing rates).

List

- Alternatives: There are no alternatives. All situation have unique consequences. The phrase "B1 is an alternative to B2" actually means that under condition A1 we get B1, while under condition A2 we get B2. See physics-based determinism, Sect. 2.5.2.
- Ambiguity: Situations in general have unique consequences. See "alternatives".
- Attention: Shift of brain states or ordinary states into the basin of attraction of certain attractors or changes of structure (e.g., in terms of eigenvalues) leading to monostability of the human pattern formation system. See Sect. 6.3.4.
- Attentional resources: See "attention"
- Behavior: BBA patterns that involve the production of muscle forces; formation of body patterns
- Choice (to choose): From a physics perspective humans are pieces of materials or man-machines (Skinner [300]) determined by laws. Neither the moon has a "choice" to spin around the earth or not to spin around the earth nor humans have a "choice" to do something or not to do something. See Sects. 2.5.2, 2.5.3, 4.6.1, 7.2.3 and 10.1.3.
- Conditioning (classical conditioning, operant conditioning): See below.
- Classical conditioning: Restructuration by means of paired forces, where one force induces a special body reaction that involves certain neuronal structures (and has been called "reflex"), see Chaps. 2 and 9.
- Operant conditioning: Restructuration by means of paired forces, where one force is reward related, see Chaps. 2 and 9.

  There is nothing mystic about restructuration of thingbeings by means of paired forces. If you hold a piece of stiff black paper and then take a pair of scissors and make a little hole into the paper, then you apply two forces to the paper (one by holding it and another one by using the pair of scissors) and, in doing so, you change the structure of the paper. Due to that change in structure, the piece of paper reacts differently to forces that act on the paper. For example, when you shine a light to one side of the paper, then the light will leak out of the little hole as a spot-light. Note that the paper has not "learned" how to produce a spot-light. Likewise, the paper has not "forgotten" how to hold back any light from penetrating it. Just like the paper that due to the application of paired forces does not "learn" or "forget" anything, animals performing in "conditioning experiments" that use paired forces do not "learn" or "forget" anything.

- Cognition: BA attractor patterns or sequences of BBA attractor patterns that do not involve muscle activity, see Sects. 5.2.2 and 7.5.3
- Compulsive: See Sect. 10.4.1. See "control" below
- Consciousness and unconsciousness: There are three types of basis patterns. Those with positive eigenvalues in the stability band, those with positive eigenvalues out of the stability band, and those with negative eigenvalues. Basis patterns of the first type have BBA attractor patterns that emerge when the ordinary state (or brain state) falls in the basin of attraction of the corresponding attractors. Basis patterns of the second and third type can only give rise to BBA

attractor patterns if conditions change such that they turn into basis patterns of the first type. However, they can be part of emerging BBA attractor patterns and they can affect the build-up and disappearance of BBA attractor patterns. This scheme can explain observations made by Freud and advocates of Freud's theory about consciousness and unconsciousness. See Sect. 7.2.2

- Control (to control and to steer): See "choice". Humans are self-organizing evolving systems that (1) form sequences or (2) are sequences or (3) are part of sequences of appearing and disappearing attractor patterns, see Sects. 4.2.4 and 10.4.1. The sequences are determined by laws just as the travel of the moon around the earth is determined by laws. Humans do not have "control" over their lives just as the moon has no "control" over its trajectory around the earth. For example, although in everyday language we say that "a person driving a car steers the car" this is an unnecessary (or even incorrect) statement. When a ball is rolling down a ramp, then (a) the ball does not "steer" its movement, (b) the ramp does not "steer" the ball movement, and (c) the gravitation does not "steer" the ball movements. Likewise, when we drive a car or walk to some place, we are like a ball rolling down a ramp. We do not "steer" or "control", where we are heading to. Likewise, we are not "steered" or "controlled" by anything.

  In the aforementioned example of the person driving a car the person evolves along a trajectory that describes a sequence of emerging attractor patterns such as sitting into the car, starting the engine, pushing the gas pedal, turning the steering wheel to the right or left, etc. The person evolves according to a sequence that was already fixed when he/she sat down in the car (or even earlier when the person was born), see Chap. 2. Humans are man-machines (Skinner [300]) or pattern formation systems that exhibit a certain dynamics that can be described in terms of certain pattern formation sequences—just like rats running in a maze towards a food source exhibit a certain dynamics that can be described in terms of certain pattern sequences. Rats turn left and right according to laws (see e.g., Sects. 6.4.3, 7.4.4, and 9.3). What they will do is determined at the beginning of the maze already—just as the future trajectory of the moon around the earth is determined at any given moment in time.

  When a person says "I have control over my car" or "I am steering my car" the person (from a physics perspective) actually means that "When I am in my car, what I do follows from certain laws and my current structure and ordinary state".

  Much of the research on "control" is actually research on cause-and-effect relationships. When a researcher talks about an entity A (e.g., a cortical area) and states that "A controls something" or "A regulates something" then the researcher actually means that "A causes something to happen". That is, A exerts (mechanical, electrical, chemical, etc.) forces on a thingbeing and those forces have an effect on that thingbeing.

- Control parameter: The phrase "control" should be avoided in science because it leads to mystification. Therefore, in this book the phrase bifurcation parameter is used.

- Decision (decision making, to decide): Bifurcations between BBA patterns. The phrase "decision" is unnecessary (in the sense that it is not needed from a science perspective) as it has been shown in various chapters of this book. Moreover, according to physics, physical systems including humans do not make "decisions". A sitting person does not "decide" to stand up (see Fig. 1.4 in Chap. 1) just as water does not "decide" to turn into ice and the moon does not "decide" to travel around the earth. See "choice" and "control".

  When you are waiting at a bus station and it is cold, then you start to sway. You do not "decide" to sway. The conditions are such that a bifurcation occurs such that your body makes a transition from a standing still pattern to an oscillatory swaying pattern. When you are a short person, then you may push yourself up to stand on tiptoes to see something better or to be able to talk to somebody (e.g., there is somebody behind a Dunkin-Donuts counter or food counter talking to you and the counter is high). You do not "decide" to stand on tiptoes. The conditions are such that you make a bifurcation from a standing with flat feet pattern to a standing on tiptoes pattern. When you marry somebody, you do not "decide" (at least from a physics perspective) to marry that person. You are in the condition to marry that person. When you buy a drink, you do not "decide" to buy a drink. You are in the condition to buy a drink.

  When you are reading these lines then in 24 h from that moment it might rain or the sun might shine. Whatever the weather will be in 24 h of that very moment when you read these lines, the weather will not "decide" to do anything. The sun will not "decide" to shine and the rain will not "decide" to fall. Likewise, in 24 h of the moment when you read these lines, you will do something. You might watch TV, you might kiss somebody. Whatever it will be what *you* will do in 24 h of that very moment when you read these lines, it will not involve a "decision" of yours. You will not "decide" to watch TV, just like the sun does not "decide" to shine. You will not "decide" to kiss somebody, just like the rain does not "decide" to fall. The concept of "decision making" mystifies human and animal beings. "Decision making" is a concept not needed in a physics perspective of human beings and animals.
- Free will: See "control", "choice", "decision". See Sect. 2.5
- Focus: See "attention"
- Forgetting: Structural change leading to a change in the reactions of humans and animals to external forces or a change in the dynamics of the brain activity of the isolated human and animal brain.

  When a car mechanic in an auto-repair shop fastens some screws and as a result of that the car engine stops shuttering, then the car mechanic changes the structure of the car engine and as a result of that change the performance of the car engine changes. The engine does not "forget" to shutter. When we relax a guitar string and it stops to produce high pitch sounds but produces low pitch sounds, then the string does not "forget" how to produce high pitch sounds. See Sect. 9.4

- Goal: The life of human individuals (and animals) is determined by their reactions to external forces and the dynamics of their isolated human brains. "Goal" is a concept outside of the physics perspective of human beings.

  "Goal" might be a very important concept when looking at humans from non-scientific perspectives (e.g. religious perspectives). Recall that physics does not make any statements about non-scientific perspectives. Taking a religious perspective the author of this book as a private person (not as a scientist) believes that it is import to ask oneself questions like "What is the purpose of my life?" and "What is the goal of my life?". See the discussion on the "purpose of life" and non-scientific perspective in Chap. 12.
- Habituation: Change in structure (restructuring)
- Illusion: See "mistake" below
- Information (as used in everyday language): External forces acting on humans. Bifurcation parameters.

  In physics the concept of "information" as used in everyday language is not needed.[1] For example a person who wants to catch a bus may look at the time on his/her watch or cell phone (mobile phone). Depending on the ratio of the remaining time before the bus arrives at the bus stop and the distance to the bus stop, the person may walk or run to the bus stop. In physics, the time to distance ratio is a bifurcation parameter. Other disciplines have referred to this ratio as "information". Physics does not need to do so. See Sect. 6.5.6.

  A laser that is pumped by energy does not pick up any "information" about the pumping. Rather, the pumping acts as bifurcation parameter. If the pumping is strong enough a bifurcation occurs and the laser produces laser light. A gas layer heated from below does not pick up any "information" about the heating. Rather, the heating acts as bifurcation parameter. If the heating is strong enough (relative to the temperature at the top of the layer), then a bifurcation occurs and the gas layer will from a convection pattern (e.g., a roll pattern).
- Information variable: Bifurcation parameter, see Sect. 6.5.6.
- Individuality: Individuality is an unnecessary and misleading concept. A point mass of 1 kg that rests in a laboratory frame accelerates in a particular way when a force of 1 N is applied to it. A point mass of 2 kg accelerates in a different way. To say that there are individual differences between the masses of 1 and 2 kg that lead to different reactions of the point masses is misleading because all mass points are subjected to the same law, which is Newton's law. That is, Newton's law determines the acceleration of all mass points irrespective of their masses.

  Humans differ with respect to their structure and ongoing brain and body state (ordinary state) just like point masses in general differ with respect to their masses. Irrespective of the differences of structure and current ordinary state, there are laws that apply to all humans and that determine how they evolve in

---

[1]Note that there are various concepts of information that differ from the concept of information as used in everyday language (e.g., Shannon information). Here, only the concept of information as typically used in casual conversations is addressed.

time just like there is Newton's law that determines the acceleration of all mass points irrespective of their masses.

In this context note that in physics there is the concept of identify. Accordingly, from the a physics perspective, humans have identity. In this book frequently the phrase "human individual" has been used. With the phrase "individual" the author of this book refers to a particular person X, that is, the author refers to the identity of that person.

- Interpretation: Reaction to external forces

As in a phrase like "perception is the process of interpreting the sensations we experience". From a physics perspective, humans exhibit reactions to external forces just like all thingbeings of the universe react to forces. The concept of a reaction to forces is sufficient to explain human life. The notion of an "interpreter" is not needed.

- Judgments: Formation of BBA attractor patterns in X2/X0 systems (excluding A2 systems) as reactions to external forces. See also "perception" below. See Sect. 5.2.2
- To judge: To react to external force by forming X2/X0 systems BBA attractor patterns
- Knowledge (to know): See Sects. 7.5.3 and 9.4
- Learning: Change of structure leading to changes in the reactions to external forces and to changes in the dynamics of brain activity in the human and animal isolated brain.

When the author of this book buys a new pair of shoes, they have a certain structure. When the author has used his shoes for half a year, they look different and have changed their structure. The author of this book also has T-shirts that change their shape and elasticity when the author uses them, on the one hand, and puts them into the washing machine from time to time, on the other hand. The shoes and T-shirts of the author change the way they react to the forces the author applies to them. This does not mean that the shoes and T-shirts have "learned" anything. When we stretch a guitar string and it changes the way it reacts to a guitar player who plays the string by producing a higher pitch sound, then this does not mean that the guitar string has "learned" to produce the higher pitch sound. See Sect. 9.4

- Long-term memory: See "memory"
- Memory: There is no physical system that can go back into the past (if there exists such as thing as a past at all). From a physics perspective, everything what happens in the next millisecond or in the next fraction of a millisecond only depends on the presence (i.e., on the now). The relevant question is: What are the circumstances of this world now?

Circumstances of the world are permanently subjected to change. For example, the shoes of the author of this book and the T-shirts change their structure when used (see "learning"). The first car of the author of this book changed its structure and got dents and scratches (see "remembering"). As a result of such structural changes, the shoes and T-shirts change the way they react to forces that the author applies to them. The car changed the way it reacted to forces. Such structural changes or the change in the reactions should not be taken as evidence that those things exhibit "memory". From a physics perspective, the structure of thingbeings (irrespective of whether they are living or non-living things) can change and this affects their relationships with the world. It is not necessary to call such structural changes "memory". In particular, it is not useful (and violates the physics perspective of the world) to mystify such changes as "memory" in human beings and animals, while saying that they are structural changes in non-living things.

While you as a reader read these lines you most probably are surrounded by air (assuming you are not reading this book in the outer space or under water). If you just move your body (e.g., your head) slightly or if you just breath in and out, then you exert forces on the molecules of the air around you. Air primarily contains nitrogen and oxygen. Those elements occur as molecules composed of two atoms (i.e., not as single atoms). Therefore, the forces that you produce while reading these lines squeeze and stretch some of the air molecules around you. Let us assume you turned your head slightly to the right and, in doing so, you squeezed an air two-atoms molecules X. After a while the distance between the two atoms of the molecule X relaxes back to its rest length. The structural changes in humans and animals due to forces acting on them are like the structural changes in air molecules produced by the reader at the moment when he/she is reading these lines. If we say that humans have "memory" because of such structural changes, then we would also need to say that the air molecules have "memory". Some of them would have "short-term" memory when the relaxation to the baseline state happens relatively quickly. Others would have "long-term" memory if the relaxation to the baseline state happens slowly. In fact, if we say that humans have the super-power of "memory" then any thingbeing (e.g., the chair you might be sitting one; the chair would "remember" your weight) would have that super-power. Physics can explain the world without introducing super-powers like the super-power "memory".

- Mistakes: Within the framework of physics-based determinism, humans evolve according to deterministic sequences. These sequences are neither "correct" or "wrong".
- Perception: Formation of BBA attractor patterns as reactions to external forces; formation of reaction patterns
- Figure-ground perception: Formation of BBA attractor patterns as reaction to figure-ground scenes or figure-ground pictures
- Letter perception: Formation of BBA attractor patterns as reaction to forces exerted by displayed letters
- Face perception: Formation of BBA attractor patterns as reaction to forces exerted by displayed faces
- To perceive: To react to external forces by forming BBA attractor patterns; to form reaction patterns; for example "to perceive a house" means to form a house reaction pattern
- Posture: Human body configuration (frequently related to a particular BBA pattern such as standing, sitting, etc.). The atoms of a piece of metal in

equilibrium are arranged in space in a particular way. A piece of metal has as much as posture as a human being.

- Priming: State-induced restructuring (i.e., restructuring by internal forces), see Sect. 7.4.1, or instruction-induced restructuring (i.e., restructuring by external forces), see Sect. 10.5.4.
- Reflex: Reaction that involves certain neuronal structures (such as the spinal cord). Although it is beyond the scope of this book, let us attempt to describe the reflex phenomenon from a physics point of view. As a working hypothesis it is proposed that the difference between a reflex and movements like walking or reaching for an object is that the reflex is produced in the human and animal *driven system*, whereas walking and the emergence of reaching movements are produced in the human and animal *self-organizing, pattern formation system*.

  Frequently, the reflex is introduced in the literature using concepts such as "voluntariness" or "being automated". These concepts contribute to the mystification of humans and animals. From a physics perspective, there is no qualitative difference between any reaction of a human or animal and any "reflex" of a human or animal with respect to the amount of involved "control". That is, any reaction of a human or animal does not involve more or less "control" than any "reflex". From a physics perspective, the distinction between "reflexes" and "non-reflexes" by saying that reflexes are "automated" and "involuntary", whereas non-reflexes are "non-automated" and "voluntary" is not needed and in fact violates the fundamental principles of physics. With respect to the degree of "control" and "voluntariness" everything what humans and animals do is just like a "reflex". Everything what humans and animals do are reactions to something that follows certain laws. See also "choice", "compulsive", "control", "decision making".
- Relation between people: Interactions between people
- Recognition: See "perception"
- Reinforcement: See "learning"
- To remember: When the author of this book was young and received his driving license, he had a Volvo that was parked in the garage of his parent's house. When trying to park in the garage, the author hit several times garage walls with the car. The car got dents and scratches. The author of this book would not say that the car dents and scratches were indicators that the car "remembered" those incidence. When the car got dents and scratches this changed the brain structure of the author of this book. In this sense, just like the car got dents and scratches the brain of the author got "dents and scratches". Due to these structural changes, when writing the text of this book, the author of this book can report about the car and the dents and scratches. This does not mean that the author of this book "remembers" anything. Those incidences, when the car hit the walls of the garage, on the one hand, changed the structure of the car, and, on the other hand, changed the structure of the author of this book. These changes, in turn, changed the way the car reacted to forces of the environment (e.g., the way how the car's body reflected light to the eyes of the author) and these changes changed the way the author reacted to the situation when sitting in front of a computer and doing

some writing. The concept of "remembering" is not needed when describing the animate and inanimate worlds from a physics perspective. In fact, it is not a useful concept because it typically is only used for humans and animals and not for cars and other non-living things. It creates a difference (and mystifies human beings and animals) where there is no difference.

- Representation: See Sect. 5.4.7

In Sect. 5.4.7 it has been clarified why "representations" are not a concept of physics. Let us approach this issue from a slightly different point of view. An ordinary pendulum swings with a particular frequency. The pendulum can be driven with an oscillatory driving force. If so, the pendulum frequency and the frequency of the driving force can be the same or different. If the frequencies are the same, using the concept of "representations" we may say that the pendulum frequency "represents" the frequency of the driving force. Likewise, a harmonic limit cycle oscillator such as the canonical-dissipative oscillator introduced in Sect. 3.6.3 has an oscillation frequency. If the limit cycle oscillator frequency happens to be the same as the frequency of an oscillatory driving force, then using the "representation" terminology, we may say that the limit cycle oscillator frequency "represents" the driving force frequency. In fact, a pendulum or a limit cycle oscillator with frequency f would "represent" all oscillatory phenomena that happen to oscillate with frequency f.

The author of this book has running shoes in his office. They are black and have a weight of 500 g. Using the "representation" terminology, the shoes represent all items that are black. They also represent all things that have a weight of 500 g. The author has a laptop with a screen that has a width of 26 cm. We could say the screen "represents" all items in this world that have a size of 26 cm. In the office of the author of this book there are numerous objects such as the running shoes and the laptop. Using "representation" terminology, these objects "represent" a huge amount of things in this world. That is, the office of the author is full of "representations". It would be fair to say that the office of the author of this book has as much "representations" as there are putatively in a human brain or an animal brain.

Cars have steering wheels. Therefore, cars have "representations" of the things in this world that are round.

Physics does not need the concept of "representations". Accordingly, cars do not have "representations". The office of the author of this book does not have "representations". The running shoes of the author do not "represent" anything. The swinging frequency of a pendulum and the oscillation frequency of a limit cycle oscillator do not "represent" anything. Humans and animals do not have "representations".

Having said that, when a swinging pendulum with a certain frequency is driven by an oscillatory force with the same frequency, then the pendulum reacts to that force in a particular way. This is the so-called resonance phenomenon. The resonance phenomenon also holds for limit cycle oscillators driven by oscillatory forces that have the same frequency as the limit cycle oscillators. Likewise, when the author of this book looks into a mirror, then the forces exerted by the face displayed in the mirror put the brain state of the author into the basin of attraction of a particular attractor. This attractor may be called the That-is-me attractor or the I attractor. Physics is about materials that exhibit certain structures and as a result of those structures react in certain ways to forces applied to them. The science of human beings and animals is just about that.

- Resources (resource allocation): See "attention"
- Self: Identity but not individuality. See "individuality"
- Self-control: See "control"
- Stimulus: From a science perspective, systems of the animate and inanimate worlds are subjected to forces. The "stimulus" concept is an unnecessary concept creating a difference, where there is no difference. See Sect. 1.7.2
- Short-term memory: See "memory"
- Social (society): From a physics perspective, "societies" are human many body systems. The concept of "society" is not needed and introduces a distinction between the animate and inanimate worlds that is not helpful to understand humans from a physics-based scientific perspective.
- Training: Change of structure (restructuring). See "learning"
- Voluntary and involuntary: See "choice", "compulsive", "control", "decision", "reflex".

13.3   My Bifurcations

In this paragraph, situations are described in which the author of this book was involved either actively or as observer. These situations were about bifurcations or state shifts into the basin of attraction of certain attractors. Therefore, the section title should actually read "My bifurcations and state shifts". For sake of brevity, it is simply called "My bifurcations".

Some events are marked with a star "*". A marked event possibly described a bifurcation. It is also possible[2] that the event was about a human pattern formation system that acted as a so-called excitable medium, whose state was shifted out of a stable fixed point. States shifted out of fixed points of excitable media typically return to their fixed points. The human reaction reported from the event would then correspond to the return to that fixed point.

2008

Tile-Walk-1 Bifurcation A pedestrian walk was plastered with gray tiles. A 6 years old girl was walking with her parents on the pedestrian walk. When being on the tiled pedestrian walk a bifurcation occurred and the girl walked in such as way to avoid stepping on the edges of the tiles (where crocodiles were assumed to hide to bite the girl).

---

[2]The author of this book tries to cheat here and tries to avoid using the word "alternatively".

Tile-Walk-2 Bifurcation The floor of a shopping center was plastered with gray tiles. Some of the tiles were colored black. As soon as the aforementioned 6 years old girl was confronted with the forces exerted by the tiled floor, a bifurcation occurred and the girl started to walk on the black tiles only.[3]

Child-Following Bifurcation When the aforementioned 6 years old girl made a gait transition to a tile-walk, the forces exerted by the girl and acting on her father shifted the eigenvalue spectrum of the father such that the tile-walk BBA pattern of the father became monostable or the forces shifted the ordinary state of the father into the basin of the attraction of the tile-walk attractor. The father followed the girl and turned into a tile-walk BBA attractor pattern.

2018

Searching Style Bifurcation There are several methods to search for a small item in a pocket. Method A is to search with the hand in the pocket, feel the shape of the items, pick a promising item out, check if it is the desired item, and repeat this procedure until the desired item is found. Method B is to empty the pocket and search visually. The author of this book typically forms a method-A searching BBA pattern. While the pattern is formed it decreases its eigenvalue. After a while a bifurcation takes place and the author bifurcates to the method-B searching BBA pattern.

Rotating-Chair Bifurcation The author of this book likes to eat fast food in McDonald's. My favorite McDonald's restaurant has bar stools that can be rotated to the left and right by about an angle of $30°$. When a teenager produces a sitting-on-the-bar-stool BBA pattern in that McDonald's restaurant, frequently, the author observes that the sitting-on-the-bar-stool BBA pattern induces another bifurcation: the teenager boy or girl starts to produce a rotating-the-stool BBA pattern. That is, the boy or girl turns the stool to the left and right in a more or less rhythmic fashion for a certain period. While producing that pattern, the corresponding eigenvalue decays such that the pattern eventually becomes unstable and disappears.

Shivering Bifurcation When the human body temperature falls below a critical temperature, humans produce muscle forces that put body parts in a movement that is referred to as shivering or shaking. The body temperature seems to act as bifurcation parameter. The shivering movement may be understood as a BBA pattern defined by a limit cycle attractor. In this context, the pattern would emerge via a Hopf bifurcation.

Nodding Bifurcation[*] The university at which the author is working has a dining hall with a salad bar. The servers who are working at the salad bar typically ask the

---

[3]The Don't-walk-in-the-hallway program initiated at the Panorama Hills School in Calgary, Canada, is aimed at changing the structure of school hallways such that school children do not simply walk in school hallways but make bifurcations to various types of gait patterns that are more active than simple walking.

students what ingredients they want for their salads. At the end, the servers ask if the order is complete. When the author of this book is asked at that final stage when everything is complete, he typically bifurcates to produce a nodding BBA pattern.

2019

Give-It-Up Bifurcation The author of this book likes to eat pancakes with jam. Opening a jar with self-made jam can be a challenge. In order to open the jam jar the lid must be rotated. The closed jam jar and the circumstances of being hungry and having a pancake in front of me make that a bifurcation occurs and the author produces a BBA pattern of rotating the lid by hand. If the lid does not move, a bifurcation occurs and the authors produces one or several tool-assisted lid-opening BBA patterns. For example, the author puts a kitchen towel between lid and his hand in order to change the friction. The author may use a knife to lift the lid to get air into the jam jar. At the end of the BBA pattern sequence of lid-opening BBA patterns the author has either opened the jam jar or a bifurcation to a give-it-up BBA pattern occurs.

Smiling Bifurcation and Shaking Head Bifurcation* A young, heterosexual couple in a vacation spot was sitting on the sidewalk after midnight and killed some time. He told his girl friend something and she reacted to it with a smile (i.e., she produced a smiling BBA pattern). Subsequently, he made a suggestion to her and she reacted to it by shaking her head sidewise as a rejection. The rejection shaking head BBA pattern acted back on the young man and caused him to produce another suggestion. That in turn induced in the woman the emergence of another shaking head BBA pattern.

Twerking Bifurcation Some teenagers and adults "like to" twerk, that is, they exhibit pattern sequences of their daily life routines in which the twerk BBA pattern frequently occurs and comes with a supplementary pattern component of positive emotions. Appropriate circumstances make those teenagers and adults to play certain pieces of modern music that in turn cause them to form twerking BBA patterns.

Pompom Wielding Bifurcation A child about 5 or 6 years old was sitting in a bus in a vacation spot. The child held in each hand a cheerleader pompom. Holding the pompoms in the hand was an unstable reference state. A bifurcation occurred and the child wielded the pompoms such that the plastic streamers or stripes swung around.

I-Am-Right-Oops-I-Am-Wrong Bifurcation In Spring 2019 the author was boarding an airplane and took his aisle seat on the right wing. The plane had ten seats in a row. Three seats on each side wing and four seats in the middle. When the author took his seat, an elderly heterosexual couple was already sitting in the row in front of the author on the right wing. The man was sitting in the middle seat, the woman in the aisle seat. The window seat was empty. Another passenger showed up and claimed the aisle seat in which the woman was sitting. The passenger was speaking in English to the woman. The woman was speaking a different native language but was able to communicate in English that the aisle seat was her seat. She produced a I-am-right BBA attractor pattern. The passenger showed his boarding card and explained to the woman the illustrations on the ceiling indicating that the aisle seat was his seat. This probably reduced the eigenvalue of the I-am-right BBA attractor pattern of the woman but was not sufficient to make the pattern unstable. The women insisted to be right. The forces exerted by the communication between the two people acted on the author and changed his eigenvalue spectrum such that sitting and listening BBA pattern of the author became unstable. The author of this book happened to speak the native language of the woman. I stood up and explained to the woman in her native language that her seat was the window seat and that the arriving passenger had the aisle seat. My explanations in her native language presumably reduced sufficiently the eigenvalue of her I-am-right BBA attractor pattern. A bifurcation occurred. The woman admitted to be wrong, directed her partner to take the window seat and took the middle seat. As a side note: the plane was only half full. After the plane had climbed to cruising altitude, the aforementioned passenger abandoned his aisle seat and took a seat in a completely empty 4-seats row in the middle of the plane. ... And they all lived happily ever after.

Move-the-Armrest Bifurcation* In 2019, the author was sitting on a 2 h ride in a tourist bus heading for a vacation spot. The author had an aisle seat on the right hand side. A middle aged Chinese woman was sitting four rows in front of the author on an aisle seat on the left hand side. The armrest of her seat was damaged. As a result of that the armrest could be moved up and down. That is, it could be rotated up and down. The woman grasped the armrest with her right hand. At that moment, grasping without doing something else became an unstable reference point. One basis pattern out of the many basis patterns that occurred at the reference point was the BBA pattern of an oscillatory up and down movement of the armrest. In fact, the circumstances (give in terms of the current structure and brain state of the woman and the forces that the armrest exerted on the woman) were such that either the system was monostable with the oscillatory armrest movement attractor being the only attractor or the system was multistable but the ordinary state of the woman happened to fall into the basin of attraction of the oscillatory armrest movement attractor. As a result, the female passenger moved up and down the armrest for a while.

Summer-Hat Spinning Bifurcation The aforementioned Chinese female passenger had a light-colored summer hat. While sitting in the bus, the activities performed by the woman (e.g., talking to her female holiday friend sitting next to her or using her cell phone) made at a certain point in time the state of having the summer hat in front of her unstable. A bifurcation occurred. She moved one hand from below into the hat and started to spin the hat around on her hand or finger. The spinning-the-hat BBA attractor pattern presumably reduced its own eigenvalue such that after a while another bifurcation occurred and the pattern disappeared.

Researcher Bifurcation The author of this book was sitting in the aforementioned tourist bus with the Chinese woman, when his brain was forming various type 1 and type 2 pattern formation sequences of the human isolated brain (i.e., the author was "thinking"). According to the pattern formation perspective of human beings, the brain structure of people working as researchers exhibits an attractor that comes with a basin of attraction to which brain states are shifted by forces of events that are work-related. For sake of simplicity, the attractor will be called the researcher attractor. In general, it might be called the my-work (or my-job) attractor. The forces exerted by the Chinese woman acted on the author and shifted the brain state into the basin of attraction of the researcher attractor of the author. The emerging researcher BBA pattern of the author gave rise to a pattern formation sequence in which the author produced a taking-a-pen BBA pattern and a making-notes BBA pattern.

String-Bag Spinning Bifurcation In the first months of 2019 the author observed two children walking on a sidewalk. One of the children, a boy (about 7 years old), was carrying a small drawstring bag. The bag had approximately half of the size of a 1 liter milk box. The bag was filled with some small items. The drawstring attached to the bag was approximately 40 cm long. Holding the bag on the string in the usual way while producing a walking BBA pattern and a talking-to-my-friend BBA pattern became unstable at a certain moment in time. The boy started to spin the bag around on a circle. The bag followed a trajectory like big wheel that was standing upright next to the boy on the sidewalk.

Hip-Swing Bifurcation Waiting in line in the airport to check-in can be boring. The author of this book was waiting with other passengers in a zig-zag line that was for checking-in and dropping off the check-in suitcases. Many passengers had their suitcases stapled on airport trolleys (airport carts). In order to move such a trolley you need to push first the trolley handle down and then you can push the trolley forward. An about 11 years old girl was waiting with her family in line and was in charge of the family trolley with the suitcases. She had both hands on the handle. The line did not move forward. She was chewing something, most likely, chewing gum. Being in a waiting state and in a doing-nothing state made that her state became unstable after a while. Having her hands on the handle and leaning with the upper body forward somewhat at that unstable reference point various basis patterns were available. One of the available basis patterns was a hip-swing pattern. The pattern won the competition with other patterns. The girl swung her hip sidewise from left to right and from right to left for a while and enjoyed herself in doing that. The hip-swing BBA pattern reduced its own eigenvalue such that after a while it became unstable and the movement stopped. The girl made another bifurcation and formed the waiting-and-chewing BBA pattern again.

Staring-at-Woman Bifurcation When I was a young boy, my family (i.e., my mother and father, my brothers and I) went to a restaurant for dinner. We sat down at a table, made our order, and were waiting for the food to come. While producing a sequence of talking and listening patterns in a human many body system, a woman entered the restaurant and was walking past our table heading towards another empty table. The woman exerted forces on my older brothers such that they formed staring-at-the-woman BBA patterns. This in turn acted on my mother and changed her eigenvalue spectrum such that she interrupted her current conversation and made a general, verbal comment on staring at women. The whole situation in turn acted on my brain structure such that when writing this book, I am describing and reporting the event. In 2019 the author happened to be in an airtrain that took passengers from the center of a city to the airport of that city. The airtrain wagon in which the author was sitting had benches on both sides of the wagon such that when the author was sitting on the one side he was looking across the wagon to the other side. Next to the author two young men (about 20 years old) were sitting. On the opposite side a woman with her two daughters were sitting. The daughters were about the same age as the young men. The three women were forming a human many body talking and listening BBA pattern including smiles and laughing patterns. They enjoyed themselves. The two men were looking at their cell phones. The playing-with-my-cell-phone patterns of the men became from time to time unstable presumable in part due to the forces of the women exerted on the men. Several times bifurcations took place in which the men looked up from their cell phones and stared at the women for a while. Note while this paragraph is called staring-at-the-woman bifurcation, it is clear that women stare at men, just like men stare at women. An extreme case in this regard are, for example, female teenager fans of Korean boys pop groups.

Four-Basis-Pattern Airplane Bifurcation In the first half of the year 2019, the author of this book was sitting in a plane flying to New York. Lights were switched off such that the passengers could sleep. The author was watching a movie on his seat screen, that is, the ordinary state of the author formed a movie-watching BBA pattern. A flight attendant formed a walking BBA pattern and walked through the aisle with a tray with drinks in plastic cups on it: water, orange juice, and apple juice. The visual world forces of the approaching flight attendant acting on the author put the eigenvalue of the movie-watching basis pattern to a negative value and in doing so created an unstable reference point. At that reference point the author as a pattern formation system exhibited a number of unstable basis patterns (i.e., basis patterns with positive eigenvalues). Among those patterns were the following four patterns: taking the water, taking the orange juice, taking the apple juice, and ignoring the offer. The author was in the condition of taking the orange juice. More precisely, either the system was monostable with a dominant positive eigenvalue of the orange juice basis pattern (i.e., the external forces of the flight attendant with the tray and the drinks shifted the eigenvalue spectrum of the author accordingly) or the system was multistable and the brain state of the author happened to be in the basin of attraction of the orange juice attractor (e.g., the brain activity related to the movie-watching BBA pattern happened to occupy in the relevant reduced amplitude space a location close to the orange juice attractor or the external forces of the drinks caused a brain state shift that shifted the brain state to such a location).

Smile-Back Bifurcation When the author walks from one building of the university campus to another building it frequently happens that the author exerts forces on students who walk towards the author. The visual world forces exerted by the author on the students make the neutral face expression state unstable and increase the eigenvalue of the smile basis pattern. As a result, the students smile to the author. The visual forces exerted by the smiling faces acting on the author, in turn, make the neutral face expression state of the author unstable and increase the eigenvalue of the smile basis pattern of the author. Consequently, a smile-back bifurcation occurs and the author smiles back.

Pedestrian-Crossing-Follower Bifurcation The university of the author of this book is a campus university in the countryside. There is not much traffic except in the morning and evening when university employees come to work or are heading home. There is a four-way crossing with traffic lights close to the building in which the author has his office. When the pedestrian signal light shows the wait sign students typically are waiting at the sidewalk even if the streets are complete empty (and I mean that literally: there is no car as far as you can see to the North, West, South and East; nevertheless students are patiently waiting). However, under such circumstances sometimes one person is stepping from the sidewalk to the street and is crossing the street despite the fact that the signal light shows the wait sign. If that happens, the visual world forces exerted by that crossing person and acting on the other students who are waiting on the sidewalk make the waiting BBA patterns of those waiting students unstable. Bifurcations take place. Walking BBA patterns emerge and the remaining students cross the street regardless of the fact that the pedestrian signal light shows the wait sign.

Detour Bifurcation In February 2019 a snow storm hit the university where the author of this book was working. University workers removed most of the snow from the university walking paths between the university buildings. The building in which the author had his office was opposite to the library building. From the main entrance of the author's building students could walk in a straight line to the main entrance of the library building. However, due to the snow storm there was a small snow field (3 m long and 1 m wide) between the two buildings. Students could cross the snow field or walk around it to the left or the right. The author was leaving his office and was heading to the library. A student was walking in front of the author, obviously heading towards the library as well. The visual world forces of the snow field acted on the student and made the walking straight to the library pattern unstable. The student turned to the right and walked around the snow field. The author was also affected by the forces exerted by the snow field. In addition, the author was affected by the forces of the detouring student walking in front of the author. As a result, the walking straight to the library pattern of the author disappeared and the author formed a detouring pattern following the student.

Pick-a-Boy and Pick-a-Girl Bifurcation Visual world forces coming from the body and particular the faces of people can affect other people in various ways. It is assumed that human individuals are affected in the same or similar way by the forces coming from the bodies and faces of other individuals that belong to a certain body type. Given a certain body type, the forces shift the brain states into the basin of attraction of a particular attractor. The BBA patterns related to such attractors can have different supplementary components. In doing so, individuals can exhibit an attractor for pretty people and another attractor for not-so-pretty people (where prettiness means in general something different for each individual such that a given face may lead to the emergence of the pretty BBA pattern for one individual but triggers the emergence of the not-so-pretty BBA pattern for another individual). Let us consider a heterosexual female teenager or adult. Let us assume the woman is dancing together with her girl friends in a discotheque and is exposed to the visual world forces exerted by the bodies of men who happen to be in the discotheque as well. If the forces of a particular male shift the brain state of the woman into the basin of attraction of the attractor for pretty people (or maybe in the basin of attraction of an attractor for "super-hot" males), then the corresponding BBA pattern emerges. The emerging BBA pattern may trigger a sequence of BBA patterns that corresponds to a flirting ritual (e.g., approaching the male person, smiling, saying hello, etc.). In short, the forces exerted by the male on the woman let the woman pick the male. Likewise, a heterosexual male can make a pick-a-girl bifurcation triggered by the forces exerted by a female on the male person. Moreover, analogous considerations can be made for individuals of all kind of sexual orientations (homosexual, bisexual, transgender, animal-oriented, object-oriented, pedophile, etc.)

Run-After-Daddy Bifurcation   In February 2019 the author observed the following scene. A father and his 6 years old son are leaving a local community center and are on the way to the parking lot of the community center. The father carries his son on the arm. Halfway to the parking lot, the father stops and puts his son down on the pedestrian path that goes to the parking lot. The boy is standing on his feet, while the father starts walking towards the parking lot. As a result, within half of a second, there is a distance between the boy and his father of about 1 m. That is, boy and father are separated. In line with the observation of the author and the human pattern formation reaction model, it is plausible to assume that the distance between the boy and his father acts as bifurcation parameter. The increasing distance or bifurcation parameter makes that the standing-still BBA pattern of the boy becomes unstable. A bifurcation takes place. The boy turns into a running-after-daddy pattern and runs into the direction of his father.

Ice-Puddle Bifurcation   The following scene took place at the aforementioned community center. It was February 2019 and temperatures had been below zero degrees Celsius. There was no snow but close to the path that connected the parking lot of the community center with the main entrance of the center there was an ice-puddle. The ice-puddle was about 1 m long and half of a meter wide. Three boys (about 8 years old) were running from the parking lot to the entrance of the community center. While the first two boys ran straight to the entrance, the third boy showed a different body dynamics. Taking a pattern formation perspective the following happened. When the boy was running to the entrance and came close to the ice-puddle, the visual world forces of the ice-puddle acted on the boy, changed the eigenvalue spectrum, and made the running BBA pattern unstable. The boy stopped and placed himself in front of the ice-puddle. The standing in front of the ice-puddle BBA pattern decreased its own eigenvalue and became unstable after a while. At that reference point presumable a basis pattern of running to the entrance and a basis pattern of stepping on the ice-puddle existed (among many other basis patterns). The basis pattern of stepping on the ice-puddle won the amplitude competition and the boy stepped on the ice-puddle.

Two-Lane-Gait Bifurcation   A boy (about 7 years old) was walking in a pedestrian zone of an European city. As such the pedestrian zone was just an asphalt street in gray color. However, in that particular section of the pedestrian zone there were two lanes or stripes in dark color (black or dark gray) that were parallel to each other. Each lane was about 10 cm wide. The two lanes were about 40–50 cm apart from each other. The boy placed his right foot on the right lane and his left foot on the left lane. Subsequently, he walked forward making sure that the feet were always on the lanes. Note that the two-lane gait may be considered as some kind of a tile-walk as described at the beginning of this section.

References

1. S.D. Aberson, J.B. Halverson, Kelvin-Helmholtz billows in the eyewall of hurricane Erin. Mon. Weather Rev. **134**, 1036–1038 (2006)
2. R.E. Adamson, Functional fixedness as related to problem solving. a repetition of three experiments. J. Exp. Psychol. **44**, 288–291 (1952)
3. R.M. Alexander, *Exploring Biomechanics: Animals in Motion* (W. H. Freeman, New York, 1992)
4. C. Allain, H.C. Cummins, P. Lallemand, Critical slowing down near the Rayleigh-Benard convection instability. Le J. de Phys. Lett. **24**, L474–L477 (1978)
5. M.C. Allen, G.R. Alexander, Gross motor milestones in preterm infants: correction for degrees of prematurity. J. Pediatr. **116**, 955–959 (1990)
6. J.R. Anderson, *Cognitive Psychology and Its Implications* (W. H. Freeman and Company, San Francisco, 1980)
7. M.C. Anderson, B.A. Spellman, On the status of inhibitory mechanisms in cognition: memory retrieval as a model case. Psychol. Rev. **102**, 68–100 (1995)
8. M.C. Anderson, E.L. Bjork, R.A. Bjork, Retrieval-induced forgetting: evidence for a recall-specific mechanism. Psychon. Bull. Rev. **7**, 522–530 (2000)
9. F. Attneave, Multistability in perception. Sci. Am. **225**, 63–71 (1971)
10. G. Avidan, M. Behrmann, Correlation between the fMRI BOLD signal and visual perception. Neuron **16**, 495–497 (2002)
11. S. Avissar, G. Schreiber, The involvement of guanine nucleotide binding proteins in the pathogenesis and treatment of affective disorders. Biol. Psychiatry **31**, 435–459 (1992)
12. R.F.W. Bader, *Atoms in Molecules: A Quantum Theory* (Oxford University Press, Oxford, 1990)
13. L.M. Baker, *Learning and Behavior* (Prentice Hall, Upper Saddle River, 2001)
14. A. Bakonyi, D. Michaelis, U. Peschel, G. Onishchukov, F. Lederer, Dissipative solitons and their critical slowing down near a supercritical bifurcation. J. Opt. Soc. Am. B **19**, 487–491 (2002)
15. G. Barbalat, M. Rouault, N. Bazargani, S. Shergill, S.J. Blakemore, The influence of prior expectations on facial expression discrimination in schizophrenia. Psychol. Med. **42**, 2301–2311 (2012)
16. S. Barbay, G. Giacomelli, F. Marin, Stochastic resonance in vertical cavity surface emitting lasers. Phys. Rev. E **61**, 157–166 (2000)
17. M. Bauer, S. Beaulieu, D.L. Dunner, B. Lafer, R. Kupka, Rapid cycling bipolar disorder - diagnostic concepts. Bipolar Disord. **10**, 153–162 (2008)

© Springer Nature Switzerland AG 2019
T. Frank, *Determinism and Self-Organization of Human Perception and Performance*, Springer Series in Synergetics, https://doi.org/10.1007/978-3-030-28821-1

18. H. Bee, *The Developing Child* (Haper and Row, New York, 1975)

19. P.J. Beek, R.C. Schmidt, A.W. Morris, M.-Y. Sim, M.T. Turvey, Linear and nonlinear stiffness and friction in biological rhythmic movements. Biol. Cybern. **73**, 499–507 (1995)

20. M. Begon, J.L. Harper, C.R. Townsend, *Ecology, Individuals, Populations and Communities* (Blackwell Scientific Publications, Boston, 1990)

21. J.W. Bendersky, *A Concise History of Nazi Germany* (Rowman and Littlefield, Chicago, 2007)

22. M. Bestehorn, R. Friedrich, Rotationally invariant order parameter equations for natural patterns in nonequilibrium systems. Phys. Rev. E **59**, 2642–2652 (1999)

23. M. Bestehorn, H. Haken, Associative memory of a dynamical system: an example of the convection instability. Z. Phys. B **82**, 305–308 (1991)

24. M. Bestehorn, R. Friedrich, H. Haken, Two-dimensional traveling wave patterns in nonequilibrium systems. Z. Phys. B **75**, 265–274 (1989)

25. J.J. Binnen, N.J. Dowrick, A.J. Fisher, M.E.J. Newman, *The Theory of Critical Phenomena* (Oxford University Press, New York, 1992)

26. O.H. Blackwood, T.H. Osgood, A.E. Ruark, *An Outline of Atomic Physics* (Wiley, New York, 1955)

27. T.A. Blondis, Motor disorders and attention-deficit hyperactivity disorder. Pediatr. Clin. North Am. **46**, 899–913 (1999)

28. H.U. Bödeker, T.D. Frank, R. Friedrich, H.G. Purwins, Stochastic dynamics of dissipative solitons in gas-discharge systems, in *Anomalous Fluctuation Phenomena in Complex Systems: Plasmas, Fluids and Financial Markets* ed. by C. Riccardi, H.E. Roman (Research Signpost, Trivandrum, 2008), pp. 145–184

29. D. Bohm, *Quantum Theory* (Dover Publications, New York, 1989)

30. L.L. Bonilla, C.J. Perez-Vicente, R. Spigler, Time-periodic phases in populations of nonlinearly coupled oscillators with bimodal frequency distributions. Phys. D **113**, 79–97 (1998)

31. Y.S. Bonneh, A. Cooperman, D. Sagi, Motion-induced blindness in normal observers. Nature **411**, 798–801 (2001)

32. C.T. Bonnet, J.M. Kinsella-Shaw, T.D. Frank, D. Bubela, M.T. Turvey, Deterministic and stochastic postural processes: effects of age, task, environment. J. Motor Behav. **42**, 85–97 (2010)

33. A. Borsellino, A. De Maroc, A. Allazetta, S. Rinesi, B. Bartollini, Reversal time distribution in the perception of ambiguous stimuli. Kybernetik **10**, 139–144 (1972)

34. C.L. Bowden et al., The efficacy of lamotrigine in rapid cycling and non-rapid cycling patients with bipolar disorder. Biol. Psychiatr. **45**, 953–958 (1999)

35. A.J. Bracken, J.G. Wood, Semiquantum versus semiclassical mechanics for simple nonlinear systems. Phys. Rev. A **73**, article 012104 (2006)

36. P.C. Bressloff, Neural networks, lattice instantons, and the anti-integrable limit. Phys. Rev. Lett. **75**, 962–965 (1995)

37. P.C. Bressloff, P. Roper, Stochastic dynamics of the diffusive Haken model with subthreshold periodic forcing. Phys. Rev. E **58**, 2282–2287 (1998)

38. P.C. Bressloff, J.D. Cowan, M. Golubitsky, P.J. Thomas, M.C. Wiener, Geometric visual hallucinations, Euclidean symmetry and the functional architecture of striate cortex. Phil. Trans. R. Soc. Lond. B **356**, 299–330 (2001)

39. P.C. Bressloff, J.D. Cowan, M. Golubitsky, P.J. Thomas, M.C. Wiener, What geometric visual hallucinations tell us about the visual cortex. Neural Comput. **14**, 473–491 (2002)

40. J. Britz, T. Landis, C.M. Michel, Right parietal brain activity precedes perceptual alternation of bistable stimuli. Cereb. Cortex **19**, 55–65 (2009)

41. P.D. Butler, S.M. Silverstein, S.C. Dakin, Visual perception and its impairment in schizophrenia. Biol. Psychiatry **64**, 40–47 (2008)

42. A.J. Capute, B.K. Shapiro, F.B. Palmer, Normal gross motor development: the influence of race, sex, and socio-economic status. Dev. Med. Child Neurol. **27**, 635–643 (1985)

43. C. Carello, A. Grosofsky, F.D. Reichel, H.Y. Solomon, M.T. Turvey, Visually perceiving what is reachable. Ecol. Psychol. **1**, 27–54 (1989)

44. O.L. Carter, J.D. Pettigrew, A common oscillator for perceptual rivalries. Perception **32**, 295–305 (2003)
45. B. Case, B. Tuller, M. Ding, J.A.S. Kelso, Evaluation of a dynamical model of speech perception. Percept. Psychophys. **57**, 977–988 (1995)
46. P. Chance, *Learning and Behavior* (Wadsworth, Belmont, 2008)
47. D. Cheyne, H. Weinberg, Neuromagnetic fields accompanying unilateral finger movements: pre-movement and movement-evoked fields. Exp. Brain Res. **78**, 604–612 (1989)
48. S. Chiangga, T.D. Frank, Stochastic properties in bistable region of single-transverse-mode vertical-surface emitting lasers. Nonlin. Phenom. Complex Syst. **13**, 32–37 (2010)
49. R. Chisholm, Human freedom and the self, in *Free Will* ed. by G. Watson (Oxford University Press, Oxford, 2003), pp. 24–35
50. D. Cho, C.L. Park, T.O. Blank, Emotional approach coping: gender differences on psychological adjustment in young to middle-aged cancer survivors. Psychol. Health **28**, 874–894 (2013)
51. S.K. Ciccarelli, J.N. White, *Psychology. An Exploration* (Pearson, New York, 2015)
52. L. Ciompi, The key role of emotions in the schizophrenia puzzle. Schizophr. Bull. **4**, 318–322 (2015)
53. M.A. Ciranni, A.P. Shimamura, Retrieval-induced forgetting in episodic memory. J. Exp. Psychol. - Learn. Mem. Cogn. **25**, 1403–1414 (1999)
54. R.J. Comer, *Abnormal Psychology* (Worth Publishers, New York, 2007)
55. M. Cook, The judgment of distance on a plane surface. Percept. Psychophys. **23**, 85–90 (1978)
56. J.T. Coyle, R.S. Duman, Finding the intercellular signaling pathways affected by mood disorder treatments. Neuron, **38**, 157–160 (2003)
57. M.C. Cross, P.C. Hohenberg, Pattern formation outside of equilibrium. Rev. Mod. Phys. **65**, 851–1112 (1993)
58. G.C. Cruywagen, P.K. Maini, J.D. Murray, Biological pattern formation on two-dimensional spatial domains: a nonlinear bifurcation analysis. SIAM J. Appl. Math. **57**, 1485–1509 (1997)
59. B.N. Cuffin, D. Cohen, Magnetic fields produced by models of biological current sources. J. Appl. Phys. **48**, 3971–3980 (1977)
60. J.A. da Silva, Scales for measuring subjective distance in children and adults in a large open field. J. Psychol. **113**, 221–230 (1983)
61. A. Daffertshofer, H. Haken, A new approach to recognition of deformed patterns. Pattern Recogn. **27**, 1697–1705 (1994)
62. J.M. Delgado-Garcia, A. Gruart, Firing activities of identified posterior inteposirus nucleus neurons during associative learning in behaving cats. Brain Res. Rev. **49**, 367–376 (2005)
63. R. Dick, *Advanced Quantum Mechanics: Materials and Photons* (Springer, Berlin, 2012)
64. F.J. Diedrich, W.H. Warren, Why change gaits? Dynamics of the walk-run transition. J. Exp. Psychol. - Hum. Percept. Perform. **21**, 183–202 (1995)
65. F.J. Diedrich, W.H. Warren, The dynamics of gait transitions: effects of grade and load. J. Motor Behav. **30**, 60–78 (1998)
66. P.A.M. Dirac, The Lagrangian in quantum mechanics. Physikalische Zeitschrift der Sowjetunion **3**, 64–72 (1933)
67. T. Ditzinger, H. Haken, Oscillations in the perception of ambiguous patterns: a model based on synergetics. Biol. Cybern. **61**, 279–287 (1989)
68. T. Ditzinger, B. Tuller, H. Haken, J.A.S. Kelso, A synergetic model for the verbal transformation effect. Biol. Cybern. **77**, 31–40 (1997)
69. T. Ditzinger, B. Tuller, J.A.S. Kelso, Temporal patterning in an auditory illusion: the verbal transformation effect. Biol. Cybern. **77**, 23–30 (1997)
70. D.G. Dotov, T.D. Frank, From the W-method to the canonical-dissipative method for studying uni-manual rhythmic behavior. Motor Control **15**, 550–567 (2011)
71. D.G. Dotov, S. Kim, T.D. Frank, Non-equilibrium thermodynamical description of rhythmic motion patterns of active systems: a canonical-dissipative approach. BioSystems **128**, 26–36 (2015)

72. J. Du, J. Quiroz, P. Yuan, C. Zarate, H.K. Manji, Bipolar disorder: involvement of signaling cascades and AMPA receptor trafficking at synapses. Neuron Glia Biol. **1**, 231–243 (2004)

73. S.L. Dubovsky, Rapid cycling bipolar disease: new concepts and treatments. Curr. Psychiatry Rep. **3**, 451–462 (2001)

74. V. Dufiet, J. Boissonade, Dynamics of Turing pattern monolayers close to onset. Phys. Rev. E **53**, 4883–4892 (1996)

75. K. Duncker, *On Problem Solving (Trans. L. S. Lees)*. Psychological Monographs, vol. 58 (1945)

76. A.K. Dutt, Turing pattern amplitude equations for a model glycolytic reaction-diffusion system. J. Mater. Chem. **48**, 841–855 (2010)

77. W. Ebeling, I.M. Sokolov, *Statistical Thermodynamics and Stochastic Theory of Nonequilibrium Systems* (World Scientific, Singapore, 2004)

78. R.E. Ecke, H. Haucke, Y. Maeno, J.C. Wheatley, Critical dynamics at a Hopf bifurcation to oscillatory Rayleigh-Benard convection. Phys. Rev. A **33**, 1870–1878 (1986)

79. B. Eckhardt, E. Ott, S.H. Strogatz, D.M. Abrams, A. McRobie, Modeling walker synchronization on the Millenium bridge. Phys. Rev. E **75**, 021110 (2007)

80. H. Einat, H.K. Manji, Cellular plasticity cascades: genes-to-behavior pathways in animal models of bipolar disorder. Biol. Psychiatry **59**, 1160–1171 (2006)

81. L. Fajutrao et al., A systematic review of the evidence of the burden of bipolar disorder in Europe. Clin. Pract. Epidemiol. Ment. Health **5**, article 3 (2009)

82. B. Fiedler, T. Gedeon, A class of convergent neural network dynamics. Phys. D **111**, 288–294 (1998)

83. P. Fitzpatrick, C. Carello, R.C. Schmidt, D. Corey, Haptic and visual perception of an affordance for upright posture. Ecol. Psychol. **6**, 265–287 (1994)

84. E.B. Foa, Cognitive behavioral therapy of obsessive-compulsive disorder. Dialogues Clin. Neurosci. **12**, 199–207 (2010)

85. T.D. Frank, Multivariate Markov processes for stochastic systems with delays: application to the stochastic Gompertz model with delay. Phys. Rev. E **66**, 011914 (2002)

86. T.D. Frank, On a mean field Haken-Kelso-Bunz model and a free energy approach to relaxation processes. Nonlin. Phenom. Complex Syst. **5**(4), 332–341 (2002)

87. T.D. Frank, Delay Fokker-Planck equations, perturbation theory, and data analysis for nonlinear stochastic systems with time delays. Phys. Rev. E **71**, 031106 (2005)

88. T.D. Frank, *Nonlinear Fokker-Planck Equations: Fundamentals and Applications* (Springer, Berlin, 2005)

89. T.D. Frank, Markov chains of nonlinear Markov processes and an application to a winner-takes-all model for social conformity. J. Phys. A: Math. Gen. **41**, 282001 (2008)

90. T.D. Frank, On a multistable competitive network model in the case of an inhomogeneous growth rate spectrum with an application to priming. Phys. Lett. A **373**, 4127–4133 (2009)

91. T.D. Frank, On the linear discrepancy model and risky shifts in group behavior: a nonlinear Fokker-Planck perspective. J. Phys. A: Math. Gen. **42**, 155001 (2009)

92. T.D. Frank, On a moment-based data analysis method for canonical-dissipative oscillator systems. Fluctuation Noise Lett. **9**, 69–87 (2010)

93. T.D. Frank, Pumping and entropy production in non-equilibrium drift-diffusion systems: a canonical-dissipative approach. Eur. J. Sci. Res. **46**, 136–146 (2010)

94. T.D. Frank, Motor development during infancy: a nonlinear physics approach to emergence, multistability, and simulation, in *Advances in Psychology Research*, ed. by A.M. Columbus, vol. 83, Chap. 9 (Nova Publ., New York, 2011), pp. 143–164

95. T.D. Frank, New perspectives on pattern recognition algorithm based on Haken's synergetic computer network, in *Perspective on Pattern Recognition* ed. by M.D. Fournier, pp. 153–172, Chap. 7 (Nova Publ., New York, 2011)

96. T.D. Frank, Rate of entropy production as a physical selection principle for mode-mode transitions in non-equilibrium systems: with an application to a non-algorithmic dynamic message buffer. Eur. J. Sci. Res. **54**, 59–74 (2011)

97. T.D. Frank, Multistable pattern formation systems: candidates for physical intelligence. Ecol. Psychol. **24**, 220–240 (2012)

98. T.D. Frank, Psycho-thermodynamics of priming, recognition latencies, retrieval-induced forgetting, priming-induced recognition failures and psychopathological perception, in *Psychology of Priming* ed. by N. Hsu, Z. Schütt, Chap. 9 (Nova Publ., New York, 2012), pp. 175–204

99. T.D. Frank, A limit cycle model for cycling mood variations of bipolar disorder patients derived from cellular biochemical reaction equations. Commun. Nonlinear Sci. Numer. Simul. **18**, 2107–2119 (2013)

100. T.D. Frank, Action flow in obsessive-compulsive disorder rituals: a model based on extended synergetics and a comment on the 4th law. J. Adv. Phys. **5**, 845–853 (2014)

101. T.D. Frank, From systems biology to systems theory of bipolar disorder, in *Systems Theory: Perspectives, Applications and Developments*, ed. by F. Miranda, Chap. 2 (Nova Publ., New York, 2014), pp. 17–36

102. T.D. Frank, Multistable perception in schizophrenia: a model-based analysis via coarse-grained order parameter dynamics and a comment on the 4th law. Univ. J. Psychol. **2**, 231–240 (2014)

103. T.D. Frank, A nonlinear physics model based on extended synergetics for the flow of infant actions during infant-mother face-to-face communication. Int. J. Sci. World **2**, 62–74 (2014)

104. T.D. Frank, Secondary bifurcations in a Lotka-Volterra model for n competitors with applications to action selection and compulsive behaviors. Int. J. Bifurcation Chaos **24**, article 1450156 (2014)

105. T.D. Frank, Domains of attraction of walking and running attractors are context dependent: illustration for locomotion on tilted floors. Int. J. Sci. World **3**, 81–90 (2015)

106. T.D. Frank, On the interplay between order parameter and system parameter dynamics in human perceptual-cognitive-behavioral systems. Nonlinear Dynamics Psychol. Life Sci. **19**, 111–146 (2015)

107. T.D. Frank, Formal derivation of Lotka-Volterra-Haken amplitude equations of task-related brain activity in multiple, consecutively preformed tasks. Int. J. Bifurcation Chaos **10**, article 1650164 (2016)

108. T.D. Frank, The Lotka-Volterra-Haken model for group buying, apples-to-apples group play and the competition and attraction of leisure activities, in *New Research in Collective Behavior*, ed. by T.D. Frank, Chap. 9 (Nova Publ., New York, 2016), pp. 139–158

109. T.D. Frank, Perception adapts via top-down regulation to task repetition: a Lotka-Volterra-Haken modelling analysis of experimental data. J. Integr. Neurosci. **15**, 67–79 (2016)

110. T.D. Frank, A synergetic gait transition model for hysteretic gait transitions from walking to running. J. Biol. Syst. **24**, 51–61 (2016)

111. T.D. Frank, Unstable modes and order parameters of bistable signaling pathways at saddle-node bifurcations: a theoretical study based on synergetics. Adv. Math. Phys. **2016**, article 8938970 (2016)

112. T.D. Frank, Determinism of behavior and synergetics, in *Encyclopedia of Complexity and Systems Science*, ed. by R.A. Meyers, Chap. 695 (Springer, Berlin, 2018)

113. T.D. Frank, P.J. Beek, A mean field approach to self-organization in spatially extended perception-action and psychological systems, in *The Dynamical Systems Approach to Cognition*, ed. by W. Tschacher, J.P. Dauwalder (World Scientific, Singapore, 2003), pp. 159–179

114. T.D. Frank, J. O'Leary, Motion-induced blindness and the spinning dancer paradigm: a neuronal network approach based on synergetics, in *Horizon in Neuroscience Research*, ed. by A. Costa, E. Villalba, vol. 35, Chap. 2 (Nova Publ., New York, 2018), pp. 51–80

115. T.D. Frank, M.J. Richardson, On a test statistic for the Kuramoto order parameter of synchronization with an illustration for group synchronization during rocking chairs. Phys. D **239**, 2084–2092 (2010)

116. T.D. Frank, A. Daffertshofer, P.J. Beek, H. Haken, Impacts of noise on a field theoretical model of the human brain. Phys. D **127**, 233–249 (1999)

117. T.D. Frank, A. Daffertshofer, C.E. Peper, P.J. Beek, H. Haken, Towards a comprehensive theory of brain activity: coupled oscillator systems under external forces. Phys. D **144**, 62–86 (2000)

118. T.D. Frank, R. Friedrich, P.J. Beek, Stochastic order parameter equation of isometric force production revealed by drift-diffusion estimates. Phys. Rev. E **74**, 051905 (2006)

119. T.D. Frank, C.E. Peper, A. Daffertshofer, P.J. Beek, Variability of brain activity during rhythmic unimanual finger movements, in *Movement System Variability*, ed. by K. Davids, S. Bennett, K. Newell (Human Kinetics, Champaign, 2006), pp. 271–305

120. T.D. Frank, M.J. Richardson, S.M. Lopresti-Goodman, M.T. Turvey, Order parameter dynamics of body-scaled hysteresis and mode transitions in grasping behavior. J. Biol. Phys. **35**, 127–147 (2009)

121. T.D. Frank, J. van der Kamp, G.J.P. Savelsbergh, On a multistable dynamic model of behavioral and perceptual infant development. Dev. Psychobiol. **52**, 352–371 (2010)

122. T.D. Frank, V.L.S. Profeta, H. Harrison, Interplay between order parameter and system parameter dynamics: considerations on perceptual-cognitive-behavioral mode-mode transitions exhibiting positive and negative hysteresis and response times. J. Biol. Phys. **41**, 257–292 (2015)

123. W.K. Frankenburg, J. Dodds, P. Archer, *Denver II Technical Manual* (Denver Developmental Materials, Denver, 1990)

124. W.J. Freeman, Tutorial on neurobiology: From single neurons to brain chaos. Int. J. Bifurcation Chaos **2**, 451–482 (1992)

125. R.W. Frischholz, F.G. Boebel, K.P. Spinner, Face recognition with the synergetic computer, in *Proceedings of the First International Conference on Applied Synergetics and Synergetic Engineering* (Frauenhofer Institute IIS., Erlangen, 1994), pp. 100–106

126. K.J. Friston, Theoretical neurobiology and schizophrenia. Br. Med. Bull. **52**, 644–655 (1996)

127. A. Fuchs, H. Haken, Pattern recognition and associative memory as dynamical processes in a synergetic system. I. Translational invariance, selective attention and decomposition of scene. Biol. Cybern. **60**, 17–22 (1988)

128. G. Gambino, M.L. Lombardo, M. Sammartino, Turing instability and traveling fronts for a nonlinear reaction-diffusion system with cross-diffusion. Math. Comput. Simul. **82**, 1112–1132 (2012)

129. G. Gambino, M.C. Lombardo, M. Sammartino, V. Sciacca, Turing pattern formation in the Brusselator system with nonlinear diffusion. Phys. Rev. E **88**, article 042925 (2013)

130. P. Genevet, S. Barland, M. Giudici, J.R. Tredicce, Stationary localized structures and pulsing structures in cavity soliton lasers. Phys. Rev. A **79**, article 033819 (2009)

131. C. Gerloff, C. Toro, N. Uenishi, L.G. Cohen, L. Leocani, M. Hallett, Steady-state movement-related cortical potentials: a new approach to assessing cortical activity associated with fast repetitive finger movements. Electroenceph. Clin. Neurophysiol. **102**, 106–113 (1997)

132. C. Gerloff, N. Uenishi, T. Nagamine, M. Hallett T. Kunieda, H. Shibasaki, Cortical activation during fast repetitive finger movements in humans: steady-state movement-related magnetic fields and their cortical generators. Electroenceph. Clin. Neurophysiol. **109**, 444–453 (1998)

133. J.J. Gibson, *The Ecological Approach to Visual Perception* (Houghton-Mifflin, Boston, 1979)

134. M.E. Gilpin, Limit cycles in competition communities. Am. Nat. **109**, 51–60 (1975)

135. L. Glass, M.C. Mackey, *From Clocks to Chaos* (Princeton University Press, Princeton, 1988)

136. S. Glucksberg, The influence of strength of drive on functional fixedness and perceptual recognition. J. Exp. Psychol. **63**, 36–41 (1962)

137. N.S. Goel, S.C. Maitra, E.W. Montroll, On the Volterra and other nonlinear models of interacting populations. Rev. Mod. Phys. **43**, 231–276 (1971)

138. N. Goldstein, *Sensation and Perception* (Wadsworth, Pacific Grove, 2002)

139. S. Gori, E. Giora, R. Pedersini, Perceptual multistability in figure-ground segregation using motion stimuli. Acta Psychol. **129**, 399–409 (2008)

140. N.J. Gotelli, *A Primer of Ecology* (Sinauer Associates, Sunderland, 2008)

141. T.D. Gould, H.K. Manji, Signaling networks in the pathophysiology and treatment of mood disorders. J. Psychosom. Res. **53**, 687–697 (2002)

142. A. Graham, *Statistics* (NTC Publishing group, Lincolnwook, 1994)
143. T.M. Griffin, R. Kram, S.J. Wickler, D.F. Hoyt, Biomechanical and energetic determinants of the walk-trot transition in horses. J. Exp. Biol. **207**, 4215–4223 (2004)
144. S. Grillner, Neurobiological bases of rhythmic motor acts in vertebrates. Science **228**, 143–149 (1985)
145. S. Grillner, P. Wallen, Central pattern generators for locomotion, with special reference to vertebrates. Annu. Rev. Neurosci. **8**, 233–261 (1985)
146. K.S. Griswold, L.F. Pessar, Management of bipolar disorder. Am. Family Phys. **62**, 1343–1353 (2000)
147. S. Grossberg, C. Pribe, M.A. Cohen, Neural control of interlimb oscillations. I. Human bimanual coordination. Biol. Cybern. **64**, 485–495 (1997)
148. S.J. Guastello, *Chaos, Catastrophe, and Human Affairs* (Lawrence Erlbaum, Mahwah, 1995)
149. J. Guckenheimer, P. Holmes, *Nonlinear Oscillations, Dynamical Systems, and Bifurcations of Vector Fields* (Springer, Berlin, 1983)
150. R. Guerra-Narbona, J.M. Delgado-Garcia, J.C. Lopez-Ramos, Altitude acclimization improves submaximal cognitive performance in mice and involves an imbalance of the cholinergic system. J. Appl. Physiol. **114**, 1705–1716 (2013)
151. H. Haken, Distribution function for classical and quantum systems far from thermal equilibrium. Z. Phys. **263**, 267–282 (1973)
152. H. Haken (ed.), *Synergetics I Cooperative Phenomena in Multi-Component Systems* (Teubner, Stuttgart, 1973)
153. H. Haken (ed.), *Cooperative Effects* (North-Holland Publ. Company, Amsterdam, 1974)
154. H. Haken, *Synergetics. An Introduction* (Springer, Berlin, 1977)
155. H. Haken, *Light - Laser Light Dynamics* (North-Holland Publ. Company, Amsterdam, 1985)
156. H. Haken, *Light: Laser Light Dynamics* (North Holland, Amsterdam, 1991)
157. H. Haken, *Synergetic Computers and Cognition* (Springer, Berlin, 1991)
158. H. Haken, *Principles of Brain Functioning* (Springer, Berlin, 1996)
159. H. Haken, *Brain Dynamics* (Springer, Berlin, 2002)
160. H. Haken, *Synergetics: Introduction and Advanced Topics* (Springer, Berlin, 2004)
161. H. Haken, G. Schiepek, *Synergetik in der Psychologie (in German)* (Hogrefe, Gottingen, 2006)
162. H. Haken, M. Wagner (eds.), *Cooperative Phenomena* (Springer, Berlin, 1973)
163. N. Herschkowitz, J. Kagan, K. Zilles, Neurobiological bases of behavioral development in the first year. Neuropediatrics **28**, 296–306 (1997)
164. N. Hirose, A. Nishio, The process of adaptation to perceiving new action capabilities. Ecol. Psychol. **13**, 49–69 (2001)
165. M. Hirsch, B. Baird, Computing with dynamic attractors in neural networks. Bio Syst. **34**, 173–195 (1995)
166. D.F. Hoyt, C.R. Taylor, Gait and energetics of locomotion in horses. Nature **292**, 239–240 (1981)
167. A. Hreljac, Effects of physical characteristics on the gait transition speed during human locomotion. Hum. Mov. Sci. **14**, 205–216 (1995)
168. A. Hreljac, R. Imamura, R.F. Escamilla, W.B. Edwards, Effects of changing protocol, grade, and direction on the preferred gait transition speed during human locomotion. Gait Posture **25**, 419–424 (2007)
169. H.C. Hsu, A. Fogel, Stability and transitions in mother-infant face-to-face communication during the first 6 months: a microhistorical approach. Dev. Psychol. **39**, 1061–1082 (2003)
170. J.D. Huppert, M.E. Franklin, Cognitive behavioral therapy for obsessive-compulsive disorder: an update. Curr. Psychiatric Rep. **7**, 268–273 (2005)
171. R.T. Hurlburt, S.A. Akhter, Unsymbolized thinking. Conscious. Cogn. **17**, 1364–1374 (2008)
172. L. Iosa, L. Gizzi, F. Tamburella, N. Dominici, Editorial: neuro-motor control and feed-forward models of locomotion in humans. Front. Hum. Neurosci. **9**, article 306 (2015)
173. P. Jena, S. Azad, M.N. Rajeevan, CMIP5 projected changes in the annual cycle of Indian monsoon rainfall. Climate **4**, article 4010014 (2016)

174. A. Jenkins, Self-oscillations. Phys. Rep. **525**, 167–222 (2013)

175. V.K. Jirsa, H. Haken, A field theory of electromagnetic brain activity. Phys. Rev. Lett. **77**, 960–963 (1996)

176. V.K. Jirsa, H. Haken, A derivation of a macroscopic field theory of the brain from the quasi-microscopic neural dynamics. Phys. D **99**, 503–526 (1997)

177. V.K. Jirsa, J.A.S. Kelso, *Coordination Dynamics: Issues and Trends* (Springer, Berlin, 2004)

178. R.S. Jope, Anti-bipolar therapy: mechanisms of action of lithium. Mol. Psychiatry **4**, 117–128 (1999)

179. B. Katz, *Nerve, Muscle and Synapse* (McGraw-Hill, New York, 1966)

180. B.A. Kay, J.A.S. Kelso, E.L. Saltzman, G. Schöner, The space-time behavior of single and bimanual movements: data and model. J. Exp. Psychol. - Hum. Percept. Perform. **13**, 178–192 (1987)

181. M.B. Keller, L.A. Baker, Bipolar disorder: epidemiology, course, diagnosis, and treatment. Bull. Menn. Clin. **55**, 172–181 (1991)

182. J.A.S. Kelso, *Dynamic Patterns - The Self-Organization of Brain and Behavior* (MIT Press, Cambridge, 1995)

183. G. Keppel, T.D. Wickens, *Design and Analysis* (Pearson Prentice Hall, Upper Saddle River, 2004)

184. S. Keri, A. Antal, G. Szekeres, G. Benedek, Z. Janka, Spatiotemporal visual processing in schizophrenia. J. Neuropsychiatry Clin. Neurosci. **14**, 190–196 (2002)

185. S.L. Kerr, J.M. Neale, Emotion perception in schizophrenia: specific deficit or further evidence of generalized poor performance? J. Abnorm. Psychol. **102**, 312–318 (1993)

186. S. Kim, T.D. Frank, Body-scaled perception is subjected to adaptation when repetitively judging opportunities for grasping. Exp. Brain Res. **234**, 2731–2743 (2016)

187. S. Kim, T.D. Frank, Correlations between hysteretic categorical and continuous judgments of perceptual stimuli supporting a unified dynamical systems approach to perception. Perception **47**, 44–66 (2018)

188. A. Kleinschmidt, C. Buchel, C. Hutton, K.J. Friston, R.S.J. Frackowiak, The neural structures expressing perceptual hysteresis in visual letter recognition. Neuron **34**, 659–666 (2002)

189. K. Koffka, An introduction to gestalt theory. Psychol. Bull. **19**, 531–585 (1922)

190. J. Kornmeier, M. Bach, Early neural activity in Necker-cube reversal: evidence for low-level processing of the gestalt phenomenon. Psychophysiology **41**, 1–8 (2004)

191. J. Kornmeier, C.M. Hein, M. Bach, Multistable perception when bottom-up and top-down coincide. Brain Cogn. **69**, 138–147 (2009)

192. S.M. Kosslyn, R.S. Rosenberg, *Psychology: the Brain, the Person, the World* (Allyn and Bacon, New York, 2001)

193. S.M. Kosslyn, W.L. Thompson, I.J. Kim, N.M. Albert, Topographical representations of mental images in primary visual cortex. Nature **378**, 496–498 (1953)

194. S.M. Kosslyn, N.M. Albert, W.L. Thompson, V. Maljkovic, S.B. Weise, C.F. Chabris, S.E. Hamilton, F.S. Buonano, Visual mental imagery activates topographically organized visual cortex: PET investigation. J. Cogn. Neurosci. **5**, 263–287 (1993)

195. S.M. Kosslyn, A. Pascual-Leone, O. Felician, S. Camposano, W.L. Thompson, W.L. Ganis, K.E. Sukel, N.M. Albert, The role of area 17 in visual imagery: convergent evidence from PET and rTMS. Science **284**, 167–170 (1999)

196. A. Kreinin, Hearing voices in schizophrenia: who's voices are they? Med. Hypotheses **80**, 352–356 (2013)

197. R. Kristeva, D. Cheyne, L. Deecke, Neuromagnetic fields accompanying unilateral and bilateral voluntary movements: topography and analysis of cortical sources. Electroenceph. Clin. Neurophysiol. **81**, 284–298 (1991)

198. J. Kriz, Synergetics in clinical psychology, in *Synergetics in Cognition*, ed. by H. Haken, M. Stadler (Springer, Berlin, 1990), pp. 393–404

199. P. Kruse, M. Stadler, T. Wehner, Direction and frequency specific processing in the perception of long-range apparent movement. Vis. Res. **26**, 327–335 (1986)

200. A.D. Kuo, The relative roles of feedforward and feedback in the control o rhythmic movements. Motor Control **6**, 129–145 (2002)
201. R.W. Kupka et al., Comparison of rapid-cycling and non-rapid-cycling bipolar disorder based on prospective mood rating in 539 outpatients. Am. J. Psychiatry **162**, 1273–1280 (2005)
202. S. Kuroda, N. Schweighofer, M. Kawato, Exploration of signal transduction pathways in cerebellar long-term depression by kinetic simulation. J. Neuroscience **21**, 5693–5702 (2001)
203. D.D. Kurylo, R. Pasternak, G. Silipo, D.C. Javin, P.D. Butler, Perceptual organization by proximity and similarity in schizophrenia. Schizophr. Res. **95**, 205–214 (2007)
204. M.D. Kuzmin, Shape of temperature dependence of spontaneous magnetization of ferromagnets: quantitative analysis. Phys. Rev. Lett. **94**, article 107204 (2005)
205. L.D. Landau, E.M. Lifshitz, *Quantum Mechanics: Non-relativistic Theory* (Pergamon Press, London, 1977)
206. M. Lavelli, A. Fogel, Interdyad differences in early mother-infant face-to-face communication: real-time dynamics and developmental pathways. Dev. Psychol. **49**, 2257–2271 (2013)
207. C.W. Leach, R. Spears, N.R. Branscombe, B. Doosje, Malicious pleasure: schadenfreude at the suffering of another group. J. Pers. Soc. Psychol. **84**, 932–943 (2003)
208. D.A. Leopold, N.K. Logothetis, Multistable phenomena: changing views in perception. Trends Cogn. Sci. **3**, 254–264 (1999)
209. D.A. Leopold, M. Wilke, A. Maier, N.K. Logothetis, Stable perception of visually ambiguous patterns. Nat. Neurosci. **5**, 605–609 (2002)
210. J.S. Leversen, M. Haga, H. Sigmundsson, From children to adults: motor performance across the life-span. PLOS One **7**, article e38830 (2012)
211. L. Li, Stability landscapes of walking and running near gait transition speed. J. Appl. Biomater. **16**, 428–435 (2000)
212. J. Liu, G. Ahlers, Spiral-defect chaos in Rayleigh-Benard convection with small Prandtl numbers. Phys. Rev. Lett. **77**, 3126–3129 (1996)
213. C.H. Liu, O.J.L. Tzeng, D.L. Hung, P. Tseng, C.H. Juan, Investigation of bistable perception with the silhouette spinner: sit still, spin the dancer with your will. Vis. Res. **60**, 34–39 (2012)
214. S.M. Lopresti-Goodman, M. Richardson, M.J. Baron, C. Carello, K.L. Marsh, Task constraints on affordance boundaries. Motor Control **13**, 69–83 (2009)
215. S.M. Lopresti-Goodman, M.T. Turvey, T.D. Frank, Behavioral dynamics of the affordance "graspable". Atten. Percept. Psychophys. **73**, 1948–1965 (2011)
216. S.M. Lopresti-Goodman, M.T. Turvey, T.D. Frank, Negative hysteresis in the behavioral dynamics of the affordance "graspable". Atten. Percept. Psychophys. **75**, 1075–1091 (2013)
217. A.J. Lotka, The growth of mixed populations. two species competing for a common food supply. J. Wash. Acad. Sci. **23**, 461–469 (1932)
218. M.H.R. Ludtmann, K. Boeckeler, R.S.B. Williams, Molecular pharmacology in a simple model system: implicating MAP kinase and phosphoinositide signalling in bipolar disorder. Semin. Cell Dev. Biol. **22**, 105–113 (2011)
219. A.S. Lunchins, *Mechanization in Problem Solving. The Effect of Einstellung*. Psychological Monographs, vol. 54(6) (1942)
220. A.C. Luo, *Regularity and Complexity in Dynamical Systems* (Springer, Berlin, 2012)
221. M.R. Lyons, A.E. West, Mechanisms of specificity in neuronal activity-regulated gene transportation. Prog. Neurobiol. **94**, 259–295 (2011)
222. A.W. MacDonald, S.C. Schulz, What we know: findings that every theory of schizophrenia should explain. Schizophr. Bull. **35**, 493–508 (2009)
223. M.C. Mackey, L. Glass, Oscillations and chaos in physiological control systems. Science **197**, 287–289 (1977)
224. P. Mackin, A.H. Young, Rapid cycling bipolar disorder: historical overview and focus on emerging treatments. Bipolar Disord. **6**, 523–529 (2004)
225. N.R.F. Maier, Reasoning in humans ii: the solution of a problem and its appearance in consciousness. J. Comp. Psychol. **12**, 181–194 (1931)
226. N.R.F. Maier, *Problem Solving and Creativity in Individuals and Groups* (Brooks/Cole, Belmont, 1970)

227. K. Mainzer, *Thinking Complexity* (Springer, New York, 1994)
228. R. Malaka, Models of classical conditioning. Bull. Math. Biol. **61**, 33–83 (1999)
229. P. Mamassian, R. Goutcher, Temporal dynamics in bistable perception. J. Vis. **5**, 361–375 (2005)
230. P. Mamassian, M.S. Landy, Observer biases in the 3d interpretation of line drawings. Vis. Res. **38**, 2817–2832 (1998)
231. M.K. Mandal, R. Pandey, A.B. Prasad, Facial expressions of emotions and schizophrenia: a review. Schizophr. Bull. **24**, 399–412 (1998)
232. L.S. Mark, Eyeheight-scaled information about affordances: a study of sitting and stair climbing. J. Exp. Psychol. - Hum. Percept. Perform. **13**, 361–370 (1987)
233. N.I. Markevich, J.B. Hoek, B.N. Kholodenko, Signaling switches and bistability arising from multisite phopsphorylation in protein kinase cascades. J. Cell Biol. **164**, 353–359 (2004)
234. J.R. Martin, G. Dezechache, D. Pressnitzer, P. Nuss, J. Dokic, N. Bruno, E. Pacherie, N. Franck, Perceptual hysteresis as a marker of perceptual inflexibility in schizophrenia. Conscious. Cogn. **30**, 62–72 (2014)
235. D. Mataix-Cols, S. Wooderson, N. Lawrence, M.J. Brammer, A. Speckens, M.L. Phillips, Distinct neural correlates of washing, checking, and hoarding symptom dimensions in obsessive-compulsive disorder. Arch. Gen. Psychiatry **61**, 564–576 (2004)
236. R.M. May, W.J. Leonard, Nonlinear aspects of competition between three species. SIAM J. Appl. Math. **29**, 243–253 (1975)
237. P. Meehl, Psychological determinism or chance: configural cerebral autoselection as tertium quid, in *Science, Mind, and Psychology: Essays in Honor of Grover Maxwell*, ed. by M.L. Maxwell, C.W. Savage (University Press of America, Lanham, 1989), pp. 211–255
238. D.J. Miklowitz, M.J. Goldstein, Family factors and the course of bipolar affective disorder. Arch. Gen. Psychiatry **45**, 225–231 (1988)
239. S. Mongkolsakulvong, T.D. Frank, Synchronization and anchoring of two non-harmonic canonical-dissipative oscillators via Smorodinsky-Winternitz potentials. Condens. Matter Phys. **20**, article 44001 (2017)
240. T.J. Mueller, A. Federspiel, A.J. Fallgatter, W.K. Strik, Eeg signs of vigilance fluctuations preceding perceptual flips in multistable illusionary motion. Neuroreport **10**, 3423–3427 (1999)
241. T.J. Mueller, T. Koenig, J. Wackermann, P. Kalus, A.J. Fallgatter, W.K. Strik, D. Lehmann, Subsecond changes of global brain state in illusory multistable motion perception. J. Neural Transm. **112**, 565–576 (2005)
242. R. Müller, D. Cerra, P. Reinartz, Synergetics framework for hyperspectral image classification, in *The International Archives of the Photogrammetry, Remote Sensing and Spatial Information Sciences 2013* (International Society for Photogrammetry and Remote Sensing, Hannover, 2013), pp. 257–262
243. J.D. Murray, *Mathematical Biology* (Springer, Berlin, 1993)
244. B. Nagler, M. Peeters, J. Albert, G. Verschaffelt, K. Panajotov, H. Thienpont, I. Veretnnicoff, J. Danckaert, S. Barbay, G. Giacomelli, F. Marin, Polarization-mode hopping in single-mode vertical-cavity surface-emitting lasers: theory and experiment. Phys. Rev. A **68**, 013813 (2003)
245. P.G. Nestor, R. Piech, C. Allen, M. Niznikiewicz, M. Shenton, R.W. McCarley. Retrieval-induced forgetting in schizophrenia. Schizophr. Res. **75**, 199–209 (2005)
246. A. Newell, H.A. Simon, *Human Problem Solving* (Prentice Hall, Englewood Cliffs, 1972)
247. A.C. Newell, J.A. Whitehead, Finite bandwidth, finite amplitude convection. J. Fluid Mech. **38**, 279–303 (1969)
248. K.M. Newell, D.M. Scully, P.V. McDonald, R. Baillargeon, Task constraints and infant grip configuration. Dev. Psychobiol. **22**, 817–832 (1989)
249. K.M. Newell, D.M. Scully, F. Tenenbaum, S. Hardiman, Body scale and the development of prehension. Dev. Psychobiol. **22**, 1–13 (1989)
250. K.M. Newell, Y.T. Liu, G. Mayer-Kress, A dynamical systems interpretation of epigenetic landscapes for infant motor development. Infant Behav. Dev. **26**, 449–472 (2003)

251. L.K. Nguyen, M.A.S. Cavadas, B.N. Kholodenko, T.D. Frank, A. Cheong, Species differential regulation of cox2 can be described by nfkb-dependent logic and gate. Cell. Mol. Life Sci. **72**, 2431–2443 (2014)

252. G. Nicolis, *Introduction to Nonlinear Sciences* (Cambridge University Press, Cambridge, 1995)

253. A. Nitzan, P. Ortoleva, J. Deutch, J. Ross, Fluctuations and transitions at chemical instabilities: the analogy to phase transition. J. Chem. Phys. **61**, 1056–1074 (1974)

254. P.L. Nunez, *Electric Fields of the Brain* (Oxford University Press, New York, 1981)

255. P.L. Nunez, *Neocortical Dynamics and Human EEG Rhythms* (Oxford University Press, New York, 1995)

256. S. Ohlsson, Information-processing explanations of insight and related phenomena, in *Advances in the Psychology of Thinking*, ed. by M. Kean, K. Gilhooley (Harvester-Wheatsheaf, London, 1992), pp. 1–44

257. F. Ortega, J.L. Garces, F. Mas, B.N. Kholodenko, M. Cascante, Bistability from double phosphorylation in signal transduction. FEBS J. **273**, 3915–3926 (2004)

258. J.H.C. Palmer, P. Gong, Associative learning of classical conditioning as an emergent property of spatially extended spiking neural circuits with synaptic plasticity. Front. Comput. Neurosci. **8**, article 79 (2014)

259. A. Pansuwan, C. Rattanakul, Y. Lenbury, D.J. Wollkind, L. Harrison, L. Rajapakse, K. Cooper, Nonlinear stability analysis of pattern formation on solid surfaces during ion-sputtered erosion. Math. Comput. Model. **41**, 939–964 (2005)

260. B. Pena, C. Perez-Garcia, Selection and competition of Turing patterns. Europhys. Lett. **51**, 300–306 (2000)

261. J.D. Pettigrew, Search for the switch: neural basis for perceptual rivalry alternations. Brain and Mind **2**, 85–118 (2001)

262. S. Poltoratski, F. Tong, Hysteresis in dynamic perception of scenes and objects. J. Exp. Psychol. - General **143**, 1875–1892 (2014)

263. H.G. Purwins, H.U. Bödeker, S. Amiranasvili, Dissipative solitons. Adv. Phys. **59**, 485–701 (2010)

264. P.B. Putfall, C. Dunbar, Perceiving whether or not the world affords stepping onto and over: a developmental study. Ecol. Psychol. **4**, 17–38 (1992)

265. M.I. Rabinovich, M.K. Muezzinoghu, I. Strigo, A. Bystritsky, Dynamic principles of emotion-cognition interaction: mathematical images of mental disorders. PLoS One **5**, e12547 (2010)

266. C. Rattanakul, Y. Lenbury, D.J. Wollkind, V. Chatsudthipong, Weakly nonlinear analysis of a model of signal transduction pathway. Nonlinear Anal. **71**, e1620–e1625 (2009)

267. J.W. Rayleigh, *Theory of Sound* (Dover, New York, 1945). First edition published 1894

268. W. Reinhard, *A Short History of Colonialism* (Manchester University Press, Nanchester, 2011)

269. M.J. Richardson, K.L. Marsh, R.M. Baron, Judging and actualizing intrapersonal and interpersonal affordances. J. Exp. Psychol. - Hum. Percept. Perform. **33**, 845–859 (2007)

270. K.R. Ridderinkhof, M.W. van der Molen, Mental resources, processing speed, and inhibitory control: a developmental perspective. Biol. Psychol. **45**, 241–261 (1997)

271. H. Risken, *The Fokker-Planck Equation: Methods of Solution and Applications* (Springer, Berlin, 1989)

272. P.A. Robinson, C.J. Rennie, J.J. Wright, Propagation and stability of waves of electrical activity in the cerebral cortex. Phys. Rev. E **56**, 826–840 (1997)

273. G. Roeper, S. Rachman, Obsessional-compulsive checking: experimental replication and development. Behav. Res. Therapy **12**, 25–32 (1976)

274. P. Romanczuk, M. Bär, W. Ebeling, B. Lindner, L. Schimansky-Geier, Active Brownian particles: from individual to collective stochastic dynamics. Eur. Phys. J. Special Topics **202**, 1–162 (2012)

275. S. Saha, D. Chant, J. Welham, J. McGrath, A systematic review of the prevalence of schizophrenia. PLos Med. **2**, article e141 (2005)

276. J.D. Salamone, A. Wisniecki, B.B. Carlson, M. Correa, Nucleus accumbens dopamine depletions make animals highly sensitive to high fixed ratio requirements but do not impair primary food reinforcement. Neuroscience **105**, 863–870 (2001)

277. G. Schiepek, G. Strunk, The identification of critical fluctuations and phase transitions in short term and coarse-grained time series: a method for real time monitoring of human change processes. Biol. Cybern. **102**, 197–207 (2010)

278. G. Schiepek, I. Tominschek, S. Karch, et al., A controlled single case study with repeated fMRI measurements during the treatment of a patient with obsessive-compulsive disorder: testing the nonlinear dynamics approach to psychotherapy. World J. Biol. Psychiatry **10**, 658–668 (2009)

279. G. Schiepek, I. Tominschek, S. Heinzel, et al., Discontinuous patterns of brain activation in the psychotherapy process of obsessive-compulsive disorder: converging results from repeated fMRI and daily self-reports. PLoS One **8**, article e71863 (2013)

280. G. Schiepek, W. Aichhorn, M. Gruber, G. Strunk, E. Bachler, B. Aas, Real-time monitoring of psychotherapeutic processes: concept and compliance. Front. Psychol. **7**, article 604 (2016)

281. R.C. Schmidt, C. Carello, M.T. Turvey, Phase transitions and critical fluctuations in the visual coordination of rhythmic movements between people. J. Exp. Psychol. - Hum. Percept. Perform. **53**, 247–257 (1990)

282. M. Schmutz, W. Banzhaf, Robust competitive networks. Phys. Rev. A **45**, 4132–4145 (1992)

283. D. Schneider, C.J. Schneider, *Slavery in America* (Facts on File, New York, 2007)

284. N. Schneiderman, I. Fuentes, I. Gormenzano, Acquisition and extinction of the classical conditioned eyelid response in the Albino Rabbit. Science **136**, 650–652 (1962)

285. J.P. Scholz, J.A.S. Kelso, G.S. Schöner, Non-equilibrium phase transitions in coordinated biological motion: critical slowing down and switching time. Phys. Lett. A **123**, 390–394 (1987)

286. G.S. Schöner, H. Haken, J.A.S. Kelso, A stochastic theory of phase transitions in human hand movement. Biol. Cybern. **53**, 247–257 (1986)

287. G. Schreiber, S. Avissar, Lithium sensitive G protein hyperfunction: a dynamic model for the pathogenesis of bipolar affective disorder. Med. Hypotheses **35**, 237–243 (1991)

288. B. Schwartz, *Psychology of Learning and Behavior* (W.W. Norton, New York, 1989)

289. F. Schweitzer, W. Ebeling, B. Tilch, Complex motion of Brownian particles with energy depots. Phys. Rev. Lett. **80**, 5044–5047 (1998)

290. F. Schweitzer, W. Ebeling, B. Tilch, Statistical mechanics of canonical-dissipative systems and applications to swarm dynamics. Phys. Rev. E **64**, 021110 (2001)

291. C.M. Schwiedrizik, C.C. Ruff, A. Lazar, F.C. Leitner, W. Singer, L. Melloni, Untangling perceptual memory: hysteresis and adaptation map into separate cortical networks. Cereb. Cortex **24**, 1152–1164 (2014)

292. L.A. Segal, The nonlinear interaction of two disturbances in thermal convection problem. J. Fluid Mech. **14**, 97–114 (1962)

293. L.A. Segal, The nonlinear interaction of a finite number of disturbances to a layer of fluid heated from below. J. Fluid Mech. **21**, 359–384 (1965)

294. R.D. Seidler, J.A. Bernard, T.B. Burutolu, B.W. Fling, M.T. Gordon, J.T. Gwin, Y. Kwak, D.B. Lipps, Motor control and aging: links to age-related brain structural, functional and biochemical effects. Neurosci. Biobehav. Rev. **34**, 721–733 (2012)

295. W. Sha, J. Moore, K. Chen, A.D. Lassaletta, C.S. Yi, J.J. Tyson, J.C. Silbe. Hysteresis drives cell-cycle transitions in Xenopus laevis egg extracts. Proc. Natl. Acad. Sci. USA **93**, 628–633 (2003)

296. M.N. Shadlen, W.T. Newsome, Motion perception: seeing and deciding. Proc. Natl. Acad. Sci. USA **93**, 628–633 (1996)

297. M.N. Shadlen, W.T. Newsome, Neural basis of a perceptual decision in the parietal cortex (area lip) of the rhesus monkey. J. Neurophysiol. **86**, 1916–1936 (2001)

298. M.L. Shik, G.N. Orlovskii, F.V. Severin, Organization of locomotor synergism. Biofizika **11**, 879–886 (1966)

299. H. Shimizu, Y. Yamaguchi, Synergetic computer and holonics: information dynamics of a semantic computer. Phys. Scripta **36**, 970–985 (1987)

300. B.F. Skinner, *Science and Human Behavior* (The Free Press, New York, 1953)

301. J.C. Slater, *Modern Physics* (McGraw-Hill, New York, 1955)

302. J.W. Smith, P. Frawley, L. Polissar, Six and twelve month abstinece rates in inpatient alcoholics treated with aversion therapy compared with matched inpatients from a treatment registry. Alcohol. Clin. Exp. Res. **15**, 862–870 (1991)

303. D. Sornette, *Critical Phenomena in Natural Science* (Springer, Berlin, 2000)

304. L. Spinelli, G. Tissoni, M. Brambilla, F. Prati, L.A. Lugiato, Spatial solitons in semiconductor microcavities. Phys. Rev. A **58**, 2542–2559 (1998)

305. H.E. Stanley, *Introduction to Phase Transitions and Critical Phenomena* (Oxford University Press, New York, 1971)

306. A.L. Stanton, S.B. Kirk, C.L. Cameron, S. Danoff-Burg, Coping through emotional approach: scale construction and validation. J. Pers. Soc. Psychol. **78**, 1150–1169 (2000)

307. A. Steinacher, K.A. Wright, Relating the bipolar spectrum to dysregulation of behavioral activation: a perspective from dynamical modelling. PLoS One **8**, article e63345 (2013)

308. L.E. Stephenson, D.J. Wollkind, Weakly nonlinear analysis of one-dimensional Turing pattern formation in activator-inhibitor/immobilizer model system. J. Math. Biol. **33**, 771–815 (1995)

309. P. Sterzer, A. Kleinschmidt, G. Rees, The neural basis of multistable perception. Trends Cogn. Sci. **13**, 311–318 (2009)

310. S.S. Stevens, On the psychophysical law. Psychol. Rev. **64**, 153–181 (1957)

311. S.S. Stevens, On the new psychophysics. Scand. J. Psychol. **1**, 27–35 (1960)

312. T.A. Stoffregen, C.M. Yang, B.G. Bardy, Affordance judgments and nonlocomotor body movement. Ecol. Psychol. **17**, 75–104 (2005)

313. S.H. Strogatz, D.M. Abrams, A. McRobie, B. Eckhardt, E. Ott, Crowd synchrony on the Millennium Bridge. Nature **438**, 43–44 (2005)

314. Y. Sumino, K. Yoshikawa, Self-motion of an oil droplet: a simple physiochemical model of active Brownian motion. Chaos **18**, 026106 (2008)

315. K. Tanaka, G.J. Augustine, A positive feedback signal transduction loop determines timing of cerebellar long-term depression. Neuron **59**, 608–620 (2008)

316. Y.B. Taranenko, I. Ganne, R.J. Kuszelewicz, C.O. Weiss, Patterns and localized structures in bistable semiconductor resonators. Phys. Rev. A **61**, article 063818 (2000)

317. G. Tesauro, Simple neural models of classical conditioning. Biol. Cyb. **55**, 187–200 (1986)

318. B. Tisdale, https://bobtisdale.wordpress.com

319. E.C. Tolman, Cognitive maps in rats and men. Psychol. Rev. **55**, 189–208 (1948)

320. J.A. Tracy, J.K. Thompson, D.J. Krups, R.F. Thompson, Evidence of plasticity in the pontocerebellar conditioned stimulus pathway during classical conditioning of the eyeblink response in the rabbit. Behav. Neurosci. **112**, 267–285 (1998)

321. F. Tretter, P.J. Gebicke-Haerter, U. an der Heiden, H.W. Mewes, C.W. Turck, Affective disorders as complex dynamic diseases: a perspective from systems biology. Pharmacopsychiatry **44**(Suppl 1), S2–S8 (2011)

322. W. Tschacher, How specific is the Gestalt-informed approach to schizophrenia. Gestalt Theory: Int. Multidiscip. J. **26**, 335–344 (2004)

323. W. Tschacher, C. Scheier, K. Grawe, Order and pattern formation in psychotherapy. Nonlinear Dynamics Psychol. Life Sci. **2**, 195–215 (1998)

324. W. Tschacher, D. Schuler, U. Junghan, Reduced perception of the motion-induced blindness illusion in schizophrenia. Schizophr. Res. **81**, 261–267 (2006)

325. B. Tuller, P. Case, M. Ding, J.A.S. Kelso, The nonlinear dynamics of speech recognition. J. Exp. Psychol. - Hum. Percept. Perform. **20**, 3–16 (1994)

326. B. Tyldesley, J.I. Grieve, *Muscles, Nerves, and Movement | Kinesiology in Daily Living* (Blackwell Science, Oxford, 1996)

327. P.J. Uhlhaas, S.M. Silverstein, The continuing relevance of Gestalt psychology for an understanding of schizophrenia. Gestalt Theory Int. Multidiscip. J. **25**, 256–279 (2003)

328. US National Oceanic and Atmospheric Administration (NOAA). https://www.nhc.noaa.gov/climo

329. J. van der Kamp, G.J.P. Savelsbergh, W.E. Davis, Body-scaled ratio as a control parameter for prehension in 5- to 9-year old children. Dev. Psychobiol. **33**, 351–361 (1998)

330. A.M. van Loon, T. Knapen, H.S. Scholte, E.S.J. Saaltink, T.H. Donner, V.A.F. Lamme, GABA shapes the dynamics of bistable perception. Curr. Biol. **23**, 823–827 (2013)

331. I. van Rooij, R.M. Bongers, W.F.G. Haselager, A non-representational approach to imagined action. Cogn. Sci. **26**, 345–375 (2002)

332. J.M. Vargas, V.J. Adesso, A comparison of aversion therapies for nailbiting behavior. Behav. Therapy **7**, 322–329 (1976)

333. F. Verhulst, *Nonlinear Differential Equations and Dynamical Systems*, 2nd ed. (Springer, Berlin, 1996)

334. V.S. Vorobev, The law of corresponding states for the entropy of rare gases. Chem. Phys. Lett. **383**, 359–361 (2004)

335. M.A. Vorontsov, A.Y. Karpov, Simulated optical patterns in a Kerr slice-feedback mirror-type experiment. J. Modern Opt. **44**, 439–446 (1997)

336. J.B. Wagman, A. Hajnal, Getting off on the right (or left) foot: perceiving by means of a rod attached to the preferred or non-preferred foot. Exp. Brain Res. **232**, 3591–3599 (2014)

337. J.B. Wagman, A. Hajnal, Task specificity and anatomical independence in perception of properties by means of wielded object. J. Exp. Psychol. - Hum. Percept. Perform. **40**, 2372–2391 (2014)

338. J.B. Wagman, C.A. Taheny, T. Higuchi, Improvements in perception of maximum reaching height transfer to increases or decreases in reaching ability. Am. J. Psychol. **127**, 269–279 (2014)

339. G.V. Wallenstein, J.A.S. Kelso, S.L. Bressler, Phase transitions in spatiotemporal patterns of brain activity and behavior. Phys. D **84**, 626–634 (1995)

340. L. Wang, B.L. Walker, S. Iannaccone, D. Bhatt, P.J. Kennedy, W.T. Tse, Bistable switches control memory and plasticity in cellular differentiation. Proc. Natl. Acad. Sci. USA **106**, 6638–6643 (2009)

341. W. Wang, Y. Lin, H. Wang, H. Liu, Y. Tan, Pattern selection in an epidemic model with self and cross diffuion. J. Biol. Syst. **19**, 19–31 (2011)

342. R.M. Warren, Verbal transformation effect and auditory perceptual mechanisms. Psychol. Bull. **70**, 261–270 (1968)

343. W.H. Warren, Perceiving affordances: visual guidance of stair climbing. J. Exp. Psychol. - Hum. Percept. Perform. **10**, 683–703 (1984)

344. F. Waters, Visual hallucinations in the psychosis spectrum and comparative information from neurodegenerative disorders and eye diseases. Schizophr. Bull. **40**(Suppl.), S233–S245 (2017)

345. W. Weidlich, Physics and social science: the approach of synergetics. Phys. Rep. **204**, 1–163 (1991)

346. R.T. Weidner, R.L. Sells, *Elementary Modern Physics* (Allyn and Bacon, Boston, 1968)

347. S.F. Weiss, *The Nazi Symbiosis* (University of Chicago Press, Chicago, 2010)

348. M. Wertheimer, Untersuchungen zur Lehre von Gestalt II (in German). Psychol. Forsch. **4**, 301–305 (1923)

349. J. Wesfreid, Y. Pomeau, M. Dubois, C. Normand, P. Berge, Critical effects in Rayleig-Benard convection. Le J. de Phys. Lett. **7**, 726–731 (1978)

350. WHO Multicentre Growth Reference Study Group. WHO Motor development study: windows of achievement for six gross motor development milestones. Acta Paediatr. Suppl. **450**, 86–95 (2006)

351. H.R. Wilson, J.D. Cowan, Excitatory and inhibitory interactions in localized populations of model neurons. Biophys. J. **12**, 1–24 (1972)

352. A.T. Winfree, *The Geometry of Biological Time*, 2 edn. (Springer, Berlin, 2001)

353. P.H. Wolff, *Behavioral States and the Expressions of Emotion in Early Infancy* (University of Chicago Press, Chicago, 1987)

354. D.J. Wollkind, L.E. Stephenson, Chemical Turing pattern formation analyses: comparison of theory with experiment. SIAM J. Appl. Math. **61**, 387–431 (2000)
355. T.S. Woodward, S. Moritz, M. Menon, Belief inflexibility in schizophrenia. Cogn. Neuropsychiatry **13**, 267–277 (2008)
356. G. Wunner, A. Pelster, *Self-organization in complex systems: the past, present, and future of synergetics* (Springer, Berlin, 2016)
357. H. Yokoi, A. Adamatzky, B.L. Costello, Excitable chemical medium controller for a robotic hand: closed-loop experiments. Int. J. Bif. Chaos **14**, 3347–3354 (2004)
358. A.A. Yudashkin, Bifurcations of steady-state solutions in the synergetic neural network and control of pattern recognition. Auto Remote Control **57**, 1647–1653 (1996)

Index

© Springer Nature Switzerland AG 2019
T. Frank, *Determinism and Self-Organization of Human Perception and Performance*, Springer Series in Synergetics,
https://doi.org/10.1007/978-3-030-28821-1

Printed by Printforce, the Netherlands